New Approaches to Identifying Structures Using Geometric Structure Analysis:

Design and Adaptation

Nikolay Nikolayevich Karabutov
MIREA-Russian Technological University, Moscow, Russia

IGI Global
Scientific Publishing
Publishing Tomorrow's Research Today

Vice President of Editorial	Melissa Wagner
Director of Acquisitions	Mikaela Felty
Director of Book Development	Jocelynn Hessler
Production Manager	Mike Brehm
Cover Design	Phillip Shickler

Published in the United States of America by
IGI Global Scientific Publishing
701 East Chocolate Avenue
Hershey, PA, 17033, USA
Tel: 717-533-8845
Fax: 717-533-7115
Website: https://www.igi-global.com E-mail: cust@igi-global.com

Library of Congress Cataloging-in-Publication Data

Names: Karabutov, Nikolay, 1950- author.
Title: New Approaches to identifying structures using geometric structure analysis / edited by Nikolay Karabutov.
Description: Hershey : IGI Global Scientific Publishing, [2026] | Includes bibliographical references and index. | Summary: "We consider an actual problem of identification theory related to the non-formalized task of evaluating the model structure. We consider an actual problem of identification theory related to the non-formalized task of evaluating the model structure"-- Provided by publisher.
Identifiers: LCCN 2024059921 (print) | LCCN 2024059922 (ebook) | ISBN 9798337308227 (hardcover) | ISBN 9798337308234 (paperback) | ISBN 9798337308241 (ebook)
Subjects: LCSH: Nonlinear systems. | System identification. | Parametric modeling.
Classification: LCC QA402.35 .N688 2025 (print) | LCC QA402.35 (ebook) | DDC 003/.1--dc23/eng/20250310
LC record available at https://lccn.loc.gov/2024059921
LC ebook record available at https://lccn.loc.gov/2024059922

British Cataloguing in Publication Data
A Cataloguing in Publication record for this book is available from the British Library.

Table of Contents

Section 3
Adaptive Identification

Preface

Implicitly, all scientific research is aimed at building models that may have a different appearance and structure. The most difficult areas are the fields where the description of processes and phenomena needs to be formalized. This area is a core of the identification theory (IT). Many IT approaches are divided into two classes: structural and parametric methods. The most developed is the parametric direction (parametric identification (PI)). We believe PI is a formed direction, at least in the class of regression models.

The second area of IT is structural identification (SI), which has not received a formalized description. There are no methods for forming the model structure, criteria for selecting the model and determining the display class for the model. The main difficulty of this direction is the impossibility of describing SI by known mathematical methods. SI is rather the intuition and experience of a researcher who studies a process or phenomenon under conditions of uncertainty. Based on the researcher's experience, there is often a class of models with accuracy up to an unknown set of parameters. In fact, the task of structural identification is reduced to PI. This is the dominant trend in TI. Most methods of structural synthesis of nonlinear models in any field are reduced to PI. This approach excludes knowledge about the nonlinear system (NS), which may be required in the tasks of formalizing NS when expanding the scope of application. There are indirect methods for estimating some structural parameters of the system. This analysis shows the complexity, versatility and relevance of the SI problem. The approach to solving the SI problem based on the construction and analysis of geometric frameworks (GF) is proposed in this book. The GF analysis formalizes the model structure for nonlinear systems (NS) and defines the PI direction for complex systems.

CHAPTER OVERVIEW

The book has the following structure. It consists of three sections. Section 1 presents an analysis of the SI problem and includes 6 chapters based on the GF analysis. In Chapter 1, we analyze approaches to SI. We consider difficulties and problems. Most SI methods are based on indirect estimates of structural parameters.

In chapter 2, a new approach to the SI system is presented. It bases the GF application in identification systems. Classes of frameworks for solving structural and parametric identification problems are proposed. Example of using GF is presented in Chapter 3.

The structural and adaptive identification issues of systems with distributed delay are discussed in Chapters 3 and 4. Chapter 3 discusses: (i) the problem of estimating the linearity of a system with distributive lag (DL), (ii) algorithms for deciding on the system structure with lag in input and output, (iii) the use of sector sets to estimate the structure and parameters of the DL.

Chapter 4 are presented methods and algorithms for adaptive identification of systems with DL. It is based on the Chapter 3 results. We analyze the impact of relationships between system variables on the quality of identification.

The SI analysis of dynamic systems based on using GF is presented in Chapter 5. Approaches are proposed for estimating the class of nonlinearities for a nonlinear dynamical system (NDS). GF construction methods have been developed for the NDS structural identification. SI methods of linear dynamical systems (LDS) using GF analysis are proposed. We are developing methods for estimating LDS structural parameters based on the construction and analysis of GF for evaluating Lyapunov exponents (LE). Structural methods are generalized to periodic systems and static systems.

In Chapter 6, we consider the identification of parametric restrictions under uncertainty. This structural parameter has not been studied under uncertainty. We are considering various ways to describe the domain of constraints. We consider various ways to describe the domain of constraints (RA) and propose GF, and we analyze GF properties and develop adaptive procedures for determining the RA parameters.

The structural identifiability (SID) methods are presented in the second part. This is an important area of SI, which is studied based on the GF method (GFM) application. In Chapter 7, the theoretical foundations of SID based on GF analysis are presented for NDS. The insignificant GF occurrence case in estimating the SID of nonlinear systems is considered. We analyze the effect of excitation constancy on SID. This is the S-synchronizability problem of the dynamic system. We propose a criterion for evaluating the identifiability. A hierarchical immersion method has been developed to SID evaluate of a system with complex nonlinearities. We show the relationship between SI and SID.

In Chapter 8, we consider SID systems with non-symmetric nonlinearities. We generalize Chapter 7 results for this case and show the relationship between h-identifiability and S-synchronizability of NDS.

SID of feedback systems with nonlinear adulterating is presented in Chapter 9. We generalize Chapter 7 and 8 results for this case. Various cases of mixing are considered in NDS. Additive mixing and the more complex case of nonlinear mixing of a function argument are studied. The SID estimates for these cases are presented.

In Chapter 4, we discuss the case of evaluating and applying Lyapunov exponents in the SI problem of the linear system. The solution to this problem depends on the ability to identifiability and detectability LE. We study these issues in chapter 10. GF is proposed to solve problem. The conditions for recoverability and detectability of LE are presented.

Section 3 contains methods for adaptive identification of complex systems under uncertainty. They are based on the results of sections 1, 2 and the φ-algorithm method. In Chapter 11, the method of φ-algorithms (FA) is presented, which we used to synthesize adaptive identification systems under constraints. The method is based on the application of Lyapunov functions and is provided the formalized solution to the problem of synthesizing the adaptive system. Resulting algorithms provide solutions for both coordinate and parametric (adaptive) control. Methods for obtaining robust FA are considered. We propose FA synthesis with the specified quality.

In chapters 11-16, the FA method and GF are used to design adaptive systems of identification complex systems. In Chapter 12, we present adaptive identification for systems with multiple nonlinearities. The identification method is based on the structural-parametric approach and the GF analysis proposed in the book. The adaptive identification system (AIS) is designed for the system a parametrised model of nonlinearity.

Adaptive identification of interconnected systems is presented in Chapter 13. The AIS synthesis method is described using the example of a two-channel system. The proposed procedures are generalized to interconnected systems.

Adaptive identification of decentralized systems (DS) is considered in Chapter 14. The DS identifiability problem has been studied. It is assumed that the nonlinearity satisfies the quadratic condition. Algorithms for parametric and signal adaptation have been designed. We show that consider functional constraints transforms an adaptive algorithm into a system in the state space. The properties of the adaptive system are studied.

Chapter 15 and 16 are contain synthesis methods of adaptive observers (AO) for nonlinear systems and linear time-varying dynamic systems. We study the identification of systems under uncertainty. Chapter 15 applies the GF and AF method. This makes it possible to implement structural and parametric stages for AO synthesis. Local identifiability conditions based on the analysis of experimental data are obtained.

Chapter 16 contains results the AO development for linear time-varying dynamic systems. Adaptive algorithms have been developed. We represent the system as an object with parametric uncertainty (PU) and we apply the φ-algorithm method. A class of multiplicative φ-algorithms is obtained. PU estimation algorithms have been developed. Properties of the adaptive system are studied.

The obtained results introduce a definite contribution to the formalized paradigm creation for the synthesis of structural identification systems.

Nikolay N. Karabutov

Prof., DTS, MIREA–Russian Technological University, Moscow, Russia

Section 1
Structural Identification

Chapter 1
Research State on Structural Identification

ABSTRACT

In this section, we consider the state of the structural identification (SI) issue. The section comprises 6 chapters. Chapter 1 reflects the state of the SI research. We analyze research from the systemic perspective. The SI problem is informal. Therefore, there are no strict methods for estimating the structure parameters. The structure as a mathematical object is not represented as a certain set of attributes. The science development has made adjustments to the classical theory of identification. We consider such aspects of SI as: 1) evaluation of the order of the system and the model; 2) structure selection of the system with lag variables; 3) assessment of nonlinearity of the system; 4) we interpret the problem of constructing the attractor for a dynamic system as the SI problem; 5) synthesis of parametric constraints for the system under uncertainty; 6) approaches to the assessment of Lyapunov exponents as a structural parameter of the system. The application of geometric structures (GF) is considered solving the SI problem.

1. INTRODUCTION

The identification problem occupies one of the central places. Parametric estimation (PE) and structure selection are the main directions of identification theory. PE is the basis of the identification theory and has now received the most complete development (Box and Jenkins, 1976; Eykhoff, 1974; Graupe, 1975; Kashyap and Rao, 1976; Ljung, 1999; Raibman and Chadeev, 1975). Structural identification (SI) has practically not been developed. This has several explanations. First, these are SI problem formalizing difficulties. They cause other problems, the solution of which many authors get on an intuitive level. This approach does not provide

DOI: 10.4018/979-8-3373-0822-7.ch001

a general method and approach to solving a wide class of SI problems. Next, we consider problems that complicate the SI problem solution.

Despite on the SI complexity, various publications devoted to the consideration and study of the problem. One of the first reviews containing the method for the structure model selection proposed by Perelman, (1983). Some aspects and approaches to structural identification are considered by Eykhoff (1974); Giannakis and Serpedin (2001); Ljung (2000); Raibman (1978). Proposed methods base on a selective parametric approach. There are approaches to choosing the model structure that have not received proper coverage in the literature, but are effective.

This work purpose is to consider some SI tasks from a systemic perspective. We consider the structure concept, which is widely interpreted by various authors. These interpretations do not always reflect the structure essence when solving applied identification problems. Next, the structural identification problem formulation considers. Difficulties arising from solving the SI problem of static systems considered. We analyze methods and approaches used to solve structural identification in various problems.

2. STRUCTURE CONCEPT IN IDENTIFICATION TASKS

The structure concept does not have a clear definition. This concept interprets widely in modern science. Each knowledge area gives its own interpretation. B. Green (2004) uses the mathematical structure (ST) concept and interprets ST as any mathematical theory. In relation to living systems, ST (Stepin, 2001) is a set of stable connections in an object. Structure) or building is the internal organization of something. The internal building related to categories of the whole and its parts.

In the automatic control theory (ACT) (Besekersky and Popov, 1975; Vinogradov, 1979), the structure is the mathematical equations set describing processes in the system. ACT uses the system concept with a variable structure (Besekersky and Popov, 1975). Here, the structure interprets as a restructuring (change) of system parameters. In control systems, the block diagram concept widely uses, which reflects the system building (composition). The data structure in computer science is an analogy of a block diagram in ACT.

The ST concept widely uses in mathematics (Greene, 2004; Stepin, 2001). This term is synonymous with various abstract concepts and categories. Often, the structure in the mathematical sense is a system of some equations. In the theory of systems (Van Gig, 1978)), ST understands as the system building, the organization of its elements.

We identify two features of the term structure interpretation. First, the structure is considered as a complex system. The structure covers many technical, social, and organizational systems. ST understands as an abstract category or set of objects used to describe and analyse complex processes and phenomena. This ST interpretation corresponds to the identification theory most adequately. We will adhere to this SI interpretation.

The identification theory is considered in the wide and narrow sense (Eykhoff, 1974). ST is a research problem in a wide sense, i.e., the choice of the form for operators or equations describing the processes in the system. Note that the choice problem of the mathematical model structure under uncertainty is far from the last solution. The SI problem complexity consists in the absence of regular methods for model structure synthesizing. Note mathematical objects describing the structural indicators of the system do not exist. It is intuitively clear that there is the structure, but SI cannot be described in a formalized language. SI difficulties considered in this paper, Karabutov (2012) analyzed.

Besides the above interpretation, ST understands as a mathematical object described by a functional mapping. Such ST has a graphical representation and describes the processes in the system in a generalized form. The specified object is called a geometric portrait defined in some space. Various authors interpret the SI concept from the perspective of applied problems (see e.g., Kerschen et al. (2006); Koh and Perry (2010); Sirca and Adeli (2012).

3. STRUCTURAL IDENTIFICATION PROBLEM STATEMENT

As noted in the introduction, the structural identification problem is intractable to formalize. We give one of the possible SI problem formulations based on the multiplicity-information approach (Karabutov (2021a).

Consider the system described by the equation

$$\begin{aligned} \dot{X}(t) &= F(X,A,t)+Bu(t), \\ y(t) &= C^T X(t)+\xi(t), \end{aligned} \tag{1.1}$$

where $X \in \mathbb{R}^m$ is the state vector, $F: \mathbb{R}^m \times \mathbb{R}^k \times \mathbb{J} \to \mathbb{R}^m$ is smooth continuously differentiable m-dimensional vector-function, $t \in \mathbb{J} \subset \mathbb{R}$, $y \in \mathbb{R}$ is output, $u \in \mathbb{R}$ is input, $A \in \mathbb{R}^k$ is parameter vector, $B \in \mathbb{R}^m$, $\xi \in R$ is piecewise continuous bounded perturbation, $C \in \mathbb{R}^m$.

A priori information

$$_a(X,_S,_S,u,\xi) \subset {}_S \cup {}_S \cup {}_a^X \cup {}_a^u \cup {}_a^\xi \tag{1.2}$$

set containing available information about the vector function $F \in {}_S$ structure, parameters $(A, B) \subset {}_{S}$, characteristics of input, output, and perturbation.

The set ${}_S$ contains the information about the operator class describing the system (1.1) dynamics, as well as some structural parameters ${}_S$. The ${}_a$ level determines the ${}_S$ cardinality. In identification tasks, the formation ${}_S$ and ${}_S$ is bases on the experience and intuition of the researcher. Given the unformalizability ${}_S$ in (1.2), the subsets ${}_{S}$, ${}_S$ cardinality can set fuzzy, and most often, is uncertain. Therefore, the SI problem solution complicates.

Experimental information

$$\mathbb{I}_o = \{u(t), y(t), t \in \mathbb{J} = [t_0, t_k]\}$$

The perturbation ξ may have a different nature (Isermann and Mönchhof (2011); Kuntsevich and Lychak (1985)). ξ is limited.

Consider the operator $\hat{F}_i(\cdot) \in \mathbb{R}^m$, which is a contender for the formation of the vector-function $F(X,A,t)$ structure in (1). Let $\hat{F}_i(\cdot) \in \mathbb{S}_S$ and is parametrized accurate to a pair $(\hat{A}_i, \hat{B}_i) \in \mathbb{A}_S \subset \mathbb{S}_S$. Apply the model:

$$\dot{\hat{X}}(t) = \hat{F}(\hat{X}, \hat{A}, t) + \hat{B}u(t).$$

Problem: use a priori $\mathbb{I}_a$ and experimental $\mathbb{I}_o$ information and estimate the vector-function F structure in (1.1) to minimise cardinality of the set $\mathbb{S}_S$

$$\arg\min_{\hat{F}} \# \mathbb{S}_S = F^* \tag{1.3}$$

The fulfilment (1.3) is equivalent to the condition

$$\arg\min_{(\hat{A}, \hat{B})} \# \mathbb{A}_S = (A^*, B^*)$$

We do not specify the class of parametric identification methods, since set $\mathbb{S}_S$ elements determine their type. The identification criterion $\# \mathbb{S}_S$ choice reflects the non-standard and complexity of the problem.

The structural identification issue is complex and requires solving various subtasks. An overview of the solution methods is given below. The structural identification issue is complex and requires solving much subtasks. We give a solution method overview below. Other statements of the SI problem know (Perelman, 1983).

The structural identification issue is complex and requires solving various subtasks. An overview of the solution methods is given below. The SI issue is complex and requires solving various subtasks (Perelman, 1983).

4. REQUIREMENTS FOR MODEL STRUCTURE

Three groups of factors are decisive for the model structure choice (Perelman, 1983):

(1) ensuring the maximum quality of the restored model;
(2) the calculation amount minimising in the model synthesis.
(3) maximum reduction of calculations during the construction of the model.

5. ON SI DIFFICULTIES OF STATIC SYSTEMS

Static frameworks (SF) (Karabutov, 2016a, 2016b) are proposed in a special structural space. SF analysis allows deciding about the nonlinearity class in static systems (SS). We explain the need to allocate the structural space for the nonlinear SS for the following reasons (Karabutov, 2016):

(i) the system output is an integrated quantity reflecting the influence of the input variable set;
(ii) the method absence of the input-output relationships identification in the nonlinearity type classification problem. Approaches to degree nonlinearity assessing of the system described by Ljung (1999), Raibman and Chadeev (1975). Nonlinearity degree estimating is the key to class choosing of nonlinear static models. The solution bases on the complex mathematical apparatus use, which does not always offer an approach to SI;
(iii) nonlinear structure a priori assignment (Boguslavsky, 2006; Graupe, 1975; Mosteller and Tukey, 1977), based on a set of existing inputs and a parametric approach, requires solving the multicollinearity problem (Draper and Smith, 2014; Johnston, 1972). The problem is complex, and the applied solution is simple. The solution is the practical implementation of the model. A simple solution is the exclusion of dependent variables, which use very used in actual control systems, but this solution closes the way to considering the SI problem. This relationship does not always understand and is not considered in existing methods;
(iv) application of parametric methods based on the polynomial (Boguslavsky, 2006; Mosteller and Tukey, 1977). The effectiveness of this approach depends on the experience and intuition of the researcher. This approach requires preliminary labour-intensive research. It does not allow determining the nonlinearity structure explicitly.

These problems explain the existing state of SI static systems. Despite the apparent simplicity of the SS mathematical description, its analysis requires the design and application of non-trivial approaches and methods (Karabutov, 2012, 2016b). Designed methods should resolve existing contradictions. Contradictions have structural form and associate with multicollinearity, correlation, lag, etc. SI problem is insoluble when the parametrization paradigm dominates. The parametric approach is a powerful tool for solving control tasks. The parametrization problem is secondary in SI tasks. This should understand on designing SI procedures under uncertainty. Only the model structure correct choice allows to go to solve the parametric estimation problem. Such the relationship is not always understood and requires the use of various procedures to compensate for emerging uncertainties. Note that collinearity (COL) is a problem that cannot solve by parametric methods. But this does not mean that COL is a "parasite" and it should not use in system the structure assessment. COL is a structural indicator. Only the information set analysis of the system is a condition for obtaining structural indicators. This is a non-trivial task. But it has the solution. Its solution got in a structural space (Karabutov, 2016a, 2016b), which does not always contain system variables. The static framework (Karabutov, 2016a) is based on generalised variables and parameters defined in this space. The analysis of changes in the framework and the assessment of its parameters allow concluding about nonlinear processes in the system.

Consider methods for evaluating the structural parameters of the system. Evaluation of structural parameters is based on the use of indirect methods.

6. MODEL ORDER ESTIMATION

Different approaches are used to estimate of the model order (Ljung, 1999). The following classification corresponds to methods:

1. Spectral characteristics study of the transfer function (TF).
2. Checking the ranks of sample covariance matrices.
3. Correlation of variables.
4. Analysis of the information matrix.
    1. *Spectral-analytical estimates.* If TF contains information about the of resonant peak magnitude, high-frequency slices, or phase shifts, then this information is the basis for the model order choose (Britenkov and Dedus, 2012; Marple, 1987; Prokhorov and Grafkin, 2010).
    2. *Checking the ranks of sample covariance matrices.* This approach bases on the fact (Ljung, 1999) that a regression model with the variables vector synthesises. Consider the matrix

$$P_q(N) = \frac{1}{N}\sum_{i=1}^{N}\Phi_{q,i}\Phi^T_{q,i} \qquad (1.4)$$

where N is data sample length, q is the evaluation of the system order. If the system input is constantly excited, then the matrix $P_q(N)$ is non-degenerate at $q \leq m$ and degenerate at $q > m$. Therefore, det $P_q(N)$ (Woodside, 1971) can use to statistical hypothesis test and decide about the system order. The relationship study between the matrix (1.4) degeneracy and the model order given by Lee (1964), and the algorithm implementing the method is proposed Woodside (1971). If there is interference, then enter a threshold for applying the matrix $P_q(N)$ to minimise the interference effect with a large amplitude. If the signal-to-noise ratio is small, then

$$\hat{P}_q(N) = P_q(N) - \sigma^2_\xi P_\xi \qquad (1.5)$$

where $\sigma^2_\xi P_\xi$ consider the effect of ξ on $P_q(N)$.

Various models are used for the object order estimate under uncertainty. Jenkins and Watts (1969) introduce criterion, which gives the model of the required order. This approach is time-consuming and imposes certain requirements on the qualification of the researcher. If the model order is too small, then smoothed spectral estimates use. If the order is too large, then the spectrum resolution increases, and appear false peaks in the spectrum. Therefore, if the autoregressive model order evaluating, then a compromise must observe between the resolution and the variance amount for classical spectral estimation methods. The model order choice (Jenkins and Watts, 1969)) is influenced by residuals of the variance. Akaike criterion (AC) (Goldenberg et al. (1990), Jenkins and Watts, 1969) is used to identify of the model order often. There are various AC modifications. First criterion allows estimating the final prediction error. The autoregressive model order choice is based on minimizing of the error average variance at each prediction step. The criterion for the autoregressive process has the form (Jenkins and Watts, 1969)

$$Q_1(m) = \hat{\sigma}_m \frac{N + (m+1)}{N - (m+1)} \qquad (1.6)$$

where N is data number, m is the model order, $\hat{\sigma}_m$ is the white noise dispersion estimation. Centred values of variables are used to estimate the error variance in (1.6). The second multiplier increases if the model order increases. Therefore, the model choice with the order m should give a minimum $Q_1(m)$. The application $Q_1(m)$ gives an understated value to the model order (Berryman, 1978; Jones, 1976).

The second AC or Akaike Information Criterion (ICA) is based on the maximum likelihood method. According to this criterion, the model order determines based on the minimisation of some information-theoretic function. If the data have a normal distribution, then the ICA has the form (Jenkins and Watts, 1969; Kay, 1999)

$$Q_2(m) = N \ln\left(\hat{\sigma}_m\right) + 2m \tag{1.7}$$

The second term in (1.7) characterises the fee for the use of additional coefficients in the model. Such representation of the criterion does not reduce the prediction error variance. Criteria (1.6) and (1.7) are asymptotically equivalent for $N \rightarrow \infty$. $Q_2(m)$ gives excellent results for ideal autoregressive processes. $Q_2(m)$ has the same problems as $Q_1(m)$.

ICA is statistically untenable (Kashyap, 1980), which the values of the model order overestimate for G2. The following criteria are used to evaluate the order of the model:

(i) Bayesian information criterion or Schwartz criterion (Wagenmakers, 2007)

$$BIC = SC = m \ln N - 2 \ln L \tag{1.8}$$

where L is the maximum likelihood function value for the estimated model;

(ii) Hannan-Quinn Information Criterion (Hannan and Quinn, 1979)

$$HQC = N \ln \frac{ECC}{N} + 2m \ln N \tag{1.9}$$

where ECC is the squared deviation sum;

(iii) There know modifications of criteria (1.6) - (1.9), which are used for the synthesis of various models.

We do not consider the interesting direction of choosing the model order for the system with lag variables (LPS). The LPS major problem is to minimise the number of model parameters and the model order. Various schemes for approximating of LPS model parameters are proposed under a priori information. Parametric schemes of I. Fischer (Malinvaud, 1980), L.M. Koyck (Johnston, 1972), S. Almon (Almon, 1965) (a changed Fischer model) use. Durbin-Watson and von Neumann statistics (Johnston, 1972) are used to verifying the of the model order estimation. Under uncertainty, the use of these schemes is associated with solving problem series. A functionally multiple approach (Karabutov, 2010) is proposed for the structural identification of discrete LPS. The LPS structure estimation (Karabutov, 2010) is based on the analysis of spatial frameworks (SF). It is shown that the distributed lag can be interpreted as nonlinearity. The second order secant for SF is the criterion for estimating on the lag length. The approach (Karabutov, 2010) is applied for choosing the model order of the system containing lags for input, output variables, and their combinations.

3. *Correlation of variables (VC).* VC is related to the choice of variables included in the model. Choice of variables indirectly affect the model order. The problem is solved by determining a correlation between the output and the candidate-variable to include in the model structure. Correlation indicators are an indicator for including lag variables in the autoregressive model. The method of canonical correlations or partial correlation (Draper and Smith, 2014; Dubrovsky et al., 2014; Hamilton, 1994) is the basis for an indirect order estimation.
4. *Analysis of the information matrix.* The information matrix $P_q(N)$ is the analysis of the primary object in the modern theory of identification. $P_q(N)$ properties influence on the parametric estimation. As shown Ljung (1999), if estimates of the model order overestimate, then the system is unidentifiable. Hence, the matrix $P_q(N)$ has an incomplete rank. Evaluation issues of the model order are based on the analysis of the information matrix (Mehra, 1974; Söderstrom, 1975; Stoica and Söderstrom, 1982; Young et al., 1980).

Remark 1. The constant excitation ($PE^S_{\alpha,\bar{\alpha}}$) is the property of the information matrix. $PE^S_{\alpha,\bar{\alpha}}$ determines the evaluating possibility of system parameters. As recent studies show (Karabutov, 2021b), the $PE^S_{\alpha,\bar{\alpha}}$ property does not guarantee the structural identification and the identifiability of a nonlinear system. $PE^S_{\alpha,\bar{\alpha}}$ should guarantee the system S-synchronizability.

Other approaches are considered for estimating of the model order by Karabutov (2016). They are not coincident with the considered paradigm. They give the indirect estimate of the system order. The S_{ey}-framework method (SFM) is used to estimating the order of a dynamic system. SFM consider below.

7. SYSTEM NONLINEARITY DEGREE

The model structure choice is based on the system linearity (nonlinearity) evaluation. The system nonlinearity degree estimation is based on the use of correlation and variance analysis (Raibman and Terekhin, 1965). In the general case, the regression between the output $Y(y)$ at time t and the input $X(s)$ at time s of the system is some curve. The nonlinear regression $M\{Y_t|X_s\}$ is the mean square deviation of this curve from some straight line. Evaluate the regression $Y(t)$ nonlinearity degree relative to $X(s)$ as

$$n_{yx}(t,s)=\sqrt{\frac{\min_{a(t,s)b(t,s)} M\{[M\{Y_t|X_s\}-(a(t,s)+b(t,s)x_s)]^2\}}{D\{Y(t)\}}} \tag{1.10}$$

$$a(t,s)=M\{X_t\}-R_{xx}(t,s)\frac{\sqrt{D\{X(t)\}}}{\sqrt{D\{X(s)\}}}M\{X_s\} \tag{1.11}$$

$$b(t,s)=R_{xx}(t,s)\frac{\sqrt{D\{X(t)\}}}{\sqrt{D\{X(s)\}}} \tag{1.12}$$

where $D\{Y(t)\}$ is variance of a random function $Y(t)$, $M\{Y_t|X_s\}$ is conditional mathematical expectation of a random function $y(t)$ relative to a random function $x(s)$, $R_{xx}(t,s)$ is normalised correlation function.

The minimum in (1.10) is reached for $a(t,s)$ and $b(t,s)$ satisfying (1.11), (1.12). We obtain the nonlinearity degree estimate from (1.10) - (1.12)

$$n_{yx}^2(t,s)=\eta_{yx}^2(t,s)-R_{yx}^2(t,s)$$

As is noted Ljung (1999), the nonlinearity degree estimation is the basis for making-decision on the structure of the model. The correlation and variance analysis development given by Billings, 1982; Haber, 1985; Hosseini et al., 2011; Malinvaud, 1969, where higher-order correlation functions used to calculate the of nonlinearity degree. The review (Demetriou and Vassiliou 2009) contains the current state of this problem.

Karabutov (2016b) proposes the approach to nonlinear static system structure choosing. It is based on the identification power evaluation. The identification power is the assessment indicator of the nonlinearity degree.

8. LPS STRUCTURAL IDENTIFICATION

Models with lag variables (LV) are used in econometrics and economics (Demetriou and Vassiliou, 2009; Dhrymes, 1971; Johnston, 1972; Malinvaud, 1969), engineering (Gershenfeld, 1999; Kailath, 1977; Karabutov, 2016b) and medicine (Armstrong, 2006; Gasparrini et al., 2010; Nelson and Schwert, 1974). The delay can have independent or dependent variables. LV considering leads to autocorrelation between variables (Boguslavsky, 2006; Haber, 1985; Malinvaud, 1969) and complicates identifying system parameters. The discrete system model structure choice is based on various schemes of approximation parameters (Almon, 1965;

Johnston, 1972; Malinvaud, 1980). This approach reduces the number of estimated parameters (Karabutov, 2016b).

The approach to structural identification based on approximation schemes of parameters for LV, widely use in econometrics. The system model with LV at input (Dhrymes, 1971) is presented as:

$$y_n = \sum_{i=0}^{k} \omega_i x_{n-i} + \xi_n \tag{1.13}$$

where ω_i are constants that are not equal to zero, x_n is an input variable, ξ_n is a random variable independent of x_i and haves zero mean and constant variance.

The choice of parametric scheme is based on a priori information. The scheme choices so that the input variable influence at time n=0 is the most significant. So, the sequence $\omega_0, \omega_1, \ldots, \omega_k$ (Malinvaud, 1980) containing the first terms of the series is increasing. If the maximum $\omega_0, \omega_1, \ldots, \omega_k$ is reached, then the sequence should decrease. I. Fischer (Fisher, 1937; Gasparrini et al., 2010; Коуск, 1954; Malinvaud, 1980) proposes changing ω_i based on a decreasing arithmetic progression

$$\omega_i = \omega_2\left(1 - \frac{i-2}{k-1}\right), \ 2 \le i \le k \tag{1.14}$$

where ω_0, ω_1 are any numbers, q>0, $\sum_i q^i = 1$. So, the task reduced to the evaluation of parameters $\omega_0, \omega_1, \omega_2$ and k. S. Almon (Armstrong, 2006; Karabutov, 2010; Nelson and Schwert, 1974) modifies I. Fischer model applied the polynomial law for coefficient variation. The Koyck scheme is widely used (Johnston, 1972; Коуск, 1954). Model of change coefficients decreasing geometric progression is presented as:

$$\omega_i = \omega_2 q^{i-2}, \ i = 2.3, \ldots, \infty \tag{1.15}$$

Schemes (1.14) and (1.15) apply to the time series if ω_i decrease with the first term. Laws (1.13), (1.14) do not work on small samples. Conditions for q^i can be interpreted as some probability distribution on a set of non-negative integers. Therefore, q^i can be formally considered as the probability assigned to the integer i. This idea gave direction to the development of parametric laws for ω_i based on probability theory of non-negative integers. The scheme with a logarithmic normal distribution proposes for q^i (Fisher, 1937). q^i interprets as the probability that a normal variable with mean μ and mean square deviation σ belongs to [ln(i), ln(i+1)]. The distribution laws assignment for q^i considers by Solow (1960). Other approaches to the construction of parametric schemes for LV are described by Dhrymes (1971); Fisher (1937); Jorgenson (1964); Theil and Stern, (1960).

Considered parametric schemes minimize the number of unknown parameters. The least squares method or its modifications is used to estimate system parameters (Demetriou and Vassiliou, 2009; Dhrymes, 1971; Johnston, 1972; Malinvaud, 1969). The model structure sets a priori and the parametric identification problem solves. An interactive algorithm (Demetriou and Vassiliou, 2008) for parameters estimating static system with LV proposes. The lag length set and parametric schemes are not applied. Maximum lag length choice based on the analysis of residues considers by Malinvaud, 1969; Solow, 1960; Yoder, 2007. The a priori uncertainty influence on the structure selection and parameters of the system has not been studied.

Yoder (2007) and Karabutov (2016a) propose the structural identification method for systems with a distributed lag under uncertainty. Method is based on the use of geometric frameworks (GF) reflecting of system properties (Karabutov, 2016a). The system linearity criterion introduces. The object linearity criterion is introduced. Algorithms are proposed for distributed lag maximum length determining. They do not require the calculation of statistics. The Durbin-Watson criterion analogue is obtained for this case. Changing laws of the system parameters do not set a priori. The lag length choice is based on the GF analysis (Karabutov, 2012b, 2013). The proposed approach is generalized to a class of autoregressive models.

9. STRUCTURAL IDENTIFIABILITY OF SYSTEMS

The identifiability problem relates to the estimate possibility of dynamic system parameters. The approach to the identifiability assessment bases on the R. Kalman's ideas (Kalman et al., 1969). Further development of these ideas gives by Elgerd (1967) and Lee (1964). R. Lee gives the following definition of identifiability.

Consider the system

$$
\begin{aligned}
X_{n+1} &= AX_n, \\
y_n &= C^T X_n,
\end{aligned}
\tag{1.16}
$$

where $X_n \in \mathbb{R}^m$is a vector state,$A \in \mathbb{R}^{m\times m}$, $y_n \in \mathbb{R}$ is the output system, $n = \mathbb{J}_n = [0, N]$ is discrete time.

Problem: determine conditions under which the system is identifiable based on

$$\mathbb{I}_o = \{y_n, n = \overline{0,N}, N < \infty\}$$

For the case $y_n \in \mathbb{R}^m$, the following sufficient and necessary conditions are given by Lee (1964).

Definition 1. The system (1.16) is identifiable if the matrix A is n-determined based on the vector X measurement.

Definition 2. The system (1.16) is called 1-identifiable if the matrix A is determined based on the measurement y.

The n-identifiability condition is satisfied if the matrix $[X_0 \quad AX_0 \quad A^2 X_0 \quad \dots \quad A^{m-1}X_0]$ is non-degenerate.

The 1-identifiability condition:

(1) System (1.16) is n-identifiable.
(2) The pair (A, C) is observable.

Lee (1964) and Karabutov (2013) consider the identifiability case when the dynamical system order is less than m.

The analysis of publications shows that the system (1.16) identifiability evaluation is performed in a parametric space. Call it IP-Identifiability (IPI). IPI is studied by many authors. The identification results (Aguirregabiria and Mira, (2007); Lee, 1964; Walter, 1982) are presented in the form used in parametric identification.

Walter (1982) introduces the of structural identifiability concepts not based on IPI. Consider two dynamic systems $S_1(U_1,Y_1,A_1)$ and $S_2(U_2,Y_2,A_2)$ with inputs U_1 and U_2, outputs Y_1,Y_2 and parameters A_1 and A_2. Systems describe by models $M_1\left(U_1, \hat{Y}_1, \hat{A}_1\right)$ and $M_2\left(U_2, \hat{Y}_2, \hat{A}_2\right)$.

Definition 3 (Walter, 1982). If condition $M_1\left(\hat{A}_1\right) \approx M_2\left(\hat{A}_2\right)$ holds for $U_1=U_2$, $Y_1=Y_2$ and $\hat{A}_1 \neq \hat{A}_2$, then the M_1, M_2 models are indistinguishable from the observed inputs and outputs.

Definition 4 (Walter, 1982). The parameter $\hat{a}_{1,i} \in \hat{A}_1$ calls structurally globally identifiable if the condition $M_1\left(\hat{A}_1\right) \approx M_2\left(\hat{A}_2\right) \Rightarrow \hat{a}_{1,i} = \hat{a}_{2,i}$ satisfies for almost any $\hat{A}_2 \in \Omega_p$, where Ω_p is parametric space.

Definition 5 (Walter, 1982). The parameter $\hat{a}_{1,i} \in \hat{A}_1$ is structurally locally identifiable if a neighbourhood $O_2\left(\hat{A}_2\right)$ exists such that $M_1\left(\hat{A}_1\right) \approx M_2\left(\hat{A}_2\right) \Rightarrow \hat{a}_{1,i} = \hat{a}_{2,i}$ follows from the condition $\hat{A}_1 \in O_2\left(\hat{A}_2\right)$ for almost any $\hat{A}_2 \in \Omega_p$.

Local identifiability is a necessary condition for global identifiability. A parameter that is structurally locally identifiable, but is not structurally globally identifiable, is called structurally globally unidentifiable. Various approaches and methods can use to verify structural identifiability (Audoly et al., 1998; Avdeenko, 2013). The concept of local parametric identifiability and local identifiability at a point introduce by Bodunov (2012) and given its theoretical justification.

Remark 2. Most of the works is devoted to identifiability do not consider the SI problem. Therefore, the structural identifiability concept does not reflect the SI problem essence. Therefore, the structural identifiability concept does not reflect the SI problem essence. This terminology actively uses in tasks of assessing identifiability. This terminology actively uses in tasks of assessing identifiability. Therefore, we use this terminology to continue analysing obtained results. Next, a concept introduces that directly relates to the structural identifiability of nonlinear systems in a structural space.

Bodunov (2012) proposes criteria for assessing the local identifiability of a linearized system (1.16). For a heterogeneous linear system, the method for estimating local identifiability based on the evaluation of the Lyapunov exponents is developed. Balonin (2010) parametric identifiability criteria introduces by Balonin (2010). Complete identifiability conditions of a linear stationary system from discrete measurements are obtained by Krasovsky (1987) and Stigter et al. (2007).

Various authors studied the IPI of nonlinear systems (see, for example, Chis et al., 2011; Krasovsky, 1987; Saccomani and Thomaseth, 2010). Approach (Stigter and Peeters, 2007) based on the system output sensitivity study is used for the identifiability analysis. The approach effectiveness shows in the identifiability study of a parameter combination. The approach effectiveness shown on the identifiability study of parameter combination. Stigter et al. (2007) obtain parametric identifiability local conditions for various measurement variants of experimental data. Joint observability and identifiability conditions of the linear stationary system are considered. The identifiability critical analysis of biological models is used by Chis et al. (2011). Models for identifiability estimation of nonlinear systems based on Taylor series expansion, identifiability tables, differential algebra consider. Models for the identifiability estimation of nonlinear systems based on Taylor series expansion, identifiability tables, differential algebra are considered. Analysis of practical identifiability (PRI) is given by Saccomani and Thomaseth (2010). PRI is based on the experimental information analysis and the differential algebra application. The approach is applied to the biology tasks.

Ayvazyan et al. (1995) consider identifiability issues of the model described by a system of simultaneous equations. The state matrix rank is the identifiability condition.

So, the model identifiability understands as the possibility of evaluating its parameters. The proposed methods are based on the information matrix non-degeneracy evaluation. Similar results are obtained in the parametric estimation theory. Checking the non-degeneracy of the matrix bases on ensuring the excitation constancy for the input and output of the system. The model structure sets a priori. Therefore, it is not always clear how to understand structural local identifiability. The structure concept widely is used in the assessment of identifiability. The nonlinear system identifiability reduces to parametric identifiability problem based on the application of various linearization methods. This approach does not consider the structural identifiability problem and does not answer the question: how to decide about the nonlinear system structure under uncertainty? PI task cannot be represented as an

SI task. The answer to this question is based on the analysis of GF and the concept of *h*-identifiability (Karabutov, 2016a).

Consider the system

$$\begin{aligned} \dot{X} &= AX + B_{\varphi}\varphi(y) + B_{u}u, \\ y &= C^{T}X, \end{aligned} \tag{1.17}$$

where $u \in \mathbb{R}, y \in \mathbb{R}$ are input and output; $A \in \mathbb{R}^{q \times q}, B_u \in \mathbb{R}^q, B_\varphi \in \mathbb{R}^q, C \in \mathbb{R}^q$; $\varphi(y)$ is a scalar nonlinear function. A is the Hurwitz matrix.

Information set

$$\mathbb{I}_o = \{u(t), y(t),\ t \in \mathbb{J} = [t_0, t_k]\}$$

Problem: estimate the system (1.17) structural identifiability (IS) based on the analysis and processing ${}_o$.

Remark 3. It is shown by Karabutov (2021b), the problems of structural identification and structural identifiability for nonlinear systems interrelate. Structural identification follows from structural identifiability.

IS estimation bases on the framework S_{ey} analysis that reflects properties of the nonlinear part (1.17). The constructing S_{ey} method describes by Karabutov (2016a, 2017). The analysis S_{ey} related to the IS problem solution for the system (1.17). Let us denote the approach described below as *h*-identifiability (HI) to emphasize its difference from IPI-identifiability.

Definition 6. Call the input *u*(*t*) representative if:

(i) The set $\mathbb{I}_o$ provides the solution to the parametric identification problem.
(ii) Input *u*(*t*) provides an informative framework $S_{ey}\left(\mathbb{I}_{N,g}\right)$.

If *u*(*t*) is representative, then S_{ey} is closed.

If *u*(*t*) is representative, then S_{ey} is closed. Denote the S_{ey} height, ask $h(S_{ey})$ where height is the distance between two points on opposite sides of the framework S_{ey}. Let $\mathbb{I}_{N,g}$ be the set for deciding about the structure of the system. Karabutov (2016a, 2017) describes its construction method.

Statement 1 (Karabutov, 2015). Let (1) the system (1.17) linear part is stable, and the nonlinearity $\varphi(\bullet)$belongs to the sector (k_0,k_1); (2) input *u*(*t*) is bounded, piecewise, continuous, and continuously excited; (3) there is $\delta_S>0$ such that $h(S_{ey})\geq\delta_S$. Then the framework S_{ey} is identifiable on the set $\mathbb{I}_{N,g}$.

Definition 7. The S_{ey} -framework satisfying conditions of statement 1 is HI.

Karabutov (2016a) considers the features of the *h*-identifiability.

Remark 4. Not every input satisfying the $PE^{S}_{\alpha,\bar{\alpha}}$ condition guarantees the system SI. In particular, the input can give the so-called insignificant S_{ey} -framework (N S_{ey}-framework) (Karabutov, 2016a).

The decision on structural identifiability (IS) bases on checking the S-synchronizability (Karabutov, 2021b) condition of the system. The verification is based on the analysis of S_{ey} structure properties.

Definition 8. Input $u(t) \in \mathbb{U}$ is called S-synchronizing system (1.17) if the definition domain D_y of the framework S_{ey} has a maximum diameter D_y on the set $\{y(t), t \in \mathbb{J}\}$.

S-synchronizability guarantees structural identifiability (SID) (h_{δ_h}-identifiability) of the system (1.17).

As definition 8 is shows, if the system (1.17) is h_{δ_h}-identifiable, then the framework S_{ey} has the maximum diameter of the region D_y. Karabutov (2016a, 2017, 2022, 2023) presented SID criteria.

10. IDENTIFICATION AND IDENTIFIABILITY OF LYAPUNOV EXPONENTS

Lyapunov exponents (LE) are widely used to analyse the qualitative behaviour of dynamical systems. LE gives behaviour estimates of systems and processes in physics (Thamilmaran et al., 2006), medicine (Porcher and Thomas, 2001), economics (Hołyst and Urbanowicz, 2000) and astronomy (Macek and Redaelli, 2000). Most often, LE is determined based on time series analysis. It is believed that a priori information about the system structure is known. We made the conclusion on structural features of the system based on the LE analysis. The emergence of systems with changing structural properties is an impetus to the development of research on LE. The focus is on calculating the largest (highest, maximum, first) LE (LLE).

Skokos (2010) presents LLE calculation overview for various classes of systems. Gencay and Dechert (1992) propose LE estimation algorithm for an unknown dynamical system. It calculates all LE.

Various algorithms are applied to the calculation of LLE for non-stationary systems based on experimental data analysis. The application of these algorithms is based on the Takens theorem (Takens, 1980). F. Takens showed that the system phase portrait (attractor) can restore (reconstructed) based on a single time series (experimental data). Therefore, the theorem is the basis for calculating various indicators of the dynamic system. LE is such the indicator. Estimate LLE obtains using the Wolf (Wolf et al., 1985) and Rosenstein methods (RWM) (Rosenstein et al., 1993). Various authors generalise and develop RWM. LLE is calculated based on logarithm and interpolation of a time series (Bespalov and Polyakhov, 2016). It is shown that the best results for stationary systems obtain by the Rosenstein method and the interpolation algorithm, and for non-stationary systems by the interpolation algorithm. A model (Bespalov and Polyakhov, 2016) containing the product of an exponent and a phase-shifted sine wave is used to compensate for the non-stationary component in the data. This procedure is not applicable for LE identification to non-stationary systems, since it removes a valuable information layer. Note that the Rosenstein method is a time-consuming procedure associated with the selection and refinement of system parameters. Golovko (2005) proposes a neural network algorithm for LLE estimating. It bases on the use of a multilayer perceptron.

Two major methods (Perederiy, 2012) are used to evaluate Lyapunov exponents on a time series. These procedures are based on a previously reconstructed attractor by the Takens method. First method determines two close trajectories in the reconstructed phase space and tracks their behaviour over a certain time interval (the Benettin algorithm (Benettin et al., 1990)). The Lyapunov exponents spectrum (LES) evaluation is obtained similarly to the LE estimation based on the original system of equations and the equations in variations. The major advantage of this method is its relative simplicity, and the disadvantage is the identifying difficulty of the full LES.

LLE plays the determining role when considering two close trajectories. Second method (Lyapunov, 1992; Dieci et al., 1997) is based on the Jacobian calculation, since the LE can define as of the Jacobi matrix eigenvalues for the system that generated the time series. The advantage of this method is the estimation of non-negative LE on the short time series, and the disadvantage is high sensitivity to noise and errors.

Takens theorem application depends on the properties of the time series (Filatov, 2013). Series properties effects on the effectiveness of the criteria used to evaluate attractor. This explains the implementing complexity of LE identification methods.

So, modifications of Rosenstein, Benetton, Wolf methods and the Takens theorem widely use to LE identification of stationary systems. The properties of the time series describing variables of the system effect on the accuracy of the obtained LE estimates. The methods give an estimate for LLE. Non-stationary systems (NSS) have their own peculiarity (Bylov et al., 1996). In particular, they contain the spectrum of Lyapunov exponents. Therefore, further modification of approaches and methods discussed above is required for NSS. Not always, criteria offer for verifying the received solutions. The LE identifiability conditions have not been studied.

Karabutov (2015b, 2018, 2023) proposes the approach to the LE identification based on the GF analysis. Frameworks describe the LE dynamics in the stationary dynamical system under uncertainty. We will discuss this approach in the following chapters.

11. IDENTIFICATION OF PARAMETRIC CONSTRAINTS IN STATIC SYSTEMS UNDER UNCERTAINTY

Parametric constraints (PC) are the basis for creating effective identification and management systems. In identification systems, the consideration of constraints reflects the actual operating conditions and guarantees the use of robust parametric estimation algorithms. Therefore, the PC is the structure element of the system.

Many authors study the application of parametric constraints in identification systems. The general method of moments (GMM) (Gagliardini et al., 2010) is used to design the model for estimating the main asset. Two kinds of restrictions impose. These are restrictions on the variable current value and on the maximum value for the variable in a certain interval. The limit on the moment set a priori. GMM (Giacomini and Rossi, 2009; Magnusson and Mavroeidis, 2012; Rossi, 2005) is used to determine the relationship between PC and parameter stability. The constraint is the equality at the moment. Bardsley (2004) studies the electromagnetism process identification problem. The problem solution given by the Levenberg-Marquardt iterative algorithm. Constraints set a priori as lower and upper bounds for parameters. The correcting problem of boundaries of constraints considered. Constraints set a priori as lower and upper bounds for parameters. The correcting problem of boundaries of constraints considered. The parametric identification problem (Palanthandalam-Madapusia et al., 2010; Van Peltb and Bernstein, 2001) is studies for the dynamic system with an ellipsoidal PC. A heuristic algorithm for solving the problem is proposed. As shown Palanthandalam-Madapusia et al. (2010), there is some indefinite quadratic constraint, depending on the level of uncontrolled noise in the finite data. Constraints are set a priori. The parameter estimation of the polynomial model, based on an ellipsoidal algorithm, is solved by Corrêa et al. (2002). Process analysis is the basis for setting constraints on parameters. Chadeev and Gusev (2012) consider a static object identification. PC domain sets a priori. The a priori defined PC area influence on the properties of the model describing electrical muscle stimulation is given by Chia et al. (1991). It is noted ignoring PC effects on results of parametric evaluation. The PC correction is proposed. It is shown that consider constraints improves the predictive properties of the model. The identification algorithm of a dynamic object with a priori set PC as the equality is proposed by Mosteller and Tukey (1977). The identification algorithm is based on the variational method of optimizing the Hamilton function. Vanli and Del Castillo, (2007) proposes the identification procedure of the system with feedback. The procedure improves parameter estimates based on the use of constraints. It is shown how to form constraints based on a priori data about the system. Hametner and Jakubek, (2011) considers the using a priori knowledge problem about an object. A priori information is presented as constraints for the nonlinear system.

Mead and Renaut (2010) studies the problem of estimating parameters used quadratic constraints. Physical assumptions and the regularization method are used to form constraints. Identification algorithms are proposed and their implementation is shown. Mazunin and Dvoinikov (2010) considers the PC estimation problem for vibration damping in nonlinear control systems. The analytical method is used to determine the constraints.

Some approaches to PC obtaining are considered by Karabutov (2008, 2013, 2016a). The method for obtaining PC is described for a dynamic linear system. It is based on the analysis of the observed information portrait. The domain of parametric constraints (PCD) is interpreted as an inequality from above on the parameter vector norm. Its assessment is based on the majority model. PC often use in the synthesis of identification systems. PCD sets a priori. The PCD construction task was not considered under uncertainty. There is a class of complex systems, and we cannot set the PCD a priori for them. This problem solution is relevant. We consider methods for obtaining a PC in the following chapters for this case.

12. APPROACHES TO CHOOSING MODEL STRUCTURE

In identification theory, approaches based on the parametrization are widely used to select the model structure. The implementation of this methodology is based on the use of various approximation schemes (Gabor, 1961; Graupe and Cline, 1975; Ivakhnenko, 1975; Parsen, 1974). The autoregressive model structure choice is based on the use of Volterra series and the group method of data handling (Ivakhnenko, 1975). Combined schemes (Isermann et al., 1974; Kashyap and Rao, 1976) are used to increasing the of decision-making efficiency.

The review (Perelman, 1983) contains the analysis of methods for selecting the model structure based on the use of training and examination signals. Decision-making is based on the application of various criteria. Note that applied approaches are used a priori information. An iterative approach (Box and Jenkins, 1976) proposes for the design of forecasting and management models. The algorithm implements structural identification on a class of models. The structure estimating problem of the autoregressive model on a defined class considers by Kashyap and Rao, 1976. Various criteria are considered for the choice of models and rules for testing hypotheses. Curve linearization method (Mosteller and Tukey, 1977) is used to select the structure of the regression model. The linearization method of static dependence is proposed in structural space (Karabutov, 2016a, 2016b). Other approaches to the SI are considered by Perelman, 1983. GF (S_{ey}-framework) analysis method (Karabutov, 2012, 2012a, 2012b, 2013, 2015a, 2020) is the most adequate approach to choosing the model structure.

13. CONCLUSION

Structural identification various aspects are considered. We show the problem is multifaceted and includes various directions. There are various approaches to solving various aspects of the structural identification problem. Most of them based on consideration of a priori information. Indirect approaches are the basis for evaluating the model structure. It makes difficult to design a common approach to solving this problem. This is because of the emergence of new subject areas (SA) that are related to identification tasks. SA analysis requires the development of novel approaches that must consider structural relationships and parameters of the system. The evaluation of new structural characteristics (for example, Lyapunov exponents) gives a new direction to the development of structural identification. This explains the significant expansion of the SI tasks that are discussed in the chapter. The versatility and difficulties of the SI problem follow from the formulation of the problem described in section 3. In addition, we wanted to analyse the SI state of system positions. Aghakouchak et al. (2000), Ghafooripour and Kiamehr (2000) contain results not reflected in the chapter.

In the following chapters of this part, we will consider formalized approaches to solving the SI problem. They are based on the use of a special class of frameworks. These structures (GF) are introduced in Section 9 and provide solutions to new directions of structural identification.

REFERENCES

Aghakouchak, A. A., Ghafouripou, A., & Kiamehr, H. (2000). An overview of system identification methods and applications. Part I: Methods of system identification and dynamic tests. *Conference: 4th international conference on coasts, ports & marine structure at: Bandar Abbas, Iran, 1*, 1-31.

Aguirregabiria, V., & Mira, P. (2007). Dynamic Discrete Choice Structural Models: A Survey. *Working Paper 297*. University of Toronto.

Almon, S. (1965). The distributed lag between capital appropriations and expenditures. *Econometrica, 33*(1), 178–196. DOI: 10.2307/1911894

Armstrong, B. (2006). Models for the relationship between ambient temperature and daily mortality. *Epidemiology (Cambridge, Mass.), 17*(6), 624–631. DOI: 10.1097/01.ede.0000239732.50999.8f PMID: 17028505

Audoly, S., D'Angio, L., Saccoman, M. P., & Cobelli, C. (1998). Global identifiability of linear compartmental models - a computer algebra algorithm. *IEEE Transactions on Automatic Control, 45*, 36–47. PMID: 9444838

Avdeenko, T. V. (2013). Identification of linear dynamical systems using concept of parametric space separators. *Automated Software Engineering, 1*(3), 16–23.

Ayvazyan, S. A., Enyukov, I. S., & Meshalkin, L. D. (1985). *Applied Statistics: Dependency Research. Reference edition*. Finansy i Statistika.

Balonin, N. A. (2010). *Identifiability theorems*. St. Petersburg: Publishing house "Polytechnic".

Bardsley, J. M. (2004). A bound-constrained Levenburg-Marquardt algorithm for a parameter identification problem in electromagnetics. http://www.math.umt.edu/bardsley/papers/EMopt04.pdf

Benettin, G., Galgani, L., Giorgilli, A., & Strelcyn, J.-M. (1980). Lyapunov characteristic exponents for smooth dynamical systems and for Hamiltonian systems; a method for computing all of them. Part. I. *Theory. Pt. II: Numerical applications. Meccanica, 15*, 9–30. DOI: 10.1007/BF02128236

Berryman, J. G. (1978). Choice of operator length for maximum entropy spectral analysis. *Geophysics, 43*(7), 1384–1391. DOI: 10.1190/1.1440902

Besekersky, V. A., & Popov, E. P. (1975). *Theory of automatic control systems* (Third edition, revised). Nauka.

Bespalov, A. V., & Polyakhov, N. D. (2016). Comparative analysis of methods for estimating first Lyapunov exponent. *Modern problems of science and education, 6*, 7.

Billings, S. A. (1982). Structure Detection and Model Validity Tests in the Identification of Nonlinear Systems. Research Report. *ACSE Report 196*, Department of Control Engineering, University of Sheffield.

Bodunov, N. A. (2012). Introduction to theory of local parametric identifiability. *Differential equations and control processes*, 1-137.

Boguslavsky, I. A. (2006). *Polynomial approximation for nonlinear estimation and control problems*. Fizmatlit.

Box, G. P., & Jenkins, G. M. (1976). *Time series analysis: Forecasting and control.* Holden-Day.

Britenkov, A. K. & Dedus, F. F. (2012). Prediction of time sequences using a generalized spectral-analytical method. Mathematical modelling. Optimal control. *Bulletin of the Nizhny Novgorod University named after N.I. Lobachevsky*, 28-32.

Bylov, F., Vinograd, R. E., Grobman, D. M., & Nemytskii, V. V. (1966). *Theory of Lyapunov exponents and its application to problems of stability*. Nauka.

Chadeev, V. M., & Gusev, S. S. (2012). Identification with restrictions. determining a static plant parameter estimate. Proceedings of the VII International Conference System Identification and Control Problems" SICPRO 'OS Moscow January 28-31, 2008. V.A. Trapeznikov Institute of Control Sciences. Moscow: V.A. Trapeznikov Institute of Control Sciences, 261-269.

Chia, T. L., Chow, P. C., & Chizeck, H. J. (1991). Recursive parameter identification of constrained systems: An application to electrically stimulated muscle. *IEEE Transactions on Biomedical Engineering, 38*(5), 429–442. DOI: 10.1109/10.81562 PMID: 1874525

Chis, O.-T., Banga, J. R., & Balsa-Canto, E. (2011). Structural identifiability of systems biology models: A critical comparison of methods. *PLoS One*, *6*(4), 1–16. DOI: 10.1371/journal.pone.0027755 PMID: 22132135

Corrêa, M. V., Aguirre, L. A., & Saldanha, R. R. (2002). Using steady-state prior knowledge to constrain parameter estimates in nonlinear system identification. *IEEE Transactions on Circuits and Systems. I, Fundamental Theory and Applications, 49*(9), 1370–1381. DOI: 10.1109/TCSI.2002.802345

Demetriou, I. C., & Vassiliou, E. E. (2008). A distributed lag estimator with piecewise monotonic coefficients. *Proceedings of the World Congress on Engineering. 2008.* V. 2. *WCE 2008, July 2 - 4, 2008, London, U.K.*

Demetriou, I. C., & Vassiliou, E. E. (2009). An algorithm for distributed lag estimation subject to piecewise monotonic coefficients. *International Journal of Applied Mathematics, 39*(1), 1–10.

Dhrymes, P. J. (1971). *Distributed Lags: Problems of estimation and formulation.* Holden-Day.

Dieci, L., Russell, R. D., & Van Vleck, E. S. (1997). On the computation of Lyapunov exponents for continuous dynamical systems. *SIAM Journal on Numerical Analysis, 34*(1), 402–423. DOI: 10.1137/S0036142993247311

Draper, N. R., & Smith, H. (2014). *Applied Regression Analysis.* John Wiley & Sons.

Dubrovsky, A. M., Mkhitaryan, V. S., & Troshin, L. I. (2003). *Multidimensional statistical methods.* Finance and Statistics.

Elgerd, O. I. (1967). *Control Systems Theory.* Mc Graw-Hill.

Eykhoff, P. (Ed.). (1974). *Modern Identification Methods.* John Wiley and Sons.

Filatov, V. V. (2013). Structural characteristics of anomalies of geophysical fields and their use in forecasting. *Geophysics, geophysical instrumentation,* 4(16), 34-41.

Fisher, I. (1937). Note on a Short-cut Method for Calculating Distributed Lags. *Bulletin de l'Institut International de Statistique, 29,* 323–327.

Gabor, D., Wiby, W. P. L., & Woodcock, R. A. (1961). A universal nonlinear filter, predictor and simulator which optimizes itself by learning processes. *Proceedings of the IEEE - Part B: Electronic and Communication Engineering, 108,* 422–438.

Gagliardini, P., Gouriéroux, C., & Renault, E. (2010). Efficient Derivative Pricing by the Extended Method of Moments. *Swiss Finance Institute Research,* 10-07.

Gasparrini, A., Armstrong, B. M., & Kenwardb, G. (2010). Distributed lag nonlinear models. *Statistics in Medicine, 29*(21), 2224–2234. DOI: 10.1002/sim.3940 PMID: 20812303

Gencay, R., & Dechert, W. D. (1992). An algorithm for the n Lyapunov exponents of an n-dimensional unknown dynamical system. *Physica D. Nonlinear Phenomena, 59*(1-3), 142–157. DOI: 10.1016/0167-2789(92)90210-E

Gershenfeld, N. (1999). *The Nature of Mathematical Modelling.* Cambridge University Press.

Ghafooripour, A., & Kiamehr, H. (2000). An overview of system identification methods and applications part ii: theory, type of tested structures, history and prospective of system identification. *4th International conference on coasts, ports & marine structure, Nov 2000 Shahid Rajaee Port Complex, Bandar Abbass*, 509-540.

Giacomini, R., & Rossi, B. (2009). Model comparisons in unstable environments. *ERID Working Paper 30*, Duke.

Giannakis, G. B., & Serpedin, E. A. (2001). A bibliography on nonlinear system identification. *Signal Processing*, *81*(3), 533–580. DOI: 10.1016/S0165-1684(00)00231-0

Goldenberg, L., Matiushkin, B. D., & Poliak, M. (1990). *Digital signal processing: Handbook*. Radio and Communications.

Golovko, V. A. (2005). Neural network methods of processing chaotic processes. Scientific session of MEPhI-2005. VII All-Russian Scientific and Technical Conference Neuro-Informatics-2005: Lectures on neuroinformatics. Moscow: MIPhI, 43-88.

Graupe, D. (1975). *Identification of systems*. Litton Educational Co.

Graupe, D., & Cline, W. K. (1975). Derivation of ARMA parameters and orders from pure AR models. *Int. J. Syst. ScL, 10*(2), 101–106.

Greene, B. R. (2004). *The fabric of the cosmos: space, time and the texture of reality*. Random House, Inc.

Haber, R. (1985). Nonlinearity tests for dynamic processes. *IFAC proceedings volumes, 18*(5), 409-414.

Hametner, C., & Jakubek, S. (2011). Nonlinear identification with local model networks using GTLS techniques and equality constraints. *IEEE Transactions on Neural Networks*, *22*(9), 1406–1418. DOI: 10.1109/TNN.2011.2159309 PMID: 21788188

Hamilton, J. D. (1994). *Time Series Analysis*. Princeton University Press. DOI: 10.1515/9780691218632

Hannan, E. J., & Quinn, B. G. (1979). The Determination of the order an autoregression. *Journal of the Royal Statistical Society. Series B, Statistical Methodology*, *41*(2), 190–195. DOI: 10.1111/j.2517-6161.1979.tb01072.x

Hastie, T., Tibshirani, R., & Friedman, J. (2001). *The elements of statistical learning*. Springer. DOI: 10.1007/978-0-387-21606-5

Hołyst, J. A., & Urbanowicz, K. (2000). Chaos control in economic model by time-delayed feedback method. *Physica A*, *287*(3–4), 587–598. DOI: 10.1016/S0378-4371(00)00395-2

Hosseini, S. M., Johansen, T. A., & Fatehi, A. (2011). Comparison of nonlinearity measures based on time series analysis for nonlinearity detection. *Modeling, Identification and Control, 32*(4), 123–140. DOI: 10.4173/mic.2011.4.1

Isermann, R., Baur, V., Bamberger, W., Kneppo, P., & Siebert, H. (1974). Comparison of six on-line identification and parameter estimation methods. *Automatica, 10*(1), 81–103. DOI: 10.1016/0005-1098(74)90012-0

Isermann, R., & Münchhof, M. (2011). *Identification of dynamic systems*. Springer-Verlag Berlin Heidelberg. DOI: 10.1007/978-3-540-78879-9

Ivakhnenko, A. G. (1975). *Long-term forecasting and management of complex systems*. Technika.

Jenkins, G. M., & Watts, D. G. (1969). *Spectral analysis and its applications*. Holden-Day.

Johnston, J. (1972). *Econometric Methods*. McGraw-Hill.

Jones, R. H. (1976). Autoregression order selection. *Geophysics, 41*(4), 771–773. DOI: 10.1190/1.1440649

Jorgenson, D. W. (1964). Minimum variance, linear, unbiased seasonal adjustment of economic time series. *Journal of the American Statistical Association, 59*(307), 681–724. DOI: 10.1080/01621459.1964.10480722

Kailath, T. (Ed.). (1977). *Linear Least-Squares Estimation,* Stroudsburg / Pennsylvania: Dowden, Hutchinson and Ross, Inc.

Kalman, R., Falb, P. L., & Arbib, M. A. (1969). *Topics in mathematical system theory*. McGraw-Hill.

Karabutov, N. (2012). About structures of state systems identification of static object with hysteresis. *Int. J. Sensing. Computing and Control, 2*(2), 59–69.

Karabutov, N. (2013). Identification of parametrical restrictions in static systems in conditions of uncertainty. *International Journal of Intelligent Systems and Applications, 5*(4), 43–54. DOI: 10.5815/ijisa.2013.04.04

Karabutov, N. (2013). Structural identification of systems with distributed lag. *International Journal of Intelligent Systems and Applications, 5*(11), 1–10. DOI: 10.5815/ijisa.2013.11.01

Karabutov, N. (2015a). Structural identification of nonlinear dynamic systems. *International Journal of Intelligent Systems and Applications, 7*(9), 1–11. DOI: 10.5815/ijisa.2015.09.01

Karabutov, N. (2015b). Structural methods of estimation Lyapunov exponents linear dynamic system. *International Journal of Intelligent Systems and Applications, 7*(10), 1–11. DOI: 10.5815/ijisa.2015.10.01

Karabutov, N. (2016a). Structural identification of dynamic systems with hysteresis. *International Journal of Intelligent Systems and Applications, 8*(7), 1–13. DOI: 10.5815/ijisa.2016.07.01

Karabutov, N. (2016b). System with distributed lag: Adaptive identification and prediction. *International Journal of Intelligent Systems and Applications, 8*(6), 1–13. DOI: 10.5815/ijisa.2017.06.01

Karabutov, N. (2017). Structural methods of design identification systems. In Uvarova, L. A., Nadykto, A. B., & Latyshev, A. V. (Eds.), *In Nonlinearity problems, solutions and applications* (Vol. 1, pp. 233–274). Nova Science Publishers, Inc.

Karabutov, N. (2021a). Geometrical frameworks in identification problem. *Intelligent Control and Automation, 12*(2), 17–43. DOI: 10.4236/ica.2021.122002

Karabutov, N. (2023). Identifiability and detectability of Lyapunov exponents in robotics. In Mellal, M. A. (Ed.), *Design and Control Advances in Robotics* (pp. 152–174). IGI Globalss Publisher of Timely Knowledge.

Karabutov, N. N. (2008). *Adaptive identification of systems: Information synthesis.* URSS.

Karabutov, N. N. (2010). Structures, fields and methods of identification of nonlinear static systems in the conditions of uncertainty," *Intelligent control and automation, 1*, 59-67.

Karabutov, N. N. (2012a). About structures of state systems identification of static object with hysteresis. International *journal sensing, computing and control,* 2(2), 59-69.

Karabutov, N. N. (2012b). Structural identification of static systems with distributed lags. *International journal of control science and engineering,* 2(6), 136-142.

Karabutov, N. N. (2016a). *Structural identification of static objects: Fields, structures, methods.* URSS.

Karabutov, N. N. (2016b). *Structural identification of systems: Analysis of information structures.* URSS.

Karabutov, N. N. (2018). *Structures in identification problems: Construction and analysis.* URSS.

Karabutov, N. N. (2020). Frameworks application for estimation of Lyapunov exponents for systems with periodic coefficients. *Mekhatronika, Avtomatizatsiya. Upravlenie, 21*(1), 3–13.

Karabutov, N. N. (2021b). *Introduction to the structural identifiability of nonlinear systems*. URSS/LENAND.

Karabutov, N. N. (2022). Identifiability and detectability of Lyapunov exponents for linear dynamical systems. *Mekhatronika, Avtomatizatsiya. Upravlenie, 23*(7), 339–350.

Kashyap, R., & Rao, A. (1976). *Dynamic stochastic models from empirical data*. Academic Press.

Kashyap, R. L. (1980). Inconsistency of the aic rule for estimation the order of autoregressive models. *IEEE Transactions on Automatic Control, 25*(5), 996–998. DOI: 10.1109/TAC.1980.1102471

Kay, S. M. (1999). *Modern spectral estimation: Theory and application*. Prentice-Hall, Inc., Englewood Cliffs.

Kerschen, G., Worden, K., Vakakis, A., & Golinval, J. (2006). Past, present and future of nonlinear system identification in structural dynamics. *Mechanical Systems and Signal Processing, 20*(3), 505–592. DOI: 10.1016/j.ymssp.2005.04.008

Koh, C. G., & Perry, M. J. (2010). *Structural identification and damage detection using genetic algorithms*. CRC Press.

Koyck, L. M. (1954). *Distributed Lags and Investment Analysis*. North-Holland Publishing Company.

Krasovsky, A. A. (Ed.). (1987). *Handbook of the theory of automatic control*. Nauka.

Kuntsevich, V. M., & Lychak, M. M. (1985). *Synthesis of optimal and adaptive control systems: A game approach*. Naukova dumka.

Lee, R. (1964). *Optimal estimation, identification, and control*. MIT Press.

Ljung, L. (1999). *System identification: Theory for the User*. Prentice Hall PTR.

Ljung, L. (2000). *System Identification Toolbox User's Guide. Computation. Visualization. Programming. Version 5*. The MathWorks Inc.

Lyapunov, A. M. (1992). *General problem of motion stability*. CRC Press. DOI: 10.1080/00207179208934253

Macek, W. M., & Redaelli, S. (2000). Estimation of the entropy of the solar wind flow. *Physical Review. E, 62*(5), 6496–6504. DOI: 10.1103/PhysRevE.62.6496 PMID: 11101986

Magnusson, L., & Mavroeidis, S. (2012). Identification using stability restrictions. https://econ.sciences-po.fr/sites/default/files/SCident32s.pdf

Malinvaud, E. (1969). Méthodes statistiques de l'économétrie. Deuxième édition. London-Paris.

Malinvaud, E. (1980). *Statistical methods in econometrics* (3rd ed.). North-Holland Publishing Co.

Marple, S. L.Jr. (1987). *Digital spectral analysis: With applications*. Prentice-Hall.

Mazunin, V. P., & Dvoinikov, D. A. (2010). Parametric constraints in nonlinear control systems of mechanisms with elasticity. *Electrical Engineering, 5*, 9–13.

Mead, J. L., & Renaut, R. A. (2010). Least squares problems with inequality constraints as quadratic constraints. *Linear Algebra and Its Applications, 432*(8), 1936–1949. DOI: 10.1016/j.laa.2009.04.017

Mehra, R. K. (1974). Optimal input signals for parameter estimation in dynamic systems. A survey and new results. *IEEE Transactions on Automatic Control, 19*(6), 753–768. DOI: 10.1109/TAC.1974.1100701

Mosteller, F., & Tukey, J. (2019). *Data analysis and regression: A second course in statistics* (1st ed.). Pearson.

Mosteller, F., & Tukey, J. W. (1977). *Data Analysis and Regression: A Second Course in Statistics*. Addison-Wesley.

Nelson, C. R., & Schwert, G. W. (1974). Estimating the parameters of a distributed lag model from cross-section data: The Case of hospital admissions and discharges. *Journal of the American Statistical Association, 69*(347), 627–633. DOI: 10.1080/01621459.1974.10480179

Palanthandalam-Madapusia, H. J., Van Peltb, T. H., & Bernstein, D. S. (2010). Parameter consistency and quadratically constrained errors-in-variables least-squares identification. *International Journal of Control, 83*(4), 862–877. DOI: 10.1080/00207170903470666

Parsen, E. (1974). Some recent advances in time series modelling. *IEEE Transactions on Automatic Control, 19*(6), 723–730. DOI: 10.1109/TAC.1974.1100733

Perederiy, Y. A. (2012). Method for calculation of Lyapunov exponents spectrum from data series. *Izvestiya VUZ. Izvestiya Vyssih Ucebnyh Zavedenij. Prikladnaya Nelinejnaya Dinamika, 20*(1), 99–104.

Perelman, I. I. (1983). Model structure selection Methodology in control objects identification. *Automation and telemechanics, 11*, 11 5–29.

Porcher, R., & Thomas, G. (2001). Estimating Lyapunov exponents in biomedical time series. [R]. *Physical Review. E, 64*(1), 010902. DOI: 10.1103/PhysRevE.64.010902 PMID: 11461218

Prokhorov, S. A., & Grafkin, V. V. (2010). *Structural and spectral analysis of random processes*. SNC RAS.

Raibman, N. S. (1978). Identification of control objects (Review). *Automation and telemechanics, 6*, 80–93.

Raibman, N. S., & Chadeev, V. M. (1975). *Construction of models of production processes*. Energiya.

Raibman, N. S., & Terekhin, A. T. (1965). Dispersion methods of random functions and their application for the study of nonlinear control objects. *Automation and telemechanics, 26*, 500–509.

Rosenstein, M. T., Collins, J. J., & De Luca, C. J. (1993). A practical method for calculating largest Lyapunov exponents from small data sets Source. *Physica D. Nonlinear Phenomena, 65*(1-2), 117–134. DOI: 10.1016/0167-2789(93)90009-P

Rossi, B. (2005). Optimal tests for nested model selection with underlying parameter instability. *Econometric Theory, 21*(5), 962–990. DOI: 10.1017/S0266466605050486

Saccomani, M. P., & Thomaseth, K. (2010). Structural vs practical identifiability of nonlinear differential equation models in systems biology. In Rogato, A., Zazzu, V., & Guarracino, M. (Eds.), *Dynamics of mathematical models in biology: Bringing mathematics to life* (pp. 31–42). Springer.

Sirca, G. F.Jr, & Adeli, H. (2012). System identification in structural engineering. *Scientia Iranica, 19*(6), 1355–1364. DOI: 10.1016/j.scient.2012.09.002

Skokos, Ch. (2010). *The Lyapunov Characteristic Exponents and Their Computation* (Vol. 790). Lect. Notes Phys. DOI: 10.1007/978-3-642-04458-8_2

Söderstrom, T. (1975). Comments on order assumption and singularity of information matrix for pulse transfer function models. *IEEE Transactions on Automatic Control, 20*(3), 445–447. DOI: 10.1109/TAC.1975.1100949

Solow, R. (1960). On a family of lag distributions. *Econometrica*, *28*(2), 393–406. DOI: 10.2307/1907729

Stepin, V. S. (Ed.). (2001). *New Philosophical Encyclopedia: In 4 volumes* / Edited by. Moscow: Mysl.

Stigter, J. D., & Peeters, R. L. M. (2007). On a geometric approach to the structural identifiability problem and its application in a water quality case study. *Proceedings of the European Control Conference 2007 Kos,* Greece, July 2-5, 2007, 3450-3456. DOI: 10.23919/ECC.2007.7068560

Stoica, P., & Söderstrom, T. (1982). On non-singular information matrices and local identifiability. *International Journal of Control*, *35*(2), 323–329. DOI: 10.1080/00207178208932896

Takens, F. (1980). Detecting strange attractors in turbulence. In D. A. Rand, L.-S. Young (Eds.), *Dynamical Systems and Turbulence. Lecture Notes in Mathematics* (366–381). Berlin: Springer-Verlag, 898.

Thamilmaran, K., Senthilkumar, D. V., Venkatesan, A., & Lakshmanan, M. (2006). Experimental realization of strange nonchaotic attractors in a quasiperiodically forced electronic circuit. *Physical Review. E*, *74*(9), 036205. DOI: 10.1103/PhysRevE.74.036205 PMID: 17025727

Theil, H., & Stern, R. M. (1960). A simple unimodal lag distribution. *Metroeconomica*, *12*(2-3), 111–119. DOI: 10.1111/j.1467-999X.1960.tb00277.x

Van Gig, J. (1978). *Applied general systems theory*. Harper & Row.

Van Peltb, H. D., & Bernstein, S. (2001). Quadratically constrained least squares identification. *Proceedings of the American Control Conference,* Arlington, VA June 25-27, 2001, 3684-3689. DOI: 10.1109/ACC.2001.946207

Vanli, O. A., & Del Castillo, E. (2007). Closed-Loop System Identification for Small Samples with Constraints. *Technometrics*, *49*(4), 382–394. DOI: 10.1198/004017007000000146

Vinogradov, I. M. (Ed.). (1979). *Mathematical Encyclopedia* (Vol. 2). Soviet Encyclopedia.

Wagenmakers, E.-J. (2007). A practical solution to the pervasive problems of p values. *Psychonomic Bulletin & Review*, *14*(5), 779–804. DOI: 10.3758/BF03194105 PMID: 18087943

Walter, E. (1982). *Identifiability of state space models*. Springer-Verlag. DOI: 10.1007/978-3-642-61823-9

Wen-Ming, S. (2010). Parameter Estimation with constraints based on variational method. *Journal Marine. Sci. Appl*, *9*, 105–108.

Wolf, A., Swift, J. B., Swinney, H. L., & Vastano, J. A. (1985). Determining Lyapunov exponents from a time series. *Physica*, *16D*(16), 285–301.

Woodside, C. M. (1971). Estimation of the order of linear systems. *Automatica*, *7*(6), 727–733. DOI: 10.1016/0005-1098(71)90010-0

Yoder, J. (2007). Autoregressive distributed lag models. *WSU Econometrics*, *II*, 91–115.

Young, P. C., Jakeman, A. J., & McMurtrie, R. (1980). An instrumental variable method for model order identification. *Automatica*, *16*(3), 281–294. DOI: 10.1016/0005-1098(80)90037-0

Chapter 2
Geometrical Framework Application in Identification Systems

ABSTRACT

In Chapter 1, we noted the lack of formalized methods of structural identification. An approach to SI based on the construction and analysis of geometrical structures (GF) was considered. In this chapter, we analyse the application areas of GF. Various classes of structures that are used in structural identification tasks are considered. We analyze approaches to SI based on the GF use, and show the effect of variables (excitation constancy) on the estimate of system structural parameters. We show how to use geometric frameworks in the example of systems with lag variables. The GF role in building attractors is explained.

1. INTRODUCTION

Many papers are devoted to conceptual questions of various approaches to creation, methods in identification problems (Chen & Narendra, 2004; Lauer et al., 2011; Pillonetto et al., 2014; Pintelon and Schoukens, 2012; Saha et al., 2009; Toth et al., 2012; Varna et al., 2008). The created theoretical constructions and approaches are the compilation of existing procedures and algorithms. They have directed to the accounting of system unique features that to give required quality to the identification system. Such the approach to the improvement of identification

DOI: 10.4018/979-8-3373-0822-7.ch002

methods is dominating (see, for example, Carino Carino et al, 2018; Nagarajaiah, 2017; Papadopoulos et al, 2017; Roettgen et al, 2017; Teh et al, 2012).

The second direction of frameworks (FR) use is associated with the analysis of the identification system quality. This direction is based on ideas of dynamic systems (DS) qualitative theory (Michel et al., 2001; Shilnikov et al., 2001; Wiggins, 2003). The mathematical mappings describing system geometrical images change are applied in this case. GF is images of attractors or phase portraits which will be formed a priori (Li and Min, 2018; Lü and Zhang, 2007; Pershin and Slipko, 2018). Attractor recover bases on the analysis of time series and Takens theorem under uncertainty. The attractor reconstruction process is very labour-consuming. The problem of identification has specificity. Therefore, the direct transfer of DS qualitative theory methods is not applicable here. GF design methods differ from the results obtained by Li and Min (2018), Lü and Zhang (2007) Pershin and Slipko (2018). The basis of the proposed approach is given by Karabutov (2018) for the problem solution structural identification for static systems. Further, got results are generalized to dynamic systems (Karabutov, (2017, 2018)). Choice of the mapping describing geometrical frameworks depends on the examined system. The GF design method is proposed by Karabutov, (2017, 2018)).

The estimation problem of Lyapunov exponents occupies a specific place. GF applied for the LE estimation differs from the frameworks used in structural identification issues. The initial data array formation for LE estimation is based on the same ideas, as the GF design for the solution of dynamic systems structural identification problem. But in this case, the subset of DS free motion trajectories (Karabutov, 2018) is allocated. The analysis of this subset allows forming the time series describing the LE change. Time series is the basis for the design of the mapping describing the LE change dynamics in a special structural space (Karabutov, (2018a, 2018c). The change calculation of LE is fulfilled by the standard formula (Bylov et al., 1966).

GF is the basis for the structural identifiability (SI) estimation of nonlinear systems. SI estimation is based on the analysis of frameworks proposed by Karabutov (2017,2018c). The property of a system S-synchronizability (Karabutov, 2018b) plays the major role in the SI analysis. The considered SI concept significantly differs from parametrial identifiability (Bodunov, 2012; Chis et al., 2011; Krasovsky, 1987; Saccomani, 2013; Stigter and Peeters, 2007).

The system structure estimation problem with the distributed lag is considered by Anishchenko et al., 2007; Karabutov, 2018b; Malinetskiyand Potapov, 2000; Rosenstein et al., 1993; Takens, 1980; Wolf et al., 1985. Its decision is based on the application of parametric schemes for the system identification with lag variables. Such a concept to the parameter estimation is based on the available a priori identification. In the following chapters, we consider the GF application to evaluate the structure of lag systems and LE. The major difference between proposed by GF and frameworks from (Li and Min, 2018; Lü and Zhang, 2007; Pershin and Slipko, 2018) consists that mathematical mapping (GF) is not postulated a priori, and is determined based on data processing.

Further, we consider the main directions of the GF application, generalizing and developing results (Karabutov, (2017, 2018c). They contain the following areas of the identification theory.

(1) Structural identification of the nonlinear system.
(2) The estimation of Lyapunov exponents.
(3) Structural identifiability of the nonlinear system.
(4) The system structure estimation with lag variables.
(5) The system phase portrait reconstruction of the time series.

2. PROBLEM STATEMENT

Consider dynamic system

$$\begin{aligned} \dot{X} &= AX + \varphi(y)I + Bu, \\ y &= C^T X, \end{aligned} \tag{2.1}$$

where $u \in \mathbb{R}$, $y \in \mathbb{R}$ are the input and the output; $A \in \mathbb{R}^{m\times m}$, $B \in \mathbb{R}^m$, $I \in \mathbb{R}^m$ $C \in \mathbb{R}^m$ are matrices of corresponding dimensions; $\varphi(y)$ is a scalar nonlinear function. A is the Hurwitz matrix. ξ is the linear combination of elements X. We suppose that $\chi=\varphi(y)$ belongs to the set

$$\chi \in F_{\varphi} = \{\upsilon_1\xi^2 \le \varphi(\xi)\xi \le \upsilon_2\xi^2,\ \xi \ne 0, 0,\ \upsilon_1 \ge 0,\ \upsilon_2 < \infty\}. \tag{2.2}$$

Therefore, further, we consider the case when $\varphi(y)$ describe by the algebraic equation.

Let the informational set be known for the system (2.1)

$$\mathbb{I}_o = \{u(t), y(t), t \in \mathbb{J} = [t_0, t_k]\}. \tag{2.3}$$

Problem: evaluate the class of nonlinear function $\varphi(y)$ in (2.1) and characteristics of the matrix A based on the data processing (2.3). Problem: evaluate the class of nonlinear function $\varphi(y)$ and characteristics of the matrix A based on the data processing (2.3).

3. GF IN DYNAMIC SYSTEMS STRUCTURAL IDENTIFICATION PROBLEM

3.1. S_{ey}-frameworks

The geometrical framework S_{ey} design is one of the main directions for the SI problem. The estimation possibility of system structural parameters defines the method of the S_{ey} design. Framework S_{ey} is derivative from a phase portrait S. S is the starting point for further research under uncertainty.

GF design approach depends on system properties and the considered problem of structure identification. The synthesis S_{ey} method is proposed Karabutov (2017) and is generalized on dynamic systems (Karabutov, (2015, 2018c)). The approach bases in the formation of a subset $\mathbb{I}_{GF}(\mathbb{I}_o)$, which allows getting a mapping for the design S_{ey}. $\mathbb{I}_{GF}$ may contain data on the transient process or the steady motion in the system. In the estimation problem of the nonlinear dynamic system (2.1) structure, at first, the set $\mathbb{I}_{GF} = \mathbb{I}_{N,g}$, which contains the information about system nonlinear properties, is formed.

The set $\mathbb{I}_{N,g}$ is identified as follows. Apply to $y(t)$ the differentiation operation and designate by the obtained variable as x_1. Determine the model

$$\hat{x}_1^l(t) = H^T[1\ u(t)\ y(t)]^T, \tag{2.4}$$

where $\hat{x}_1^l$ is the estimation of the linear component in x_1 on the time gape $\mathbb{J}_g = \mathbb{J} \setminus \mathbb{J}_{tr}$ corresponding to the steady motion in the system (2.1); $H \in \mathbb{R}^3$ is the vector of model (2.4) parameters; $\mathbb{J}_{tr}$ is the time gap corresponding to the transient process in the system. Determine by the vector H having applied at least square method.

Find the forecast for the variable x_1 used in the model (2.4), and form the error $e(t) = \hat{x}_1^l(t) - x_1(t)$. $e(t)$ depends on the nonlinearity $\varphi(y)$ in the system (2.1). So, we obtain the set $\mathbb{I}_{N,g} = \left\{y(t), e(t)\, t \in J_g\right\}$. Next; we apply the designation $y(t)$ assuming that $y(t) \in \mathbb{I}_{N,g}$.

Construct the phase portrait S and GF S_{ey} described by functions $\Gamma:\{y\} \rightarrow \{y'\}$, $\Gamma_{ey}:\{y\} \rightarrow \{e\}$. S_{ey} is the basis for the further analysis and the identification system design. The framework S_{ey} should have specified properties (Karabutov, 2017). Properties of structural identifiability and S-synchronizability (Karabutov, 2018b) are the basic for S_{ey}. The obtained GF correctness sign is a regularity of its presentation and the condition $|e(t)| > \delta_e$ for $\forall t \geq t_q > t_r$, where t_{tr} is the end time of the transient process, $\delta_e > 0$ is some number. The application of the described approach gives how significant S_{ey}, and insignificant NS_{ey} frameworks ($S_{ey} = NS_{ey}$). Decision-making on the significance of S_{ey} is based on the results obtained by Karabutov (2017). The framework NS_{ey} is results non-fulfilment of the condition S-synchronizability

(SS) of the system (2.1). S-synchronizability of the system (2.1) (framework S_{ey}) depends on fulfilment of the excitation constancy condition for the input $u(t)$. The significance S_{ey} estimation algorithm is based on the sector set properties analysis for S_{ey} (Karabutov, 2018c) if the SS-condition of is made.

Definition 1. The framework S_{ey} is called the regular if the condition S-synchronizability is satisfied for the system (2.1).

The example of the regular framework S_{ey} is presented for the system with a static hysteresis in Figure 1. If the function $\varphi(y)$ has a complex law of change, then applying the approach described above gives the false S_{ey} or NS_{ey} framework.

Figure 1. Frameworks S and S_{ey} for the second order system (2.1) with static hysteresis

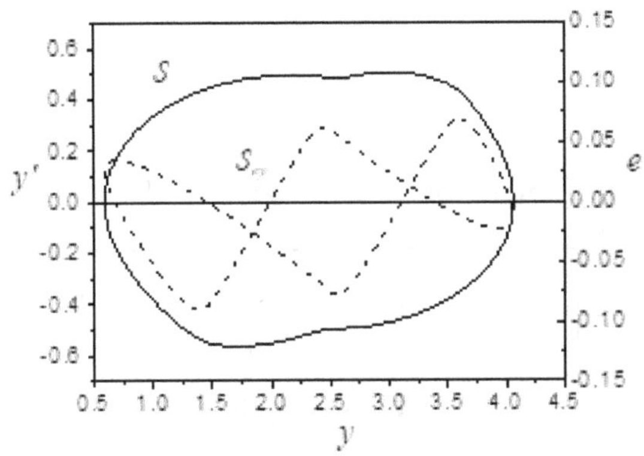

In Figure 2, the framework is shown for the system RC-OTA generated the chaotic oscillator:

$$\begin{aligned} \ddot{x} - 0.1\dot{x} + x &= \varphi \\ \dot{\varphi} &= 10(-\varphi + \text{sgn}(x + \text{sgn}(\varphi))) \end{aligned} \tag{2.5}$$

RC-OTA is applied to the design of electronic and control systems (Achho, 2013).

Obtaining of the regular framework gives to the application of the hierarchical immersion method (Karabutov, 2018c) in state-space. This method provides the model (2.4) structure choice for each layer of hierarchy. The example of the regular framework S_{ey} for the system (2.5) is presented in Figure 3. The regular framework S_{ey} example for the system (2.5) is shown in Figure 3.

Figure 2. Frameworks S, S_{ey} for the second order system (2.5) with dynamic hysteresis

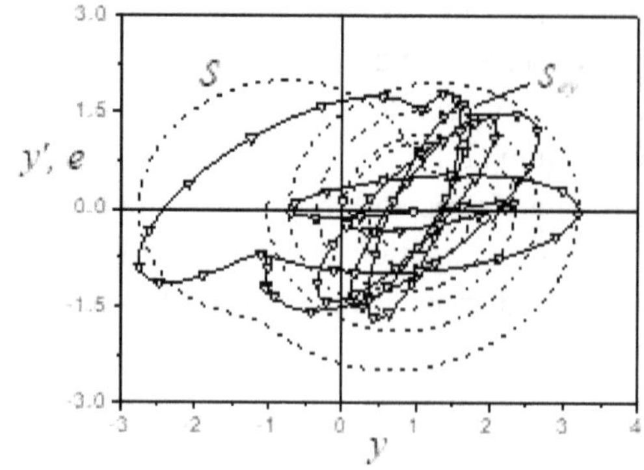

Figure 3. Regular structure for the system (2.5)

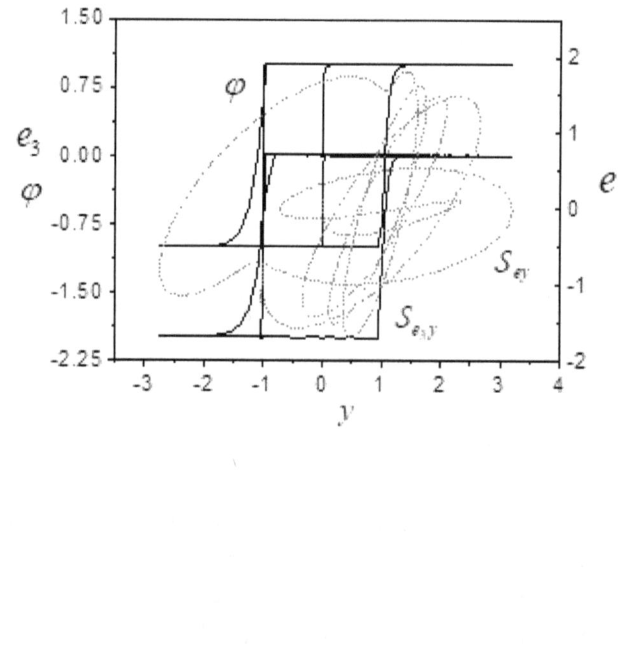

Figure 4. Framework S_{ey}, phase portrait and z

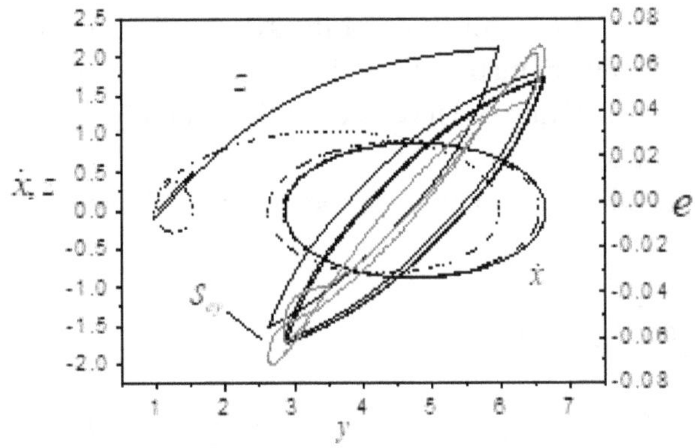

Example 1. Consider a mechanical system with Bouc-Wen hysteresis (Gao et al., 2019)

$$
\begin{aligned}
& m\ddot{x} + c\dot{x} + F(x,z,t) = f(t),\ y = x, \\
& F(x,z,t) = \alpha kx(t) + (1-\alpha)kdz(t), \qquad (2.6) \\
& \dot{z} = d^{-1}(a\dot{x} - \beta|\dot{x}||z|^{n}sign(z) - \gamma\dot{x}|z|^{n})
\end{aligned}
$$

where $m>0$ is weight; $c>0$ is damp; $F(x,z,t)$ is the restoring force; $d>0$, $n>0$, $k>0$, $\alpha\in(0,1)$, $f(t)$ is exciting force; α,β,γ are some numbers. Denote by the system (2.1) - (2.3) as SBW.

Parameters in the SBW-system are equal: $d=a=m$, $n=1.5$, $\beta=0.5$, $\alpha=1.5$, $k=0.6$, $m=1$, $c=2$. The exciting force is $f(t)=2-2\sin(0.15\pi t)$.

Model (2.4) has the form: $\hat{\dot{x}} = -0.199x + 0.471f$. The application of the proposed method gives S_{ey}-framework (Figure 4). Ranges of definition z and $\dot{x}$ are matched. Analysis of the S_{ey}-structure shows that the system (2.6) is nonlinear.

3.2. $S_{k_{s,\rho}}$, $SK_{\Delta k^{i}_{s,\rho}}$-frameworks

Another class of frameworks $S_{k_{s,\rho}}$ (Karabutov, (2015, 2018b)) is designed based on the system (2.1) general solution analysis. $S_{k_{s,\rho}}$ applies to the structure choice of the system (2.1) linear part. This task differs from the problem considered above. Therefore, mappings allowing deciding should have other form (Karabutov, 2015). They are based on the analysis of the Lyapunov exponent dynamics change. Apply the model:

$$\widehat{X}_q(t) = \widehat{A}_q W(t) \forall t \in \mathbb{J}_q \tag{2.7}$$

to the particular solution estimation of the system (2.1) on the output y, where $\widehat{A}_q \in \mathbb{R}^{2\times 2}$ is the parameter matrix, $W = [u\ u']^T$, $\widehat{X}_q \in \mathbb{R}^2$ is the estimation of the system output and its derivative. The choice of the interval $\mathbb{J}_q \subset \mathbb{J}$ depends on the system (2.1) properties.

Further, we obtain the estimation for the system (2.1) general solution on the basis $\widehat{X}_q$:

$$\widehat{X}_g(t) = X(t) - \widehat{X}_q(t) \quad \forall t \in J_g$$

where $\widehat{X}_g(t) = \left[\hat{y}_g(t)\ \dot{\hat{y}}_g(t)\right]^T$. This approach can be generalized to the case $m>2$. Functions

$$\rho(\hat{y}_g) = \rho_g = \ln\left|\hat{y}_g(t)\right| \ \forall t \in \widehat{\mathbb{J}}_g \subset \mathbb{J}_g$$

$$k_s(t,\rho) = \frac{\rho(\hat{y}_g)}{t} \tag{2.8}$$

are the basis for the mapping describing $S_{k_{s\rho}}$, where $\widehat{\mathbb{J}}_g = \left[t_0, \bar{t}\right]$ is determined based by the LE theory (Bylov et al., 1996). $K_s(t,\rho)$ is the basis for the Lyapunov exponent calculation.

Remark 1. The framework $S_{k_{s\rho}}$ use simplifies the choice of the upper bound for time at the calculation of LE.

Analyse sets

$$\begin{aligned} \mathbb{I}_{k_s} &= \left\{k_s\left(t, \rho\left(\hat{y}_g(t)\right)\right), t \in \widehat{\mathbb{J}}_g\right\}, \\ \mathbb{I}_{k'_s} &= \left\{k_s\left(t, \rho\left(\dot{\hat{y}}_g(t)\right)\right), t \in \widehat{\mathbb{J}}_g\right\} \end{aligned} \tag{2.9}$$

for the LE determination.

Introduce framework $S_{k_{s\rho}}$ described by the mapping $\Gamma_{k_{s\rho}} : \mathbb{I}_{k_s} \to \mathbb{I}_{k'_s}$ on sets $\mathbb{I}_{k_s}, \mathbb{I}_{k'_s}$. The framework $S_{k_{s\rho}}$ reflects the change dynamics of indexes depending on LE. Consider also the function describing the first difference $k_s(t, \rho(\dot{\hat{y}}_g(t)))$ change:

$$\Delta k'_s(t) = k_s\left(t, \rho\left(\dot{\hat{y}}_g(t+\tau)\right)\right) - k_s\left(t, \rho\left(\dot{\hat{y}}_g(t)\right)\right) \tag{2.10}$$

where $\tau > 0$.

We describe this approach in more detail in chapters 5 and 6.

4. STRUCTURAL IDENTIFIABILITY OF NONLINEAR DYNAMICS SYSTEM

In the previous section, we note that the nonlinear DS structure estimation depends on the system identifiability.

Many publications (see for example, Bodunov, 2012; Chis et al., 2011; Saccomani, 2013) and Chapter 1 are devoted to the parametric identifiability problem of dynamic systems. The structural identifiability of nonlinear dynamic systems is reduced to the parametric identifiability based on various approximation methods application (Chis et al., 2011; Saccomani, 2013; Stigter and Peeters, 2007). Modifications of proposed approaches are considered in case when not all system parameters can be identified.

Karabutov (2018b) consider structural identifiability (IS) in the following aspect: determine by conditions under which the nonlinear system structure estimation is possible under uncertainty. The solution to this problem we present for the system (2.1) (Karabutov, 2018b) when the nonlinear function $\varphi(y)$ satisfies the condition (2.2). Decision-making is based on the analysis of the framework S_{ey}, which describes the system (2.1) behaviour in the steady state. It is showed that the system should satisfy to h-identifiability property.

The major results and definitions of IS are discussed in the next chapters of this part.

5. SYSTEM ATTRACTOR RECONSTRUCTION

Reconstruction (restoration) of the phase portrait (PP) or a system attractor can be performed based on time series analysis. The proof of this approach is given by Takens (1980), and the practical application is based on Wolf and Rosenstein algorithms (Rosenstei et al., 1993; Wolf et al., 2003). This problem can be interpreted as the system structure restoration task in the phase space. Many authors (see e.g., Anishchenko et al., 2007; Malinetskiy and Potapov, 2000; Pecora et al., 2007) have studied this problem. Proposed procedures are heuristic (Malinetskiy and Potapov, 2000). The PP construction depends on the choice of reconstruction optimum parameters. The main issue is the choice of the time delay for new variables based on the time series. To solve this problem, various approaches (see e.g., Bradley and Kantz, 2015; Pecora et al., 2007) are used: the autocorrelation and cross-correlation, the choice of the attractor shape, the method of the neighbour, and also the prediction statistics based on various models. Recommendations about the choice of the delay value estimation method are not provided. The complexity and the variety of considered objects explains this. The second problem is the quality criteria choice

(Malinetskiy and Potapov, 2000) for the estimation of the PP reconstruction. Unfortunately, this problem has not obtained the last solution.

The the attractor dimension choice (Anishchenko et al., 2007; Buzug and Pflster, 1992; Liebert et al., 1991) is also an important task. An attempt to resolve the existing problems is made by Pecora et al., (2007). It is shown that these statistics can be applied to the attractor creation of multidimensional systems. The further analysis problem of constructed attractor is not completed at this stage. The designed attractor does not always satisfy requirements of the researcher. Therefore, various methods (Anishchenko et al., 2007; Deshpande et al., 2010) applies to get the smooth mapping.

Identification dynamic systems based on the obtain of state variables is considered by Anishchenko et al., (2007) and Malinetskiy and Potapov, (2000). This issue is discussed in the review (Aguirre and Letellier, 2009) in more detail. Various approximation methods are applied for state estimation of dynamic systems. Identification methods are based on interpolation procedures decomposition of a nonlinear function on the specified basis, the application of a spline-functions and neural networks, and also many other approaches (Aguirre and Letellier, 2009).

Remark 2. As Malinetskiy and Potapov, (2000) note, considered identification methods are not always effective. Therefore, the major role is played by heuristics, the experience of the researcher and the prior information. This remark is true also for PP restoration methods (Anishchenko et al., 2007). At first, the data approximation is performed on the class of basic functions. Next, the phase portrait is designed the equivalent of the initial system. Further, unknown parameters are introduced in the obtained model that properties improve of the mapping. Various heuristics and procedures are applied, using additional information on the system. Obtained models are very unwieldy for the application. As it is noted Malinetskiy and Potapov, (2000), the use of the complex models is not always justified in practical applications.

6. SYSTEM STRUCTURE CHOICE WITH LAG VARIABLES

Models with distributed lags (DL) are widely applied in various areas (Armstrong, 2006; Dhrymes, 1971; Johnston, 1972; Malinvaud, 1980; Solow, 1960). Independent and dependent variables can have the delay. The DL accounting activates autocorrelation between variables (Armstrong, 2006) and the parameter identification process complicates. Various schemes of approximation parameters at DL (Dhrymes, 1971; Johnston, 1972) apply (see Chapter 1), considering a prior information. Such the approach reduces the estimated parameter number of the system. The least-squares procedure and its modifications apply to the parameter estimation. Methods for the maximum length lag choice are considered. Statistics based on the analysis of residuals (Armstrong, 2006; Solow, 1960) are the basis of the applied approaches. The Akai criteria and Bayes information criteria are used for decision-making on the model structure. The identification of the system structure and parameters is not examined under uncertainty.

Approximation scheme choice of parameters is bound by performing labour-consuming calculations under uncertainty. Karabutov (2012) applies the approach to the structure DL choice based on the analysis of properties framework $S^{v}_{k,e}$. Therefore, previously considered methods are not applicable to its analysis. The structure estimation of the system with DL is based on the analysis of secants (Karabutov, 2012).

Further, the estimation method of the DL system structure based on LE identification is stated. This method is the development of the approach described in subsection 2.2. The direct transfer of results (Karabutov, (2015 2018b) on the considered system class is impossible, since these systems have the specifics.

Consider the system

$$y_n = A^T U_n + B^T X_n + \xi_n, \tag{2.11}$$

where $y_n \in \mathbb{R}$ is output; $U_n \in \mathbb{R}^k$ is input vector, which elements are limited extremely nondegenerate functions; $n \in \mathbb{J}_N = [0, N]$ is discrete time, $N<\infty$; $X_n \in \mathbb{R}^m$,

$$X_n = X(u_{i,n} \in U_n) = \left[u_{i,n-1}, u_{i,n-2}, \ldots, u_{i,n-m}\right]^T$$

is the vector of DL on $u_{i,n} \in U_n$; $A \in \mathbb{R}^k$, $B \in \mathbb{R}^m$ are constant parameter vectors; $\xi_n \in \mathbb{R}$ is external disturbance, $|\xi_n|<\infty$ for all $n \in \mathbb{J}_N$.

Let the informational set $\mathbb{I}_o$ containing the data on measured inputs and output on the interval $\mathbb{J}_N \mathbb{J}_N$ has the form

$$\mathbb{I}_o = \left\{U_n, y_n, n \in \mathbb{J}_N\right\}. \tag{2.12}$$

Problem: estimate the vector X_n dimension based on $\mathbb{I}_o$ analysis.

Remark 3. We consider the case of lags on the input U_n. If the output y_n contains lags, then the proposed approach allows to estimate the DL structure and in this case.

Analyse the effect of $u_{j,n}$, $j = \overline{1,k}$ on the output y_n. Determine by determination coefficient $r^2_{u_j y}$ for everyone $u_{j,n-1}$. Introduce the number $\delta>0$. Find such j that $r^2_{u_j y} \geq \delta$ satisfied and designate $i=j$. So, the element of the vector U_n is determined. Form the vector $\overline{U}_n \in \mathbb{R}^{m-1}$, which does not contain the element $u_{i,n}$, for the lag estimation on $u_{i,n}$ and apply model

$$\hat{y}_n = \overline{B}^T \overline{U}_n \tag{2.13}$$

where $\tilde{B}^T \in \mathbb{R}^{m-1}$ is the parameter vector.

The system (2.13) is not dynamic in the standard sense.

Assumption 1. Let the system (2.13) contain the variable $\pi_n = u_{j,n}$, which changes on the dynamic law

$$S_\pi : \pi_n = \sum_{i=1}^{h} \alpha_i \pi_{n-i} + \kappa \zeta_n, \tag{2.14}$$

where α_i, κ are some numbers; $h<\infty$; $\zeta_n \in \mathbb{R}$ is some limited function for all $n \in \mathbb{J}_N$.

Let the system (2.16) be stable, i.e. $\alpha_i<1$.

Definition 2. The systems (2.13) have π-steady state or π -state if such $j\geq 1$ exists that the variable $u_{j,n} \in U_n$ satisfies the equation (2.16).

Allocate the transient process (the system (2.16) general solution) for the application of LE to the S_π-system. Localise in (2.13) the space, which the variable $\pi_n=u_{j,n}$ belongs, and π -steady state eliminate on the interval $\mathbb{J}_g = \mathbb{J}_N \backslash \mathbb{J}_{\pi,N}$, where the interval corresponding to the π -state in the S_π -system.

Consider the set $\mathbb{I}_o$ (2.14). Apply the model (2.15) on the interval $\mathbb{J}_{\pi,N}$, where $\mathbb{J}_{\pi,N}$ choose so that the coefficient of determination was maximal between $\hat{\hat{y}}_n$ and $\hat{y}_n$. Next, calculate the error $e_n = y_n - \hat{\hat{y}}_n$. Note that the variable e_n contains information about $u_{j,n}$.

Now the analysis reduced to the study of the discrete dynamic system properties with the output e_n. We obtain the system S D$_e$ that is a prototype of the system (2.14).

The problem is reduced to the LE estimation based on the set $\mathbb{I}_e = \{e_n, n \in \mathbb{I}_g\}$ analysis (Karabutov, 2021). This problem is an analogue of the task for restoring the dynamical system attractor on $\mathbb{I}_e$. Reconstruction is based on the Takens theorem (Takens, 1980) application. F. Takens has proved that the new row $e_{d,n}$ based on lagging values e_n gives to the PP reconstruction problem solution. The row $e_{d,n}$ describes the change the dynamics of the derivative variable e_n. Many procedures are proposed for the choice of the delay interval (Anishchenko et al., 2007). It is supposed that trajectories of the dynamic system belong to the smooth manifold. It is necessary to notice that the delay interval choice problem did not obtain the last solution. Heuristic procedures, algorithms of approximation and smoothing time series are used in practical applications. A priori information is important. After obtaining of the set $\{e_n, \; e_{d,n}, n = \overline{1, n_k}\}$, we solve the problem of the PP building which also is nontrivial (Anishchenko et al., 2007; Liebert et al., 1991; Malinetskiy and Potapov, 2000).

Remark 4. Smoothing algorithms are used in the attractor reconstruction problem. Smoothing procedures application to set $\mathbb{I}_e$ elements at the LE identification for the system (2.13) can lead to the loss of valuable information. Residual errors caused by disturbances ξ_n in (2.13) effect on properties of obtained LE estimations.

Use the formula (2.8) for the calculation of Lyapunov exponents. Considering remark 4, the detection of LE, but not their values, we will evaluate.

Consider analogues of frameworks $S_{k_{s,\rho}}$, $SK_{\Delta k'_{s,\rho}}$ and $LSK_{\Delta k'_{s,\rho}}$ defined at $t_n = n\tau$, where τ is the data measurement interval. Introduce the function

$$b_n = \begin{cases} 1, \ if \ \Delta k'_{s,n} \geq 0 \\ -1, \ if \ \Delta k'_{s,n} < 0 \end{cases} \tag{2.15}$$

where $b_n = b(b\tau)$, $\Delta k'_{s,n} = \Delta k'_s(n\tau)$.

Theorem 1 (Karabutov, 2018b). *If the function* b_n changes the sign h times on the interval $[t_0, t^*] \subset \vec{\mathbb{J}}_g$ $(t^* \leq \bar{t})$, that the system (2.14) has the order h.

Consider frameworks

$$SK_{\Delta k'_s} \to \mathbb{I}_{k_s} \times \mathbb{I}_{\Delta k'_s}, LSK_{\Delta k'_s} \to \mathbb{I}_{k_s} \times \left(\mathbb{I}_{\Delta k'_s}\right)$$

where sets $\mathbb{I}_{k_s}$ and $\mathbb{I}_{k'_s}$ have form (2.9).

We show (Karabutov, 2018b) if the theorem 1 conditions are made, then local minima of the framework $SK_{\Delta k'_s}$ correspond to LE estimations of the system (2.14) in space $(k_s, \Delta k'_s)$.

Theorem 2 (Karabutov, 2018b). *If conditions of the theorem 1 are satisfied and the framework* $SK_{\Delta k'_s}$ described by the function $\Gamma_{\Delta k'_s} : \mathbb{I}_{k_s} \to \mathbb{I}_{\Delta k'_s}$ has local minima on the plane $(k_s, \Delta k'_s)$, then S_π-system have π-state.

The local minima quantity corresponds to the lag structure of the system (2.11) on the variable $u_{j,n}$.

So, we have shown that the discrete informational set $\mathbb{I}_o$ modification based on the approach (Karabutov, 2018b) allows extending the GF application method to systems with the distributed lags of input variables. Consider the LE identifiability problem for this case.

Let the vector U_n is limited constantly excited

$$PE_\alpha : U_n U_n^T \geq \alpha I_k \tag{2.16}$$

for some $\alpha > 0$ and $\forall n \geq 0$ on the interval $\mathbb{J}_N$, where $I_k \in \mathbb{R}^{k \times k}$ is the unit matrix.

If (2.16) is satisfied, then we will write $U_n \in PE_\alpha$. As shown in section 5, the fulfilment (2.16) is sufficient for the S_π-system π-state estimation. Parameters of the model (2.12) are identifiable if $\widetilde{U}_n = U_n \backslash u_{i,n}$ and $\widetilde{U}_n \in PE_{\overline{\alpha}}$, $\overline{\alpha} > 0$. S_π-system with π-state corresponds to the system (2.12). Let the framework $LSK_{\Delta k'_{s,\rho}}$ and the function b_n, which on the interval $[t_0, t^*] \subset \vec{\mathbb{J}}_g$ changes the sign h of times exist. Then the system has h Lyapunov exponents (set M_{S_π}). Therefore, S_π-system is identifiable on the set M_{S_π} Lyapunov exponents. So, it is a true statement.

Theorem 3. *Let: (i) the* $U_n \in PE_\alpha$; (ii) the vector $\tilde{B}^T \in R^{m-1}$ of the model (2.13) is identifiable with $\tilde{U}_n \in PE_\alpha$; (iii) the framework $LSK_{\Delta k'_{s,\rho}}$ and the function b_n (2.15) satisfying theorems 1 conditions exist; 4) the S_π-system (2.16) has the π-state. Then the dynamic S_π-system (2.16) corresponding to the system (2.11) is identifiable on the Lyapunov exponent set.

Figure 5. System (2.11) phase portrait of the with k=3 and h=2

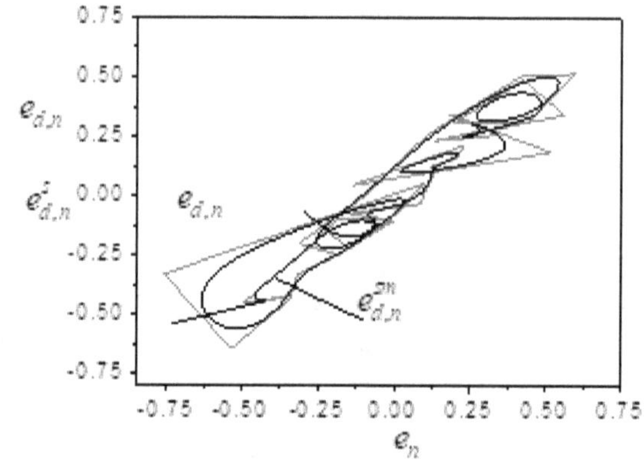

Example 2. Consider the system (2.11) with k=3 and h=2, A=[0,7; 3; 3,5]^T, B= [0,4; 0,45]^T, $X_n = \left[u_{1,n-1}, u_{1,n-2}\right]^T$. $u_{1,n}$ is obtained as the system (2.14) output with the input ζ_n, distributed to the normal law with the zero average and final dispersion. The set $\square_o$ (2.12) is generated for $n\in$[1; 60]. The analysis of the set $\square_o\square_o$ is shown that lags are had by the variable $u_{1,n}$. Time series $\{e_n\}_{n=\overline{1;60}}$, $\{e_{d,n}\}_{n=\overline{1;60}}$ are formed. Apply the model (2.13)

$$\hat{y}_n = [3;3.52]\left[u_{2,n};u_{3,n}\right]^T + 7.35 \tag{2.17}$$

which is obtained based on LSM for $n\in$[30; 60]. The determination coefficient of the model (2.17) is 0.99.

The system (2.14) phase portrait and its smoothed analogue (variable $e^{sm}_{d,n}$) are shown in Figure 5.

Figure 6. Lyapunov exponents estimations

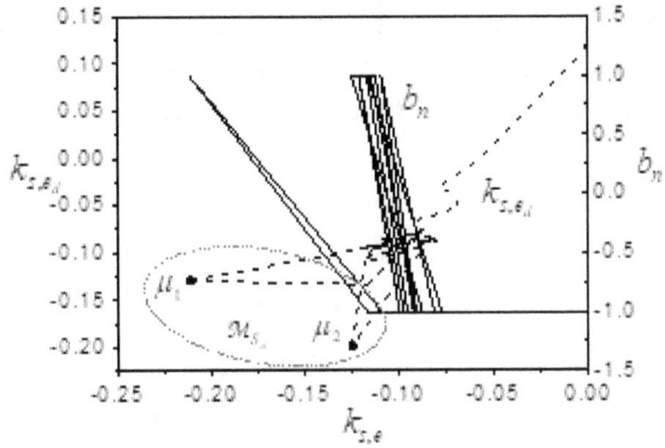

Figure 5 is presented, processes in the S_π-system are non-smooth. Results of the lag structure estimation are presented in Figure 7, where frameworks $S_{k_{s,\rho}}$ and $LSK_{\Delta k'_{s,\rho}}$ are shown. Designations in Figure 6: μ_1,μ_2 are estimations of Lyapunov exponents, k_s are calculated on the basis of (2.8)

$$k_{s,e,n} = \frac{\rho(e_n)}{n\tau}, \quad k_{s,e_d,n} = \frac{\rho(e_{d,n})}{n\tau}$$

The set M_{S_π} of Lyapunov exponent is showed also in Figure 6. The analysis of results shows that the system (2.14) describing the change $u_{1,n}$ has the order 2.

Example 3. Consider the control system for supplying cars to the Vladivostok transport hub (Russia). We consider the case of 6 cars giving from railway tracks on berth tracks. The maximum capacity of the hub is 175 cars. Let N_4 be the number of cars from the railway; N_5 be the number of cars received on the railway lines of the port. Determine ω=N_5–N_4. The variable ω reflects the current status of a hub and influences on the giving process of cars. The mathematical model for decision-making has the form

$$\hat{N}_{5,n} = f(\hat{N}_{5,n-1}, N_{4,n}, \omega_n), \tag{2.18}$$

where $\hat{N}_{5,n}$ is a model output at the moment *n*. The model (2.18) structure is described by an autoregressive equation of the first order. Apply the approach stated above and evaluate the effect ω. The system (2.14) has the first order to ω. Apply algorithms from section 3.2 and evaluate the auto-regression order. The model (2.18) has the form:

$$\widehat{N}_{5,n} = 1.06\widehat{N}_{5,n-1} - 0.13\omega_{n-1} - 0.08N_{4,n} - 4.59 \tag{2.19}$$

The determination coefficient of the model (2.19) is 0.964. The simulation showed good predictive properties of the model (2.19).

So, modelling results confirm efficiency of the proposed approach to the lag structure estimation of the system (2.11).

7. CONCLUSION

We consider the concept "framework" (FR) use in identification problems and show that this concept is widely applied in parametrical estimation problems. FR can be interpreted as a frame, a structure, the system, a platform, the concept, the basis, the system of approaches. It is shown that the framework is used in two directions: (i) the paradigm for integrate methods, approaches or procedures; (ii) the mapping describing in the generalised form processes and properties in the system. The second direction is closer to methods which are applied in the qualitative theory of dynamic systems. We interpret this approach as the method based on the analysis of virtual geometrical frameworks. The major difference of GF from the approaches, applied in the theory of chaos, accidents and so on, consists that mathematical mapping (structure) is not postulated a priori, and it is determined based on the experimental data processing. The geometrical framework is the analysis primary object. They allow the decision about properties and features of the system. The present analysis contains virtual GF application directions in structural identification. We consider five areas of the identification theory where GF is applicable:

1. Structural identification of the nonlinear system.
2. Lyapunov exponent estimation.
3. Structural identifiability of the nonlinear system.
4. The system phase portrait reconstruction on the time series.
5. The system structure estimation with lag variables.

We show Lyapunov exponents can apply to the decision-making on static systems structure with lag variables. The structural identifiability estimation of the system with lag variables is given.

REFERENCES

Achho, L. (2013). Hysteresis Modeling and Synchronization of a Class of RC-OTA Hysteretic-Jounce-Chaotic Oscillators. *Universal journal of applied mathematics, 1*(2), 82-85. DOI: 10.13189/ujam.2013.010207

Aguirre, L. A., & Letellier, C. (2009). Modeling nonlinear dynamics and chaos: A Review. *Mathematical Problems in Engineering, 2009*(1), 1024–1235. DOI: 10.1155/2009/238960

Anishchenko, V. S., Astakhov, V., Neiman, A., Vadivasova, T., & Schimansky-Geier, L. (2007). *Nonlinear dynamics of chaotic and stochastic systems. nonlinear dynamics of chaotic and stochastic systems. Tutorial and modern developments* (2nd ed.). Springer., DOI: 10.1007/978-3-540-38168-6

Armstrong, B. (2006). Models for the relationship between ambient temperature and daily mortality. *Epidemiology (Cambridge, Mass.), 17*(6), 624–631. DOI: 10.1097/01.ede.0000239732.50999.8f PMID: 17028505

Bodunov, N. A. (2012). Introduction to the theory of local parametrical identifiability. *Differential equations and management processes*, 2.

Bradley, E., & Kantz, H. (2015). Nonlinear time-series analysis revisited. *arXiv:1503.07493v1*. doi: DOI: 10.1063/1.4917289

Buzug, T., & Pfister, G. (1992). Optimal delay time and embedding dimension for delay-time coordinates by analysis of the global static and local dynamical behaviour of strange attractors. *Physical Review A, 45*(10), 7073–7084. DOI: 10.1103/PhysRevA.45.7073 PMID: 9906779

Bylov, F., Vinograd, R. E., Grobman, D. M., & Nemytskii, V. V. (1966). *Theory of Lyapunov exponents and its application to problems of stability*. Nauka.

Carino, J. A., Delgado-Prieto, M., Iglesias, J. A., Sanchis, A., Zurita, D., Millan, M., Ortega Redondo, J. A., & Romero-Troncoso, R. (2018). Fault Detection and Identification Methodology Under an Incremental Learning Framework Applied to Industrial Machinery. *IEEE Access : Practical Innovations, Open Solutions, 99*, 1–9. DOI: 10.1109/ACCESS.2018.2868430

Chen, L., & Narendra, K. S. (2004). Identification and Control of a Nonlinear Discrete-Time System Based on its Linearization: A Unified Framework. *IEEE Transactions on Neural Networks, 15*(3), 663–673. DOI: 10.1109/TNN.2004.826206 PMID: 15384554

Chis, O.-T., Banga, J. R., & Balsa-Canto, E. (2011). Structural identifiability of systems biology models: A critical comparison of methods. *PLoS One*, *6*(4), 1–16. DOI: 10.1371/journal.pone.0027755 PMID: 22132135

Deshpande, A., Chen, Q., Wang, Y., Lai, Y.-C. G., & Do, Y. (2010). Effect of smoothing on robust chaos. *Physical Review. E*, *82*(2), 026209. DOI: 10.1103/PhysRevE.82.026209 PMID: 20866895

Dhrymes, P. J. (1971). *Distributed Lags: Problems of Estimation and Formulation*. Holden-Day.

Gao, H., Jézéquel, L., Cabrol, E., & Vitry, B. (2019). Characterization of a Bouc-Wen model-based damper model for automobile comfort simulation. *HAL* Id: hal-02188563.

Iqbal, A., Aftab, S., Ullah, I., Saeed, M. A., & Husen, A. (2019). (). A classification framework to detect DoS attacks. *International Journal of Computer Network and Information Security*, *11*(9), 40–47. DOI: 10.5815/ijcnis.2019.09.05

Johnston, J. (1972). *Econometric methods* (2nd ed.). McGraw-Hill Book Company.

Karabutov, N. (2015). Structural methods of estimation Lyapunov exponents linear dynamic system. *International Journal of Intelligent Systems and Applications*, *7*(10), 1–11. DOI: 10.5815/ijisa.2015.10.01

Karabutov, N. (2017). Structural Methods of Design Identification Systems. In Uvarova, L., Nadykto, A. B., & Latyshev, A. V. (Eds.), *Nonlinearity: Problems, Solutions and Applications* (Vol. 1, pp. 233–274). Nova Science Publishers Inc.

Karabutov, N. (2018a). About Lyapunov exponents identification for systems with periodic coefficients. *International Journal of Intelligent Systems and Applications*, *10*(11), 1–10. DOI: 10.5815/ijisa.2018.11.01

Karabutov, N. (2018b). About structural identifiability of nonlinear dynamic systems under uncertainty. *Global Journal of Science Frontier Research: (A). Physics and Space Science*, *18*(11), 51–61.

Karabutov, N. (2018c). *Frameworks in identification problems: Design and analysis*. URSS/Lenand.

Karabutov, N. (2021). Geometrical Framework Application Directions in Identification Systems. [Review]. *International Journal of Intelligent Systems and Applications*, *13*(2), 1–20. DOI: 10.5815/ijisa.2021.02.01

Karabutov, N. N. (2012). Structural identification of static systems with distributed lags. *International Journal of Control Science and Engineering*, *2*(6), 136–142. DOI: 10.5923/j.control.20120206.01

Krasovsky, A. A. (Ed.). (1987). *Reference book on automatic control theory*. Nauka.

Kurtoglu, T., & Tumer, I. Y. (2008). A Graph-Based Fault Identification and Propagation Framework for Functional Design of Complex Systems. *Journal of Mechanical Design*, *130*(5), 051401–051408. DOI: 10.1115/1.2885181

Lauer, F., Bloch, G., & Vidal, R. (2011). A continuous optimization framework for hybrid system identification. *Automatica*, *47*(3), 608–613. DOI: 10.1016/j.automatica.2011.01.020

Li, C., Min, F., & Li, C. (2018). Multiple coexisting attractors of the serial–parallel memristor-based chaotic system and its adaptive generalized synchronization. *Nonlinear Dynamics*, *94*(7191), 1–22. DOI: 10.1007/s11071-018-4524-3

Liebert, W., Pawelzik, K., & Schuster, H. G. (1991). Optimal embedding of chaotic attractors from topological considerations. *Europhysics Letters*, *14*(6), 521–526. DOI: 10.1209/0295-5075/14/6/004

Lü, J., & Zhang, S. (2007). Controlling Chen's chaotic attractor using backstepping design based on parameters identification. *Physics Letters. [Part A]*, *32*(4), 148–156. DOI: 10.1016/S0375-9601(01)00383-8

Malinetskiy, G. G., & Potapov, A. B. (2000). *Modern problems of nonlinear dynamics*. Editorial URSS.

Malinvaud, E. (1980). *Statistical methods in econometrics* (3rd ed.). North-Holland Publishing Co.

Michel, A., Wang, K., & Hu, B. (2001). *Qualitative theory of dynamical systems*. CRC Press., DOI: 10.1201/9780203908297

Nagarajaiah, S. (2017). Sparse and low-rank methods in structural system identification and monitoring. *X International Conference on Structural Dynamics (EURODYN 2017). Procedia Engineering*, *199*, 62–69. DOI: 10.1016/j.proeng.2017.09.153

Papadopoulos, P. N., Guo, T., & Milanović, J. V. (2017). Probabilistic Framework for Online Identification of Dynamic Behavior of Power Systems with Renewable Generation. *IEEE Transactions on Power Systems*, *33*(1), 45–54. DOI: 10.1109/TPWRS.2017.2688446

Pecora, L. M., Moniz, L., Nichols, J., & Carroll, T. L. (2007). A Unified approach to attractor reconstruction. *Chaos (Woodbury, N.Y.), 17*(1), 013110. DOI: 10.1063/1.2430294 PMID: 17411246

Pershin, Y. V., & Slipko, V. A. (2018). Dynamical attractors of memristors and their networks. *arXiv: 1808.07947.*

Pillonetto, G., Dinuzzo, F., Chen, T., De Nicolao G., & Ljung, L. (2014). Kernel methods in system identification, machine learning and function estimation: A survey. *Automatica, 50*(3), 657-682. doi.: DOI: 10.1016/j.automatica.2014.01.001

Pintelon, R., & Schoukens, J. (2012). *System Identification. A Frequency Domain Approach* (2nd ed.). Published by John Wiley & Sons, Inc., DOI: 10.1002/9781118287422

Roettgen, D., Allen, M. S., Kammer, D., & Mayes, R. L. (2017). Substructuring of a Nonlinear Beam Using a Modal Iwan Framework, Part I: Nonlinear Modal Model Identification. *Dynamics of Coupled Structures, 4*, 165–178. DOI: 10.1007/978-3-319-54930-9_15

Rosenstein, M. T., Collins, J. J., & De Luca, C. J. (1993). A practical method for calculating largest Lyapunov exponents from small data sets Source. *Physica D. Nonlinear Phenomena, 65*(1-2), 117–134. DOI: 10.1016/0167-2789(93)90009-P

Saccomani, M. P. (2013). Structural vs Practical identiability in system biology. International Work-Conference on Bio informatics and Biomedical Engineering. IWBBIO 2013. *Proceedings. Granada, 18-20*(March), 305–313.

Saha, B., Goebel, K., Poll, S., & Christopherson, J. (2009). Prognostics methods for battery health monitoring using a Bayesian framework. *IEEE Transactions on Instrumentation and Measurement, 58*(2), 291–296. DOI: 10.1109/TIM.2008.2005965

Shilnikov, L. P., Shilnikov, A. L., Turaev, D. V., & Chua, L. O. (2001). *Methods of qualitative theory in nonlinear dynamics. Part II.* World Scientific., DOI: 10.1142/4221

Solow, R. (1960). On a family of lag distributions. *Econometrica, 28*(2), 393–406. DOI: 10.2307/1907729

Stigter, J. D., & Peeters, R. L. M. (2007). On a geometric approach to the structural identifiability problem and its application in a water quality case study. *Proceedings of the European Control Conference 2007 Kos,* Greece, July 2-5 2007, 3450-3456. DOI: 10.23919/ECC.2007.7068560

Takens, F. (1980). Detecting strange attractors in turbulence," *Dynamical systems and turbulence. Lecture notes in mathematics* / Eds. D.A. Rand, L.-S. Young. Berlin: Springer-Verlag, *898*, 366-381.

Teh, C. Y., Kerk, Y. W., Tay, K. M., & Lim, C. P. (2018). On modelling of data-driven monotone zero-order TSK fuzzy inference systems using a system identification framework. *IEEE Transactions on Fuzzy Systems*, *26*(6), 3860–3874. DOI: 10.1109/TFUZZ.2018.2851258

Toth, R., Sanandaij, B. M., Poolla, K., & Vincent, T. L. (2012). Compressive system identification in the linear time-invariant framework," *50th IEEE Conference on Decision and Control and European Control Conference (CDC-ECC) Orlando, FL, USA, December 12-15, 2011*, 783-790.

Varna, A. L., Swaminathan, A., & Wu, M. (2008). A Decision Theoretic Framework for Analyzing Binary Hash-based Content Identification," *DRM '08 Proceedings of the 8th ACM workshop on Digital rights management*, 67-76. DOI: 10.1145/1456520.1456532

Wiggins, S. (2003). Introduction to applied nonlinear dynamical systems and chaos. Springer Science & Business Media, 4(5). .DOI: 10.1007/b97481

Wolf, A., Swift, J. B., Swinney, H. L., & Vastano, J. A. (1985). Determining Lyapunov exponents from a time series. *Physica*, *16D*(16), 285–301.

Chapter 3
Structural Identification of Systems With Distributed Lag

ABSTRACT

We consider approaches and methods for evaluating the structure of distributed lag systems. A method for estimating the order of a distributed lag system (SDL) is proposed. It is based on the construction and analysis of geometric structures (GF). We describe the design GF method. An approach to estimating the system linearity degree based on the analysis of secants for GF is suggested. Decision on the structure of the lag variable is made based on the analysis of the second-order secant field for geometric structures. We show the lag can be interpreted as a non-linear element of the system. The cases of lags in the input and output variables of the system are considered.

1. INTRODUCTION

Models with the distributed lags are widely applied in the econometrics and economy (Demetriou and Vassiliou, 2009; Harvey and Dhrymes, 1971; Johnston, 1976; Malinvaud, 1980), the engineering (Gershenfeld, 1999; Kailath, 1977; Karabutov, 2009) and medicine (Armstrong, 2006; Nelson and Schwert, 1974; Gasparrini et. al, 2010). Both independent and dependent variables can have lags. The distributed lag (DL) account activates autocorrelation between variables (Johnston, 1976; Malinvaud, 1980; Harvey and Dhrymes, 1971) and complicates the process of identification parameters. To identification of parameters of system, various models apply for DL approximation parameters. Such an approach reducing number of estimated parameters of the system. Various schemes for approximating parameters are proposed

DOI: 10.4018/979-8-3373-0822-7.ch003

(see Fisher model (Fisher, 1937; Malinvaud, 1980), Koyck scheme (Коуск, 1954; Johnston, 1976), Almon (1965), and other approaches (Harvey and Dhrymes, 1971; Fisher, 1937; Theil and Stern, 1960; Solow 1960) and method least-squares method (LSM) (Demetriou and Vassiliou, 2009; Harvey and Dhrymes, 1971; Johnston, 1976; Malinvaud, 1980). The a priori information effect on the dependence evaluation between DL is given by Jorgenson (1964). The considered models minimize the number of unknown parameters. In these works, the model structure is postulated a priori. Approaches (Malinvaud, (1980; Solow, 1960; Yoder, 2007) based on the analysis of residues are proposed. Hsiao (2003) and Tsay (2005) study methods for estimating DL system parameters with a priori specified model structure.

The a priori information effect received from the empirical data (Carter and Zellner, 2004) consider on the structure model choice for DL. Polynomial schemes for estimation parameters are offered. Lansing (2002) proposes an algorithm for estimating the parameters of the DL system. The inflation process based on autoregressive models describes by Duffee (2003). Lag length choice is based on a step-by-step increase in delays and the model adequacy estimation. The Akaike criterion and Bayesian information criteria are applied, and the solution is made on model structure. The combination of modelling methods is widely used in the analysis of statistical data and the DL structure selection (for example, Alessi et al., 2008; Campos et al., 2005; Castle et al., Hansen, 2022). In most works, the shortcomings of existing statistics are considered and the scope of their application is shown.

The t approach (Karabutov, 2012) is offered to structural identification of DL systems with an irregular input. It is based on the analysis of static structures (virtual portraits), reflecting nonlinear properties of the system in a special structural space (Karabutov, 2011).

It is based on the analysis of static structures (visual portraits), reflecting nonlinear properties of the system in the special structural space (Karabutov, 2011). Decision-making algorithms on the DL maximum length are offered. Algorithms do not demand statistical criteria calculation. The analogue of Durbin-Watson criterion is offered for a considered case. We are developing the approach proposed by Karabutov (2012). Decision-making criteria on DL length are described in input and output variables of the system. It is shown that for structural identification of DL systems is not applicable sector sets by (Karabutov, 2010, 2012). The approach is offered to parametric identification of the system under uncertainty.

2. PROBLEM STATEMENT

Consider system

$$y_n = A_n^T U_n + \xi_n \tag{3.1}$$

where $y_n \in \mathbb{R}$ is an output; $U_n \in \mathbb{R}^k$ is input vector which elements are limited; $n \in \mathbb{J}_N = [0,N]$ is discrete time, $N<\infty$; $X_n = X\left(u_{i,n} \in U_n, y_n\right) \in \mathbb{R}^m$ is a DL vector on $u_{i,n} \in \mathbb{U}_n$ and y_n; $A \in \mathbb{R}^k, B \in \mathbb{R}^m$ are vectors of constant parameters; $\xi_n \in \mathbb{R}$ is a perturbation, $|\xi_n|<\infty$ for all $n \in \mathbb{J}_N$.

We supposed to $u_{i,n} \in \mathbb{U}_n$ $(i = \overline{1,k})$ and $\xi_n \in \mathbb{R}$ are irregular functions of time. The set of the measured values

$$\mathbb{I}_o = \left\{y_n, U_n \forall n \in \mathbb{J}_N\right\} \tag{3.2}$$

and mapping Γ_o: $\{U_n\} \times \{y_n\}$ $\forall n \in \mathbb{J}_N$ describing an observable informational portrait (Karabutov, 2009).

Problem: estimate of linearity degree and vector X_n dimension based on analysis of the system (3.1) structure and $\mathbb{I}_o, \Gamma_o$.

3. ESTIMATION OF LINEARITY DEGREE SYSTEM (3.1)

We consider the narrowing of the observed information portrait $\Gamma_o^{u_i} \subset \Gamma_o\big|_{u_i \in \mathbb{U}}$, and for each $\Gamma_o^{u_i}$, we construct a secant

$$\overline{\gamma}(y, u_i) = a_i u_{i,n}$$

where a_i is some real number.

Introduce set on (3.2) set of secants for Γ_o

$$\mathbb{S}\left(U,y\right) = \left\{\overline{\gamma}(y,u_i) \forall i = \overline{1,k}\right\}$$

Definition 1 (Karabutov, 2009). Field of structures S_S name a set of maps $\overline{\gamma}(y,u_i):\left\{u_i\right\} \rightarrow \{y\}$ $\forall i = \overline{1,k}$ on Euclidean plane E $S_S = \mathbb{S}\left(U,y\right)$.

Designate $\Phi_n = \left[\overline{\gamma}(y_n, u_{1,n}), \overline{\gamma}(y_n, u_{2,n}), \ldots, \overline{\gamma}(y_n, u_{k,n})\right]^T$ and consider the equation

$$\hat{y}_n = \Psi^T \Phi_n \tag{3.3}$$

where vector $\Psi \in \mathbb{R}^k$ define by the least-squares method (LSM). Estimation Ψ is based on suppositions made for U_n in section 2.

Completeness of system (3.1) on S_S follows from statement.

Theorem 1 (Karabutov, 2006). Consider the vector of informative variables $U_n \in \mathbb{R}^k$ and $S_S \subset \mathbb{S}(U,y)$ for (3.1). Then the field of structures S_S for system (3.1) is full, if

$$\sum_{i=1}^{k} \psi_i = 1, \tag{3.4}$$

where ψ_i is i-th element of vector Ψ in (3.4).

The theorem 1 gives sufficient conditions of linearity (nonlinearity, collinearity) systems (3.1) on $S_S \subset \mathbb{S}(U,y)$. If the condition (3.4) is fulfilled, that field S_S is full. Hence, S_S is a linear span of an output system (3.1). Otherwise, make a solution in the presence of nonlinearity or collinearity in the system (3.1).

Let $\sum_{i=1}^{k} \psi_i = \chi$. Magnitude $\Delta(\Psi)=\chi-1$ is the system (3.1) nonlinearity level in parametric space $P_P = (A,B)$. Since DL causes multicollinearity in (3.1), $\Delta(\Psi)$ takes small values.

4. SET FOR ESTIMATION OF DISTRIBUTED LAG STRUCTURE

Consider a case, when $X_n=X(u_{i,n})$. Apply the approach offered Karabutov (2011). Generate set for an estimation of structure (3.1). Introduce a variable $s_n = I^T U_n$, where $I \in \mathbb{R}^k$ is a unit vector. Apply model $\hat{y}_{s,n} = \hat{a}_s s_n$. Find parameter $\hat{a}_s \in \mathbb{R}$ as:

$$\arg\min_{\hat{a}_s} (\hat{y}_{s,n} - y_n)^2 = \hat{a}_s^* \tag{3.5}$$

$e_n = \hat{y}_{s,n} - y_n$ is a variable which contains the data about the structure of a system (3.1) lag. $X(u_{i,n})$ use as argument variable $u_{i,n} \in \mathbb{U}_n$, which ensures maximum for the coefficient of determination r^2_{e,u_i} between e and u_i.

As argument $X(u_{i,n})$ use variable $u_{i,n} \in \mathbb{U}_n$, which ensures maximum value of coefficient of determination between e and u_i. It is true and for y_n. As shown Karabutov (2011), set $\{e_n, u_{i,n}\}$ does not allow to solve a problem of structural identification. From the statement follows that the index in the Holder inequality (Shilov, 1973) is close to zero for e_n. Therefore, introduce a coefficient of structural properties (CSP) (Karabutov, 2009)

$$k_{e,u_i,n} = k_s(e,u_i,n) = \frac{e_n}{u_{i,n}} \tag{3.6}$$

Generate set

$$\mathrm{I}_k = \mathrm{I}_k\left(k,e\right) = \left\{e_n, k_n, n \in \mathbb{J}_N = [0,N]\right\},\ k_n \stackrel{df}{=} k_{e,u_i,n}$$

Definition 2. $P_s = (k,e)$ is the structural space of the system (3.1) allowing to identify of vector X_n structure.

Generate $\left\{k_q^v\right\}$, where $k_q^v = k_s\left(e,u_i,q\right), q \in \mathbb{J}_N^v = \left[0,N\right]$. As to every k_q^v corresponds value e_q^v receive $\mathrm{I}_k^v = \left\{e_q^v, k_q^v, q \in \mathbb{J}_N^v\right\}$.

In P_s, mapping $\Gamma_{e,k}^v : \left\{k_q^v\right\} \to \left\{e_q^v\right\}$and structure $S_{k,e}^v$ define corresponding to it. Now estimate structure X_n based on the analysis I_k^v and $S_{k,e}^v$.

Such approach well works at an estimation of structure of nonlinear static systems (Karabutov, 2011). For systems with the DL, approach demands modification.

5. DECISION-MAKING ON LAG LENGTH IN $X\left(u_{i,n}\right)$

Set $\mathrm{I}_{e,u} = \left\{e_n, u_{i,n}\right\}$ contains uncertainty N_L and N_{Lg}. N_L is an incomplete account of the linear component system (3.1). N_{Lg} reflects the of disturbance DL effect. Construct the secant for e_n for an exception N_L (Karabutov, 2012)

$$\overline{\gamma}(e,u_i) = \alpha_0 + \alpha_1 u_{i,n},$$

where α_0, α_1 фку positive number.

Introduce new variable $\varepsilon_n = e_n - \overline{\gamma}\left(e,u_i\right)$ which do not contain N_L. Analyze the set $\{\varepsilon_n, u_{i,n}\}$ for estimation N_{Lg}. A preliminary estimation of the maximum delay m is got using the algorithm CD_ρ (Karabutov, 2012).

We apply the set-functional approach to structural identification. Therefore, known methods of the length lag choice, based on statistical criteria (see Section 1), we do not consider.

Apply the theorem 1 for the estimation of vector X_n independence elements. As the set $\mathrm{I}_{e,u}$ analysis is inefficient, apply results of Section 4 and generate set:

$$\mathrm{I}_{\varepsilon,k_X} = \left\{\varepsilon_n, k_{\varepsilon,x_{i,n}}, i = \overline{1,m} \forall n \in \mathbb{J}_N\right\} \tag{3.7}$$

where $k_{\varepsilon,x_i,n}$ calculate based of (3.6) and designate $e_n = \varepsilon_n$, $x_{i,n} \in X_n$.

Introduce transformation $\Gamma_{\varepsilon,k_{\varepsilon,x_i}} : \left\{k_{\varepsilon,x_{i,n}}\right\} \to \left\{\varepsilon_n\right\}$ on $\mathrm{I}_{\varepsilon,k_X}$. $\Gamma_{\varepsilon,k_{\varepsilon,x_i}}$ corresponds structure $S_{k_{\varepsilon,x_i},\varepsilon}$ in space $P_{s,\varepsilon} = \left(k_{\varepsilon,x_{i,n}}, \varepsilon\right)$. Construct secants for $S_{k_{\varepsilon,x_i},\varepsilon}$

$$\overline{\gamma}\left(\varepsilon_n, k_{\varepsilon,x_{i,n}}\right) = \beta_i k_{\varepsilon,x_{i,n}} \quad \forall i = \overline{1,m_*} \tag{3.8}$$

where β_i are defined as the problem (3.5) solution; m_* is a value m received by algorithm CD_ρ.

Generate a vector

$$\Phi_{\varepsilon,n} = \left[\bar{\gamma}\left(\varepsilon_n, k_{\varepsilon,x_{1,n}}\right), \bar{\gamma}\left(\varepsilon_n, k_{\varepsilon,x_{2,n}}\right), \ldots, \bar{\gamma}\left(\varepsilon_n, k_{\varepsilon,x_{m_*n}}\right)\right]^T \tag{3.9}$$

and apply model $\hat{\varepsilon}_n = \Psi_\varepsilon^T \Phi_{\varepsilon,n}$ to forecasting ε_n, where the vector $\Psi_\varepsilon \in \mathbb{R}^{m_*}$ define based on results Section 3.

Theorem 2 (Karabutov, 2012). *Let (i) field of secants is constructed for* ε_n

$$S_\varepsilon = \left\{\bar{\gamma}\left(\varepsilon_n, k_{\varepsilon,x_{i,n}}\right)(\forall i = \overline{1,m_*}) \,\&\, (n \in \mathbb{J}_N)\right\}$$

on set the $\mathbb{I}_{\varepsilon,k_x}$; (ii) the model $\hat{\varepsilon}_n = \Psi_\varepsilon^T \Phi_{\varepsilon,n}$ is applied to forecasting ε_n. Then the vector $X_n \in R^{m_*}$ is an element of the system (3.1), if

$$\sum_{i=1}^{m_*} \psi_{\varepsilon,i} \neq 1,$$

where m_* *is defined by algorithm* CD_ρ, $\psi_{\varepsilon,i} \in \Psi_\varepsilon$.

Remark 1. As ε_n contains the information on the vector X_n influence on statistical treatment U_n, ξ_n use analogue of criterion Durbin-Watson criterion (Karabutov, 2012)

$$d = \frac{\sum_{n=1}^{N} \varepsilon_n^2}{\sum_{n=1}^{N} e_n^2}$$

System (3.1) nonlinearity level, considering DL for X_n, is defined as:

$$\Delta(\Psi_\varepsilon) = \sum_{i=1}^{m_*} \psi_{\varepsilon,i} - 1$$

Theorem 3. *Consider the secants* S_ε *set on* $\mathbb{I}_{\varepsilon,k_x}$, *and the secant* $\bar{\gamma}\left(e, k_{e,u_i}\right) = \alpha_1 k_{e,u_i}$ *for* $S_{k_{e,u},e}$. *Let coefficients of determinations are known for secants* $r^2_{\varepsilon,k_{x_j}}, j = \overline{1,m}$ *and* $r^2_{e,k_{u_i}}$. *Then the vector* $X\left(u_{i,n}\right) \in \mathbb{R}^m$ *is of the structure system (3.1) element, if* $\forall z = \overline{1,m}$

$$\left|r^2_{\varepsilon,k_{x_z}} - r^2_{e,k_{u_i}}\right| \leq \delta_x,$$

where $\delta_x \geq 0$ is set value.

DL interprets as a nonlinear component of the system (3.1) at the structure analysis. It follows from the theorem 1. Therefore, develop the procedure, deciding on the uncertainty N_{Lg} class. We offer the method of the DL estimation, based on the development of the approach by Karabutov (2012).

Note that there is the dependency between $u_{i,n}$ and $u_{i,n-j}$ $(j \geq 1)$. Therefore, the application of linear secants $\overline{\gamma}$ can appear inefficient for decision-making. Consider set:

$$\mathbb{I}_{e,k_x} = \left\{ e_n, k_{e,x_{\rho,n}}, \rho = \overline{1,m} \forall n \in \mathbb{J}_N \right\}, \tag{3.10}$$

where $x_{\rho,n} \in X(u_{i,n})$.

Arrange $\left(e_n, k_{e,x_{\rho,n}}\right)$ in ascending order for each $\rho = \overline{1,m}$ and get the set

$$\mathbb{I}^v_{e,k_{x_\rho}} = \left\{ e^v_{\rho,q}, k^v_{e,x_{\rho,q}}, q \in J^v_N \right\},$$

where the index ρ in $e^v_{\rho,q}$ is introduced to underline dependence e^v_q from $k^v_{e,x_{\rho,q}}$.

Corresponding set $\mathbb{I}^v_k$ construct for u_i. Consider mappings:

$$\Gamma^v_{e,e_\rho} : \left\{ e^v_q \right\} \to \left\{ e^v_{\rho,q} \right\}, \ \rho \geq 1 \tag{3.11}$$

which describe structures S^v_{e,e_ρ}. For every S^v_{e,e_ρ} construct, the secant

$$\overline{\gamma}_{2,\rho}\left(e^v_q, e^v_{\rho,q}\right) = a^v_{\rho,0} e^v_q + a^v_{\rho,1} \left(e^v_q\right)^2 \tag{3.12}$$

define and the coefficient of determination (CD) r^2_{2,e^v,e^v_ρ}.

Theorem 4. *Let for system (3.1) with $X\left(u_{i,n} \in \mathbb{U}_n\right) \in \mathbb{R}^m$: (i) structures S^v_{e,e_k}, described by mappings (3.11), sets $\mathbb{I}^v_{e,k_x}$ and $\mathbb{I}^v_k$ are constructed; (ii) secants $\overline{\gamma}_{2,\rho}\left(e^v_q, e^v_{\rho,q}\right)$ are defined for S^v_{e,e_ρ}. If the secant $\overline{\gamma}_{2,\rho}\left(e^v_q, e^v_{\rho,q}\right)$ for S^v_{e,e_ρ} $(\rho \geq 1)$ satisfies the condition:*

$$r^2_{2,e^v,e^v_\rho} \geq \delta_v,$$

where $\delta_v > 0$ is some set number, then $x_{\rho,n} = u_{i,n-\rho}$ is the element of system (3.1) structure.

Theorem 4 defines DL length for variable $u_i \in \mathbb{U}$. The decision accepts based on the interdependence analysis between e_n and its forecast on variables $x_{\rho,n}$.

The decision on the lag for u_i can be received based on the data analysis in space $P_s = (k,e)$.

Consider set $\mathbb{I}^v_{k_i} = \{e^v_q, k^v_{i,q}, q \in \mathbb{J}^v_N\}$ and define mapping $\Gamma^v_{e,k_i}: \{k^v_{i,q}\} \rightarrow \{e^v_q\}$, where $k^v_{i,q}$ is CSP on the input u_i. Set the structure S^v_{e,k_i} on the basis Γ^v_{e,k_i}. Define secants for S^v_{e,k_i} and $S^v_{e,k_{e,x_\rho}}$

$$\bar{\gamma}_2(e^v_{i,q}, k^v_{i,q}) = \bar{\gamma}_{2,i} = a^v_{i,0} k^v_{i,q} + a^v_{i,1} (k^v_{i,q})^2 \tag{3.13}$$

$$\bar{\gamma}_2(e^v_{x_\rho,q}, k^v_{x_\rho,q}) = \bar{\gamma}_{2,x_\rho} = \tilde{a}^v_{\rho,0} k^v_{x_\rho,q} + \tilde{a}^v_{\rho,1} (k^v_{x_\rho,q})^2 \tag{3.14}$$

where $a^v_{i,0}$, $a^v_{i,1}$, $\tilde{a}^v_{\rho,0}$, $\tilde{a}^v_{\rho,1}$ are LSM estimations. Coefficients of determinations $r^2_{2,\bar{\gamma}_{2,i},e^v}$, are known for $\bar{\gamma}_2(e^v_{i,q}, k^v_{i,q})$ and $\bar{\gamma}_2(e^v_{x_\rho,q}, k^v_{x_\rho,q})$. Designate average values $x_{\rho,n}$ as $\bar{x}_{\rho,n}$.

Theorem 5. *Let for system (3.1) with $X(u_{i,n} \in U_n) \in \mathbb{R}^m$ in space $P_s = (k,e)$: (i) structures S^v_{e,k_i} and S^v_{e,k_i}, described by mappings $\Gamma^v_{e,k}$ and Γ^v_{e,k_i}, on sets $\mathbb{I}^v_{e,k_{x_\rho}}$, $\mathbb{I}^v_{k_i}$ are seed; (ii) secants (3.13), (3.14) are defined for S^v_{e,k_i} and $S^v_{e,k_{e,x_\rho}}$. If the secant $\bar{\gamma}_2(e^v_{x_\rho,q}, k^v_{x_\rho,q})$ for $S^v_{e,k_{e,x_\rho}}$ ($\rho \geq 1$) satisfies the condition*

$$\left| r^2_{2,\bar{\gamma}_{2,x_\rho},e^v_1} - r^2_{2,\bar{\gamma}_{2,i},e^v_1} \right| \leq \delta_\rho \tag{3.15}$$

$$\left| \tilde{a}^v_{\rho,1} - \bar{x}_{\rho,n} \right| \leq \pi_\rho \tag{3.16}$$

where $\delta_\rho > 0$ and π_ρ are some set numbers, then $x_{\rho,n} = u_{i,n-\rho}$ is the structure element of the system (3.1).

The theorem 5 proof is based that frameworks S^v_{e,k_i} and $S^v_{e,k_{e,x_\rho}}$ describe variable change e^v_q. Therefore, the decision except one inclusion in the system (3.1) structure for variable $x_\rho \in X_n$, which ensures given level of determination coefficient. $r^2_{2,\bar{\gamma}_{2,\rho},e^v}$ should belong to the interval (3.15). If values $r^2_{2,\bar{\gamma}_{2,\rho},e^v}$G2 are approximate, then use restriction (3.16) for a received estimation of average $x_{\rho,n} = u_{i,n-\rho}$. Other approaches to the choice of DL length are proposed by Karabutov (2012).

3. CHOICE OF LENGTH LAG FOR $X(y_n)$

Let $X_n = [x_{1,n}, x_{1,n}, \ldots, x_{l,n}]^T$ in system (3.1), where $x_{i,n} = y_{n-i}$, $i = \overline{1,l}$. Apply algorithm CD_ρ for a prior estimation l. Choice DL length method is proposed Karabutov (2012). It is based on the application of the theorem 1. As show modelling results, the decision on DL structure for y_n not always is possible on linear secants for $S^v_{e,k}$. Therefore, we describe the approach based on the application of theorems 4, 5.

Consider space $P_{s,e} = (e^v_q, e^v_{i,q})$, where e^v_q, $e^v_{i,q}$ are elements of sets

$$\mathbb{I}^{v}_{e,k_y} = \left\{ e^{v}_{q}, k^{v}_{e,y_q}, q \in \mathbb{J}^{v}_{N} \right\}$$

$$\mathbb{I}^{v}_{e,k_{x_i}} = \left\{ e^{v}_{i,q}, k^{v}_{e,x_{i,q}}, q \in \mathbb{J}^{v}_{N} \right\}, \ i \geq 1$$

Introduce mapping $\Gamma^{v}_{e,e_i}: \left\{ e^{v}_{q} \right\} \rightarrow \left\{ e^{v}_{i,q} \right\}, i \geq 1$ for $\mathbb{I}^{v}_{e,k_y}, \mathbb{I}^{v}_{e,k_{x_i}}$. The construction method of mappings is described in Section 5. Γ^{v}_{e,e_i} corresponds the structure S^{v}_{e,e_i}. Construct the secant for every S^{v}_{e,e_i}

$$\overline{\gamma}_{2,i}\left(e^{v}_{q}, e^{v}_{i,q}\right) = a^{v}_{i,0} e^{v}_{q} + a^{v}_{i,1} \left(e^{v}_{q}\right)^{2} \tag{3.17}$$

and define CD $r^{2}_{2,e^{v},e^{v}_{i}}$. Apply the theorem 4 for the DL length estimation y_n.

Consider space $P_{s,k_{x_i},e} = \left(k^{v}_{e,x_i}, e^{v}_{i}\right)$. Define mapping $\Gamma^{v}_{e,k_{x_i}}: \left\{ k^{v}_{x_i q} \right\} \rightarrow \left\{ e^{v}_{q} \right\}$ and a virtual portrait $S^{v}_{e,k_{e,x_i}}$ on set $\mathbb{I}^{v}_{k_{x_i}}$, where $k^{v}_{x_i q}$ is CSP on an input x_i. Consider set

$$\mathbb{I}^{v}_{e,k_y} = \left\{ e^{v}_{q}, k^{v}_{e,y,q}, q \in \mathbb{J}^{v}_{N} \right\}$$

introduce mapping $\Gamma^{v}_{e,k_{e,y}}: \left\{ k^{v}_{e,y,q} \right\} \rightarrow \left\{ e^{v}_{q} \right\}$ and structure $S^{v}_{e,k_{e,y}}$. Apply (3.13), (3.14) and get secants

$$\overline{\gamma}_{2}\left(e^{v}_{q}, k^{v}_{e,y,q}\right) = \overline{\gamma}_{2} = a^{v}_{y,0} k^{v}_{e,y,q} + a^{v}_{y,1} \left(k^{v}_{e,y,q}\right)^{2} \tag{3.18}$$

$$\overline{\gamma}_{2}\left(e^{v}_{i,q}, k^{v}_{x_i q}\right) = \overline{\gamma}_{2,i} = \tilde{a}^{v}_{i,0} k^{v}_{x_i q} + \tilde{a}^{v}_{i,1} \left(k^{v}_{x_i q}\right)^{2} \tag{3.19}$$

where $a^{v}_{y,0}$, $a^{v}_{y,1}$, $\tilde{a}^{v}_{i,0}$, $\tilde{a}^{v}_{i,1}$ are some numbers.

We have CD $r^{2}_{2,\overline{\gamma}_2,e^{v}}$, $r^{2}_{2,\overline{\gamma}_{2,i},e^{v}}$ for $\overline{\gamma}_{2}\left(e^{v}_{q}, k^{v}_{e,y,q}\right)$ and $\overline{\gamma}_{2}\left(e^{v}_{i,q}, k^{v}_{x_i q}\right)$. Designate average values $x_{i,n}$ as $\overline{x}_{i,n}$. Apply the theorem 5 and receive the of DL length estimation for y_n.

The offered approach applies in a case

$$X_n = X\left(u_{i,n} \in U_n, y_n\right) \in \mathbb{R}^{m}$$

Apply Lyapunov exponents to the DL estimation (Lyapunov, 1992).

4. ABOUT SECTOR SETS

Applied the analysis of sector sets (SS) (Karabutov, 2010) we get the decision on the nonlinearity class. Despite complexities, construction SS allows solving problems of structural identification for nonlinear static systems. Extending this approach to systems with DL is difficult. Explain it to that DL results from the decision of difference equations. The resulting solution is only some approximation of the variable, which is the basis for obtaining the delay. Therefore, SS will differ in structural space based on variable and its lags. It does not allow design effective algorithms for decision-making. DL presence on variables y_n is the sign of decrease CD between input variables and the system output.

5. ABOUT ESTIMATION OF SYSTEM (3.1) PARAMETERS AT DL

The identification of parameters in the system with DL is realized on a priori given models for parameters (see Section 1). Such the approach is based on number minimization of estimated parameters under multicollinearity. It is the dominating approach. Authors offer various variations of approaches within a given of identification concept (Demetriou and Vassiliou, 2014). It is based on estimations of the CSP system on the input set. The coefficient of structural properties is a parameter estimation of the considered DL. Here, we give development of the proposed approach.

We accept the received estimations of an average CSP for admissible estimations DL, and form the vector $\breve{B} \in \mathbb{R}^m$. Introduce some limited vector $\Theta \in \mathbb{R}^h$, where $h<m$, $\Theta \in \Omega_\Theta$. In particular, at $h=1$,

$$\theta \in \Omega_\theta = \left\{\theta \in \mathbb{R}, \theta \in \left[1-\varepsilon_h, 1+\varepsilon_h\right]\right\}$$

where $\varepsilon_h \geq 0$. Write a second component in (3.1) as:

$$B^T X_n = \theta \breve{B}^T X_n \tag{3.20}$$

Apply LSM and receive an estimation θ.

If the model adequacy with the estimation θ is unsatisfactory, then choose $h>1$. Consider the vector Θ and execute division X_n into parts that they contained non-singular components. Let $X_n = \left[X_{1,n}^T, X_{2,n}^T\right]^T$. Then $\breve{B} = \left[\breve{B}_1^T, \breve{B}_2^T\right]^T$ and $\Theta=[\theta_1,\theta_2]^T$, $D(\Theta)=\theta_1 I_{h_1} \dotplus \theta_2 I_{h_2}$, where $h_1+h_2=h$, $D(\Theta)$ is a diagonal matrix from the vector Θ, $I_{h_1} \in R^{h_1 \times h_1}$ is unity matrix, $\dotplus$ is the sum direct of matrixes.

The model for $B^T X_n$ write as

$$B^T X_n = \left[\theta_1 \breve{B}_1^T, \theta_2 \breve{B}_2^T\right]\left[X_{1,n}^T, X_{2,n}^T\right]^T$$

Apply LSM to the estimation W2. Estimate adequacy of the received model. If the forecasting error is big, then change the vector Θ dimension. Execute the identification process again.

Remark 2. The vector Θ is possible to identify, applied by adaptive φ-algorithm (Karabutov, 2006), considering restriction $\Theta \in \Omega_{\Theta}$.

6. EXAMPLES

Consider system (3.1) with $A = [1.5; 2; 2.8]^T$, $B = [0.7; 0.6]^T$, $X_n \in \mathbb{R}^2$, $X_n = X(u_{1,n})$. $u_{i,n} \in U_n \in \mathbb{R}^3$ are limited stochastic functions, $\xi_n \in \mathbb{R}$ is a random variable with a zero expectancy and a final dispersion, $|\xi_n| \leq 0.3$. $u_{1,n}$ is the difference equation solution of the second order. Apply the theorem 1 and receive, that the system is nonlinear. $\Delta(\Psi)=0.03$. Value $\Delta(\Psi)$ is the linear interdependence (multicollinearity) sign of the system. Generate set for lag estimation

$$\mathbb{I}_{e,u} = \left\{ e_n, u_{i,n} \left(\forall i = \overline{1,3}\right) \& \left(\forall n \in \mathbb{J}_N\right) \right\} \tag{3.21}$$

for lag estimation.

The analysis $\mathbb{I}_{e,u}$ showed that the $u_{1,n}$ contains the lag. Apply algorithm CD_ρ and receive a prior estimation for lag length m. We have $m^*=2$ for $u_{1,n}$. Apply the theorem 4 to validation of the received result.

Generate sets $\mathbb{I}^v_{e,k_x}$, $\mathbb{I}^v_k$, consider mappings (3.11) and structures S^v_{e,e_ρ}, where $\rho \geq 1$. Define secants $\overline{\gamma}_{2,\rho}\left(e^v_q, e^v_{\rho,q}\right)$ and r^2_{2,e^v,e^v_ρ} for S^v_{e,e_ρ}:

$$\overline{\gamma}_{2,1}\left(e^v_q, e^v_{1,q}\right) = 0.975\, e^v_q - 0.0133 \left(e^v_q\right)^2, \quad r^2_{2,e^v,e^v_1} = 0.948$$

$$\overline{\gamma}_{2,2}\left(e^v_q, e^v_{2,q}\right) = 0.967\, e^v_q - 0.019 \left(e^v_q\right)^2, \quad r^2_{2,e^v,e^v_2} = 0.934$$

$$\overline{\gamma}_{2,3}\left(e^v_q, e^v_{3,q}\right) = 0.954\, e^v_q - 0.0038 \left(e^v_q\right)^2, \quad r^2_{2,e^v,e^v_3} = 0.91$$

Figure 1. Change y_n and ε_n

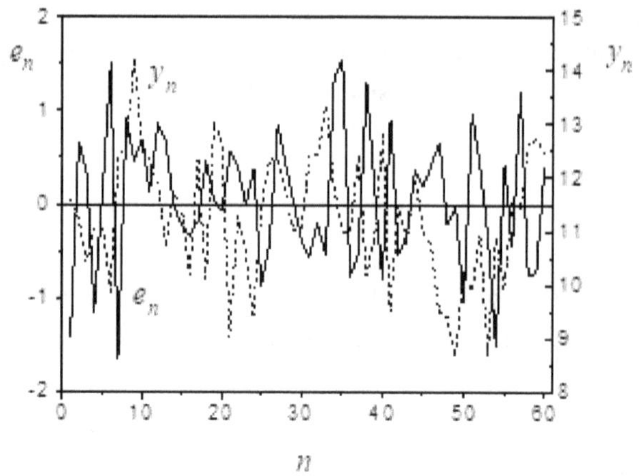

Let δ_v=0.93. Apply the theorem 4 and receive m^*=2. Figure 1 shows the delay effect on $u_{1,n}$.

Apply theorem 5 to check of the received estimation.

To check of the received estimation, apply theorem 5. Define secants (3.13), (3.14) and CD

$$\overline{\gamma}_2\left(e^v_{1,q}, k^v_{1,q}\right) = \overline{\gamma}_{2,1} = 2.13k^v_{i,q} - 0.793\left(k^v_{1,q}\right)^2, \; r^2_{2,\overline{\gamma}_{2,1},e^v_1} = 0.971$$

$$\overline{\gamma}_2\left(e^v_{x_1,q}, k^v_{x_1,q}\right) = \overline{\gamma}_{2,x_1} = 1.93k^v_{x_1,q} + 0.48\left(k^v_{x_1,q}\right)^2, \; r^2_{2,\overline{\gamma}_{2,x_1},e^v_{x_1}} = 0.964$$

$$\overline{\gamma}_2\left(e^v_{x_2,q}, k^v_{x_2,q}\right) = \overline{\gamma}_{2,x_2} = 1.76k^v_{x_2,q} + 0.22\left(k^v_{x_2,q}\right)^2, \; r^2_{2,\overline{\gamma}_{2,x_2},e^v_{x_2}} = 0.928$$

$$\overline{\gamma}_2\left(e^v_{x_3,q}, k^v_{x_3,q}\right) = \overline{\gamma}_{2,x_3} = 1.69k^v_{x_3,q} + 0.053\left(k^v_{x_3,q}\right)^2, \; r^2_{2,\overline{\gamma}_{2,x_3},e^v_{x_3}} = 0.922$$

Average values for u_1 and x_p: $\overline{u}_1 = 2.026, \overline{x}_1 = 2.011, \overline{x}_2 = 1.996, \overline{x}_3 = 1.98$. Let δ_ρ=0.043, π_ρ=0.02. Check-up conditions (3.15), (3.16). They are performed for lags x_1, x_2. So, we get estimates of the DL length for u_1.

In Figure 2, results of the theorem 5 application are shown. They confirm the conclusion about DL influence as some equivalent nonlinearity in the system. This conclusion receives from the analysis of structure S^v_{e,k_1} secant $\overline{\gamma}_{2,1}$ change.

For $\overline{\gamma}_{2,x_3}$, we have almost linear dependence that confirms the lag x_3 elimination in the model.

Consider system (3.1) with $X_n = [x_{1,n}, x_{2,n}]^T, x_{1,n} = y_{n-1}, x_{2,n} = u_{1,n-1}$. Performance the set (3.21) analysis and apply algorithm CD_ρ to the DL estimation. Analysis results give lags on y_n and $u_{1,n}$. m^*=1 for these variables. Apply the theorem 4 and confirm this conclusion.

Figure 2. Structures $S^v_{a,k}$ and secants $\bar{\gamma}_{2,x}$

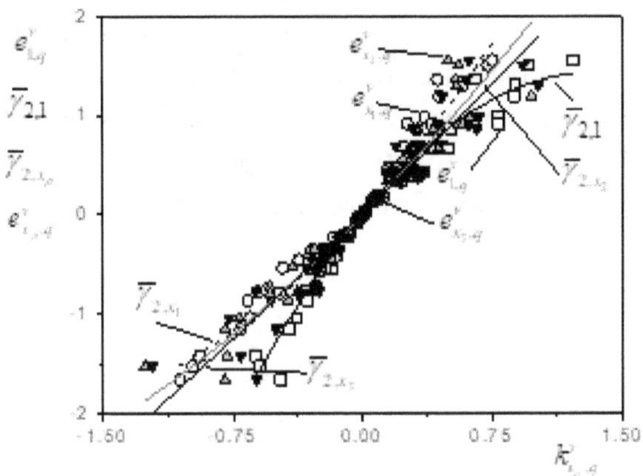

Define secants (3.13), (3.14) and CD

$$\bar{\gamma}_2\left(e^v_{1,q}, k^v_{1,q}\right) = \bar{\gamma}_{2,1} = 1.909\,k^v_{i,q} - 0.567\left(k^v_{1,q}\right)^2, \quad r^2_{2,\bar{\gamma}_{2,1},e^v_1} = 0.933$$

$$\bar{\gamma}_2\left(e^v_{x_2,q}, k^v_{x_2,q}\right) = \bar{\gamma}_{2,x_2} = 2.0\,k^v_{x_2,q} + 0.771\left(k^v_{x_2,q}\right)^2, \quad r^2_{2,\bar{\gamma}_{2,x_2},e^v_{x_2}} = 0.965$$

$$\bar{\gamma}_2\left(e^v_{x_{2,2},q}, k^v_{x_2,q}\right) = \bar{\gamma}_{2,x_{2,2}} = 1.694\,k^v_{x_{2,2},q} + 0.115\left(k^v_{x_{2,2},q}\right)^2, \quad r^2_{2,\bar{\gamma}_{2,x_{2,2}},e^v_{x_{2,2}}} = 0.946$$

where $x_{2,2,q} = u_{1,q-2}$.

Average values for $u_{1,n}, x_{2,n}$ and $x_{2,2,n}$: $\bar{u}_1 = 2.028$, $\bar{x}_2 = 2.023$, $\bar{x}_{2,2} = 2.0$. Secants $\bar{\gamma}_{2,1}, \bar{\gamma}_{2,x_2}$ are confirmed lag presence in $u_{1,n}$. The secant $\bar{\gamma}_{2,x_{2,2}}$ is almost linear (for a linear secant $\bar{\gamma}_{1,x_{2,2}}$ $r^2_{2,\bar{\gamma}_{1,x_{2,2}},e^v_{x_{2,2}}} = 0.944$). The condition (3.16) is not true for $\bar{\gamma}_{2,x_{2,2}}$ with π_ρ=0.15. Therefore, $x_{2,2,n}$ is not an element of the system (3.1) structure. Apply the statement (Karabutov, 2012) to explain the relatively high $r^2_{2,\bar{\gamma}_{1,x_{2,2}},e^v_{x_{2,2}}}$. Similarly, we perform the DL identification for y_n.

So, modelling results confirm the efficiency of offered algorithms and methods.

7. CONCLUSION

We offer is functional-set approach to the structural identification of discrete systems with the distributed lag. The decision on DL structure accepts based on the analysis of static frameworks. It is shown that the DL can be considered as nonlinearity. We apply second order secants to decision-making on lag length. It is shown that the linearity degree of the second order secant is the criterion for decision on the distributed lag. The approach to parametric identification is offered for the system with DL. We do not do any assumptions on the law of change parameters. The approach is based on tentative estimations for model parameters and the analysis of the system structural properties. These estimations are specified based on the identification of auxiliary factors (multiplicative parameters). Results of modelling have confirmed working capacity of offered methods and algorithms.

REFERENCES

Alessi, L., Barigozzi, M., & Capasso, M. (2008). A robust criterion for determining the number of static factors in approximate factor models. *Working paper series*, 903.

Almon, S. (1965). The distributed lag between capital appropriations and expenditures. *Econometrica*, *33*(1), 178–196. DOI: 10.2307/1911894

Armstrong, B. (2006). Models for the relationship between ambient temperature and daily mortality. *Epidemiology (Cambridge, Mass.)*, *17*(6), 624–631. DOI: 10.1097/01.ede.0000239732.50999.8f PMID: 17028505

Campos, J., Ericsson, N. R., Hendry D. F. (2005). General-to-specific modeling: An overview and selected bibliography. *International finance discussion papers, 2005*(838). DOI: 10.17016/ifdp.2005.838

Carter, R. A. L., & Zellner, A. (2004). The arar error model for univariate time series and distributed lag models. *Studies in Nonlinear Dynamics and Econometrics*, *8*(1), 2–2. DOI: 10.2202/1558-3708.1132

Castle, J. L., Doornik, J. A., & Hendry, D. F. (2010). *Evaluating automatic model selection. (Discussion paper series, 474) (working Paper)*. Economists Online.

Demetriou, I. C., & Vassiliou, E. E. (2009). An algorithm for distributed lag estimation subject to piecewise monotonic coefficients. *International Journal of Applied Mathematics*, *39*(1), 1–10.

Demetriou, I. C., & Vassiliou, E. E. (2014). On Distributed-Lag Modeling algorithms by r-convexity and piecewise monotonicity. In Rassias, T., Floudas, C., & Butenko, S. (Eds.), *Optimization in Science and Engineering. Springerб 115-140*. DOI: 10.1007/978-1-4939-0808-0_6

Duffee, G. (2006). R. (2005). Term structure estimation without using latent factors. *Journal of Financial Economics*, *79*(3), 507–536. DOI: 10.1016/j.jfineco.2005.03.007

Fisher, I. (1937). Note on a Short-cut Method for Calculating Distributed Lags. *Bulletin de l'Institut International de Statistique*, *29*, 322–327.

Gasparrini, A., Armstrong, B., & Kenwardb, M. G. (2010). Distributed lag non-linear models. *Statistics in Medicine*, *29*(21), 2224–2234. DOI: 10.1002/sim.3940 PMID: 20812303

Gershenfeld, N. (1999). *The Nature of Mathematical Modelling*. Cambridge University Press.

Hansen, B. E. (2022). *Econometric*. Princeton University Press.

Harvey, A. C., & Dhrymes, P. J. (1971). *Distributed Lags: Problems of Estimation and Formulation*. Holden-Day., DOI: 10.2307/2987844

Hsiao, C. (2003). *Analysis of Panel Data*. Cambridge University Press. DOI: 10.1017/CBO9780511754203

Johnston, J. (1976). *Econometric methods* (2nd ed.). McGraw-Hill Book Company., DOI: 10.2307/2988145

Jorgenson, D. W. (1964). Minimum variance, linear, unbiased seasonal adjustment of economic time series. *Journal of the American Statistical Association, 59*(307), 681–724. DOI: 10.1080/01621459.1964.10480722

Kailath, T. (1977). *Linear Least-Squares Estimation*. Dowden, Hutchinson and Ross, Inc.

Karabutov, N. (2012). Structural identification of static systems with distributed lags. *International journal of control science and engineering, 2*(6), 136-142. DOI: 10.5923/j.control.20120206.01

Karabutov, N. N. (2006). *Adaptive identification of systems. Information synthesis*. URSS, Librokom.

Karabutov, N. N. (2009). *Structural identification of systems: the analysis of informational structures*. URSS, Librokom.

Karabutov, N. N. (2010). Structures, fields and methods of identification of nonlinear static systems in the conditions of uncertainty. *Intelligent control and automation, 1*, 59-67. DOI: 10.4236/ica.2010.12007

Karabutov, N. N. (2011). *Structural identification of static plants: Fields, structures, methods*. URSS, Librokom.

Karabutov, N. N. (2012). Adaptive algorithms for structural parameters identification of single-valued nonlinearities of static systems with a vectorial input. *Int. J. Sensing. Computing and Control*, 2(1), 1–11.

Koyck, L. M. (1954). *Distributed Lags and Investment Analysis*. North-Holland Publishing Company.

Lansing, K. J. (2002). Real-time estimation of trend output and the illusion of interest rate smoothing. *FRBSF economic. RE:view*, 18–34.

Lyapunov, A. M. (1992). *General problem about a movement stability*. CRC Press.

Malinvaud, E. (1980). *Statistical methods in econometrics* (3rd ed.). North-Holland Publishing Co.

Nelson, C. R., & Schwert, G. W. (1974). Estimating the parameters of a distributed lag model from cross-section data: The case of hospital admissions and discharges. *Journal of the American Statistical Association, 69*(347), 627–633. DOI: 10.1080/01621459.1974.10480179

Shilov, G. (1973). *Mathematical analysis.* MIT Press.

Solow, R. (1960). On a family of lag distributions. *Econometrica, 28*(2), 393–406. DOI: 10.2307/1907729

Theil, H., & Stern, R. M. (1960). A simple unimodal lag distribution. *Metroeconomica, 12*(2-3), 111–119. DOI: 10.1111/j.1467-999X.1960.tb00277.x

Tsay, W.-J. (2005). The long memory autoregressive distributed lag model and its application on congressional approval. *Institute of Economics, Academia Sinica.* DOI: 10.1016/j.electstud.2009.06.004

Yoder, J. (2007). Autoregressive distributed lag models. *WSU Econometrics, II*, 91–115.

Chapter 4
Adaptive Identification of Systems With Distributed Lags

ABSTRACT

In this chapter, we apply the results got in Chapter 3. The adaptive identification problem is studied under uncertainty and perturbations. Adaptive algorithms for identification of system parameters with lags in the output variable are proposed, and convergence conditions are obtained. The case is considered when the number of tuning model parameters can be reduced. Here, we apply adaptive multiplicative algorithms (AMA) for the system with lags in the input variable. The convergence of the AMA is proved. We consider the identification of system parameters in the presence of relationships between variables. The case is studied linear relationship between the system output and the disturbance. Using adaptive algorithms presented above does not allow to get true estimates of parameters. To solve the problem, it is proposed to evaluate the current disturbance. The relevant procedures are described, and their effectiveness is proven.

1. PROBLEM STATEMENT

Consider the system described by the equation

$$y_n = A^T U_n + B^T X_n + \xi_n \tag{4.1}$$

where $y_n \in \mathbb{R}$ is the output; $U_n \in \mathbb{R}^k$ is the input vector whose elements are piecewise bounded function of time; $n \in {}_N = [0,N]$ is the discrete time variable $X_n = X(u_{i,n}, y_n) \in \mathbb{R}^m$ is the vector of distributed lags regarding $u_{i,n} \in U_n$,

DOI: 10.4018/979-8-3373-0822-7.ch004

y_m, $X_n = \left[X^T_{u_{i,n}}, X^T_{y_n}\right]^T$, $A \in \mathbb{R}^k$, $B \in \mathbb{R}^m$ is the vector of constant parameters, and $\xi_n \in \mathbb{R}$ is the external perturbation, $|\xi_n|<\infty$ for all $n \in {}_N$.

For (2.1), the set of measured values

$$_o = \{y_n, U_n \ \forall n \in {}_N\} \tag{4.2}$$

and the mapping Γ_o:$\{U_n\}\times\{y_n\}$ $\forall n \in {}_N$ that describes the observable information portrait (see Karabutov, (2016; 2013) are known.

Problem: estimate the set parameters for the system (2.1) based on the analysis $_o$, Γ_o.

First, we consider the case, where:

$$X_{i,n} = X_{u_{i,n}} = \left[u_{i,n-1}, u_{i,n-2}, \ldots, u_{i,n-m}\right]^T \in \mathbb{R}^m$$

i.e., only lags regarding the input variables $u_{i,n} \in U_n$ are considered. Assume that the structure of the lag regarding $u_{i,n}$ has already been determined by the approach proposed by Karabutov (2013).

2. ADAPTIVE IDENTIFICATION OF SYSTEM (4.1) WITH LAG IN $X_{u_{i,n}}$

2.1. Identification of Parameters in Absence of Relationship Between Lags

For the identification of the parameter of the system (4.1), we apply the model

$$\hat{y}_n = \hat{A}^T_{n-1} U_n + \hat{B}^T_{n-1} X_{i,n} \tag{4.3}$$

where $\hat{y}_n \in \mathbb{R}$ is the output of the model, $\hat{A}_{n-1} \in \mathbb{R}^k$ and $\hat{B}_{n-1} \in \mathbb{R}^m$ are vectors of adjustable parameters in (4.3).

Introduce the error of the forecast output of the system (4.1)

$$e_n = \delta A^T_{n-1} U_n + \delta \hat{B}^T_{n-1} X_{i,n} + \xi_n \tag{4.4}$$

where $\delta A_n = A_n - \hat{A}_n$, and $\delta B_n = B_n - \hat{B}_n$ are parametric errors.

Consider the Lyapunov function (LF) $V_{e,n} = e_n^2$. We find an adaptive algorithm from the stability condition for the system (4.1), (4.3):

$$\Delta V_{e,n} = V_{e,n} - V_{e,n-1} \leq 0$$

Differentiating ΔV_n regarding δA_n and δB_n, we obtain:

$$\delta A_n = \delta A_{n-1} - \Gamma_A e_n U_n \tag{4.5}$$

$$\delta B_n = \delta B_{n-1} - \Gamma_B e_n X_{i,n} \tag{4.6}$$

where $\Gamma_A \in \mathbb{R}^{k\times k}$ and $\Gamma_B \in \mathbb{R}^{m\times m}$ are diagonal matrices with $\gamma_{ii,A}>0$, $\gamma_{jj,B}>0$. The matrices Γ_A, and Γ_B provide the convergence of the algorithms (4.5) and (4.6) in the presence of perturbations.

From (4.5) and (4.6), we obtain tuning adaptive algorithms of the parameters of model (4.3):

$$\hat{A}_n = \hat{A}_{n-1} + \Gamma_A e_n U_n \tag{4.7}$$

$$\hat{B}_n = \hat{B}_{n-1} + \Gamma_B e_n X_{i,n} \tag{4.8}$$

Introduce notation

$$\delta H_n = \left[\delta A_n^T, \delta B_n^T\right]^T,\ K_n = \left[U_n^T, X_{i,n}^T\right]^T,\ \Gamma = \Gamma_A \dotplus \Gamma_B,$$

where $\dotplus$ means the direct sum of matrices. Then for δH_n we obtain the algorithm:

$$\delta H_n = \delta H_{n-1} - \Gamma e_n K_n \tag{4.9}$$

Assumptions: 1. The vector $K_n \in \mathbb{R}^{k+m}$ is extremely nondegenerate, bounded, and

$$\alpha_0 I_{(k+m)} \leq K_n K_n^T \leq \alpha_1 I_{(k+m)} \tag{4.10}$$

where α_0 and α_1 are some nonnegative numbers and $I_{k+m} \in \mathbb{R}^{(k+m)\times(k+m)}$ is the identity matrix.

2. $\|K_n\|^2 \leq \beta^2 < \infty$ for $\forall n \geq 0$.

The convergence of algorithms (4.7) and (4.8) is implied by the following assertion (Karabutov et al., 2016).

Theorem 1. *Assume that the following conditions hold:*

(1) the perturbation ξ_n in (4.1) is bounded: $|\xi_n| \leq \upsilon$ for all $n \geq 0$, where $\upsilon \geq 0$ is some number;

(2) there exist functions $V_{e,n}$, and $V_{\delta,n} = \|\delta H_n\|^2$, where " · " is a vector norm;

(3) $\tilde{\gamma} I_{k+m} \leq \Gamma \leq \bar{\gamma} I_{k+m}$, where $\tilde{\gamma} = \lambda_{\min}(\Gamma)$and $\bar{\gamma} = \lambda_{\max}(\Gamma)$are the minimal and maximal eigenvalues of the matrix Γ;

(4) the vectors U_n and $X_{i,n}$ are piecewise continuous functions of n satisfying the assumptions 1 and 2.

Then all trajectories of the system (4.1), (4.3), (4.5), (4.6) are bounded and the following estimates are valid:

$$V_{\delta,n} \leq \frac{\kappa v^2}{1-\mu}, \quad V_{e,n} \leq \frac{\kappa_\xi v^2}{\kappa_e}$$

if $\tilde{\gamma}\alpha_0 \leq 1 + \bar{\gamma}^2\beta^4$, $0<\kappa_e<1$, where $\|\Gamma\| \leq \pi$,

$$K_n^T \Gamma K_n \leq \vartheta < \infty \;, \; \left|K_{n-1}^T \Gamma K_n\right| \leq \tilde{\vartheta} < \infty \;, \; \mu = 1 - \tilde{\gamma}\alpha_0 + \bar{\gamma}^2\beta^4$$

$$\beta^2\left(\frac{\pi^2}{\rho} + 2\bar{\gamma}^2\right) = \kappa, \;\; \kappa_e = 1 + \vartheta - 2\tilde{\vartheta} - \theta, \;\; \kappa_\xi = \vartheta + \tilde{\vartheta}$$

Theorem 1 implies that the algorithms (4.7) and (4.8) converge if the vectors U_n and $X_{i,n}$ are bounded and extremely nondegenerate vector-valued functions, i.e., the condition (4.10) is valid for them. The limit properties of estimates for parameters and error e_n are depend on the power of the perturbation ξ_n. If condition (4.10) does not hold, then the estimation of adaptive system can be performed by the approach proposed Karabutov (2016).

Consider the information power (see Karabutov, 2011) of the perturbation ξ_n:

$$\Xi_\xi = \frac{1}{N}\sum_{i=1}^{N}\xi_i^2 \tag{4.11}$$

If we treat ξ_n as a random variable with zero mean, then Ξ_ζ coincides with the variance of ξ_n, and Theorem 1 implies the well-known result on the asymptotic estimate for the parameters of system (4.1).

Now we consider the case where condition (4.10) does not hold. As is known, in this case, various a priori precondition on laws of the change of the elements of the vector B in (4.1) are used. Such preconditions do not always correspond to actual parameters of the system. Therefore, it is necessary to compensate for the parametric perturbations that arise; the existing approaches have been discussed in the Introduction. Below, we present an approach that does not require such assumptions.

2.2. Multiplicative Adaptive Algorithm

Now we present an approach that allows one to reduce the number of adjustable parameters of the model described by system (2.1) for the case $X_{i,n} = X_{u_{i,n}}$. This method gives the a posteriori estimates for the vector H_B introduced in Sec. 3.1. These estimates serve as a basis for constructing a model of the system. To obtain the adequate mathematical model, the final operational development of the basic model is performed.

Assume that the method finds the number m of lagged variables suggested (Karabutov, 2013). For constructing the vector H_B, we use the following algorithm AH_B.

Algorithm AH_B

1. Apply the least square method (LSM) and find the parameters for secants y_n by $x_{j,n} \in X_{i,n}$, $u_{i,n} \in U_n$:

$$\gamma_{y,x_{j,n}} = a_{y,j,0} + a_{y,j,1} x_{j,n}, \quad 1 \le j \le m$$

$$\gamma_{y,u_{l,n}} = a_{y,l,0} + a_{y,l,1} u_{l,n}, \quad 1 \le l \le k$$

and the determination coefficients $r^2_{y,x_{j,n}}$, $r^2_{y,u_{l,n}}$.

2. Introduce the variable $S_n = I^T U_n$ and construct the model

$$\hat{y}_{s,n} = \hat{a}_{s,0} + \hat{a}_{s,1} s_n$$

whose parameters are calculated by LSM.

3. Find the variable $\varepsilon_n = \hat{y}_{s,n} - y_n$, the secants for it:

$$\gamma_{\varepsilon,x_{j,n}} = a_{\varepsilon,j,0} + a_{\varepsilon,j,1} \varepsilon_n, \quad 1 \le j \le m$$

and the determination coefficients $r^2_{\varepsilon,x_{j,n}}$.

4. Construct the vector $\tilde{H}_B$ by using the sets $\{a_{\varepsilon,j,1}\}$, $\{a_{y,j,1}\}$, $\{a_{y,l,1}\}$. The following rule performs the selection of coefficients of lagged variables, including $u_{i,n}$:

If $r^2_{y,x_{j,n}} \le r^2_{\varepsilon,x_{j,n}}$, then $a_{\varepsilon,j,1} \in H_B$, otherwise $a_{y,j,1} \in H_B$. This rule is also applied for the choice of the coefficient of $u_{i,n}$.

5. Construct the model $y_{B,n} = \tilde{H}_B^T K_n$ and make a forecast y_n.

6. Calculate the means values $\bar{y}_n$ and $\bar{y}_{B,n}$ for y_n, $y_{B,n}$ and information powers Ξ_y and Ξ_{y_B} by (4.11).
7. Introduce the numbers $\varphi \geq 0$ and $\varphi_B \geq 0$. If

$$\left(|\bar{y} - \bar{y}_B| \leq \varphi\right) \& \left(\left|\Xi_y - \Xi_{y_B}\right| \leq \varphi_B\right) \tag{4.12}$$

then we set $H_B = \tilde{H}_B$ and take H_B as the base vector.

8. If (4.11) does not hold, then carry out the correction of the elements $\tilde{H}_B$.

For system (2.1), we consider the adaptive multiplicative model

$$\hat{y}_{B,n} = \hat{\eta}_{n-1} H_B^T K_n, \tag{4.13}$$

where $\hat{\eta}_n \in \mathbb{R}$ is an adjustable parameter of the model and $\hat{y}_{B,n}$ is the output of the model; the vector $H_B \in \mathbb{R}^{k+m}$ is defined by the algorithm AH_B.

Denote by $e_{B,n} = y_n - \hat{y}_{B,n}$ the forecast error of the output of system (4.1). Consider the Lyapunov function $V_{e_B,n} = e_{B,n}^2$.

From the stability condition of system (4.1), (4.13) we obtain the tuning algorithm for the multiplicative parameter $\hat{\eta}_n$:

$$\hat{\eta}_n = \hat{\eta}_{n-1} + \gamma_\eta e_{B,n} H_B^T K_n = \hat{\eta}_{n-1} + \gamma_\eta e_{B,n} \kappa_n \tag{4.14}$$

where $\gamma_\eta > 0$ is the parameter that provides the convergence of (4.13), $\kappa_n = H_B^T K_n$.
It is easy to see that algorithm (4.14) provides the minimization of the criterion

$$V_{\delta,B,n} = \|H - \hat{\eta}_n H_B\|^2$$

We write the expression for the forecast error $e_{B,n} = y_n - \hat{y}_{B,n}$ as:

$$e_{B,n} = \delta H_{B,n-1}^T K_n + \xi_n$$

where $\delta H_{B,n-1} = H - \hat{\eta}_{n-1} H_B$.

We present the convergence conditions for the algorithm (4.14).

Theorem 2 (Karabutov, 2016). Assume that the following condition hold:

(1) exists a vector $H_B < \infty$ got by the algorithm AH_B;
(2) assumptions 1 and 2 hold.

Algorithm (4.14) converges if

$$\gamma_\eta \leq \frac{e_{B,n}^2 - \upsilon^2}{e_{B,n}^2 \beta^2}$$

and the estimate

$$V_{\delta,n} \leq \frac{\kappa_\upsilon \upsilon^2}{\gamma_\eta(\alpha_0 - \gamma_\eta \beta^4)} \tag{4.15}$$

is valid if $\alpha_0 > \gamma_\eta \beta^4, \gamma_\eta(\alpha_0 - \gamma_\eta \beta^4) \leq 1,$

$$\kappa_\upsilon = \frac{\gamma_\eta \beta^2(1 + 2\gamma_\eta(\alpha_0 - \gamma_\eta \beta^4))}{(\alpha_0 - \gamma_\eta \beta^4)}.$$

Theorems 1 and 2 imply that if the input K_n satisfies condition (4.10), then algorithms (4.7) and (4.8) are given valid estimates for parameters under the absence of perturbations ξ_n on the set of initial condition. This is because the adaptive procedure can be considered as difference equations of first order regarding the adjustable parameters of the model (4.4). This is also valid for the model (4.13) with the algorithm (4.14).

In actual conditions, the adaptive algorithm (4.14) under the action of ξ_n provides the fulfilment of the condition $|e_{B,n}| \leq \delta_\eta(\xi_n)$, where δ_η>0. If this condition does not hold, then the model (4.13) must be transformed, for example, to the form

$$\hat{y}_{B,n} = [\hat{\eta}_{1,n-1}, \hat{\eta}_{2,n-1}]^T H(H_B) K_n$$

where $H(H_B) \in \mathbb{R}^{2\times(k+m)}$ is a matrix constructed H_B.

In this approach, we speak of the minimization of the forecast error: it is impossible to minimize the function

$$V_{\delta,B,n} = \|H - \hat{\eta}_n H_B\|^2$$

here. The application of this approach is expedient, where a high accuracy of estimates is not required.

2.3. Identification of Parameters in Presence of Relations Between Lags

Where there is a relationship between lagged variables, the method of the principal component (Karabutov, 2009) is used for the identification of parameters of the system.

The method of principal components comprises transformations of input variables that eliminate relations between them. It requires some preliminary preparations and is not always applicable in actual conditions.

Remark 1. For the construction of regression and auto-regression models, the method of adaptive expectations can be used (see Chow, 2011). It is based on a forecast of the input variable obtained by exponential smoothing filter (first-order recursive filter) and is irrelevant to adaptation. Below, we use an approach based on the transformation of input lag variables that eliminate relations between them. As new variables, we use the differences of lagged variables $x_{i,j,n}$ and $x_{i,j+1,n}$:

$$z_{i,n} = x_{i,j,n} - x_{i,j+1,n}$$

where $x_{i,j,n}$ are components of the vector $X_{i,n}$.

Remark 2. Some authors (see, e.g., Malinvaud, 1980) apply this approach to eliminating relations between lags of the dependent variable y_n. This is fair for estimating parameters by the method of the last squares, but the model obtained is not dynamical. This should be considered when choosing appropriate methods of identification.

The choice of the components of the vector $X_{i,n}$ to be transformed depends on properties of the system. Usually, lagged variables with the larger value of the lag are transformed.

Here, the technique of adaptive identification is exactly the same as in Sec. 2.1.

3. ADAPTIVE IDENTIFICATION OF SYSTEMS WITH LAGS WITH REGARDING TO $X_{y,n}$

Let $X_n = X_{y,n}$, where $X_n = X_{y,n}$, $X_{y,n} = [y_{n-1}, \ldots, y_{n-k}]^T$. This means that the auto-regression equation describes the system (2.1). Assume that the perturbation $\xi_n \in \mathbb{R}$is linearly related to the output y_n.

For the identification of parameters of the system (4.1), we apply the model

$$\hat{y}_{\xi,n} = \hat{A}^T_{n-1} U_n + \hat{B}^T_{n-1} X_{\hat{y},n} + \hat{\xi}_n \tag{4.16}$$

where $\hat{\xi}_n \in \mathbb{R}$ is the estimate of perturbation, $\hat{A}_n$ and $\hat{B}_n$ are the vectors of adjustable parameters, and $X_{\hat{y},n} = \left[\hat{y}_{\xi,n-1}, \ldots, \hat{y}_{\xi,n-k}\right]^T$ is the vector constructed by the output for $\hat{y}_{\xi,n}$.

Assume that, for the system (4.1), the information set is available $_o$.

It is required to estimate the set of parameters for the system (4.1) and obtain the estimate of the perturbation ξ_n based on the analysis of $_o$ by the model (4.1).

First, we show how to establish a relation between y_n and ξ_n. For this, we use the following modification of the procedure proposed by Karabutov (2013).

Algorithm AR_ξ

1. Introduce the variable $s_n = I^T U_n$ and construct the model

$$\hat{y}_{s,n} = \hat{a}_{s,0} + \hat{a}_{s,1} s_n$$

whose parameters are found by using the method of least squares.

2. Find the variable $\varepsilon_n = \hat{y}_{s,n} - y_n$, secants for it:

$$\gamma_{\varepsilon,x_{i,n}} = a_{\varepsilon,i,0} + a_{\varepsilon,i,1}\varepsilon_n, \quad 1 \le i \le m$$

and the determination coefficients $r^2_{\varepsilon,x_{i,n}}$.

3. Specify the level $\delta_x > 0$. Choosing $x_{i,n} \in X_n$ for which the condition $r^2_{\varepsilon,x_{i,n}} \ge \delta_x$ holds, construct the vector $\bar{X}_n$ (by the condition of the problem, it must coincide with X_n).
4. Construct the model $\hat{y}_{\varepsilon,n} = A_\varepsilon^T \bar{X}_n$ for ε_n and find the error $\varepsilon_{\varepsilon,n} = \varepsilon_n - \hat{y}_{\varepsilon,n}$.
5. Approximate $\varepsilon_{\varepsilon n}$ by the dependence $\hat{\varepsilon}_{\varepsilon,n} = \omega_{1,\varepsilon} y_n + \omega_{0,\varepsilon}$, where the coefficients $\omega_{0,\varepsilon}$, $\omega_{1,\varepsilon}$ are found by the method described above. Find r^2_{ε,y_n}. If the inequality:

$$r^2_{\varepsilon,y_n} \ge \delta_{\varepsilon,y}$$

holds, where $\delta_{\varepsilon,y} > 0$, then there exists a relation between y_n and ξ_n.

Under uncertainty, to account for the existing relationship, we construct an observer for ξ_n. Denote by $e_{\xi,n}$ the forecast error y_n by the model (4.16):

$$e_{\xi,n} = y_n - \hat{y}_{\xi,n}$$

We set

$$\hat{\xi}_n = \hat{\chi}_{n-1} e_{\xi,n-1}$$

and rewrite (4.16) in the form

$$\hat{y}_{\xi,n} = \hat{A}^T_{n-1} U_n + \hat{B}^T_{n-1} X_{\hat{y},n} + \hat{\chi}_{n-1} e_{\xi,n-1} \tag{4.17}$$

where $\hat{\chi}_{n-1}$ is the tuning parameter.

This approach applies, where the range of the error is constant. Since in the adaptation process, the error varies $e_{\xi,n}\varepsilon_n$ contains distorted information on the perturbation.

Therefore, as an information carrier for $\hat{\xi}_n$ we take the variables ε_n or $\varepsilon_{\varepsilon n}$. We rewrite (4.17) as:

$$\hat{y}_{\xi,n} = \hat{A}^T_{n-1} U_n + \hat{B}^T_{n-1} X_{\hat{y},n} + \hat{\chi}_{n-1} \varepsilon_n \tag{4.18}$$

The equation for the forecast error y_n has the form:

$$e_{\xi,n} = \delta A^T_{n-1} U_n + \delta B^T_{n-1} X_{\hat{y},n} + \delta \xi_n \tag{4.19}$$

where $\delta \xi_n = \xi_n - \hat{\chi}_{n-1} \varepsilon_n$

Applying the second Lyapunov method with the function $V_{e_\xi,n} = e^2_{\xi,n}$, we obtain the following adaptation algorithms:

$$\hat{A}_n = \hat{A}_{n-1} + \Gamma_A e_{\xi,n} U_n \tag{4. 20}$$

$$\hat{B}_n = \hat{B}_{n-1} + \Gamma_B e_{\xi,n} X_{y,n} \tag{4.21}$$

$$\hat{\chi}_n = \hat{\chi}_{n-1} + \gamma_\chi e_{\xi,n} \varepsilon_n \tag{4.22}$$

where Γ_A, Γ_B, and γ_χ are matrices that provide the convergence of these algorithms.

From (4.20), (4.21), and (4.22) we can easily deduce equations for δA_n and δB_n. These functions satisfy Eq. (4.9) in which $e_n = e_{\xi,n}$. For $\delta\hat{\chi}_n = \chi - \hat{\chi}_n$ we have:

$$\delta\chi_n = \delta\chi_{n-1} - \gamma_\chi e_{\xi,n} \varepsilon_n \tag{4.23}$$

where χ is a number such that $\xi_n = \chi\varepsilon_n$. By the choice of ε_n, condition (4.23) is not always valid, but so far, we assume that this relation holds.

Theorem 3. Assume that the conditions of Theorem 1 and the condition $|\varepsilon_n| \le \mu < \infty$ $\forall n \ge 0$ hold. Algorithms (4.20), (4.21), and (4.22) converge if

$$2\pi \ge \bar{\gamma}^2 \beta^2 + \gamma_\chi^2 \mu^2$$

where $\pi = \min(\bar{\gamma}, \gamma_\chi)$.

Proof. Consider the Lyapunov function $V_{\delta,n} = \|\delta H_n\|^2 + (\delta\chi_n)^2$. For $\Delta V_{\delta,n}$ we obtain

$$\Delta V_{\delta,n} = -2e_{\xi,n}\left(\delta H_{n-1}^T \Gamma K_n + \gamma_\chi \delta\chi_{n-1}\varepsilon_n\right) + e_{\xi,n}^2 K_n^T \Gamma^2 K_n + e_{\xi,n}^2 \gamma_\chi^2 \varepsilon_n^2$$

Using the conditions of Theorem 3, we transform $\Delta V_{\delta,n}$ to the form

$$\Delta V_{\delta,n} \leq -2\pi e_{\xi,n}^2 + e_{\xi,n}^2 \bar{\gamma}^2 \beta^2 + e_{\xi,n}^2 \gamma_\chi^2 \mu^2$$

For the convergence of algorithms (4.90) and (4.22) it is necessary that

$$2\pi \geq \bar{\gamma}^2\beta^2 + \gamma_\chi^2\mu^2$$

The theorem proves. ■

Now we consider the case where the condition $\xi_n = \chi_{\varepsilon n}$ does not hold. This situation can appear in many times. Therefore, we will assume that:

$$\delta\xi_n = \xi_n - \hat{\chi}_{n-1}\varepsilon_n = \delta\chi_{n-1}\varepsilon_n + \varepsilon_\chi \tag{4.24}$$

where $|\varepsilon_\chi| \leq \mu_\chi$ and $\mu_\chi \geq 0$ is the certain number.

ε_χ characterizes the uncertainty that results from unreliable information about the disturbance. It is easy to see that in this case the properties of the identification system can be estimated by Theorem 2. Because of the presence of the signal uncertainty in (4.24) and hence in (4.19), we choose the value of the parameter γ_χ to minimize the error $e_{\xi,n}$:

$$\min_{\gamma_\chi} e_{\xi,n}^2 \rightarrow \gamma_\chi^*$$

Thus, we have presented methods of adaptive identification of discrete systems with lagged variables in the presence of perturbations. Various assumptions about perturbations can be imposed; we show how these assumptions affect the properties of adaptive algorithms and identification systems.

4. RESULTS OF SIMULATION

Consider system (4.1) with $A = [2.5, 2.4, 2]^T$ and $B = [0.8, 0.4]^T$; $U_n \in \mathbb{R}^3$; $u_{i,n}$ are random variables with finite expectation and variance; $X_n = \left[u_{1,n-1}, u_{1,n-2}\right]^T$; the perturbation ξ_n is a random function with zero expectation and a finite variance, $|\xi_n| \leq 0.3$.

The input $u_{i,n}$ is obtained as a solution of a difference equation with a random input whose expectation is nonzero and variance is finite.

Figure 1. Tuning of parameters for the system under the constraint $x_{i,n}$

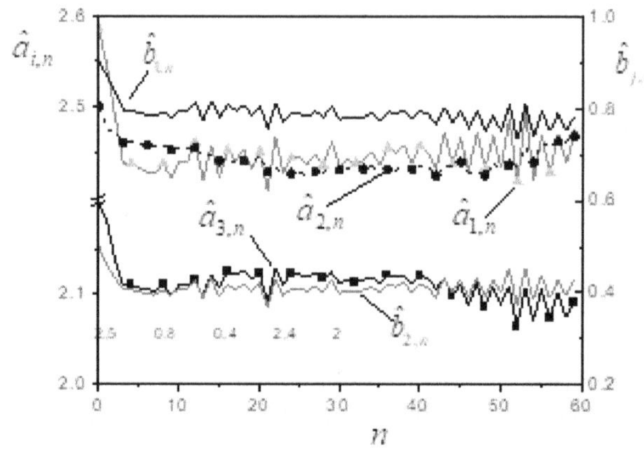

In Figure 1, we show adaptive identification with ξ_n=0. Because of the presence of a relation between the elements of the vector X_n, the convergence of adaptive algorithms (4.7) and (4.8) is not observed, but trajectories of the identification system remain bounded.

Figure 2. Results of model (4.21) identification

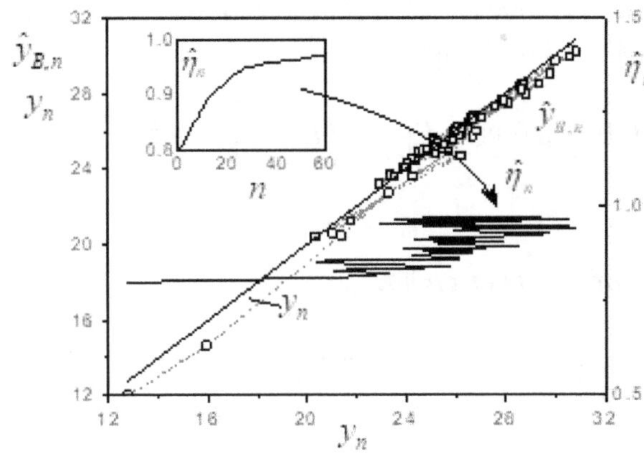

In Figure 2, the results of the tuning of the multiplicative parameter $\hat{\eta}_{n-1}$ and the adequacy of the model (4.13) are presented. For this purpose, we apply the approach described in Sec. 2.2. By the algorithm AH_BB, we find the vector $H_B= [A_B, B_B]^T$, where $A_B= [2.48, 2.05, 1.8]^T$, $B_B= [0.83, 0.7]^T$.

In (4.12) $\bar{y} = 25, \bar{y}_B = 26.1, r^2_{y,y_B} = 0.98, \Xi_y=660, \Xi_{y_B} = 691, \varphi=1.5, \varphi_B=35$.

Further, we apply algorithm (4.14) with $\gamma_\eta=0.0008$. The character of the variation of $\hat{\eta}_{n-1}$ can be explained by the compensation of appearing parametric perturbations because of the choice of the vector H_B. The maximal forecast error $e_{\eta,n}$ on the range of y_n does not exceed 2.5%.

Figure 3. Adjustment of parameters for model (4.3)

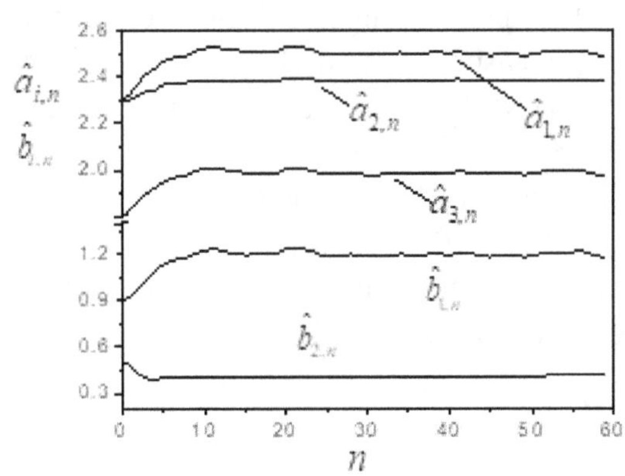

For improving the quality of the parametric identification, we applied the approach described in Sec. 2.3. Instead $x_{2,n}$, we used the variable $z_{1,n} = x_{1,n} - x_{2,n}$. The results of calculations by the adaptive algorithms (4.7) and (4.8) are presented in Figure 3. The matrices Γ_A and Γ_B in (4.7) and (4.8) are equal:

$$\Gamma_A = diag(0.005; 0.008; 0.008), \ \Gamma_B = diag(0.02; 0.008)$$

Figure 4. Variation of the criterion $\|\delta H_n\|^2$

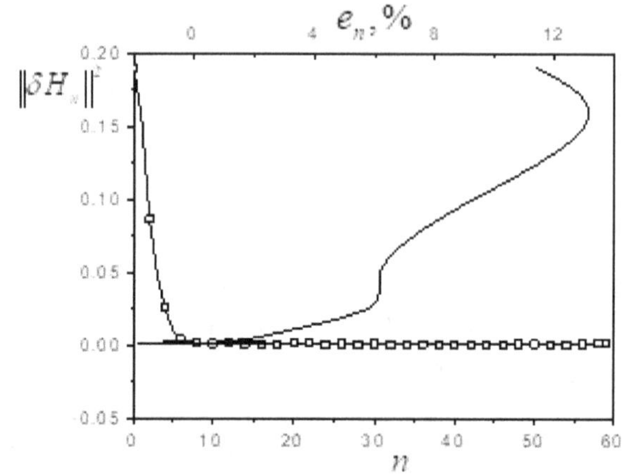

The norm variation of the parametric error

$$\|\delta H_n\|^2 = \|\delta A_n\|^2 + \|\delta B_n\|^2$$

is shown in Figure 4. The approach suggested allows one to find valid estimates of the system. Maximal forecast error for $n \geq 10$ does not exceed 2%. The choice validity of components for the vector $X_{i,n}$ in (4.3) based on determination coefficients between $z_{1,n}$ and $u_{1,n}$, $x_{1,n}$: $r^2_{z_1,u_1} = 0$, $r^2_{z_1,x_1} = 0.028$.

Now we consider the system (4.1) with the lag by y_n: $X_{y,n} = y_{n-1}$. We assume that $\xi_n = \xi(y_n)$. Take $A = [2; 2.6; 1.8]^T$, $B = b = 0.4$.

For the identification of the system, we apply the model (4.18) with the modified adaptive algorithms (4.20), (4.21), and (4.22):

$$\widehat{A}_n = \widehat{A}_{n-1} + \frac{\Gamma_A e_{\xi,n} U_n}{\|U_n\|^2 + y_{n-1}^2 + \varepsilon_n^2} \tag{4.25}$$

$$\hat{b}_n = \hat{b}_{n-1} + \frac{\Gamma_B e_{\xi,n} y_{n-1}}{\|U_n\|^2 + y_{n-1}^2 + \varepsilon_n^2} \tag{4.26}$$

$$\hat{\chi}_n = \hat{\chi}_{n-1} + \frac{\gamma_\chi e_{\xi,n} \varepsilon_n}{\|U_n\|^2 + y_{n-1}^2 + \varepsilon_n^2} \tag{4.27}$$

Modification does not introduce any changes to the properties of the procedures (4.22), (4.23), and (4.24), but allows one to simplify the choice of the matrices Γ_A, Γ_B, and γ_χ:

$$\Gamma_A = 0.36 I_3, \ \Gamma_B = 0.06, \ \gamma_\chi = 0.15$$

Figure 5. Tuning of parameters for model (4.18)

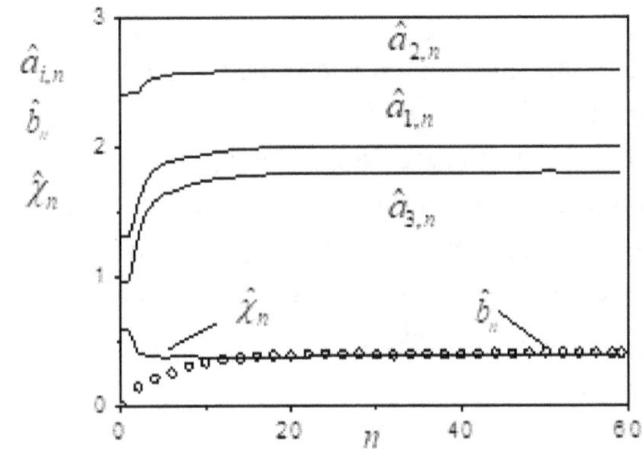

The results of calculations by algorithms (4.25), (4.26), and (4.27) are presented in Figure 5. In Figure 6, we show the estimate of the perturbation ξ_n and the variation of the norm of the parametric discrepancy $\|\delta A_n\|^2$. The forecast error $e_{\xi,n}$ does not exceed 6% starting from $n \geq 30$.

Figure 6. Perturbation estimates and the norm variation of parametric residual

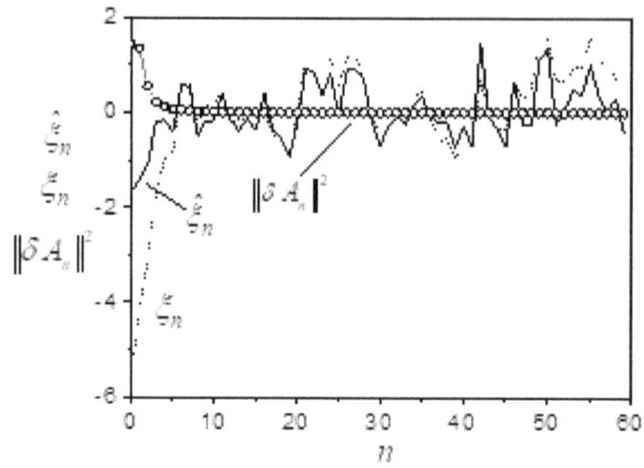

Now we consider the same system in the absence of any relation between y_n and ξ_n. In the model (4.19), instead of $X_{\hat{y},n} = \hat{y}_{n-1}$, we take $X_{y,n} = y_{n-1}$, i.e., we apply the regression model instead of auto-regression. This approach is fundamental in the analysis of lagged models and allows for the use of econometric techniques based on LSM. Application of the adaptive algorithms proposed above encounters with the problem of the dependence between variables and does not allow one to get valid estimates for parameters of the system.

5. CONCLUSION

In this chapter, we propose adaptive algorithms for the identification of discrete systems with lagged variables. For the case with lags of input variables, we design adaptive algorithms and prove their convergence. Models with multiplicative parameters are proposed. This reduces the number of estimated parameters. We develop the procedure of choosing of basic parameters for this model. If lagged variables are interconnected, we propose a transformation which allows one to exclude this mutual influence of variables. We analyze of adaptive algorithms for the systems described by equations with the lagged output variable.

REFERENCES

Chow, G. C. (2011). Usefulness of adaptive and rational expectations in economics, *CEPS Working Paper No. 221*.

Karabutov, N. (2011). *Structural identification of static objects: Fields, structures, methods*. Librocom Book House.

Karabutov, N., & Feklin, V. (2016). Adaptive identification of systems with distributed lags. *Journal of Mathematical Sciences*, *216*(5), 649–666. DOI: 10.1007/s10958-016-2925-8

Karabutov, N. N. (2009). Structural identification of systems: analysis of information portraits. URSS. DOI: 10.13140/2.1.1843.2324

Karabutov, N. N. (2013). Structural identification of static systems with distributed lags. *Int. J. of Control Science and Engineering*, 2(6), 136–142. DOI: 10.5923/j.control.20120206.01

Karabutov, N. N. (2013). Structural identification of systems with distributed lag. *International Journal of Intelligent Systems and Applications*, *5*(11), 1–10. DOI: 10.5815/ijisa.2013.11.01

Karabutov, N. N. (2016). *Adaptive Identification of Systems: Information Synthesis*. URSS.

Malinvaud, E. (1980). *Statistical Methods in Econometrics*. North-Holland., DOI: 10.1604/9780444854735

Chapter 5
Structural Identification of Dynamic Systems

ABSTRACT

In this chapter, we present structural identification methods based on the analysis of geometric frameworks (GF). Linear and nonlinear dynamical systems (NDS) are considered. The method of constructing GF for a different class of systems is described. The method for estimating of nonlinearity class based on GF analysis is proposed. We describe an approach for evaluating the NDS structure. Algorithms for estimating the class of nonlinearity have been developed. We propose the method of structural-frequency analysis to decide about structural properties of the system. Procedures are developed for constructing GF for evaluating Lyapunov exponents (LE) for stationary systems and systems with periodic coefficients (PKS). We introduce the concept of an almost periodic function in the sense of Bohr and give its application to SPK. The method for constructing GF for SPK has been developed.

1. INTRODUCTION

Structural identification (SI) problem is solved based on the application of the traditional paradigm. Paradigm is based on the application of the parametric approach and some other procedures (neural network and genetic algorithms) and intuition. These procedures are variant to the parametric approach. Analysis shows that the mathematical frameworks describing a state of nonlinear systems as portraits and mappings, in the SI theory practically were not applied. The state analysis of researches in the SI field is given Karabutov (2013) and chapter 1. The main resume (Karabutov, 2013): direct (parametric) methods are not applicable for the analysis of SI problems under uncertainty. Therefore, the design of novel approaches is base for the SI problem solution. We refer to geometrical (structural) methods of synthesis identification systems. The development of such the paradigm demands to revise of

DOI: 10.4018/979-8-3373-0822-7.ch005

existing approaches and creation of new informational and structural approaches. This approach should obtain the framework of the nonlinear dynamic system (NDS).

We propose (Karabutov, (2015a) the approach giving the SI problem solution to dynamic systems. Two concepts are the basis of approach:

1) introduction of the new variables characterizing framework of the system;
2) geometric framework (GF) construction reflecting the NDS.

Bases of the proposed structural identification approach (SIA) are stated Karabutov (2006). The further development and generalization SIA on a class of nonlinear static systems is given Karabutov (2009, 2011). We introduce concepts which are necessary for understanding the SIA. Formation methods of an auxiliary informational set, which is basic for the solution to the construction framework S_{ey}, are given. The GF is the basis for the design of methods for the estimation of SI and structural identifiability NDS. Criteria are represented for decision-making about the class of the functions describing NDS. Algorithms of SI nonlinear dynamic system are considered. We show how the SIA can apply to the estimation of the framework of linear DS.

2. PROBLEM STATEMENT OF SI NDS

Consider dynamic system

$$\begin{aligned} \dot{X} &= F(X,U,A,t), \\ Y &= F_Y(X,U,A,t), \end{aligned} \tag{5.1}$$

where $U \in \Omega_U \subset U \subseteq \mathbb{R}^m$, $Y \in Y \subset \mathbb{R}^n$ are an input and output, $X \in X \subseteq \mathbb{R}^q$ is state vector, $Y \subseteq X \subseteq \mathbb{R}^n$, $F:\mathbb{R}^q \times \mathbb{R}^m \times \mathbb{J} \to \mathbb{R}^q$ is smooth continuously differentiable on X and A a vector function, $A \in R^{q\times q}$ the matrix of parameters, $t \in \mathbb{J} \subset \mathbb{R}$, $F_Y:\mathbb{R}^q \times \mathbb{R}^m \times \mathbb{J} \to \mathbb{R}^n$ is a function specifying a mode of formation the system output.

Set of the measured data

$$\mathbb{I}_o = \mathbb{I}(U,Y) = \left\{\widetilde{U} \in \mathbb{R}^m, \widetilde{Y} \in \mathbb{R}^n, t \in \mathbb{J}\right\}, \tag{5.2}$$

observed on a time gap $\mathbb{J}$.

Problem: decide about a structure operator F in (5.1) based on the analysis and processing of the set $\mathbb{I}_o$.

3. STRUCTURAL METHODS IN IDENTIFICATION OF NDS

3.1. Approach to Identification of Class Nonlinearities

State the approach to structural identification for the system (5.1) with a selected linear part

$$\begin{aligned} \dot{X} &= AX + \varphi(y)I + Bu, \\ y &= C^T X, \end{aligned} \tag{5.3}$$

where $u \in \mathbb{R}$, $y \in \mathbb{R}$ are input and output; $A \in \mathbb{R}^{q \times q}$, $B \in \mathbb{R}^q$, $I \in \mathbb{R}^q$ $C \in \mathbb{R}^q$ are matrixes; $\varphi(y)$ is some scalar nonlinear function. A is Hurwitz matrix.

Different assumptions can become about the function structure of the function $\chi = \varphi(y)$. They are determined by a level of the a priori information.

Assumption 1

$$\chi \in F_\varphi = \{\varphi(\xi)\xi \geq \xi^2,\ \xi \neq 0,\ \varphi(0) = 0\} \tag{5.4}$$

where $\xi \in \mathbb{R}$ is input nonlinear element.

Assumption 2

$$\chi \in F_\varphi = \{\gamma_1 \xi^2 \leq \varphi(\xi)\xi \leq \gamma_2 \xi^2,\ \xi \neq 0,\ \varphi(0) = 0,\ \gamma_1 \geq 0,\ \gamma_2 < \infty\} \tag{5.5}$$

where $\xi \in \mathbb{R}$ is the combination of elements X.

The set I_o for system (5.3) has the form

$$\mathbb{I}_o = \{u(t), y(t), t \in \mathbb{J} = [t_0, t_k]\} \tag{5.6}$$

Problem: estimate the class of nonlinear functions $\varphi(y)$ in (5.3) and characteristics the matrix A based on the data processing (5.6).

Apply the informational synthesis (IS) (Karabutov, (2006, 1009) to the problem solution. IS realizes the following steps.

1. Obtaining of a set $\mathbb{I}_{N,g}$ containing the data about the nonlinear part of the system (5.3).
2. GF construction reflecting nonlinear properties of the system (5.3).
3. The algorithm design for estimate of the class nonlinearity F_φ.
4. The procedure design of nonlinearity identification in the class F_φ.
5. The method design of decision-making about a nonlinearity form in (5.3).
6. The estimation of a dimension and a spectrum of eigenvalues the matrix A.

These steps are based on construction and the analysis GF. Form and properties GF depend on properties of the system (5.3).

3.2. Formation of Set $I_{N,g}$

The construction method of set $I_{N,g}$ is based on results Karabutov, 2015a. Apply differentiation operation to $y(t)$ and designate the goat variable as x_1. The consideration x_1 gives the expansion of the informational set I_o: $I_{ent} = \{I_o, x_1\}$.

Remark 1. If variables u,y are measured with an error, then apply the procedure of filtering or smoothing.

Consider the first step IS. Generate the subset $I_g \subset I_{ent}$ corresponding to the particular solution of the system (5.3) (to the steady state). We form set I_g, except for the data I_{tr}. I_{tr} contains the information about the transient process in the system. Then $I_g = I_{ent} \backslash I_{tr}$. Apply mathematical model:

$$\hat{x}_1^l(t) = H^T[1 \; u(t) \; y(t)]^T \tag{5.7}$$

to the selection of the linear component x_1. The variable x_1 is defined in the interval $\mathbb{J}_g = \mathbb{J} \backslash \mathbb{J}_{tr}$. Here $H \in \mathbb{R}^3$ is a vector of parameters.

Determine the vector H as:

$$\min_H Q(e)\Big|_{e=\hat{x}_1^l - x_1} \rightarrow H_{opt}.$$

where $Q(e)=0.5e^2$.

Find the prediction for the variable x_1 basis on the model (5.7) $\forall t \in I_g$ and form the error $e(t) = \hat{x}_1^l(t) - x_1(t)$. $e(t)$ depends on a nonlinearity $\varphi(y)$. Then we obtain the set $I_{N,g} = \{y(t), e(t) t \in \mathbb{J}_g\}$, which we use at the second stage IS. Apply the designation $y(t)$, supposing that $y(t) \in I_{N,g}$.

Remark 2. The structure choice of the model (5.7) is one stage SI. Results of modelling show that the model (5.7) applies in identification systems of objects with static nonlinearities. The choice problem of the model (5.7) structure demands the further research for other classes of nonlinearities. Next, we consider the example of the problem solution for one class of systems.

The further problem solution of SI is based on the analysis of frameworksS_{ey}, S_{ek}, reflecting the state of the system (5.3).

3.3. Frameworks S_{ey}, S_{ek}

Geometric frameworks have been proposed by Karabutov (2006) for the analysis and synthesis of identification systems. Development and generalization of GF on static systems is given by Karabutov (2013). We describe the approach to construction GF for dynamic systems.

Let S is a phase portrait of the system (5.3) described by a function $x_1 = f_S(y)$. We study the phase portrait of the system (5.3) in space $P_{ye} = (y, e)$. P_{ye} is the structural space. The framework S_{ey} described by function $\Gamma_{ey}: \{y\} \rightarrow \{e\}\ \forall t \in \mathbb{J}_g$, is the phase portrait of a nonlinear part of the system (5.3). S_{ey} must be closed. It is a characteristic property S_{ek} of dynamic systems. We will also use the S_{ek}-framework, which is described by the function Γ_{ek}: $\{k_s(t)\} \rightarrow \{e(t)\}$, where $k_s(t) \in \mathbb{R}$ is a coefficient of structural properties (Karabutov, 2006) $k_s(t) = \frac{e(t)}{y(t)}$.

Example 1. Consider the system (5.3) second order with

$$\varphi(y) = \begin{cases} 2.5, \ if((y-d) > \bar{\upsilon}) \ \& \ (y' > 0), \\ \alpha(y-d), \ if((y-d) \le \bar{\upsilon}) \ \& \ (y' > 0), \\ 1.5, \ if((y-d) \le \underline{\upsilon}) \ \& \ (y' > 0), \\ 2.5, \ if(y > \bar{\upsilon}) \ \& \ (y' \le 0), \\ \alpha y, \ if(y < \bar{\upsilon}) \ \& \ (y' \le 0), \\ 1.5y, \ if(y < \underline{\upsilon}) \ \& \ (y' \le 0), \end{cases} \tag{5.8}$$

$A = \begin{bmatrix} 0 & 1 \\ -3 & -4 \end{bmatrix}$, $y(0)=x(0)=3.5$, $y'(0)=x'(0)=2$, $u(t)=5-5\sin(0.1\pi t)$, $y=x$.

Show phase portrait S and the framework S_{ey} describing the steady state of the system in Figure 1. We see (Figure 1) that the framework S is virtually not informative. The framework S_{ey} confirms the nonlinearity of the system. S_{ey} is designed based on the model (5.7) with $H[0.22\ 0.23\ \text{-}0.59]^T$. Show framework S_{ek} in Figure 2.

Figure 1. Frameworks for system (5.3) of second order with the function φ(y) (5.8)

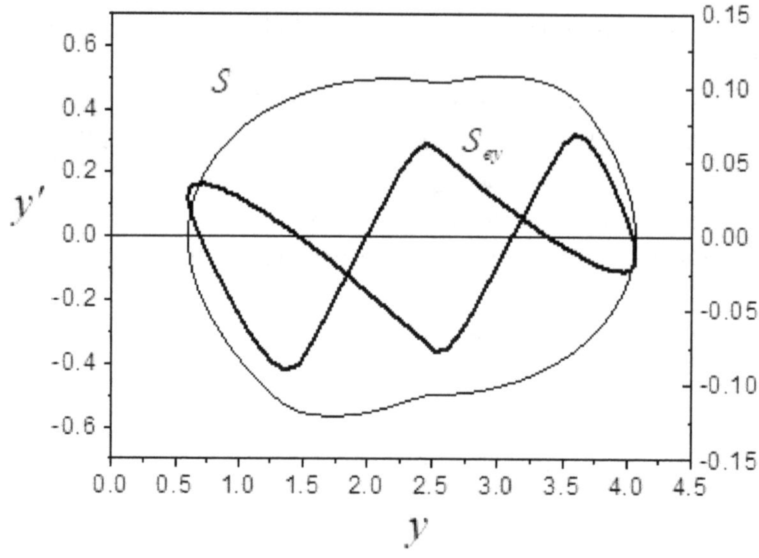

Figure 2. Framework S_{ek}

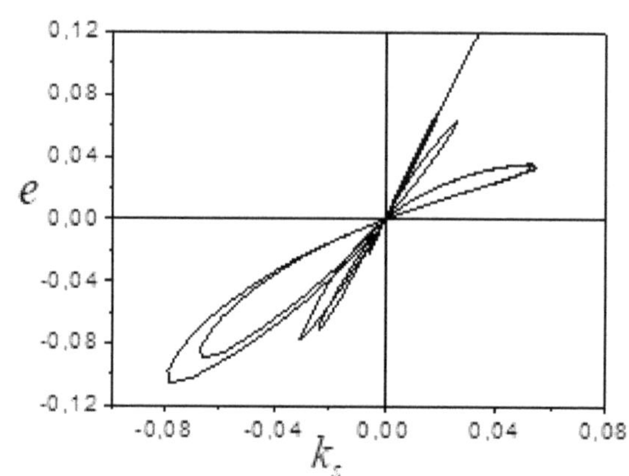

We see that the form and properties S_{ey} influence on the identifiability NDS. Therefore, we will consider properties of the set $\mathbb{I}_{N,g}$ on which the framework S_{ey} is specified.

3.4. About Properties $\mathbb{I}_{N,g}$. Structural Identifiability of System (5.3)

Consider the properties of set $\mathbb{I}_{N,g}$ ensuring the SI problem solution. At first, the set $\mathbb{I}_o$ should ensure the parametric identification problem solution for the model (5.7). This means that the input $u(t)$ should be non-degenerate (constantly excited) at an interval $\mathbb{J}$. Second, the input $u(t)$ should ensure informative framework $S_{ey}\left(\mathbb{I}_{N,g}\right)$ (or S_{ek}), and decision-making about nonlinear properties of the system (5.3). We call such an input representative.

Let the framework S_{ey} is closed. This to mean that the area S_{ey} is not null. Designate height S_{ey} as $h\left(S_{ey}\right)$. $h\left(S_{ey}\right)$ is the distance between two points of the framework S_{ey} on opposite sides.

Statement 1 (Karabutov, 2015a). *Let: (i) the linear part of the system (5.3) is stable, and nonlinearity satisfies to the condition (5.5); (ii) the input u(t) is restricted, piecewise, continuous, and constantly excited; 3) exists* $\delta_S>0$*such that* $h\left(S_{ey}\right) \geq \delta_S$. *Then the framework* S_{ey} *is identified on the set* $\mathbb{I}_{N,g}$.

Next, we suppose that S_{ey} has the specified property. Framework S_{ey} having the specified properties is *h*-identified.

Features of concept *h*-identifiability.

1. *h*-identifiability (HI) is a concept not parametric, but SI.
2. The demand for the parametric identifiability is the basis for HI.
3. HI makes more rigid demands for the system input.

Feature 3 means that the bad input can satisfy of the constancy excitation condition. Such input can give a so-called insignificant S_{ey}-structure (NS_{ey}-framework). But the NS_{ey}-structure can be *h*-identified. Under uncertainty, the insignificance property gives the identification of nonlinearity, atypical for an examined system.

Consider existence conditions of NS_{ey}-structures. Consider a class of nonlinear functions to which homotopy operation is applicable. The homotopy (Choquet, 1964) is the operation of obtaining of one part of a geometrical figure from another part based on its rotation and extension about a certain point on a plane (y,e).

Consider the framework S_{ey}. Let$S_{ey} = F^l_{S_{ey}} \cup F^r_{S_{ey}}$, where $F^l_{S_{ey}}, F^r_{S_{ey}}$ are left and right fragments S_{ey}. Determine for $F^l_{S_{ey}}, F^r_{S_{ey}}$ secants

$$\gamma^l_S = a^l y, \quad \gamma^r_S = a^r y \tag{5.9}$$

where a^l, a^r are parameters computed using the least-squares method (LSM).

Theorem 1 (Karabutov, 2017). Let: 1) the framework S_{ey} is HI; 2) the framework S_{ey} has the form $S_{ey} = F^l_{S_{ey}} \cup F^r_{S_{ey}}$, where $F^l_{S_{ey}}, F^r_{S_{ey}}$ are left and right fragments S_{ey}; 3) secants for $F^l_{S_{ey}}, F^r_{S_{ey}}$ have the form (5.9). Then S_{ey} is NS_{ey}-structure, if $|a^l - a^r| > \delta_h$, where $\delta_h > 0$.

Remark 3. The theorem 1 can be proved based on a homotopy of sets (Choquet, 1964). Estimate proximity of sets $F^l_{S_{ey}}, F^r_{S_{ey}}$. The approach based on the method of secants is simpler in implementation.

Remark 4. NS_{ey}-structures are characteristic for systems with multiple-valued nonlinearities. They are inadequate application result of the input.

Definition 1. If the framework S_{ey} is HI and the condition $|a^l - a^r| \leq \delta_h$ is satisfied, then S_{ey} is structurally identified (STI) or h_{δ_h}-identified.

Let the framework S has m features. We understand function f features as continuum loss on some interval I^j_y, flex points, or extremes. These features are sing on nonlinearity of the function.

Apply the model (5.7) and construct the framework S_{ey} in the space P_{ye}.

Definition 2. Model (5.7) is *SM*-identifying if the framework S_{ey} is h_{δ_h}-identified.

Theorem 2. Let: (i) the input $u(t)$ is constantly excited and ensures HI of the system (5.3); (ii) the phase portrait S of the system (5.3) contains m features; (iii) the S_{ey}-structure is h_{δ_h}-identified and contains the fragments corresponding to features of the phase portrait S. Then model (5.7) is *SM*-identified.

Proof (Karabutov, 2017). The phase portrait S is h_{δ_h}-identifying. Model (5.7) is linear. Input $u(t)$ is constantly excited and ensures h_{δ_h}-identifiability of the system (5.3). If the model (5.7) is adequately and describes the change in the linear part of the system (5.3) the variable e reflects features S. The of the model (5.7) adequacy is determined by the set structure of the equation (5.7) on set $I_o\left(t \in J_g\right)$, where J is the definition domain S. Since the S_{ey}-structure is HI, and the equation (5.7) gives the mode for it getting, the model (5.7) is adequate and, hence, *SM*-identifying. ■

The theorem 2 shows if model (5.7) is not *SM*-identifying, then it is necessary to the model (5.7) structure change or the informational set for its construction.

Consider one more approach to the estimation of h_{δ_h}-identifiability of the system (5.3).

Problem: choose the integral index, allowing to decide about the no significant of the S -structure.

Approaches (Feder, 1988) based on a principle of a covering, are proposed for the dimension (capacity) estimation of an attractor in DS and fractals. The simplified index is a topological dimension (TD). TD estimates framework geometry, and it does not reflect its interior features. Attractors and fractals are often inhomogeneous. Inhomogeneity reflects nonuniformity of distribution points on the framework (fractal). Framework inhomogeneity estimations obtain based on parameters reflecting properties of the system. The inhomogeneity reason is the different probabilities of filling geometrically equal elements of the fractal. Inhomogeneity is a mismatch between probabilities of fractal filling the specified objects and geometrical sizes of corresponding areas (Bozhokin and Parshin, 2001). Such inhomogeneous fractal objects name multifractals. S_{ey}-structures of dynamic systems with multiple-valued nonlinearities are the example of inhomogeneous frameworks.

Various coverage indicators (correlation aspect, information aspect, etc.) are used, and require time-consuming calculations. They do not always give an estimation of a geometrical difference of fragments. We introduce the integral estimation of GF performance. It is the distribution function of the variable *e*. Such the approach is eliminated a priori assumption concerning covering GF with local objects. The approach is based on a method of the structurally-frequency analysis (SFA) which is considered in subsection 3.7.

Consider the system (5.3). Obtain for it the framework S_{ey}. Fulfil the fragmentation S_{ey}. Fragmentation conditions depend on the form S_{ey}. We have $S_{ey} = F^l_{S_{ey}} \cup F^r_{S_{ey}}$, where $F^l_{S_{ey}}, F^r_{S_{ey}}$ are left and right parts of the framework S_{ey}. Functions describe fragments $F^l_{S_{ey}}, F^r_{S_{ey}}: \{e^l\} \subseteq \{e\}, \{e^r\} \subseteq \{e\}$. Construct frequency distribution functions (histogram) H^l, H^r for $F^l_{S_{ey}}, F^r_{S_{ey}}$. Obtain distribution integral functions IH^l, IH^r based on H^l, H^r. Let $\mathrm{I}_H = \{i\Delta e,\ i = \overline{1,k}\}$ is a definition range of functions H^l, H^r. Describe the range of values of functions IH^l, IH^r by vectors:

$$L(IH^l) = \left[IH^l_1, IH^l_2, \ldots, IH^l_k\right]^T,\ R(IH^r) = \left[IH^r_1, IH^r_2, \ldots, IH^r_k\right]^T$$

Here k is the number of the pockets specified on I_H, Δe is a magnitude of the pocket.

Apply the model $\hat{R} = a_H L(IH^l)$ and determine the parameter a_H by LSM.

The model is adequate, if parameter $a_H \in O(1)$, where $O(1)$ is a neighborhood 1. If the condition $a_H \in O(1)$ is true, the system (5.3) is h_{δ_h}-identified and $S_{ey} \neq NS_{ey}$. Otherwise, the framework S_{ey} is insignificant.

So, the statement is correct.

Statement 2. Let for system (5.3): 1) the framework S_{ey} is obtained; 2) the framework S_{ey} has the form $S_{ey} = F^l_{S_{ey}} \cup F^r_{S_{ey}}$, where $F^l_{S_{ey}}, F^r_{S_{ey}}$ are fragments S_{ey} defined on set $\{y(t)\}$; 3) functions H^l, H^r and cumulative frequency functions IH^l, IH^r are got for $F^l_{S_{ey}}, F^r_{S_{ey}}$; 4) the function $R(IH^r) = a_H L(IH^l)$ is known. Then the system (5.3) is h_{δ_h}-identified, if $a_H \in O(1)$.

Introduce D_h-dimension of the system (5.3) by analogy to fractals.

Definition 3. The system (5.3) has dimension $D_h = a_H$ if it is h_{δ_h}-identified.

Definition 3 shows that the dimension of a structurally identified system is approximate to 1.

Example 2. Consider the example 1 of subsection 3.1. Let $u(t) = 3 - 2\sin(0.1\pi t)$. Obtain structures S, S_{ey} (Figure 3).

Figure 3. Structurally no identified system of the second order with nonlinearity (5.8)

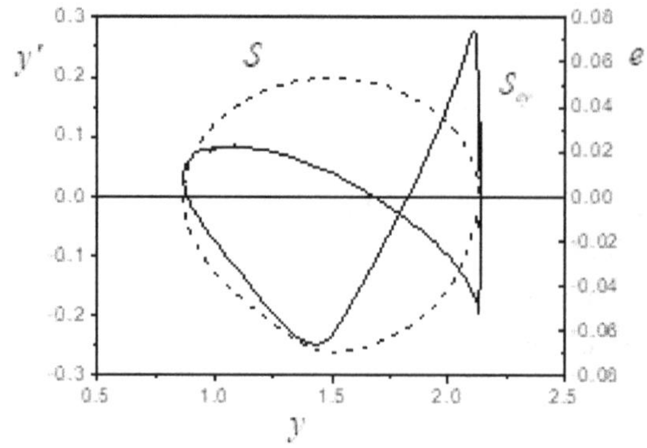

Figure 4. Defectiveness

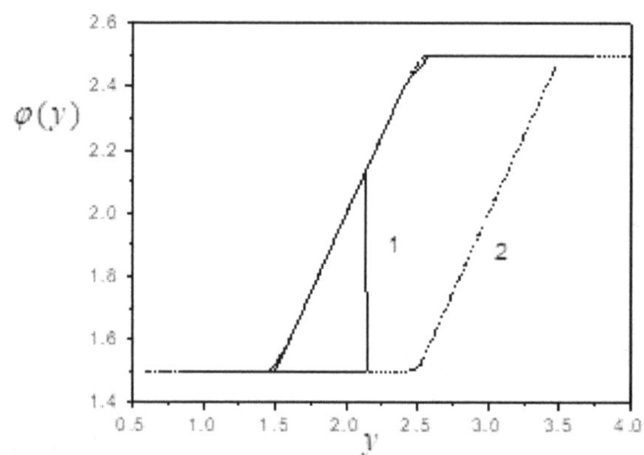

Figure 3 shows that frameworks S and S_{ey} are asymmetrical concerning the straight line y=1.5. Frameworks S, S_{ey} are h-identified, but they are not h_{δ_h}-identified. Asymmetry S_{ey} shows the difference between the left and the right parts S_{ey}. So, the system is structurally not identifiable. S_{ey} is the example NS_{ey}-structure.

Divide the framework S_{ey} in the straight line y=1.5 into two fragments $F^l_{S_{ey}}, F^r_{S_{ey}}$. Secants (5.9) are determined for fragments $F^l_{S_{ey}}, F^r_{S_{ey}}$ with a^l= -0.038, a^r=0.041. Let δ_h=0.02. Apply the theorem 1 and obtain that S_{ey} is NS_{ey}-structure. Confirmation of the function $\varphi(y)$ insignificance is shown in Figure 4. In the Figure 4, curves 1 and 2 are a defective and the reference hysteresis $\varphi(y)$.

Let $u(t)$= 5–5sin(0.1πt). Get frameworks and show them in Figure 1. We see that the system is h_{δ_h}-identified. We apply model (5.7) with H= [0.22 0.23 -0.59] ^{T} to get the S_{ey}-structure. So, $S_{ey} \neq NS_{ey}$. This conclusion confirms the model is *SM*-identifying.

Use the SFA to check results. Get function IH^l, IH^r for $F^l_{S_{ey}}, F^r_{S_{ey}}$ (Figure 5).

Figure 5. SFA application

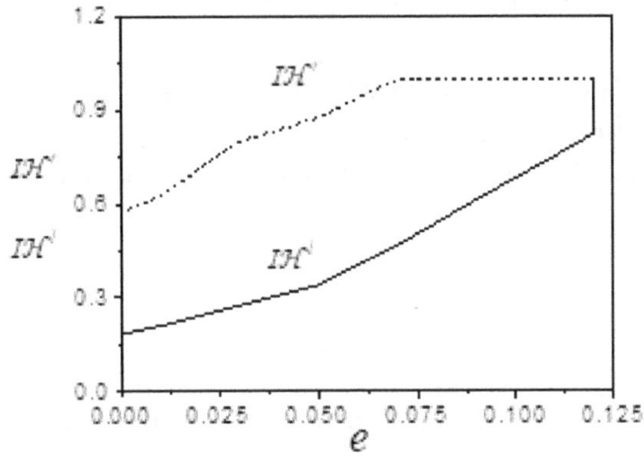

The factor a_H of the model $\widehat{R} = a_H L(IH^l)$ (Figure 5(b)) is 1.447. The coefficient of determination is 0.9. This inference confirms Figure 5 (5.a). We see the difference between IH^l, IH^r. Hence, the SFA confirms that $S_{ey} = NS_{ey}$.

Go now to the estimation of the class for nonlinearity $\varphi(y)$. It is one SI problem.

3.5. Estimation of Nonlinearity Class

Consider classes of single-valued F_{ov} and multiple-valued F_{mv} nonlinearities. These classes contain many nonlinear functions. We will describe the approach to identification of the nonlinearity class (Karabutov, 2013, 2015a) based on the analysis of sector sets.

Fulfil the framework S_{ey} fragmentation using a subset $\square_{\varphi} \subseteq \square_{N,g}$. $\square_{\varphi}$ reflects the change of the function $\chi = \varphi(y)$ in the structural space P_{ye}. The subset obtaining problem is non-trivial and depends on the a priori information. Consider the case of the a priori information absence. The problem is reduced to the selection of the change interval $y \in \square_{\varphi}$, on which we see features of function χ. Select the subset $\square_{\varphi}$, based on the framework S_{ey} change analysis. The fragment $FR_{\varphi} \subset S_{ey}$ corresponds $\square_{\varphi}$. Use fragments $FR_{\varphi} \subset S_{ey}$ in which features of the change S_{ey} are reflected.

Select the set of candidates $\mathrm{I}_y^j \subset \mathrm{I}_\varphi (j = \overline{1,s})$ based on the analysis $FR_\varphi \subset S_{ey}$. $\mathbb{I}_y^j$ can contain not associated with features of the nonlinearity of subsets. We name the set $\mathbb{I}_y^j$ informative (Karabutov, 2009) if it allows detecting these features.

Statement 3. *Let on some interval $\mathbb{I}_y^j$ of functions*

$$\left(\mu_e^j(t)=\frac{de(t)}{dy}\right) \vee \left(\mu_{k_s}^j(t)=\frac{dk_s(t)}{dy}\right) \ \forall j = \overline{1,s} \tag{5.10}$$

is not continuous, or in their behavior, there are features. Then the interval $\mathbb{I}_y^j$ is the informative.

Such the approach is effective under uncertainty. The specified informativeness property depends on h-identifiability of the framework S_{ey}. Exclude uninformative subsets $\mathbb{I}_y^j$ got as the result of analyzing changes μ_e^j, $\mu_{k_s}^j$. Include on remained $\mathbb{I}_y^j$ in composition $\mathbb{I}_\varphi$

$$\mathbb{I}_\varphi = \overset{i}{\cup}\mathbb{I}_\varphi^i\left(\mathbb{I}_y^i\right), \ \nu \le s$$

So, the informational set $\mathbb{I}_\varphi \subseteq \mathbb{I}_{N,g}$ is obtained. Go to the estimation of a nonlinearity class based on the analysis I_φ.

Consider the fragment $FR_\varphi^i \subset S_{ey}$ defined on $\mathbb{I}_\varphi^i$ for $\exists i \ge 1$. Construct a sector set for FR_φ^i (Karabutov, 2013). Apply LSM and determine $\mathbb{I}_\varphi^i$ the secant:

$$\overline{\gamma}_i = \overline{\gamma}(y(t)) = a_i y(t) + b_i \tag{5.11}$$

Calculate the mean value $\overline{y}_i$ for $y(t)$ on $\mathbb{I}_y^i \subset \mathbb{I}_\varphi$. Let $\overline{y}_i$ is a center FR_φ^i on $\mathbb{I}_y^i \subset \mathbb{I}_\varphi$. Draw a perpendicular from the point $\overline{y}_i$ before intersection with $\overline{\gamma}_i$ on the plane (y,e). Set $c_i>0$ and in the point $\alpha = \left(\overline{y}_i, e(\overline{y}_i)\right)$ construct straight lines:

$$\overline{\gamma}_{i,-} = a_{i,-} y(t) + b_i, \ \overline{\gamma}_{i,+} = a_{i,+} y(t) + b_i$$

in the point $\alpha = \left(\overline{y}_i, e(\overline{y}_i)\right)$, where $a_{i,+(-)} = a_i \pm c_i$. Let $\mathrm{Sec}_{a_i}\left(FR_\varphi^i\right) = \left(\overline{\gamma}_{i,-}, \overline{\gamma}_{i,+}\right)$ is the sector set for FR_φ^i and

$$\mathrm{Sec}\left(FR_\varphi^i\right) = \mathrm{Sec}_{\alpha,l}\left(FR_\varphi^i\right) \cup \mathrm{Sec}_{\alpha,r}\left(FR_\varphi^i\right)$$

where $\mathrm{Sec}_{\alpha,l}\left(FR_\varphi^i\right)$, $\mathrm{Sec}_{\alpha,r}\left(FR_\varphi^i\right)$ are subsets $\mathrm{Sec}\left(FR_\varphi^i\right)$ located more to the left and more to the right of the point α.

Construct secants

$$\overline{\gamma}_{i,l} = a_{i,l} y(t) + b_{i,l}, \quad \overline{\gamma}_{i,r} = a_{i,r} y(t) + b_{i,r} \tag{5.12}$$

for each of parts $FR^i_{\varphi,l(r)}$ of the fragment FR^i_φ, belonging $\text{Sec}_{\alpha,l}\left(FR^i_\varphi\right)$ and $\text{Sec}_{\alpha,r}\left(FR^i_\varphi\right)$.
Let exists $_i$>0 such that

$$\left|a_{i,l} - a_i\right| \leq \delta_i, \left|a_{i,r} - a_i\right| \leq \delta_i \tag{5.13}$$

Theorem 3 (Karabutov, 2013). *Let for system (5.3) be in space* $P_{ye} = (y, e)$*:*

(I) frameworks $FR^i_{\varphi,r}$*,* $FR^i_{\varphi,l}$ *described by mappings* $\Gamma_{ey,l(r)}: \{y\}_{i,l(r)} \to \{e\}_{i,l(r)}$*, and secants (5.11) corresponding to it are obtained;*

(ii) the fragment FR^i_φ *has the secant (5.12), where* $\{y\}_{i,l(r)} \subset \mathrm{I}^i_\varphi$*,* $\{e\}_{i,l(r)} \subset \mathrm{I}^i_\varphi$*.*

Then: (a) function $\varphi(y) \in F_{ov}$ *if conditions (5.13) are satisfied; (b) function* $\varphi(y) \in F_{mv}$ *if conditions (5.13) are not held.*

The theorem 3 shows if conditions (5.13) are satisfied, then the Holder-Lipchitz condition is fair for $\varphi(y)$ and homotopy operation applies to sectors $\text{Sec}_{\alpha,l}\left(FR^i_\varphi\right)$, $\text{Sec}_{\alpha,r}\left(FR^i_\varphi\right)$.

Remark 5. Frameworks S_{ey} and S_{ek} are usable for estimation of nonlinearity parameters.

3.6. Estimation of Nonlinearity Structure

The SI problem of nonlinear systems under uncertainty is complicated. The general approach is not developed for its solution. Each class of nonlinearities has the features. They reflect in the change of trajectories. Detection of these features gives the analysis S_{ey} or S_{ek} under uncertainty. We describe procedures for the of structure $\varphi(y)$ estimation despite the specified complexities. They are based on a phenomenological analysis S_{ey}. The nonlinearity class influences on the function $\varphi(y)$ identification procedure. Identification is performed in the structural space $P_{ye} = (y, e)$. This feature also influences on the design of structure estimation procedure. We give the development of results got Karabutov (2015a).

Class F_{ov}

We introduce a series of concepts (Karabutov, 2015a) and propose decision-making criterion. Then we synthesize the function $\varphi(y)$ structure satisfying to this criterion.

Let $\varphi(y) \in F_{ov}$. Construct for ith fragment $FR^i_{\varphi,k} \subset S_{ek}$ the secant described by the polynomial

$$\bar{\gamma}^i_{k,p}(e, k_s) = \bar{\gamma}^i_{k,p} = a^i_{0,p} + \sum_{j=1}^{p_i} a^i_{1,j,p} k^j_s, \tag{5.14}$$

where $k_s(t)=e(t)/y(t)$.

Determine a polynomial (5.14) order from the condition

$$\max_j r^2_{e\overline{\gamma}^i_{k,p}}(j) \to p_i^* \tag{5.15}$$

where $r^2_{e\overline{\gamma}^i_{k,p}}$ is the coefficient of determination between e and $\overline{\gamma}^i_{k,p}$.

Consider the class of elementary functions F_{P_f}. Based on $P_{f_i} \in F_{P_f}$, construct the secant for the fragment $FR^i_{\varphi,k} \subset S_{ek}$

$$\overline{\gamma}^i_{k,P_{f_*}}\left(e, k_{s,i}\right) = \overline{\gamma}^i_{k,P_{f_*}}(t) == a^i_{k,P_{f_*}} k_{s,n}(t, P_{f_n}),\ n = \overline{1, m_{F_{P_f}}} \tag{5.16}$$

and compute the coefficient of determination $r^2_{ek,i,P_{f_*}}$, where $k_{s,i}(t, P_{fi}) = e(t)/P_{f_i}(t)$, $m_{F_{P_f}} = \# F_{P_f}$ is a cardinal number of the class F_{P_f}.

Let

$$\delta^2_* = r^2_{ek,i,P_{f_*}}\left(p_i^*\right) \tag{5.17}$$

Theorem 4 (Karabutov, 2013). *Let for the system (5.3) in the space* $P_{k_s e} = \left(k_s, e\right)$*:*

(i) the framework S_{ek} and the secant described by the equation (5.14) are constructed;

(ii) the polynomial (5.14) order satisfies the condition (5.15);

(iii) the class F_{P_f} of elementary functions is given;

(iv) the set of secants (5.16) for S_{ek} is constructed based on functions $P_{f_n} \in F_{P_f}$, $n = \overline{1, m_{F_{P_f}}}$, and the coefficient of determination r^2_{ek,i,P_f} determined.

Then the function $\varphi(y)$ structure coincides with the function $P_{f_n} \in F_{P_f}$ structure, if

$$\left| r^2_{ek,P_{f_*}} - \delta^2_* \right| \leq \Delta,$$

where $\Delta \geq 0$ is some value, δ^2_* is determined from the condition (5.17).

Algorithm AF_{ov} for the structure $\varphi(y)$ choice on the class F_{ov}.

1. Construct framework S_{ey}.
2. Perform the fragmentation S_{ey}: $S_{ey} = \bigcup_i FR^i_\varphi$.
3. Determine the structure S_{ey} based on the analysis of fragments $FR^i_\varphi \subseteq S_{ey}$ and construct the sector $\mathrm{Sec}\left(FR^i_\varphi\right)$ for $FR^i_\varphi \subseteq S_{ey}$.
4. Estimate structural identifiability of the system on S_{ey}.
5. Find the solution in the class F_{ov}. If $\varphi(y) \in F_{ov}$, then the step 6 fulfills, otherwise, stops.
6. Go into structural space $P_{ke} = \left(k_s, e\right)$ and construct framework.
7. In S_{ek}, select the fragments $FR^i_{\varphi,k}$ corresponding $FR^i_\varphi \subseteq S_{ey}$.
8. Consider the fragment $FR^i_{\varphi,k} \subset S_{ek}$, $i \geq 1$ on S_{ek}.
9. Determine the secant $\overline{\gamma}^i_{k,p}$ (5.14), and parameter p_i based on the condition (5.15).

10. Get a class of the elementary single-valued functions F_{P_f}.
11. Construct the set of secants (5.16) on F_{P_f} for $FR^i_{\varphi,k} \subset S_{ek}$ and compute the coefficients of determination $r^2_{ek,i,P_{f_i}}$, $n = \overline{1, m_{F_{P_f}}}$.
12. Apply the theorem 4 and obtain for the function $\varphi(y)$ structure.

Remark 6. The presented approach for the secant choice is the straightening method (SM) (Karabutov, 2009). The SM similar to this method applied Mosteller and Tukey (1977).

Remark 7. We determine the form and parameters $\varphi(y)$ based on the of framework S_{ek} analysis. Explain it to that the framework S_{ek}, unlike S_{ey}, gives real estimations of parameters for the function $\varphi(y)$. The application substantiation of the framework S_{ek} is given Karabutov (2011).

Class F_{mv}

Consider a class of multiple-valued functions F_{mv}. Use frameworks S_{ey}, S_{ek} for the problem solution.

Algorithm $A\,F_{mv}$ of the function $\varphi(y)$ structure choice in the class F_{mv}.

1. Construct framework S_{ey}.
2. Estimate on S_{ey} h_{δ_h}-identifiability of the system (5.3). If the system (5.3) is structurally identified, then go to the step 3, otherwise, end the algorithm.
3. Divide S_{ey} into fragments $FR^i_{\varphi} \subseteq S_{ey}$, $i \geq 1$.
4. Construct the sector $\mathrm{Sec}\left(FR^i_{\varphi}\right)$ for FR^i_{φ}. Get the solution for the class F_{mv}. If $\varphi(y) \in F_{mv}$, then the step 5 fulfills, otherwise, end of the algorithm.
5. Get a preliminary decision on the $\varphi(y)$ function form based on the S_{ey} analysis.
6. Go into the structural space $P_{ke} = \left(k_s, e\right)$ and construct the framework S_{ek}.
7. Estimate properties and parameters of the function $\varphi(y)$ based on the analysis of fragments $FR^i_{\varphi,k} \subset S_{ek}$, $i \geq 1$. Apply the secant method in space P_{ke}.
8. For $FR^i_{\varphi,k}$, find the secant $\overline{\gamma}_{i,k}$ and the coefficient of determination $r^2_{e,\overline{\gamma}_{i,k}}$ corresponding to it.
9. Make the solution on a significance of obtained parameters of the function $\varphi(y)$ on $r^2_{e,\overline{\gamma}_{i,k}}$.

Remark 8. The proposed approach allows to get parametric estimations for $\overline{\gamma}_{i,k}$ as some subset belonging to the informational set $\mathbb{I}_o$. The level of got values is one more adequacy indicators of the proposed approach.

Remark 9. The algorithm, based on the analysis of frameworks S_{ey} and S_{ek}, does not allow identifying any class of nonlinearities. The decision on the algorithm $A\,F_{mv}$ completeness or its modification, we made based on the detailed analysis of the framework S_{ey}. Results of work (Karabutov, 2012b) confirm this conclusion.

Remark 10. The system phase portrait S can give the additional information on the system class. But this information is not such a representative. The framework S_{ey} is gives the accented reflection of system properties.

So, we propose algorithms for the nonlinearity type identification of the system (5.3). Now perform a check of obtained estimations. The first approach is based on the parametric identification. It is a very labor-consuming stage and is not always simple realized under uncertainty. We propose the structurally-frequency analysis method (SFAM). SFAM (Karabutov, 2015a) is proposed for verification of dynamic system structural identification results with complicated piecewise constant nonlinearity.

3.7. SFAM

SFA is based on the nonlinearity representation as a structurally-frequency chart (SFC). This diagrams class is proposed in (Karabutov, 2015a). SFC is the method of decision-making under uncertainty. The frequency analysis is the development of the histogram method applied in statistics. We interpret the appearance frequency of the certain level signal as the indicator of existing properties in function. It is the fundamental difference between SFC and the accepted statistical processing of frequency.

SFC establishes the relation between changes of argument y and a response (function $\varphi(y)$) on this effect. The frequency of distribution y can increase in some areas. It is the emersion indicator in a system of a non-standard (nonlinear) regime. Distribution functions y and $\varphi(y)$ have the equal definition domain. Therefore, the growth of frequency n_y should confirm features of change $\varphi(y)$. These properties reflect the SFC.

Remark 11. SFC can have the original form. Distribution laws y, $\varphi(y)$ depend on properties of the nonlinear system. As $y(t)$ is a continuous function almost for $\forall t \geq 0$, then n_e (distribution $\varphi(y)$) has to a maximum in areas $\mathbb{J}_y^M$. $\mathbb{J}_y^M$ corresponds to features of processes in the system. SFC can have several such areas.

Consider SFC as one more form of dynamic systems frameworks at the decision-making level. We consider these distributions as S-distributions to distinguish them from statistics distributions.

Figure 6. Frameworks for second order system with nonlinearity (5.18)

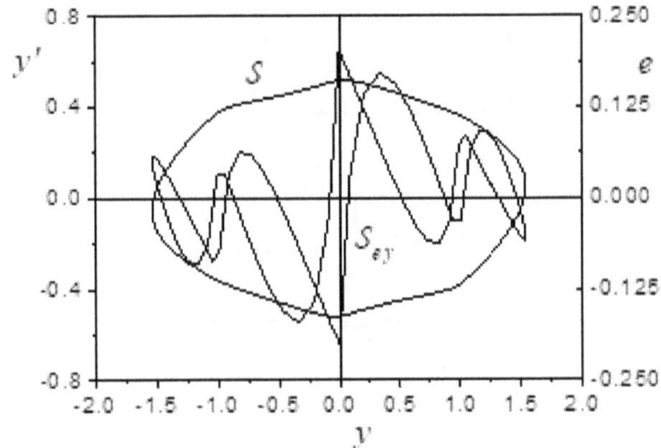

Example 3. Consider the system (5.3) of the second order with parameters $A = \begin{bmatrix} 0 & 1 \\ -3 & -4 \end{bmatrix}$, $y(0)=x(0)=2$, $y'(0)=x'(0)=1$, $u(t)=3\sin(0.1\pi t)$, $y=x$,

$$\varphi(y) = \begin{cases} 2, & if\ y > 1, \\ \text{sign}(y), & if\ y \le 1, \\ -2, & if\ y < -1. \end{cases} \tag{5.18}$$

where sign(y) is the sign function.

Show frameworks describing the steady state in Figure 6. We see (Figure 6) that the system is h_{δ_h}-identified. The system has features in points $y=-1$, $y=0$, $y=1$. These points are sign change points of a derivative (see S_{ey}-structure).

Identification of structural parameters of the system is considered Karabutov (2015a). Consider the decision-making about nonlinearity. Apply the SFAM. Results show in Figure 7. We see the SFC has two levels. The distribution function $f(y)$ for the variable y is presented at the top level. The distribution function $f(e)$ is shown at the lower level (the y-axis is on the right). $f(e)$ has a complicated appearance, and levels of its values are correlated with change $f(y)$. We represent at this level result decision-making on $\varphi(y)$. We present the restored configuration $\varphi(y)$ as $\varphi_e(y)$. Show arrows switching points off function $\varphi(y)$. Represent the fragment of the top piece of the framework S_{ey} on the plane (y,e).

Figure 7. SFAM application

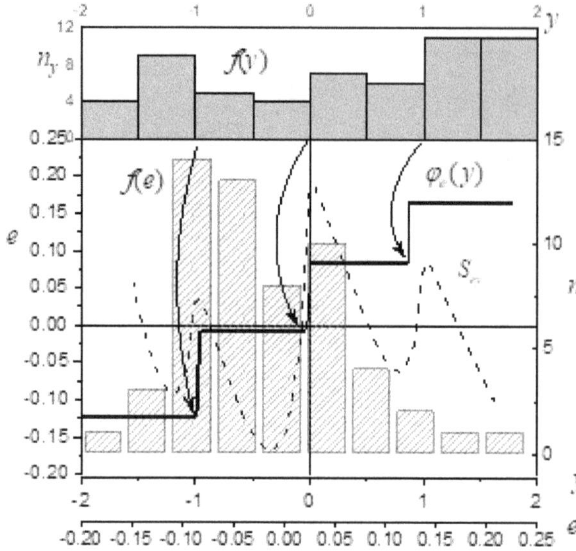

So, we conclude SFAM allows getting the form $\varphi(y)$ based on the analysis of functions $n_e(e)$, $n_e(e)$ and the framework S_{ey}. The approach is demonstrative but demands an expert knowledge of the object domain. Therefore, the combination of a priori knowledge and SFAM allows obtaining the finished solution to the problem.

Remark 12. The SFA applies to the solution of various problems, the analysis of dynamic systems. The example 3 shows that the correct interpretation of results expands decision-making options in systems of structural identification. SFC are applicable also in identification system of eigenvalues dynamic system (see next).

Remark 13. Frameworks S_{ey} also S_{ek} solve different classes of problems. S_{ey} is applied at the decision-making level about the nonlinearity form and system structural identifiability. The framework S_{ek} is applied for the estimation of structural parameters for nonlinearity.

Frameworks $RK_{\upsilon r}$ and $K_{\upsilon\pm}$ (Karabutov, 2012a) also apply for the estimation of the nonlinearity structure. Frameworks are proposed for the hysteresis estimation in a static system. Binary transformation $B_{\pm}$ is introduced for the analysis of these frameworks. $B_{\pm}$ recovers the hysteresis structure as a logic function. Results of modelling show that $RK_{\upsilon r}$, $K_{\upsilon\pm}$ and the transformation $B_{\pm}$ apply for structural identification of DS.

We note that the model (5.7) choice is the complicated problem. Show how to choose the model (5.7) for an obtaining of the framework S_{ey} and to identify the nonlinearity form.

Example 4. Consider the system (5.3) of the second order with $\mathbb{I}_o = \{y(t), u(t), t \in \mathbb{J}\}$. Phase portrait S is presented in Figure 8. We saw that processes in the system had complicated character.

Figure 8. Frameworks S and S_{ey}

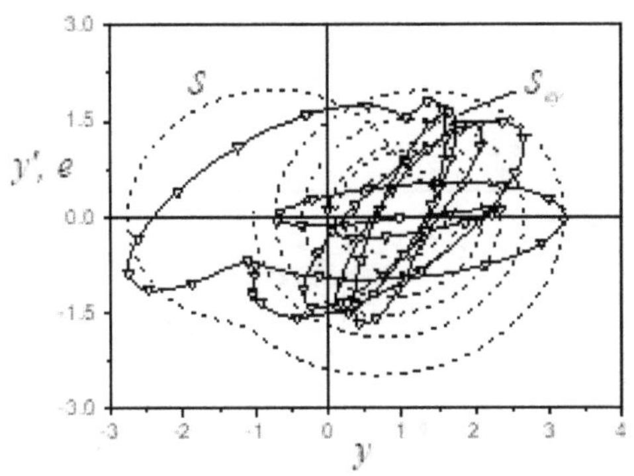

Construct the model (5.7) and generate the variable $e(t)$ for $t \in [0;7]$ s. Model (5.7) has the form $\hat{x}_1^l(t) = [-0.12; 14.68; 0.1][1\ u(t)\ y(t)]^T$. Coefficient of determination $r_1^2 = 0.97$. Construct the framework S_{ey} and show to it in Figure 8. The framework S_{ey} has the complicated form. We could not make the solution on the function $\varphi(y)$ class by S_{ey}. Consider two hypotheses.

1. Model (5.7) is inadequate to the system.
2. Model (5.7) are not applicable for the specific class of systems.

We reject the hypothesis 1, as the coefficient of determination is very high. Hence, the hypothesis 2 is true. The model (5.7) has a more complex structure. Suppose a differential equation describes that nonlinearity (compare S_{ey} and S). Examine this hypothesis.

Apply numerical differentiation and determine the second derivative $y(t)$. Designate the obtain variable as x_2. Consider the model:

$$\hat{x}_2^l(t) = H_2^T[1\ u(t)\ \dot{y}(t)]^T$$

Let the variable x_2 is more informative than x_1. Apply LSM and get:

$$\hat{x}_2^l(t) = [0.961; 1.91; -0.982][1\ u(t)\ y(t)]^T, \quad r_2^2 = 0.987 \tag{5.19}$$

The model (5.19) is obtained for $t \in [0;7.8]$ s. Introduce the variable $e_2(t) = \hat{x}_2^l(t) - x_2(t)$.

In Figure 9, we show frameworks reflecting the state of the identification system. Present in Figure 9 the framework S_{e_2y} described by mapping $\Gamma_{e_2y}: \{y\} \to \{e_2\}$.

Figure 9. Frameworks S, S_{e_2y}

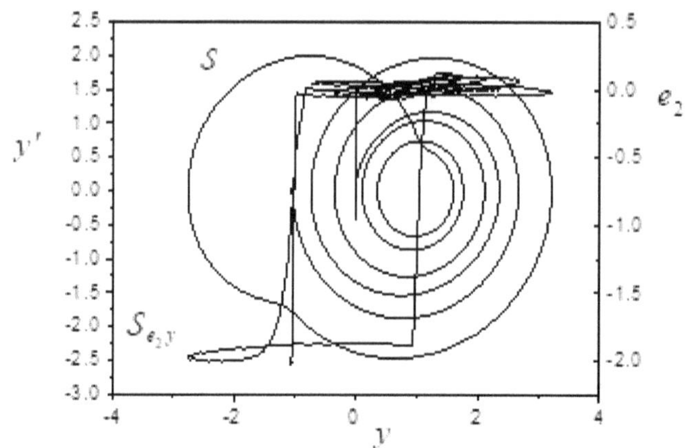

Figure 9 allows is decided on the nonlinearity of the system. The solution gives to the analysis S_{e_2y}. S_{e_2y} is the modification of the ideal hysteresis. We see that the variable e_2 have small fluctuations in the saturation range. Analyze the nature of these fluctuations.

Assumption 3. Fluctuations are the system property or impact of uncertainty which inadequacy of the model (5.19) gives.

Apply model to verification of the assumption 3

$$\hat{x}_{2(3)}^l(t) = H_3^T \left[1\ u(t)\ y(t)\ x_1(t)\right]^T \tag{5.20}$$

Get the estimation H_3 on the time gap [0; 7.8]s. Vector H_3= [0.99; 1.06; -1; 0.06]^T, $r_2^2 = 0.996$. Let $e_3(t) = \hat{x}_{2(3)}^l(t) - x_2(t)$. In Figure 10, we show to the framework S_{e_3y} describing the change of the hysteresis $\varphi(y)$.

Figure 10. Hysteresis $\varphi(y)$ and its estimation $e_3(y)$

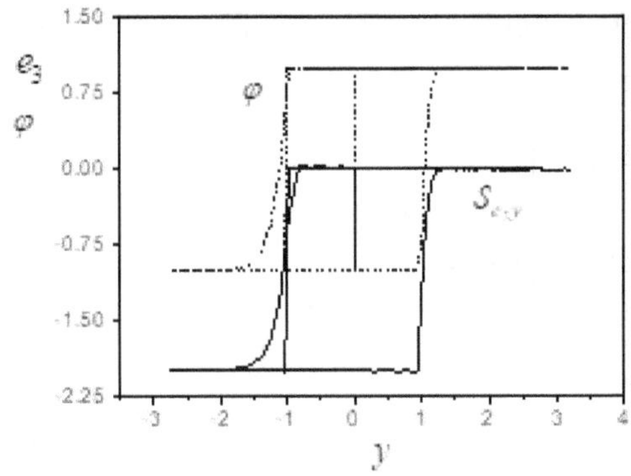

Compare S_{e_2y}, S_{e_3y} and obtain what fluctuations of the framework S_{e_2y} is the effect of the model (5.19) inadequacy in the saturation range. The model (5.20) eliminates these fluctuations. In Figure 10, we show to change of initial nonlinearity $\varphi(y)$. S_{e_3y} and $\varphi(y)$ are congruent. The offset results from the transition in the structural space. We do not consider the choice problem of a time gap for identification of models (5.7), (5.19), (5.20). The interval choice depends on properties of the system and is defined at the modeling stage. Our goal is choice of frameworks allowing to solve the problem of the nonlinearity class estimation. Further researches of this task are given Acho, (2013). They show that the examined nonlinearity is RC-OTA chaotic oscillator (Acho, 2013) described by the system of equations:

$$\begin{aligned} \ddot{x} - 0.1\dot{x} + x &= \varphi, \\ \dot{\varphi} &= 10(-\varphi + \operatorname{sgn}(x + \operatorname{sgn}(\varphi))), \\ y &= x, \end{aligned}$$

Example 4 shows that the equation (5.7) choice for obtaining of the framework S_{ey} is a complex task. NDS Properties define by the detail level of the model (5.7). If nonlinearity is described by the DS, then the framework S_{ey} based on the model (5.7) has the complex form. Equation form complicates the decision-making on the structure of the nonlinearity. It is the sign of a new dependence search for the identification of S_{ey} or its analogues. We understand analogues as virtual frameworks which determine relations between new variables. It is the main inference which gives to the example 4.

4. SI OF LINEAR DYNAMIC SYSTEMS

We consider to the structural identification approach of linear dynamic systems (LDS) (Karabutov, 2015b).

4.1. Problem Statement

Consider linear dynamic system

$$\begin{aligned} \dot{X} &= AX + BU, \\ Y &= CX + DU, \end{aligned} \tag{5.21}$$

where $X \in \mathbb{R}^m$ is a state vector, $U \in \mathbb{R}^k$, $Y \in \mathbb{R}^n$ are input and output, $A \in \mathbb{R}^{m\times m}$, $B \in \mathbb{R}^{m\times k}$, $C \in \mathbb{R}^{n\times m}$, $D \in \mathbb{R}^{n\times k}$.

The experimental information for (5.21) has the form

$$\mathbb{I}_o = \{Y(t), U(t), t \in \mathbb{J} = [t_0, t_1]\} \tag{5.22}$$

Write a solution to the system (5.21) as

$$X(t) = X(t_0, U, t) \tag{5.23}$$

where χ is the operator uniquely defined by matrixes A,B.

Obtain the system (5.21) solution with $X_0 = X(t_0)$

$$X(t) = X_g(t) + X_q(t)$$

where $X_q(t)$ is a particular solution (5.21) with $U \in \mathrm{I}_o$, $X_g(t)$ is a general solution (5.21) with $U(t)=0$.

We designate the general solution (5.21) with $X_0 = X_0(Y_0) \in \mathbb{I}_o$ as $X_g(X_0, t)$.

Problem: find solution estimations $X_g(t) = X_g(X_q, X_0, t)$ on the set I_o and evaluate the spectrum of eigenvalues and the system (5.21) order.

4.2. Estimation $X_g(t)$

Consider the system (5.21) with $Y=y$, $y \in R$, $U=u$, $u \in R$, and $m=2$. Apply approach (Karabutov, 2009). Generate a set $\{X_g(t)\}$ and get $\{X(t)\} \backslash \{X_q(t)\}$. Present $\mathbb{I}_o$ as:

$$\mathrm{I}_o = \mathrm{I}_o^q(J_q) \cup \mathrm{I}_o^g(J_g)$$

where $\mathbb{J}_q \cup \mathbb{J}_g = \mathbb{J} \subseteq \mathbb{R}$; $\mathbb{I}_o^q$, $\mathbb{I}_o^g$ are sets which contain the information about X_q and X_g.

Find the particular solution estimation for the system (5.21) based on the analysis $\mathbb{I}_o^q(\mathbb{J}_q)$. As $x_1 = y \in \mathbb{R}$ apply the differentiation operation to the variable y to obtaining the component $x_2 = \dot{x}_1$ for the vector $X \in \mathbb{R}^2$. Designate $x_2 = \dot{y}$.

Statement 4 (Karabutov, 2009). *The model*:

$$\hat{X}_q(t)=\hat{A}_q W(t) \quad \forall t \in \mathbb{J}_q \tag{5.24}$$

gives the estimate $X_q(t)$ on the set $\mathbb{I}_o^q$, where $\hat{A}_q \in \mathbb{R}^{2\times 2}$ is the matrix of parameters, $W = [u\ u']^T$.

Model (5.24) properties depend on the choice of an interval $\mathbb{J}_q \subset \mathbb{J}$. We determine the estimation $\hat{X}_q(t)$ for the system (5.21) particular solution $X_q(t)$ on the set $\mathbb{I}_o^g$, applied the model (5.24). The estimation of the general solution is

$$\hat{X}_g(t)= X(t)-\hat{X}_q(t) \quad \forall t \in J_g$$

where $\hat{X}_g(t)= \left[\hat{y}_g(t)\ \hat{\dot{y}}_g(t)\right]^T$.

The proposed approach can be generalized to the multidimensional case. Next, we consider the system (5.21) with one input u and one output y.

The model (5.24) applies to the certain restrictions imposed on the input $u(t)$.

The input $u(t)$ should be constantly excited and limited, and the system (5.21) is h-identified.

Next, we use Lyapunov exponents (LE) and frameworks as function LE for structural identification of LDS.

Remark 14. The considered problem has the specificity. Issue solution demands the introduction of frameworks which differ from S_{ey}. These frameworks are based on LE and a coefficient of structural properties (CSP).

4.3. LE. CSP

Apply LE (Bylov et al., 1966) to the estimation of eigenvalues spectrum for LDS (5.21). LE use as the system stability criterion. Determine LE for a real-valued function $h(t)$ as:

$$\chi[h]= \overline{\lim_{t\to\infty}}\frac{\ln|h(t)|}{t} \tag{5.25}$$

where $\overline{\lim\limits_{t\to\infty}}$ is a limit superior. $\chi_i\,(i = \overline{1,m})$ is the non-zero solution of the system (5.21) coinciding with actual parts of eigenvalues λ_i of a matrix A.

Let the general solution $X_g(t)$ estimation $\forall t\in J_g$ for the system (5.21) is known and the system is stable. Apply (5.25) to $\hat{y}_g(t)$:

$$\chi\left[\hat{y}_g\right] = \overline{\lim_{t\to\bar{t}}}\frac{\ln\left|\hat{y}_g(t)\right|}{t} \tag{5.26}$$

where $\bar{t} \in J_g$ is the maximum value t on the interval $\mathbb{J}_g \subset \mathbb{J}$.

$\chi\left[\hat{y}_g\right]$ is the largest Lyapunov exponent (LLE). If the limit (5.25) exists, $\chi\left[\hat{y}_g\right]$ is the estimation of the largest eigenvalue for the matrix A. So, $\chi\left[\hat{y}_g\right]$ is an index of stability degree the system (5.21). If m=2, for $\dot{\hat{y}}_g$ obtaining

$$\chi\left[\dot{\hat{y}}_g\right] = \overline{\lim_{t\to\bar{t}}}\frac{\ln\left|\dot{\hat{y}}_g\right|}{t}$$

Consider the index

$$\eta[h] = \underline{\lim_{t\to\infty}}\frac{\ln|h(t)|}{t}$$

where $\underline{\lim\limits_{t\to\infty}}$ is a bottom limit. $\eta[h]$ is the Perron bottom index (Izobov, 2006). Next, we describe the estimation method $\eta[h]$.

The LE application idea is proposed in (Karabutov, 2009) for SI problems. The approach is based on the analysis of the coefficient of structural properties (CSP).

Introduce the index

$$\rho\left(\hat{y}_g\right) = \rho_g(t) = \ln\left|\hat{y}_g(t)\right| \quad \forall t \in \hat{\mathbb{J}}_g \subset \mathbb{J}_g$$

$\hat{\mathbb{J}}_g = \left[t_0, \bar{t}\right]$ determine based on (5.26).

Consider a system with the input t and the output $\rho\left(\hat{y}_g\right)$. Define CSP as:

$$k_s(t,\rho) = \frac{\rho\left(\hat{y}_g\right)}{t} \tag{5.27}$$

$k_s(t,\rho)$ is a basis for computing of the index $\chi[\hat{y}_g]$ on the interval $\hat{\mathbb{J}}_g$.

Next, we describe the application $k_s(t,\rho)$ for the LE estimation.

So, we show the relationship between $\chi\left[\hat{y}_g\right]$ and $k_s(t,\rho)$. These indexes define on the informational set $\mathbb{I}_\rho = \left\{\rho\left(\hat{y}_g(t)\right), t \in \hat{\mathbb{J}}_g\right\}$.

Consider the set

$$\mathbb{I}_g(\hat{y}_g,t) = \{\hat{y}_g(t), t \in \mathbb{J}_g\} = \mathbb{I}_{\hat{X}_g} \setminus \{\dot{\hat{y}}_g(t), t \in \mathbb{J}_g\}$$

which contains the data about the change variable of $\hat{y}_g$ on the interval $\mathbb{J}_g$.

Assumption 4. The system (5.21) is stable, i.e. $\mathrm{Re}(\lambda_i) \leq 0$, $i = \overline{1,m}$, where $\lambda_i \in \sigma(A)$ is ith eigenvalue of matrix A.

Problem: estimate the spectrum $\sigma(A)$ of eigenvalues for the matrix A and the system (5.21) order based on the set $\mathbb{I}_g$ analysis.

4.4. Frameworks for Estimation LE

Consider sets

$$\mathbb{I}_{k_s} = \{k_s(t, \rho(\hat{y}_g(t))), t \in \mathbb{J}_g\},$$

$$\mathbb{I}_{k'_s} = \{k_s(t, \rho(\dot{\hat{y}}_g(t))), t \in \mathbb{J}_g\}$$

Define mapping $S_{k_{s,\rho}} \to \mathbb{I}_{k_s} \times \mathbb{I}_{k'_s}$. The framework $S_{k_{s,\rho}}$ reflects the dynamics of change parameters depending on LE. Consider the function:

$$\Delta k'_s(t) = k_s(t, \rho(\dot{\hat{y}}_g(t+\tau))) - k_s(t, \rho(\dot{\hat{y}}_g(t)))$$

where $\tau > 0$.

Generate the set $\mathbb{I}_{\Delta k'_s} = \{\Delta k_s(t, \rho(\dot{\hat{y}}_g(t))), t \in \mathbb{J}_g\}$ and introduce the framework $SK_{\Delta k'_{s,\rho}} \to \mathbb{I}_{k_{s,\rho}} \times \mathbb{I}_{\Delta k'_{s,\rho}}$. Consider corresponding $SK_{\Delta k'_{s,\rho}}$ the mapping:

$$LSK_{\Delta k'_{s,\rho}} \to \mathbb{I}_{k_{s,\rho}} \times \mathbb{B}(\mathbb{I}_{\Delta k'_{s,\rho}})$$

where $\mathbb{B}(\mathbb{I}_{\Delta k'_{s,\rho}}) \subset \{-1;1\}$. Define elements $b(t) \in \mathbb{B}(\mathbb{I}_{\Delta k'_{s,\rho}})$ as

$$b(t) = \begin{cases} 1, & \text{if } \Delta k'_s(t) \geq 0, \\ -1, & \text{if } \Delta k'_s(t) < 0, \end{cases} \quad t \in \hat{\mathbb{J}}_g$$

Remark 15. We can choose for some class of systems boundary a limit superior in (5.26) based on the analysis $S_{k_{s,\rho}}$.

Remark 16. We define the choice values for the function $b(t)$ by convenience of its graphical analysis. $b(t)$ is possible to specify as the binary set {0 1}.

4.5. Estimation of System Order

Consider a criterion for the order estimation of system (5.21) based on analysis of the framework $LSK_{\Delta k'_{s\rho}}$ properties (Karabutov, 2015b).

Theorem 5. If the function $b(t)$ changes the sign m–1 of times in the interval $[t_0, t^*] \subset \mathbb{J}_g$ $(t^* \leq \bar{t})$, then the system (5.21) has the order m.

Proof of Theorem 5. Consider a case of simple real roots. The system (5.21) is stable. Let the framework S, described functions $\Gamma_S:\hat{y}_g \to \dot{\hat{y}}_g$, be the phase portrait for system (5.21) with output $y \in \mathbb{R}$. System eigenvalues are arranged in decreasing order $\lambda_1 > \lambda_2 > \ldots > \lambda_m$. The phase trajectory changes an inclination angle at the prevalence of each component $e^{\lambda_i t} \in \hat{y}_g$ on some subinterval $\mathbb{J}_{\lambda_i} \subset \mathbb{J}$. This process occurs exactly m–1 times. As function $\rho(\dot{\hat{y}}_g(t))$ depends on $\dot{\hat{y}}_g(t)$, then each such change transform to the change of properties of structure $LSK_{\Delta k'_{s\rho}}$.

We give one more proof of the theorem 6. Consider the system (5.21) with a stable matrix A. Eigenvalues matrix A are simple real and are arranged as $\lambda_1 > \lambda_2 > \ldots > \lambda_m$. Then (Bylov et al., 1966) solution of the system (5.21) is linean $L^m(t)$, which is a part of linear space $L^m(t)$. As the system is linear, for $L^{l_i}(t)$ fair and embedding

$$0 \equiv L^{l_0}(t) \subset L^{l_1}(t) \subset \ldots \subset L^{l_n}(t) \equiv L^m(t)$$

which we name a pyramid, where l_i is the dimension $L^{l_i}(t)$.

Difference between two neighboring lineans:

$$dL_k = L^{l_0}(t) \subset L^{l_k}(t) \backslash L^{l_{k-1}}(t)$$

is a pyramid step. The step has weight $m_k = l_k - l_{k-1}$. l_k is a number of basic solutions which contains $L^{l_k}(t)$.

$L^{l_1}(t)$ contains an index λ_m on an initial slice of time. The solution passes to the pyramid step $L^{l_2}(t)$ at increase t. The index λ_{m-1} dominates on $L^{l_2}(t)$. It calls a change of properties of the function $b(t)$. Further, we apply an induction on steps of a pyramid to obtaining of all space $L^m(t)$. ■

The theorem 5 is fair for multiple roots and simpler eigenvalues of the matrix A system (5.21).

So, the application of structural methods allows obtaining the estimation for the order of systems (5.21). Theorem 5 has sufficient character and is applicable not to all types of roots. If conditions of the theorem 5 not are fulfilled, then use the algorithm of decision making (Karabutov, 2006).

Remark 17. The algorithm for the estimation of the order of the system under indeterminacy is proposed Karabutov, (2009). It based on the factor analysis CSP the system (5.21). The approach described here is simple for application.

4.6. Decision-Making About Type of Roots

The approach for decision-making about a type of roots is proposed by Karabutov, (2009). It is based on the analysis of a variation in the function $d(t) = \rho(\hat{y}_g)/\rho(\dot{\hat{y}}_g)$. If the system has imaginary or complex eigenvalues, then apply other approaches and additional criteria. We propose the approach based on the analysis of special frameworks.

As shown Karabutov (2015b), system eigenvalues affect the framework S form. Therefore, we introduce a special framework for the estimation the spectrum of the system. The framework of analysis gives to estimations of numbers λ_i.

Consider frameworks $SK_{k^i_{s,\rho}} \to \mathrm{I}_{k_s} \times \mathrm{I}_{k^i_s}$ and $LSK_{\Delta k'_{s,\rho}}$, where i designate ith derivative $\hat{y}_g(t)$,

$$\mathrm{I}_{k^i_s} = \left\{k_s\left(t,\rho\left(\hat{y}^{(i)}_g(t)\right)\right), t \in \mathbb{J}_g\right\}$$

The LLE $\chi_l\left[\hat{y}_g\right]$ is located in the left part of the framework $SK_{k^i_{s,\rho}}$. All other indexes $\chi_i\left[\hat{y}_g\right]$ coincide with Perron bottom indices $\eta_i\left[\hat{y}_g\right]$ and are located to the right from $\chi_l\left[\hat{y}_g\right]$. $\chi_i\left[\hat{y}_g\right]$ correspond to local minima on $SK_{k^i_{s,\rho}}$. The disposition of minima on $SK_{k_{s,\rho}}$ coincides with the change of sign $LSK_{\Delta k'_{s,\rho}}$. We obtain the set $M_{LE} = \left\{\hat{\lambda}_1, \hat{\lambda}_2, \ldots, \hat{\lambda}_m\right\}$, where $\hat{\lambda}_i$ is the estimation ith eigenvalue λ_i of the matrix A system (5.21).

Theorem 6 (Karabutov, 2012b). If the system (5.21) is stable and has simpler eigenvalues, then frameworks $SK_{k^i_{s,\rho}}$, $i = \overline{1,m}$ contain the information on LE.

We consider the stable system (5.21). So, we decide only for negative elements of I_{k_s}.

Apply the method of SFC to confirmation of the obtained solution (see section 3.7). Construct for each informational set I_{k_s}, $\mathrm{I}_{k^i_s}$ histograms $H_{k_s}, H_{k^i_s}$ showing distribution of got indexes $\chi_i\left[\hat{y}_g\right]$. The function $H_{k^i_s}$ maximum value is the index $\chi_l\left[\hat{y}_g\right]$, and the minimum values are $\chi_i\left[\hat{y}_g\right]$. These results are one more confirmation of LE classification for the DS.

Remark 18. We analyze properties of the framework $LSK_{\Delta k'_{s,\rho}}$ for the detection of false LE in H_{k_s}, $H_{k^i_s}$.

4.7. Procedure of Finding LE

Proposed methods and algorithms are the approach based on the SI of LDS. In generalizing the obtained results, we propose the SI procedure. Procedure is based on the computing LE. The proposed approach gives the estimation of type LE and the dimension evaluation for LDS.

Procedure CLE.

1. Generate sets for computing LE. Use results obtained in section 4.2. Apply statement 4 and obtain the variable $X_g(t)$.
2. Construct the phase portrait for the system (5.21) and check the h-identifiability.
3. Determine the system coefficient of structural properties on the set $▯_g(\hat{y}_g, t)$.
4. Construct frameworks $LSK_{\Delta k'_{s,\rho}}$, $SK_{\Delta k'_{s,\rho}}$.
5. Estimate the system order. Analyze $LSK_{\Delta k'_{s,\rho}}$ and apply the theorem 5.
6. Analyze the function $d(t)$ and make the solution on the type of roots for the dynamic system (5.21).
7. Fulfil check of the correctness of obtained estimations based on the construction of a frequency distributions H_{k_s}, $H_{k'_s}$.

Example 5. Consider the system (5.21) with parameters

$$A = \begin{bmatrix} 0 & 1 & 0 \\ 0 & 0 & 1 \\ -6 & -11 & -6 \end{bmatrix}$$

$\sigma(A) = (-1; -2; -3)$, $y(0) = y''(0) = 1$, $y'(0) = 2$, $u(t) = 3 + 1\sin(0.1\pi t)$, $y = x_1$.

Apply the approach proposed in section 4.2, and get the set $\{\hat{y}_g(t), 0 \le t \le 16\text{s}\}$. Generate sets $▯_{\Delta k'_s}$, $▯_{k_s}$, $▯_{k^i_s}$, $i=1,2$ based on processing $\hat{y}_g(t)$. Apply the theorem 5 and present estimation results of the system order and eigenvalues for the matrix A (see Figure 11).

Figure 11. Estimation of order and eigenvalues

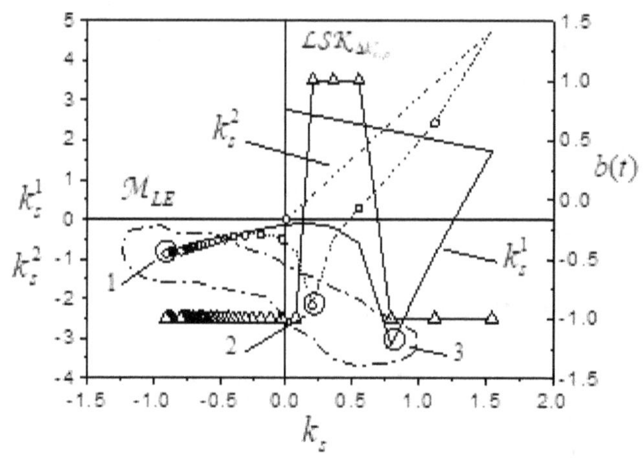

We present set M_{LE} and estimations λ_i, i=1,2,3. We see that the framework LSK_{Δ} $_{k'_{s,\rho}}$ changes a sign twice. The system has the third order. Designations in Figure 11:

$$k_s = k_s(t,\rho), k_s^1 = k_s\left(t,\rho\left(\hat{\dot{y}}_g(t)\right)\right), k_s^2 = k_s\left(t,\rho\left(\hat{\ddot{y}}_g(t)\right)\right).$$

Construct graphs of distributions $\hat{\lambda}_i$ (Figure 12) for correctness check of solutions. In Figure 12, we show the frequencies f_i for k_s, k_s^1, k_s^2 at specified intervals and the set M_{LE}. As the system is stable, we consider frequencies f_i only for negative values k_s^i. Get following set $M_{LE} = \{-0.9;-2.16;-3.16\}$ of eigenvalues the system. Except value –4 based on the remark 18.

Figure 12. Distribution of simple roots for third order system

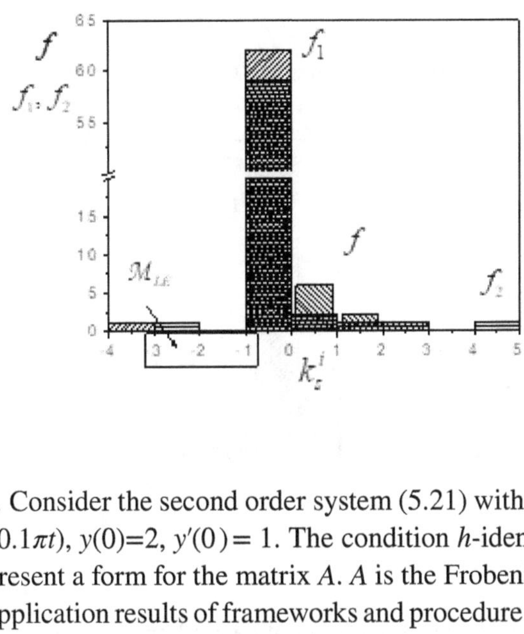

Example 6. Consider the second order system (5.21) with $\sigma(A)$= (-2i; 2i). Input $u(t)$= 3+2sin(0.1πt), y(0)=2, y'(0)= 1. The condition h-identifiability is fulfilled. We do not represent a form for the matrix A. A is the Frobenius matrix.

Show the application results of frameworks and procedure CLE in Figure 13. We see that the system does not have real roots. The system order is equal to two. The framework $SK_{k'_{s,\rho}}$ has oscillations near coordinate origin. Oscillations diminish with decreasing k_S. It is a sign of imaginary roots. The LLE on a bar graph (Figure 14) is equal to zero. It coincides with the initial assumption. The estimation method of an imaginary part of roots describes Karabutov, (2011).

Figure 13. Estimation of imaginary roots

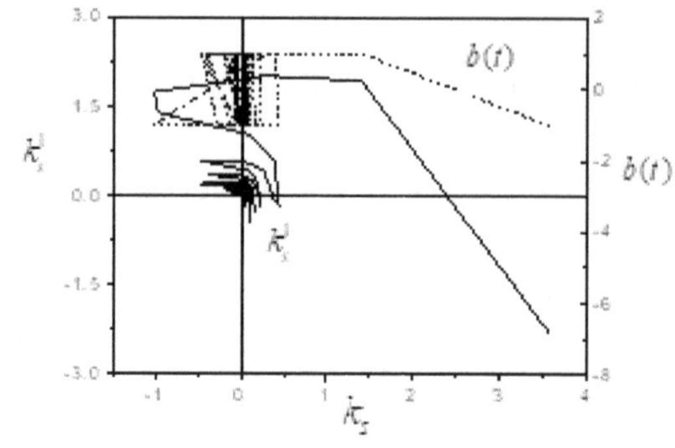

Figure 14. Distribution of imaginary roots for second order system

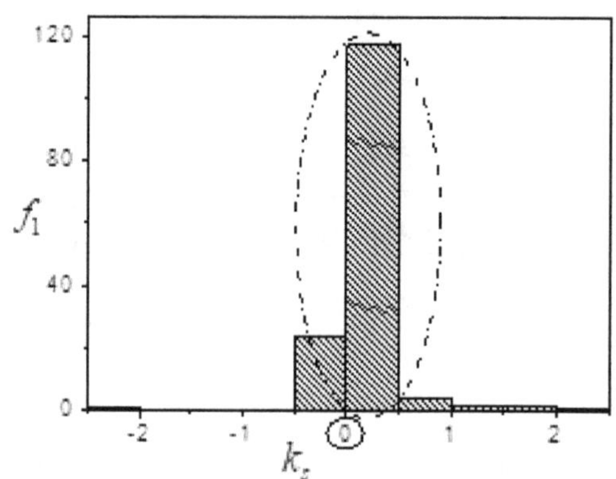

Example 7. Consider the system (5.21) of second order with

$$A = \begin{bmatrix} 0 & 1 \\ -2 & -2 \end{bmatrix},$$

$\sigma(A)= (-1+i, -1-i)$, $u(t)=5+2\sin(0.1\pi t)$, $y(0)=2$, $y'(0) = 1$.
Identification results are presented in Figures 15 and 16.

Figure 15. Frameworks for system order estimation with complex roots

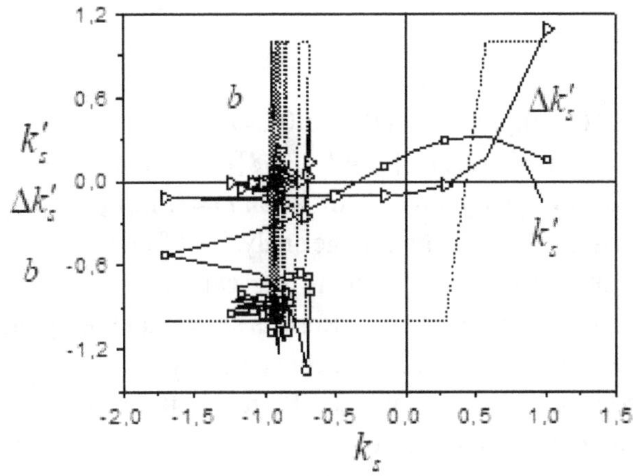

The variation $b(t)$ analysis (Figure 15) gives the estimation of the system order: 2. We observe oscillations with decreasing amplitude around the value $k_S(t,\rho)$=-1. $k_S(t,\rho)$=-1 is the real part value of the root. So, the framework $LSK_{\Delta k'_{s\rho}}$ confirms presence of complex roots.

The function f of distribution values k_S is shown in Figure 16. f confirms the complex root and shows the oscillations near value –1.

Figure 16. Distribution of imaginary roots

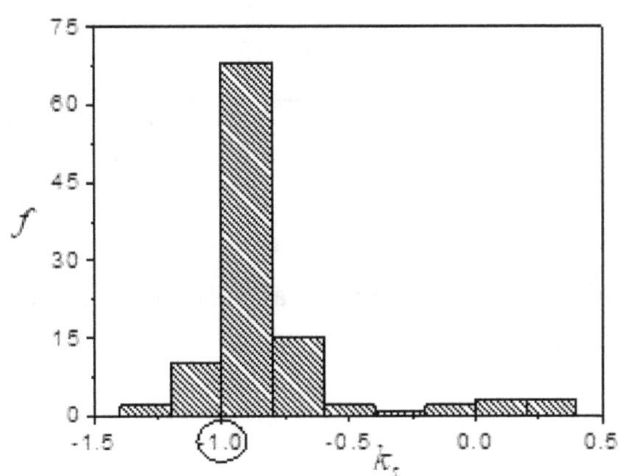

Example 8. Consider the system (5.21) of second order with

$$A = \begin{bmatrix} 0 & 1 \\ a_1(t) & a_2(t) \end{bmatrix}$$

$a_1(t)$= -3+0.2sin(0.02πt), $a_2(t)$= -4+0.3sin(0.04πt).

Spectrum is $\sigma(A)$= ([-1.32; -0.81], [-3.47; -2.38]). The analysis of a variation in the function $b(t)$ gives the system order estimation m=2. In Figure 17, we use the same designations as in Figure 11. The change analysis of functions $k_s^1(t)$, $k_s^2(t)$ show two estimations are got for LLE. We obtain the interval estimation [-0.89; -1.14] for λ_i. The mean value λ_1 on the interval is –1. Results show that the pyramid $L^2(t)$ has no fixed steps. It does not allow obtaining a constant value of λ_i for pyramid steps. The structurally-frequency method gives the Perron bottom index as -3.68. This estimation is underrated. But the proposed method confirms its existence. We have the explanation for the got results. The offered method demands a further study of the non-linear systems.

Figure 17. Structures for an estimation of an order non-stationary system

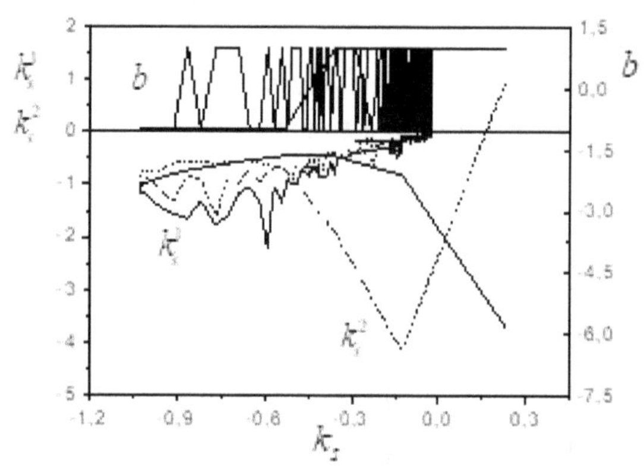

So, modeling results confirm workability and efficiency of proposed structural methods and procedures of LDS identification.

5. SI OF SYSTEMS WITH PERIODIC COEFFICIENTS

Consider the linear dynamic system with periodic coefficients (PCS)

$$\begin{aligned} \dot{X} &= A(t)X + BU, \\ Y &= CX + W_U U, \end{aligned} \tag{5.28}$$

where $X \in \mathbb{R}^m$ is state vector, $U \in \mathbb{R}^k, Y \in \mathbb{R}^n$ are input and output, $W_U \in \mathbb{R}^{n\times k}$, $C \in \mathbb{R}^{n\times m}$, $A(t) \in \mathbb{R}^{m\times m}$, $B \in \mathbb{R}^{m\times k}$.

Assumption 5. $A(t)$ is a continuous Frobenius matrix and limited

$$\|A(t)\| \leq \alpha_A \tag{5.29}$$

where $\alpha_A>0$, $\|\cdot\|$ is norm of a matrix.

Assumption 6. $A(t)$ is almost periodic

$$A_i(t) = A_i(t-\tau_i) \tag{5.30}$$

i.e. the subsequence which is uniformly convergent on all axes to some almost periodic matrix $\overline{A}(t)$ can be chosen from any sequence.

Assumption 7. $A(t)$ is Hurwitz matrix for almost all $t\geq 0$.

Experimental information

$$\mathbb{I}_o = \{Y(t), U(t), t \in \mathbb{J} = [t_0, t_1]\} \tag{5.31}$$

The solution of the system (5.28)

$$X(t) = X(t_0, U, t) \tag{5.32}$$

where χ is the operator determined by matrixes A, B.

Obtain from (5.32) the solution of the system (5.28) at $X_0 = X(t_0)$

$$X(t) = X_g(t) + X_q(t) \tag{5.33}$$

where $X_q(t)$ is a particular solution (5.28) with $U \in I_o$, $X_g(t)$ is a general solution with $U(t)=0$ at the unknown $X_0 \in \mathbb{I}_o$.

Problem: perform the I_o analysis and evaluate the general solution $X_g(t) = X_g(X_q, X_0, t)$, the eigenvalues for the A2 matrix and the order of the system.

The methods described below are based on the results get above in this section.

5.1. Almost Periodic Functions to Bohr

Consider a class of almost periodic functions to Bohr.

Definition 4 (Bohr, 2018). The numerical set $\Xi=\{\xi\}$ is relatively dense on the real axis $-\infty<x<\infty$, if such a number $l>0$ exists that each segment $a\leq x\leq a+l$ of length l contains at least one element of our set, i.e., at any a we have $[a, a+l] \cap \Xi \neq 0$.

Definition 5 (Bohr, 2018). The number $T=T_f(\delta)$ is called almost the function $f(x)$ period accurate within δ (or δ- almost the period or δ-shift) if inequality:

$$|f(x+T_f)-f(x)| < \delta, \quad \delta > 0 \tag{5.34}$$

fairly for any $x\in(-\infty,\infty)$.

Definition 6 (Bohr, 2018). Function $f(x)\in(-\infty,\infty)$ is called almost periodic in Bohr sense (BF-function) if relatively dense almost set T_f of the function $f(x)$ exists accurately within δ, i.e. such positive number $l=l(\delta)$ exists that any segment $[a,a+l]$ contains, at least, one number T_f for which it is fair

$$|f(x+T_f)-f(x)| < \delta \text{ for } x \in (-\infty ,\infty),$$

where δ is any positive number.

$\hat{y}_g(t)$ belongs to exponential-sinusoidal function class. Therefore, the condition (5.34) is not fulfilled. Provide the belonging $\hat{y}_g(t)$ to BF-functions. Perform following operations.

Consider some point $t \in \mathbb{R}$ and its neighborhood O_t. Determine the average value $\hat{y}_g(t)$ for $\forall t \in O_t$

$$\alpha = \bar{\hat{y}}_{g,O_t} = \frac{1}{N_t}\sum_i \hat{y}_i$$

where N_t is the point quantity on O_t, $t_i \in O_t$ is the current coverage of the interval O_t with a step τ.

For $t \in \mathbb{R}$ belonging to the neighborhood $O_{t+T_{\hat{y}_s}}$ obtain:

$$\pi = \bar{\hat{y}}_{g,O_{t+T_{\hat{y}_s}}} = \frac{1}{N_{t+T_{\hat{y}_s}}}\sum_i \hat{y}_i$$

Definition 7 (Karabutov, 2018). The function $\hat{y}_g(t) \in (-\infty, \infty)$ is called almost periodic in Bohr sense ($BF_{\alpha\pi}$-function) if relatively dense almost T_f set of the function $\hat{y}_g(t)$ exists accurately within δ, i.e. such positive number $l=l(\delta)$ exists that any segment $[a,a+l]$ contains, at least, one number T_f for which it is fair:

$$\left|\frac{\hat{y}_g(t+T_f)}{\pi} - \frac{\hat{y}_g(t)}{\alpha}\right| < \delta \text{ for } t \in [0,\infty).$$

5.2. Frameworks for PCS

We apply the approach to evaluating the properties of PCS described in section 4 and consider the characteristics of PCS. For PCS, theorem 5 holds.

Note the local minima location on $SK_{k^i_{s,\rho}}$ coincides with areas D_{sk} of the framework $SK_{\Delta k^i_{s,\rho}}$. The analysis D^i_{sk} gives the set M_{LE} containing LE estimations for the system (5.21). The cardinal number M_{LE} cannot coincide with the number of LE for the system. M_{LE} characterizes the set of lineanes for system (5.21).

Perform the time $\bar{t}$ choice in (5.25) based on framework $SK_{\Delta k^i_{s,\rho}}$ analysis.

Remark 19. The framework $SK_{\Delta k^i_{s,\rho}}$, where $i>1$, can be applied for decision-making on LE.

The proposed approach gives the estimation for the smallest Lyapunov exponent $\eta_i[\hat{y}_g]$. It is the difference between this approach and procedures proposed in the literature. If the framework $SK_{\Delta k^i_{s,\rho}}$ contains a single abrupt change of the value, then it is a sign that $SK_{k^i_{s,\rho}}$ contained the estimation $\eta_m[\bullet]$. As $|\eta_m[\cdot]| > |\chi_i[\cdot]|$, where $i = \overline{1, m-1}$, $\eta_m[\bullet]$ is designated as κ_m and calls the upper bound of the smallest LE. Explain this that the corresponding linear $L^m(t)$ has the minimum definitional domain. Function $\hat{y}_g(t)$ on $L^m(t)$ is not $\alpha\pi$-almost periodic. So, α,π parameters quickly decrease. Therefore, the condition (5.34) is not satisfied, and lineal $L^m(t)$ contains only one value which corresponds to κ_m. So, fairly

Statement 5. Let (i) the system (5.21) satisfy assumptions 5, 6 and 7; (ii) the framework $SK_{\Delta k^i_{s,\rho}}$ contains the point M in which value sharply changes. Then the global minimum corresponds to the point M on the framework $SK_{k^i_{s,\rho}}$, and the global minimum value is the upper bound κ_m for the smallest $\eta_i[\hat{y}_g]$.

Remark 20. As the decision is accepted based on several frameworks $SK_{k^i_{s,\rho}}$ ($i \geq 1$), choose the upper bound from $\kappa_{m,i}$ and designate it as κ_m.

The set M_{LE} is formed based on minimum $SK_{k^i_{s,\rho}}$ analysis and the remark 19.

Consider the definition problem of the domain which belong the set M_{LE}. Find an admissible domain for M_{LE} and the number defining mobility of the highest LE. This domain is limited to the indicator κ_m from below. Determination of the specified parameters is realized on the interval $[0,\bar{t}]$ where $\bar{t}$ is chosen from (5.26). Realize domain choice for χ_1 based on framework $SK_{\Delta k^i_{s,\rho}}$ analysis. The area $D^1_{sk} \subset SK_{\Delta k^i_{s,\rho}}$ is the indicator $_1$ existence. The fragment $V^1_{k^i_{s,\rho}}$ on the framework $SK_{k^i_{s,\rho}}$, which changes to the interval $J^1_{k^i_{s,\rho}}$, corresponds to the area $\dot{D}^1_{sk}$. Therefore, $\chi_1 \in J^1_{k^i_{s,\rho}}$. The admissible mobility boundary of the higher indicator χ_1 is determined as

$$\chi_1 \leq \sup J^1_{k^i_{s,\rho}}. \tag{5.35}$$

The inequality (5.35) gives the admissible boundary (boundary of mobility (Karabutov, 2018) for the change in χ_1 under uncertainty. Mobility areas for χ_1 (i>1) are determined similarly.

Consider criteria for the estimation of set M_{LE} elements and (5.25). The concept of the adequacy accepted in parametrical identification theory is inapplicable in this case. Majority of publications (see chapter 1) are devoted to the calculation of LE. Problem checks questions of obtained estimations not considered. Such theoretical indicators as durability attainability (Bylov et al., 1966) do not give in to validate under uncertainty. Further, we offer the method (Karabutov, 2018) to verify adequacy for obtained LE estimations. The method is based on the analysis of frameworks proposed above.

Consider the framework $S_{\hat{y}_g,\hat{\hat{y}}_g}$ described by the function $f_{\hat{y},\hat{\hat{y}}}:\hat{y}_g \rightarrow \hat{\hat{y}}_g$ in space $R_y = \left(\hat{y}_g, \hat{\hat{y}}_g\right)$. As the system (5.21) satisfies A1-A3, $S_{\hat{y}_g,\hat{\hat{y}}_g}$ contains areas which reflect $\alpha\pi$-almost periodic behavior of the system. As the system (5.21) satisfies assumptions 5, 6 and 7, $S_{\hat{y}_g,\hat{\hat{y}}_g}$ contains areas, which reflect $\alpha\pi$-almost periodic behavior of the system. Consider frameworks $SL_{\Delta k^i_{s,\rho}}$ and $SL_{k^i_{s,\rho}}$ which are described by functions

$$f_{SL_{k^i_{s,\rho}}}:\hat{y} \rightarrow k^i_{s,\rho}, f_{SL_{\Delta k^i_{s,\rho}}}:\hat{y} \rightarrow \Delta k^i_{s,\rho} \tag{5.36}$$

Definition 8. Estimations of Lyapunov exponents χ_i are χ-adequate in space $\mathbb{R}$ if their definition ranges coincide with $\alpha\pi$-almost periodicity areas of the framework $S_{\hat{y}_g,\hat{\hat{y}}_g}$.

Introduce fragments $D^j_{sl} \subset SL_{\Delta k^i_{s,\rho}}$ (j≥1) by analogy with $D_{sk} \subset SK_{\Delta k^i_{s,\rho}}$ in the space R_y. Designate the definitional domain D^j_{sl} as dom D^j_{sl}.

Theorem 7. Consider fragment D^j_{sl} definitional domains of the framework $SL_{\Delta k^i_{s,\rho}}$. If fragment D^j_{sl} coincides with $\alpha\pi$-almost periodicity areas of the framework $S_{\hat{y}_g,\hat{\hat{y}}_g}$, then estimations χ_i are χ-adequate to areas $\alpha\pi$-almost periodicity for $S_{\hat{y}_g,\hat{\hat{y}}_g}$.

Proof of the theorem 7. Function $f_{\hat{y},\hat{\hat{y}}}$ contains not fewer m areas D^j_{sk} according to the theorem 5. Frameworks $SK_{\Delta k^i_{s,\rho}}$ and $SL_{\Delta k^i_{s,\rho}}$ have an identical range of values. This statement is correct also for D^j_{sk}, D^j_{sl}. D^j_{sk}, D^j_{sl} set by LE area change. Congruence of value ranges D^j_{sk}, D^j_{sl} follows from equality of definition ranges D^j_{sk}, D^j_{sl}. If the definition range for fragment $D^j_{sl} \subset SL_{\Delta k^i_{s,\rho}}$ and region $\alpha\pi$-almost periodicity $S_{\hat{y},\hat{\hat{y}}}$ coincide, then at the existence of the dependence between functions $f_{SL_{k^i_{s,\rho}}}$ and

$f_{SL_{\Delta k_\varphi}}$, we get some element of set M_{LE} has corresponded to $\alpha\pi$-almost periodic region $S_{\hat{y}_g,\hat{\dot{y}}_g}$. Therefore, estimations χ_i are -adequate in space R_y. ■

5.3. Examples

Consider the system (5.28). The phase portrait is showed in Figure 18. Input is $u(t)$= 5+2sin(0.2πt). We see that the system has fluctuations.

Figure 18. Phase portrait

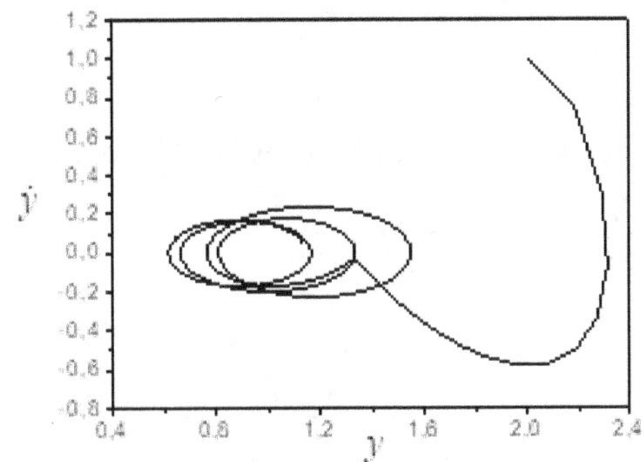

Apply the method described in section 4.2 and obtain of the general solution for $y(t)$. The model (5.24) has the form

$$\hat{y}_q(t)=\hat{A}_q^T[1\ u(t)\ \dot{u}(t)]^T,\ \hat{A}_q = [0,302;0,189;-0,203]^T.$$

The determination coefficient of the model (21) is 0.99.

Construct the phase portrait (PP) in the space R_y for the check of system (5.21) belong to PCS. PP is showed in **Figure** 19.

Figure 19. System phase portrait in space R_γ

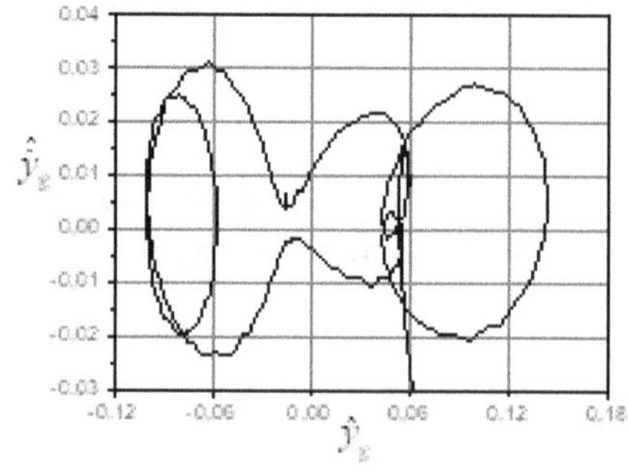

LE identification results are shown in Figures 20. Change of frameworks $SK_{\Delta k'_{s,\rho}}$ and $S_{k^1_{s,\rho}}$ is presented in Figure 20.

Figure 20. LE set

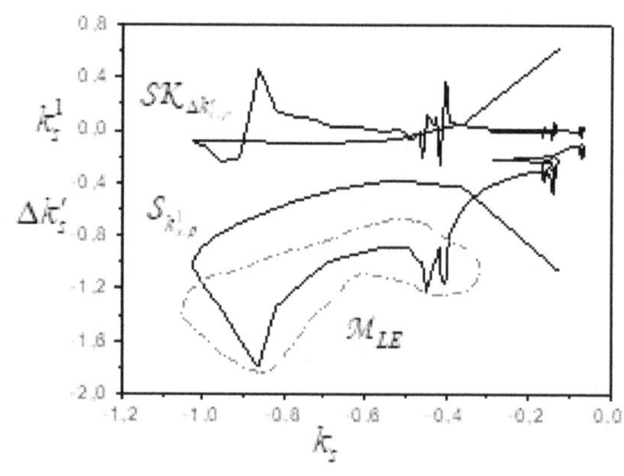

We determine by LE set: $M_{LE} = \{-1.8, [-1.21, -0.88]\}$ based on the analysis $SK_{\Delta k'_{s,\rho}}$. Apply the theorem 5 and get that the system order is 2. The upper estimation for the smallest LE is κ_m = -1.8. The mobility admissible limit of the higher indicator χ_1 is -0.8. We have another LE set on the interval $k_s \in$[-0.2, -0.13].

χ-adequacy estimation results of the LE set are presented in Figures 21 and 22. In Figure 21, the function f_{sk} regions D^i_{sk} coincide with $\alpha\pi$-almost periodicity regions of the framework $S_{k^1_{s,\rho}}$ that χ-adequacy of LE estimations confirms, and $R_\Delta = (\hat{y}_g, \Delta k'_{s,\rho})$.

Figure 21. χ-adequacy estimation results in spaces R_y and R_{Δ}

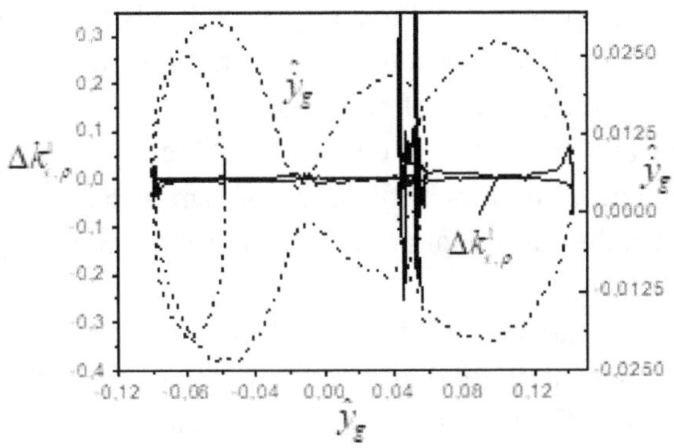

Figure 22. χ-adequacy estimation results in spaces R_y and R_{k1}

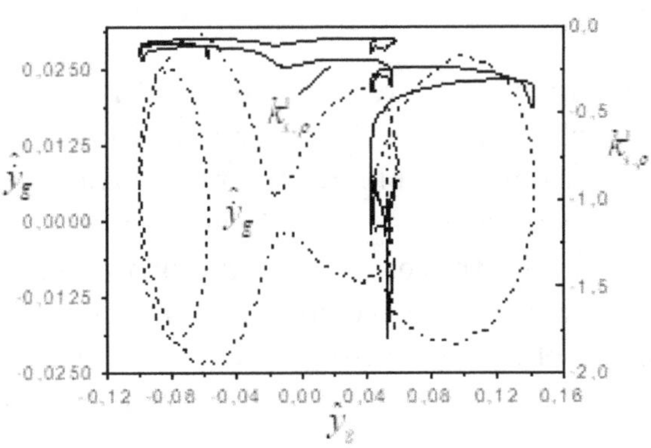

Figure 22 represents χ-adequacy results of LE estimations in spaces R_y and $R_{k^1} = \left(\hat{y}_g, k^1_{s,\rho}\right)$. They are correlated with results presented in Figure 21.

Karabutov (2018) considers other approaches to LE estimation for PCS.

6. STRUCTURAL METHODS FOR NONLINEAR STATIC SYSTEMS

Ideas and the methods described in Section 1 and 3 have been verified for the first time on static systems (SS) (Karabutov, 2011). SS have the specificity. They do not have an equilibrium state. Therefore, the framework S_{ey} is ineffective in these conditions. Often, inputs SS are irregular (random). Therefore, it is difficult to decide on S_{ey}. Development is obtained by SI methods based on the analysis of S_{ek}-frameworks. Data ordering procedure in SS on the coefficient of structural properties applies to effect elimination of an irregularity and not measured noises. We have considered the above value S_{ek} in decision-making. S_{ek} gives to the solution for the estimation problem of nonlinearity structural parameters (Karabutov, 2012b). The secant method is the basis of analysis S_{ek}. The Set of secants specifies the field of frameworks SS.

Major advantage of the secant method is decision-making on a class of linear functions (secants). Secant method gives to the problem identification solution of parametric restrictions SS under uncertainty (Karabutov, 2018). The special class of frameworks, specified in the space between system outputs and model outputs, is applied to the estimation of parametric restrictions. The secant method applies for estimation of the linearity SS.

7. CONCLUSION

We give to generalization and development of the SIA-approach on NDS structural identification problems. The basis SIA-approach is GF S_{ey}, reflecting properties of the NDS. We consider methods of formation of an auxiliary informational set. This set is a basis for construction S_{ey}. Models are proposed for the formation of the geometric framework S_{ey}. The construction problem with such models is actual. The model choice defines properties and structure of the nonlinear dynamic system. We consider examples of the model structure choice obtained in the analysis of system properties. As there is the NDS set, a general approach does not exist in the construction of such models.

GF is the base for the design of structural identifiability methods for NDS. Structural identifiability defines a possibility of the framework design for nonlinearity. We propose criteria for the structural identifiability estimation. These criteria determine demands for the form S_{ey}. Emersion conditions of no significant frameworks S_{ey}, which properties of structural identifiability it not have, are considered. We consider emersion of no significant frameworks S_{ey}, which properties of the structural identifiability it not has.

We propose criteria for decision-making about a class of functions describing nonlinearity in the dynamic system. Classes of single-valued and multiple-valued nonlinearities are considered. Algorithms and methods are proposed for the structure estimation of single-valued and multiple-valued nonlinearities. The check of obtained structural solutions is fulfilled based on the method of the structurally-frequency analysis. The method applies under uncertainty. We apply also frameworks S_{ek}. S_{ek} give to the solution of such problems as the choice of the nonlinearity form and its structural parameters.

We show geometric frameworks apply for structure estimation solutions and problems of linear dynamic systems. The application of GF for the LDS analyzes is shown. Identification of the order and spectrum eigenvalues for the system is possible using GF. Geometric frameworks describe LE change. Geometric frameworks describe LE change and apply for the system structure estimation. We use frequency charts (FC) for the spectrum check of eigenvalues for the system. FC is a special case of structurally-frequency analysis.

We generalize proposed approaches to static nonlinear systems (SNS). GF features are noted for SNS. We show that the analysis of S_{ek}-frameworks plays a major role. S_{ek} give the structural identification problem solution in a special space.

So, the GF method allows solving various problems of structural identification. The available a priori and experimental information are used to develop GF construction. Geometric frameworks of construction are a complex problem under uncertainty. The GF form defines system structural features in the identification problem. GF correctness is verified based on the analysis of the geometrical image.

REFERENCES

Acho, L. (2013). Hysteresis modelling and synchronization of a class of RC-OTA hysteretic-jounce-chaotic oscillators. *Universal Journal of Applied Mathematics, 1*(2), 82-85 14 41.

Bohr, H. (2018). *Almost Periodic Functions*. Dover Publications, Incorporated.

Bozhokin, S. V., & Parshin, D. A. (2001). *Fractals and Multifractals*. Moscow-Izhevsk: Scientific Publishing Centre \. *Regular and Chaotic Dynamics*.

Bylov, B. F., Vinograd, R. E., Grobman, D. M., & Nemytsky, V. V. (1966). *Theory of Lyapunov indexes and its application to stability problems*. Nauka.

Choquet, G. (1964). *L' enseignement de la geométrie*. Hermann.

Feder, J. (1988). *Fractals*. Plenum Press.

Izobov, N. A. (2006). *Introduction in theory of Lyapunov indexes*. BGU.

Karabutov, N. (2012a). About structures of state systems identification of static object with hysteresis. *International journal of sensing, computing & control, 2*(2), 59-69.

Karabutov, N. (2012b). Structural identification of static processes with hysteresis nonlinearities in civil engineering. *Journal of civil engineering and science, 1*(4), 22-29.

Karabutov, N. (2015a). Structural identification of nonlinear dynamic systems. *International Journal of Intelligent Systems and Applications*, *7*(9), 1–11.

Karabutov, N. (2015b). Structural methods of estimation Lyapunov exponents linear dynamic system. *International Journal of Intelligent Systems and Applications*, *7*(10), 1–11.

Karabutov, N. (2016). Structural Identification of dynamic systems with hysteresis. *International Journal of Intelligent Systems and Applications*, *8*(7), 1–13.

Karabutov, N. (2017). Frameworks in Problems of Structural Identification Systems. *International Journal of Intelligent Systems and Applications*, *9*(1), 1–19.

Karabutov, N. (2018). About Lyapunov exponents identification for systems with periodic coefficients. *International Journal of Intelligent Systems and Applications*, *11*, 1–10.

Karabutov, N. N. (2006). *Adaptive identification of systems: Informational synthesis*. URSS.

Karabutov, N. N. (2011). *Structural identification of static objects: Fields, structures, methods*. Librokom.

Karabutov, N. N. (2013). Structural identification of nonlinear static system on basis of analysis sector sets. *International Journal of Intelligent Systems and Applications*, *6*(1), 1–10.

Karabutov, N. N. (2009). *Structural Identification of Systems. The Analysis of Information Structures*. URSS/ Librokom.

Mosteller, F., & Tukey, J. (1977). *Data analysis and regression: A second course in statistics*. Addison-Wesley Publishing Company.

Chapter 6 Identification of Parametric Restrictions Under Uncertainty

ABSTRACT

In this chapter, we present estimate methods of area parametric constraints (APR). They are based on the dominance indicator analysis of the model output over the system output and a special parameter that sets the acceptable level of dominance errors. Various cases of APR representation are considered. We propose an adaptive algorithm for estimating the upper bound for APR. Various vector norms and corresponding algorithms for constructing the area of parametric constraints are proposed. We develop geometric frameworks and apply them to decide about APR. We introduce a field of secants for geometric frameworks and decide on APR.

1. INTRODUCTION

Parametric restrictions (PR) are a basis for designing of effective identification and control systems. In systems of identification, the consider PR reflects actual conditions and to realize robust algorithms of a parameter estimation.

Various publications are devoted to the application of PR. The fixed asset price model identification (Gagliardini and Gourieroux, 2011) is based on the method of moments (MM). Two types of restrictions imposed at the moment are considered. These are restrictions on variable current value or on maximum value for a variable on some interval. Restriction on the moment is set a priori. MM is applied for a correlation define between PR and stability of parameters (Giacomini and Rossi, 2009; Magnusson and Mavroeidis, 2014; Rossi, 2005). Restriction is set as equality for the moment. Restrictions are set a priori. Bardsley (2004) considers the estimation

DOI: 10.4018/979-8-3373-0822-7.ch006

of electromagnetism parameters under physical restrictions on parameters. Iteration algorithm Levenburg-Marquardt applies to its solution. The area PR (APR) is set a priori as the lower and upper boundaries for parameters. The correction problem of set boundaries for restrictions is considered. PR considers as a parametric ellipse (Palanthandalam-Madapusia et al., 2010; Van Pelt and Bernstein, 2001). The heuristic algorithm for a solution to an issue PR is offered. With finite data, it was shown (Palanthandalam-Madapusia et al., 2010) that there is some indefinite quadratic constraint, depending on the noise. Restrictions are set a priori the identification of a static object with a priori set area APR is analyzed by Corrêa et al. (2002). A priori set APR (Chia et al., 1991) considers in the construction problem of auto-regression. It is noticed that ignoring PR effects on results of the identification. Case of correction of restrictions is considered. The synthesis method of identification algorithm (Wen-Ming, 2010) is proposed for a dynamic object with a priori PR as equality. The improvement of the model properties for a closed system based on a priori constraint is given by Vanli and Del Castillo (2007). It is shown how to generate restrictions for the object. A priori knowledge about the object (Hametner and Jakubek, 2011) is the basis for the PR formation for a nonlinear system. The estimate problem of system parameters based on consider square constraints is studied by Mead and Renaut (2010). Physical knowledge and a regularization method are used for construction of restrictions. The estimation PR issue (Mazunin and Dvojnikov, 2010) is considered for the damping of oscillations in nonlinear control systems of mechanisms with elasticity. The system model is set a priori. Some approaches to obtaining PR are proposed by Karabutov (2006, 2009a). The estimate for APR Ω_A, given as the inequality for the parameter vector, is based on the majority model.

So, we see PR is often used at the synthesis of identification systems. Usually, the APR is formed a priori. Algorithms for Ω_A correction have a heuristic character. The construction Ω_A problem was not considered under uncertainty. There is a class of complex systems for which APR estimation is a priori unfeasible. Therefore, the solution to the APR problem for this class of systems is actual.

The approach to Ω_A construction is offered for static systems under uncertainty by Karabutov (2009a). Various Ω_A description methods are considered. They are based on the domination concept and the analysis of special structures for linear static systems (LSS), introduced by Karabutov (2009a). Next, we summarize the PR identification methods proposed by Karabutov (2011) for static systems. Adaptive algorithms of construction APR are offered. Construction Ω_A methods based on the analysis of a field of secants for LSS are described. Multiple-geometrical methods based on the analysis of an observable informational portrait of the system are considered. The multiple-geometrical methods based on the analysis of a system observable informational portrait (OIP) are considered.

2. INTERVAL FORM ESTIMATION Ω_A

2.1. Problem Statement

Consider system

$$y_n = A_n^T U_n + \xi_n \tag{6.1}$$

where $A_n \in \Omega_A \subset \mathbb{R}^k$ is a vector of parameters belonging to the limited area:

$$\Omega_A = \{A_n \in \mathbb{R}^k : \varphi_{b,i} \leq a_{i,n} \leq \varphi_{t,i}, (\varphi_{b,i}, \varphi_{t,i}) < \infty, a_{i,n} \in A_n\} \tag{6.2}$$

with unknown parameters $\varphi_{b,i} \in \Phi_b \in \mathbb{R}^k$, $\varphi_{t,i} \in \Phi_t \in \mathbb{R}^k$, $n \in \mathbb{J}_N = [0, N]$ is discrete time, $N < \infty$, $U_n \subset \mathbb{R}^k$ is a vector of an input with $u_{i,n} \in U_n$, $\xi_n \in \mathbb{R}$ is the limited perturbation.

Information $\mathbb{I}_o = \mathbb{I}_o(y_n, U_n)$ is known and OIP $\Gamma_0 : \{U_n\} \times \{y_n\}$ is constructed for (6.1). Introduce vector $\Phi_t = \mathrm{col}(\varphi_{t,1}, \varphi_{t,2}, \ldots, \varphi_{t,k})$ which majority vector A_n in (6.1). Consider the special case Ω_A corresponding to an upper bound PR

$$\Omega_{A,t} = \{A_n \in \mathbb{R}^k : a_{i,n} \leq \varphi_{t,i}, \ \varphi_{t,i} < \infty, \ a_{i,n} \in A_n\} \tag{6.2}$$

Apply mathematical model to estimation $\Phi_t \in \Omega_{A,t}$

$$\hat{y}_{t,n} = \widehat{\Phi}_t^T U_n \tag{6.3}$$

Problem: obtain estimations for vector Φ_t in (6.2) based on the analysis $\mathbb{I}_o$, mapping Γ_0, and the model (6.3) such that

$\hat{y}_{t,n} > y_n$ for almost $\forall n \geq 0$ (6.4)

We understand inequality $y_n \leq \hat{y}_{t,n}$ as domination for almost $\forall n \in \mathbb{J}_N$.

2.2. Method of Solution

It is possible to apply some approaches to the problem solution. State the approach (Karabutov, 2009b) based on the check of a domination condition, and use secondary criterion as average value of study variables.

Consider projections of OIP Γ_0 to planes (u_i, y) and construct secants $\bar{\gamma}_y(u_i) \in \mathbb{R}$ using the least mean square.

$$\bar{\gamma}_y(u_i) = \bar{a}_i u_i + \bar{b}_i, i = \overline{1,k} \tag{6.5}$$

with areas of definition and values: $\mathrm{dom}\left(\overline{\gamma}_y(u_i)\right) \in \mathbb{J}_{u_i}$, $\mathrm{rng}\left(\overline{\gamma}_y(u_i)\right) \in \mathbb{J}_y$, where $J_{u_i} \in \mathbb{R}$, $\mathbb{J}_y \in \mathbb{R}$ are intervals of a modification for variables u_i, y; $\overline{a}_i, \overline{b}_i$ are some numbers.

Let $\widehat{\varphi}_{t,i,0} = \overline{a}_i$. Then we have the approached estimation for area $\Omega_{A,t}$:

$$\Omega^0_{A,t} = \left\{ A_n \in \mathbf{R}^k \left| A_n \leq \widehat{\Phi}_{t,0} \right. \right\}$$

with $\widehat{\Phi}_{t,0} = [\widehat{\varphi}_{t,1,0}\ \widehat{\varphi}_{t,2,0} \cdots \widehat{\varphi}_{t,k,0}]^T$.

We understand inequality $A \leq \widehat{\Phi}_{t,0}$ as single-item. The index 0 is an elementary level of the correction $\widehat{\Phi}_t$ for the model (6.3) and does not coincide with the step of the interval change $\mathbb{J}_N$.

Apply model (6.3) with $\widehat{\Phi}_{t,0}$, define prediction $\widehat{y}_{t,0,n}$ of an output y_n and coefficient of determination $r^2_{\widehat{y}_{t,0}y}$.

Check a condition of domination (DC) $\widehat{y}_{t,0,n} \succ y_n$. If DC is fulfilled for almost all n, suppose $\Omega_{A,t} = \Omega_{A,t,0}$, and process of definition APR finish. Otherwise, we apply of correction $\Omega_{A,t,0}$ algorithm for condition (6.4).

2.3. Correction $\Omega_{A,t,0}$ Algorithm

Calculate average values

$$\overline{y} = \frac{1}{N}\sum_{n=0}^{N} y_n, \overline{\widehat{y}}_{t,0} = \frac{1}{N}\sum_{n=0}^{N} \widehat{y}_{t,0,n}$$

and define variable

$$\overline{\varepsilon}_{t,0} = \overline{y} - \overline{\widehat{y}}_{t,0} \exists \widehat{\Phi}_{t,0} \tag{6.6}$$

Lemma (Karabutov, 2011). $\overline{\varepsilon}_{t,0} \neq 0$.

$\overline{\varepsilon}_{t,0}$ shows how much variable $\widehat{y}_{t,0,n}$ does not come nearer to domination area over y_n. If $\overline{\varepsilon}_{t,0} \leq 0$, then $\widehat{y}_{t,0} \succ y$ almost $\forall n \in \mathbb{J}_N$. Generate the vector of corrections for $\widehat{\Phi}_{t,0}$ based on $\overline{\varepsilon}_{t,0}$. Define magnitudes:

$$\overline{\varphi}^{\varepsilon}_{t,i,n} = \frac{\overline{\varepsilon}_{t,0}}{u_{i,n}} \forall n \in \mathbb{J}_N \tag{6.7}$$

and calculate their average values $\overline{\varphi}^{\varepsilon}_{t,i} = \overline{\varphi}^{\varepsilon}_{t,i}\left(\overline{\varphi}^{\varepsilon}_{t,i,n}\right)$. Generate a vector of corrections

$$\Delta \Phi_{t,0} = \left[\overline{\varphi}^{\varepsilon}_{t,1}\ \overline{\varphi}^{\varepsilon}_{t,2} \cdots \overline{\varphi}^{\varepsilon}_{t,k}\right]^T \tag{6.8}$$

Algorithm for vector $\widehat{\Phi}_{t,0}$ correction

$$\widehat{\Phi}_{t,1} = \widehat{\Phi}_{t,0} + \Gamma_t \Delta \Phi_{t,0} \tag{6.9}$$

where $\Gamma_t \in \mathbb{R}^{k \times k}$ is a diagonal matrix with $\gamma_{t,ii} > 0$. Apply model (6.3) and find prediction $\hat{y}_{1,n}$ for $y_n \ \forall n \in \mathbb{J}_N$. Determine $\overline{\hat{y}}_{t,1}$ and $r_{\hat{y}_{t,1}y}$. If $\overline{\hat{y}}_{t,1} \geq \overline{y}$ and

$$\left| r_{y\hat{y}_{t,1}} - r_{y\hat{y}_{t,0}} \right| \leq \Delta_r, 0 < \Delta_r < \infty$$

then suppose $\Omega_{A,t} = \Omega_{A,t}(\widehat{\Phi}_{t,1})$ and process of construction APR finish. Otherwise, adaptation of vector $\widehat{\Phi}_t$ continues and applies algorithm (6.9). Here Δ_r is some set magnitude.

Let the initial value of vector $\widehat{\Phi}_t$ be defined based on secants (6.5), and its correction in the equation (6.3) is realized according to (6.8), (6.9)

Statement 1. Let the initial value of vector $\widehat{\Phi}_t$ is defined based on secants (6.5), and its correction in the equation (6.3) is realize according to (6.8), (6.9). Then the estimation of APR:

$$\Omega_{A,t} = \left\{ A_n \in \mathbb{R}^k : a_{i,n} \leq \varphi_{t,i}, \varphi_{t,i} < \infty, a_{i,n} \in A_n \right\} \tag{6.10}$$

is valid if the condition of domination $\hat{y}_{t,1,n} \succ y_n$ is satisfied for almost $\forall n \in \mathbb{J}_n$, and

$$r_{\hat{y}_{t,1}y} \in \Omega_r = \left\{ r_{\hat{y}_{t,1}y} \in \Omega_r : \left| r_{\hat{y}_{t,1}y} - r_{\hat{y}_{t,0}y} \right| \leq \Delta_r, \ 0 < \Delta_r < \infty \right\} \tag{6.11}$$

The condition (6.11) superimposes restriction on a class of models (6.3) under uncertainty. Value $r_{\hat{y}_{t,0}y}$ is the area Ω_r center.

Remark 1. In (6.7), the current (at time $n \in \mathbb{J}_N$) offset value $\varepsilon_{t,0,n} = y_n - \hat{y}_{t,0,n}$ can be used instead $\overline{\varepsilon}_{t,0}$.

Remark 2. If the system is stochastic, then decision-making requires setting an acceptable level of dominance. (see indicator $P_{\succ\hat{y}}$).

Algorithm tuning of vector $\widehat{\Phi}_{t,m}$

$$\widehat{\Phi}_{t,m+1} = \widehat{\Phi}_{t,m} + \Gamma_t \Delta \Phi_{t,m}, \tag{6.12}$$

where $\Delta \widehat{\Phi}_{t,m} = \Delta \widehat{\Phi}_{t,m}(n)$ is formed based on application to the procedure stated above.

Evaluate the quality of the PR identification system. Introduce quantitative criterion. Consider that $\hat{y}_{t,n} > y_n$ almost on all interval $\mathbb{J}_N$, if probability:

$$P_{>\hat{y}_t} = P(\hat{y}_{t,n} \geq y_n) \geq p_>$$

$$\forall n \in \mathbb{J}_N, \tag{6.13}$$

where $p_> > 0$ is the magnitude.

Definition 1. Vector$\widehat{\Phi}_{t,m} \in \mathbb{R}^k$ is optimal if gives an acceptable estimate $\Omega_{A,t}$ and provides a maximum of the indicator$P_{>\hat{y}_t}$.

Statement 2. The model (6.3) output with vector $\widehat{\Phi}_{t,m} \in \mathbb{R}^k$ corresponding to an optimum majority estimation for area $\Omega_{A,t}\left(\widehat{\Phi}_{t,m}\right)$, has a maximum informational performance.

Proof. Consider acceptable outputs of model (6.3) for which conditions of the statement 1 are satisfied, and the domination indicator satisfies (6.13). Therefore, we believe the majority estimate for $\Omega_{A,t}$ is optimal if

$$\max_{k \in \mathbb{N}_m} P\left(\hat{y}_{t,k,n} \geq y_n\right) = P_{>\hat{y}_t}(k_*) \forall n \in \mathrm{J}_N \tag{6.14}$$

where $\mathbb{N}_m \subseteq \mathbb{Z} \subset \mathbb{R}$ is a set of integers for which it is fulfilled (6.13), $k_* \in \mathbb{N}_m$ is the iteration step on $\widehat{\Phi}_{t,m}$ satisfying (6.14). Fulfilment (6.14) corresponds to such output of model (6.3) for which we receive a maximum value of an informational performance (Karabutov, 2009a)

$$\Xi_y = \frac{1}{N}\sum_{i=1}^{N}\hat{y}_i^2. \blacksquare$$

Algorithm (6.12) refers to a class finite-converging.

Theorem 1 (Karabutov, 2013). *The system of identification (6.1), (6.3) and (6.12) with an acceptable vector of parameters $\widehat{\Phi}_{t,m} \in \Omega_{A,t}$ has the limited trajectories, if for matrix $\Gamma_t \in \mathbb{R}^{k \times k}$ in algorithm (6.12) the inequality is fulfilled*

$$\frac{\lambda_{t,\max}^2\left(\Gamma_t\right)}{\lambda_{t,\min}\left(\Gamma_t\right)} \leq \frac{2\left(\delta A_{t,m,n}\right)^T \Delta \Phi_{t,m}}{\left\|\Delta \Phi_{t,m}\right\|^2};$$

(ii) exists acceptable vector. Then the system (6.1), (6.3) and (6.12) has limited trajectories, where $\lambda_{t,\min}\left(\Gamma_t\right)$, $\lambda_{t,\max}\left(\Gamma_t\right)$ are minimum and maximum eigenvalues of matrix Γ_t, $\delta A_{t,m} = A_n - \widehat{\Phi}_{t,m}$.

Optimal value of matrix Γ_t

$$\Gamma_t = D(\delta A_{t,m,n}) D^{-1}(\Delta \Phi_{t,m})$$

where D is a vector $\delta A_{t,m,n}$ transformation operator in a diagonal matrix.

Similarly, define the lower boundary for area Ω_A (6.2). Here, the domination (6.4) condition writes as $\hat{y}_{b,n} \prec y_n$ for almost $\forall n \geq 0$, where $\hat{y}_{b,n} \in \mathbb{R}$ is the model output

$$\hat{y}_{b,m,n} = \widehat{\Phi}^T_{b,m} U_n \tag{6.15}$$

with vector of parameters $\widehat{\Phi}_{b,m} \in \mathbb{R}^k$. Algorithm of tuning $\widehat{\Phi}_{b,m}$

$$\widehat{\Phi}_{b,m+1} = \widehat{\Phi}_{b,m} - \Gamma_b \Delta \widehat{\Phi}_{b,m} \tag{6.16}$$

Misalignment on receives from (6.7) and (6.8) considering remarks 1, 2. In (6.15) $\Gamma_b \in \mathbb{R}^{k \times k}$ is the diagonal matrix with $\gamma_{b,ii} > 0$. Properties of algorithm (6.16) receive from an analogue of the theorem 1.

2.4. Adaptive Algorithms for Estimation of Parameters $\Omega_{A,t}$

We propose algorithm for tuning of vector $\widehat{\Phi}_{t,n}$ in (6.3)

$$\widehat{\Phi}_{t,n+1} = \begin{cases} \widehat{\Phi}_{t,n} - \gamma(\hat{y}_n - y_n) U_n, if\ (\hat{y}_n \prec y_n) \vee (r_{\hat{y}y} \notin \Omega_r), \\ \widehat{\Phi}_{t,n}, \quad if (\hat{y}_n \succ y_n) \wedge (r_{\hat{y}y} \in \Omega_r), \end{cases} \tag{6.17}$$

where $\gamma \geq 0$ is the parameter ensuring algorithm convergence.

Now specify the condition of a belonging to the model (6.3) output area

$$\Omega_{>} = \{ y_n \in \mathbb{R}, \hat{y}_n \in \mathbb{R} : \hat{y}_n \succ y_n \text{ almost for } \forall n \in J_N \}.$$

Apply indicator $P_{>\hat{y}_t}$ (6.13) to decision-making on algorithm (6.17) work quality.

The algorithm (6.17) belongs to the class for finite-converging, and (6.13) is the decision-making condition for estimate $\Omega_{A,t}$.

Theorem 2. *Let (i) vector $U_n \in \mathbb{R}^k$ in (6.1) satisfy the constant excitation condition, (ii) $\widehat{\Phi}_{t,0}$ in (6.17) obtain based on secants. Then the algorithm (6.17) allows receiving the vector $\widehat{\Phi}_{t,n} \in \Omega_{A,t}$ admissible estimation, ensuring performance of target inequality $\hat{y}_n \geq y_n$ if:*

$$0 < \gamma \leq \frac{2}{\|U_n\|^2}$$

and the relative number of algorithm errors on interval J_N *is*

$$m \leq \frac{\left\|\delta \Phi_{t,0}\right\|^2}{\delta_y^2}\vartheta^2$$

where $\vartheta^2 = \max_n \|U_n\|^2, \delta\Phi_{t,0} = \widehat{\Phi}_{t,0} - \Phi_t$.

Proof. Write the algorithm (6.17) as

$$\delta\Phi_{t,n} = \delta\Phi_{t,n-1} - \gamma(\hat{y}_n - y_n)U_n,$$

where $\delta\Phi_{t,n} = \widehat{\Phi}_{t,n} - \Phi_t$.

Consider Lyapunov function $V_{\Phi,n} = \|\delta\Phi_{t,n}\|^2$. Receive for $\Delta V_{\Phi,n} = V_{\Phi,n} - V_{\Phi,n-1}$

$$\Delta V_{\Phi,n} = -2\gamma e_n \delta\Phi_{t,n-1}^T U_n + \gamma^2 e_n^2 \|U_n\|^2$$

where $e_n = \hat{y}_n - y_n$.

The algorithm (6.22) converges at

$$0 < \gamma \leq \frac{2}{\vartheta^2}$$

where $\vartheta^2 = \max_n \|U_n\|^2$.

Optimal value γ defines from condition $\partial\Delta V_{\Phi,n}/\partial\gamma = 0$: $\gamma_{opt} = \vartheta^{-2}$. At such γ_{opt}, when domination (6.4) ($|e_n| \geq c$) condition performances, receive:

$$\Delta V_{\Phi,n} \leq -\frac{e_n^2}{\vartheta^2} \leq -\frac{c^2}{\vartheta^2}, c > 0.$$

The number of algorithm (6.17) errors in the final interval J_N is

$$m \leq \frac{\delta\Phi_{t,0}^2}{c^2}\vartheta^2. \blacksquare$$

Some modifications of algorithm (6.17) are also possible. In particular, it is possible to demand performance of δ-domination condition (Karabutov, 2009a)

$$\hat{y} \underset{\delta_y}{\succ} y \Leftrightarrow (\hat{y} - y) \geq \delta_y, \delta_y > 0 \tag{6.18}$$

Notice that the condition (6.18) toughens requirements to algorithm (6.17). A payment for it is the value Δ_r magnification in (6.11). There is a limit δ_y above which the δ-domination condition is not realized.

2.5. Example

Consider system (6.1) with $A = [0.3\ -0.8\ 1]^T$ and additive limited perturbation $\xi_n \in \mathbb{R}, |\xi_n| \le 0.3$. Elements $U \in \mathbb{R}^3$ are the random functions having final expectation and a variance. It is necessary to estimate area PR (6.2).

We show vector Φ_t identification results (Figure 1) for $\Omega_{A,t}$ applying system (6.3), (6.7) and (6.12). Here, the vector A_n for system (6.2) is also shown. Estimations of the least method squares (LMS) for vector A: $\widehat{A}_n = [0.38\ -0.7\ 0.96]^T$. Vectors $\widehat{\Phi}_{t,1}$ and $\widehat{\Phi}^*_{t,1}$ define by algorithm (6.12) with $\Gamma_t = 0.3 I_3$ and $\Gamma_t = I_3$, where $I_3 \in \mathbb{R}^{3\times3}$ is identity matrix. Values $r_{\hat{y}y}$ are: $r_{\hat{y}_{t,0}y} = 0.94$, $r_{\hat{y}_{t,1}y} = 0.935$, $r^*_{\hat{y}_{t,1}y} = 0.939$.

According to the statement 1, received estimations are admissible. Final indicators for obtained estimates are equal (Figure 2): $P_{>\hat{y}_{t,0}} = 8\%$, $P_{>\hat{y}_{t,1}} = 55\%$, $P^*_{>\hat{y}_{t,1}} = 100\%$. The same results are given by an informational portrait (Figure 3) in space $(y_n, \hat{y}_{t,n})$. The lower boundary for Ω_A is $\widehat{\Phi}_b = [0.36\ -0.81\ 0.915]^T$.

Figure 1. Vector Φ estimation

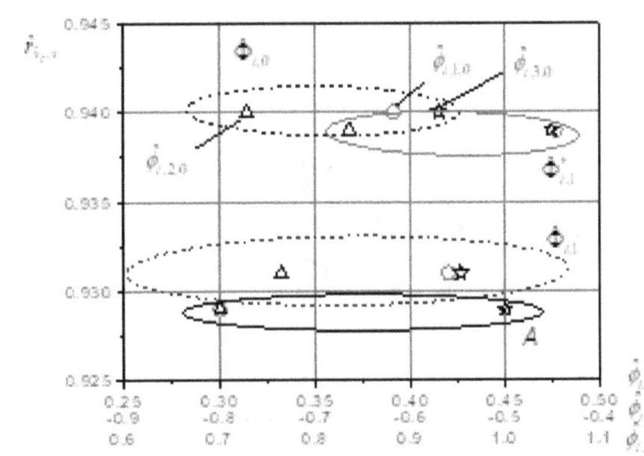

Figure 2. Variable $\hat{y}$ domination at various values of vector $\widehat{\Phi}_t$

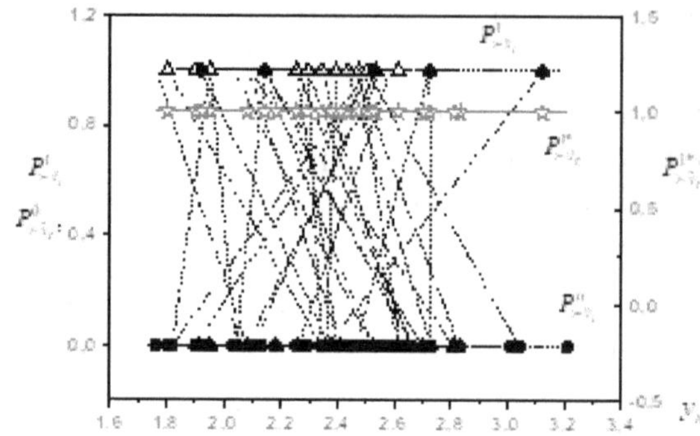

Figure 3. Output of model (6.3) in space $(y_n, \hat{y}_{t,n})$

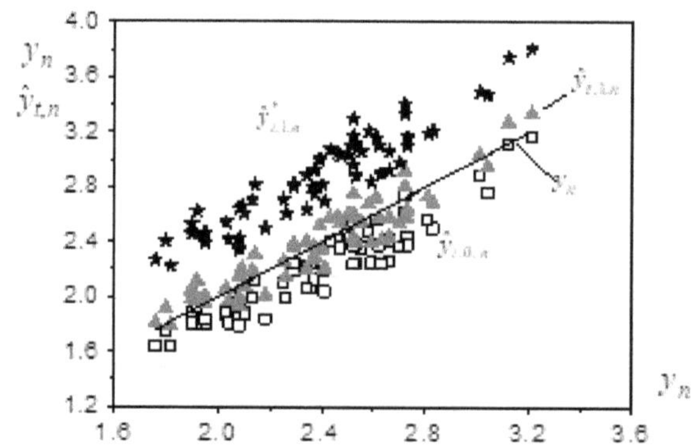

Results of the algorithm (6.17) work are shown in Figure 4 when estimating the area $\Omega_{A,t}$ of the object (6.1) with $A = [0.3 \;\; -0.8 \;\; 1]^T$. Initial estimation $\widehat{\Phi}_{t,0}$ for vector Φ_t in (6.3) is $[0.39 \;\; -0.77 \; 0.93]^T$ received on the analysis basis of the field of secants. $P_{>\hat{y}} = 0.77$, and $p_{>} = 0.6$ for considered case. Values of cross-correlation coefficients $r_{y\hat{y}_0} = 0.94$ and $r_{y\hat{y}} = 0.939$ are confirmed that the received value for Φ_t is admissible, and $r_{y\hat{y}} \in \Omega_r$. Tuning the model (6.3) parameters using the algorithm (6.15) is shown in Figure 5.

In Figure 5, level lines are shown for $r_{\hat{y}y}$ the line with a circle reflects the modification $r_{\hat{y}y}$, and the line with quadrant is the parameter $\widehat{\varphi}_2$ modification.

Figure 4. Results of adaptive algorithm (6.17) work

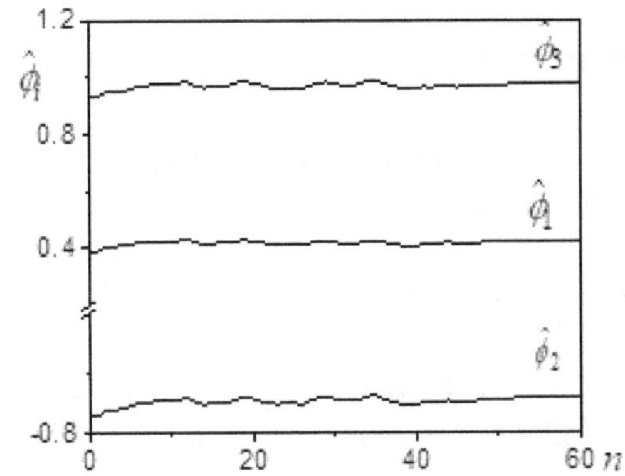

Figure 5. Results of algorithm (6.17) work in space $\left(\hat{\varphi}_1, \hat{\varphi}_2, r_{y\hat{y}}\right)$

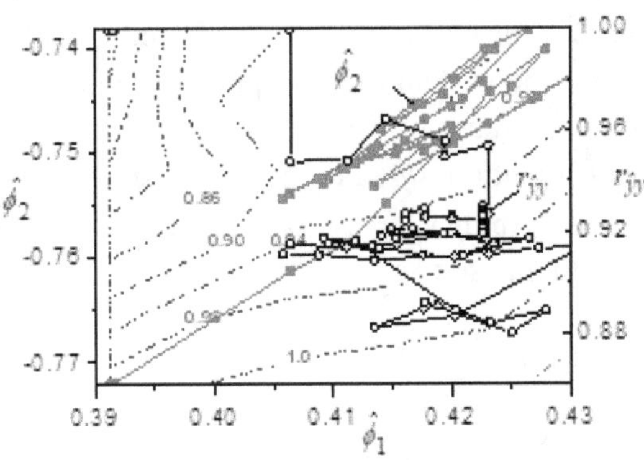

As noted in section 2.D, there is a limit δ_y above which the δ-dominance condition is not satisfied. The indicator is $P_{\succ_\delta \hat{y}} = 0.8$ for the algorithm (6.17) and condition δ-domination (6.18), and algorithm (6.17) gives $P_{\succ \hat{y}} = 0.6$ at $\delta_y = 0.1$. Adequacy of the received models by criterion $P_{\succ}$ with usual and δ-domination: $P_{\succ \hat{y}} = 0.65$, $P_{\succ_\delta \hat{y}} = 0.8$.

So, results of modelling confirm efficiency of the offered approach.

3. PR AREA ESTIMATION BASED ON FIELD ANALYSIS OF SECANTS

3.1. Approach Description

Reduce the problem solution from section 2, using the analysis of secants. We have the set of experimental data:

$$\mathbb{I}_o = \{y_n \in \mathbb{R}, U_n \in \mathbb{R}^k, n \in \mathbb{J}_N\}.$$

For the estimation of vector $\Phi_t = \mathrm{col}(\varphi_{t,1}, \varphi_{t,2}, \ldots, \varphi_{t,k})$, setting an upper bound of area Ω_A (6.2), apply following the approach. Construct field of secants (Karabutov, 2011) for (6.1)

$$S_S(y) = \{\bar{\gamma}(y, u_i)\ \forall i = \overline{1,k}\}$$

where $\bar{\gamma}(y, u_i) = a_{0,i} + a_{1,i}u_i; a_{0,i}, a_{1,i}$ are some real numbers. $\bar{\gamma}(y, u_i)$ approximates an observable informational portrait $\Gamma_o: \{U_n\} \times \{y_n\}$ of system (6.1).

Define the cross-correlation coefficient $r_{y\bar{\gamma}_i}$ for $\bar{\gamma}(y, u_i)$. Accept parameters $a_{1,i}$ as initial estimations $\widehat{\Phi}_{t,0}$ for vector $\widehat{\Phi}_t$ in (6.3). Using model (6.3) with $\widehat{\Phi}_{t,0}$, we obtain the prognosis for the system (6.1) output. Designate the received variable as $\hat{y}_t(S)$. Calculate $r_{y\hat{y}_t(S)}$, averages for $\hat{y}_t(S)$ and y.

If $\bar{\hat{y}}_t(S) < \bar{y}$, then estimation $\widehat{\Phi}_{t,0}$ improves. Consider model $\hat{y}_n = \widehat{A}_L^T U_n$ and apply LMS to estimate $\widehat{A}_L$. Apply $\hat{y}_n = \widehat{A}_L^T U_n$ and define the prediction y_n for system (6.1). Generate informational set:

$$\mathbb{I}_{LSM} = \{\hat{y}_{n,L} \in \mathbb{R}, U_n \in \mathbb{R}^k, n \in \mathbb{J}_N\}, \hat{y}_{n,L} \triangleq \hat{y}_n.$$

Construct a new field of structures

$$S_S(\hat{y}_L) = \{\bar{\gamma}(\hat{y}_L, u_i)\ \forall i = \overline{1,k}\}\ (\hat{y}_L, u_i) \in \mathbb{I}_{LSM},$$

where $\bar{\gamma}(\hat{y}_L, u_i) = \bar{\gamma}_{i,L} = \hat{a}_{0,i} + \hat{a}_{1,i}u_{i,n}$.

For each secant $\bar{\gamma}_{i,L}$ define $r_{y\bar{\gamma}_{i,L}}$. If the condition

$$r_{y\bar{\gamma}_{i,L}} \geq r_{y\bar{\gamma}_i}\ \forall i = \overline{1,k},$$

is satisfied, then specifies vector $\widehat{A}_{t,1}$ based on coefficients $\hat{a}_{1,i}$ for secants $\bar{\gamma}_{i,L}$. Otherwise, appropriate value from $\widehat{\Phi}_{t,0}$ to corresponded element of vector $\widehat{A}_{t,1}$.

Apply model $\hat{y}_{n,L*} = \hat{A}^{T}_{t,1} U_n$ to the prediction of output y_n. The prediction y_n, received from $\hat{y}_{n,L*} = \hat{A}^{T}_{t,1} U_n$, designate as $\hat{y}_{n,L,t}$. Calculate $\overline{\hat{y}}_{n,L,t} = \overline{\hat{y}}_{n,L,t}\left(\hat{y}_{n,L,t}\left(S\left(\hat{y}_{n,L*}\right)\right)\right)$ and $r_{y\hat{y}_{n,L,t}(S)}$. If conditions

$$\left(\overline{\hat{y}}_{n,L,t} > \overline{y}\right) \& \left(\left|r_{y\hat{y}_{n,L,t}(S)} - r_{y\hat{y}_{t}(S)}\right| \leq \delta_r\right)$$

are satisfied, where $\delta_r > 0$, then we accept $\hat{A}_{t,1}$ as the upper estimation Ω_A: $\hat{\Phi}_t = \hat{A}_{t,1}$. Otherwise, fulfil correction of vector $\hat{\Phi}_t$, using the adaptive procedure described in the previous section. We use I_{LSM} as an informational set.

The described approach applies to find of the lower boundary Ω_A. Use the condition of domination (CD) $\hat{y}_t(S) \prec y$. If CD is not fulfilled, then apply model $\hat{y}_{b,n} = \hat{\Phi}^{T}_{b,n-1} U_n$ and analogue of algorithm (6.17).

3.2. Example

Consider plant (6.1) with vector $A = [0.3 \ -0.8 \ 1]^T$. As the initial estimation $\hat{\Phi}_t$ in (6.3), we use vector $\hat{\Phi}_{t,0} = [0.39 \ -0.77 \ 0.93]^T$ received from the analysis of secants $S_S(y)$. $\overline{\hat{y}}_t(S) = 2.2$, $\overline{y} = 2.39$. As $\overline{\hat{y}}_t(S) < \overline{y}$, the CD with $\hat{\Phi}_{t,0}$ is not fulfilled. The estimation $\hat{A}_{t,1}$ for vector $\hat{\Phi}_t$:

$$\hat{A}_{t,1} = [0.396 \ -0.76 \ 1.012]^T.$$

As $r_{y\overline{y}_{i,L}} \geq r_{y\overline{y}_i}$, $\hat{\Phi}_t = \hat{A}_{t,1}$. Domination condition $\overline{\hat{y}}_{n,L,t} > \overline{y}$ is satisfied, and $r_{y\hat{y}_t(S)} = 0.94$, $r_{y\hat{y}_{n,L,t}(S)} = 0.94$, $\overline{\hat{y}}_{n,L,t} = 2.48$. Indicator of domination is $P_{>\hat{y}_{L,t}} = 0.7$. Adequacy results for the received estimation $\Omega_{A,t}$ are shown in Figure 6. Labels in Figure 6: y is a straight line, $\hat{y}_{n,L}$ is a circle, $\hat{y}_t(S)$ is a rhomb, $\hat{y}_{n,L,t}$ is a quadrate.

As $\hat{y}_t(S) \prec y$, we accept vector $\hat{\Phi}_b = \hat{\Phi}_0$ as the lower boundary for the area of restrictions Ω_A. Indicator of domination is $P_{>\hat{y}_b} = 0.98$ for model with $\hat{\Phi}_b$.

Figure 6. Area (6.2) estimation results

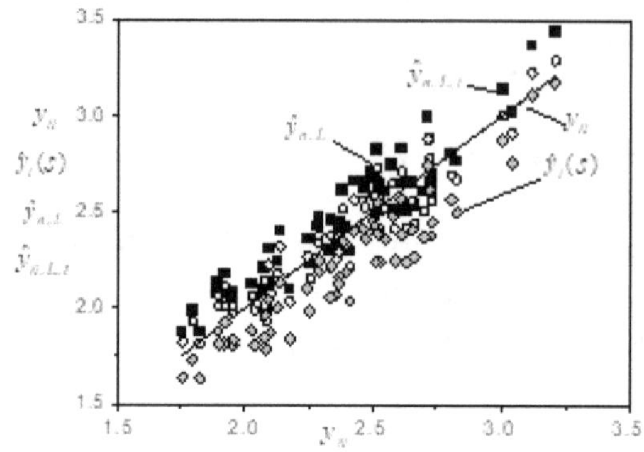

So, the estimation method APR (6.2) for system (6.1) based on the analysis secants is proposed. Sometimes, it is necessary to apply a two-level field of secants to consider set $\mathbb{I}_o$ features of the system (6.2). It is true for systems which input is irregular.

4. PR AREA AS RESTRICTION ON MODIFICATION NORM OF PARAMETERS

4.1. Approach to Estimation Ω_A

Consider system

$$y_n = A_n^T U_n + \xi_n \tag{6.18}$$

where vector $A_n \in \mathbb{R}^k$ belongs a priori to an unknown area

$$\Omega_A = \{A_n \in \mathbb{R}^k : \alpha_b \le \|A_n\| \le \alpha_t, 0 \le (\alpha_b, \alpha_t) < \infty\} \tag{6.19}$$

$U_n \in \mathbb{R}^k$ is an input vector, $y_n \in \mathbb{R}$ is an output, is the limited perturbation. The informational set

$$\mathbb{I}_o = \mathbb{I}_o(y_n, U_n, n \in \mathbb{J}_N)$$

Problem: estimate α_b, α_t of area (6.19) based on the set analysis of secants for the system (6.18).

The considered problem substantially differs from the statement stated in the previous sections. The area Ω_A representation as (6.19) does not allow applying the approaches offered earlier. It explains nonlinearity of a considered problem as on parameters and informational variables. The main thing here is a suitable vectorial norm (VN) choice. VN depends on properties of vector U_n and its correlation with output y_n. Depending on the applied norm, we received various estimations for Ω_A. As shown Karabutov (2006),

$$\hat{\alpha}_b \leq \|A_n\| \leq \hat{\alpha}_t, \tag{6.20}$$

where $\hat{\alpha}_b = \frac{\max_{n \in_N} |y_n|}{\max_{n \in_N} \|U_n\|}$, $\hat{\alpha}_t = \frac{\max_{n \in J_N} |y_n|}{\min_{n \in J_N} \|U_n\|}$.

As is shown (Karabutov, 2006) the upper bound for the vector A_n norm is overestimated. Explain it both by the applied approach and by the impact ξ_n. Therefore, estimation $\hat{\alpha}_t$ specify in the identification process. If y_n and U_n are measured with errors, then smoothed values are used in (6.20).

Apply the norm function (Lancaster, 1969)

$$\rho(X) = \max_i |x_i|, x_i \in \mathbb{X}, \mathbb{X} \in \mathbb{R}^k. \tag{6.21}$$

Application $\rho(X)$ to (6.20) gives the estimation for the element maximum of the vector A_n. To decision-making on adequacy of the received estimations, we apply a domination procedure.

Consider model

$$\hat{\rho}_{\mu,n} = \hat{\alpha}_\mu \rho_\mu(U_n), n \in \mathbb{J}_N, \tag{6.22}$$

where $\hat{\rho}_{\mu,n}$ $(\mu = t, b)$ is an upper or lower estimation of norm $\rho(y_n)$, $\rho_\mu(U_n)$ is an estimation for $\rho(U_n)$.

Define $\hat{\rho}_{\mu,n}$ and checkup condition $\hat{\rho}_{\mu,n} \succ \rho(y_n)$ based on evaluation for indicator $P_{\succ \hat{\rho}_\mu}$. We correct of current estimations in (6.20) for receiving the set level of domination. Apply the approach described in Section 2.

The following stage is the dependence definition between $\hat{\alpha}_\mu$ and a corresponding element of vector A_n. For this purpose, study correlations between $\rho(U_n)$ and $|u_{i,n}|$ on all interval $\mathbb{J}_N$.

4.2. Approach Modification

There are two methods for correcting estimates in (6.20). The first approach is based on the analysis of the coefficient modification of structural properties (Karabutov, (2006, 2011). The second approach is based on adaptive approach application. Decision-making criterion has the form:

$$P_{>\hat{\rho}_{t,s}} = P(\hat{\rho}_{t,s} \geq \rho_s(y_n)) \geq p_{t>} \forall n \in \mathrm{J}_N \tag{6.23}$$

where $p_{t>} > 0$, $p_{b>} > 0$ are some set magnitudes.

Get evaluations for (6.22) and apply δ-domination condition

$$\hat{\rho}_{t,n} - \rho(y_n) \geq \delta_y \tag{6.24}$$

where $\hat{\rho}_{t,n}$, $\hat{\rho}_{b,n}$ is model (6.22) output, $\delta_y \geq 0$.

Note that we consider condition (6.24) to reduce the algorithm sensitivity to perturbations. Initial estimates $\hat{\alpha}_\mu$ in (6.22) use the values got from (6.20). Apply adaptive model

$$\hat{\rho}_{\mu,n} = \hat{\alpha}_{\mu,n-1}\rho(U_n), \mu = t, b \tag{6.25}$$

φ-algorithm (Karabutov, 2006) for parameter $\hat{\alpha}_{\mu,n}$ tuning

$$\bar{\hat{\rho}}_{\mu,n} = F(\hat{\rho}_{\mu,n}) \tag{6.26}$$

$$\hat{\alpha}_{\mu,n} = \begin{cases} \hat{\alpha}_{\mu,n-1} - \gamma_\mu(\bar{\hat{\rho}}_{\mu,n} - \bar{\rho}(y_n))\rho(U_n), & if\ s(\mu)\bar{e}_{\mu,n} > \delta_y, \\ \hat{\alpha}_{\mu,n-1}, & if\ s(\mu)\bar{e}_{\mu,n} \leq \delta_y, \end{cases} \tag{6.27}$$

where F is an operator of an average (smoothing), $\bar{\hat{\rho}}_{\mu,n}$ is an average value of the model (6.25) output on $[0, n]$, $\bar{e}_{\mu,n} = \bar{\hat{\rho}}_{\mu,n} - \bar{\rho}(y_n)$, $\bar{\rho}(y_n)$ is average value $\rho(y_n)$ on interval J_N, $\gamma_\mu \in \mathbb{R}$ is the positive parameter ensuring convergence of algorithm,

$$s(\mu) = \begin{cases} 1, & \mu = t, \\ -1, & \mu = b. \end{cases}$$

The procedure (6.27) has the form of averaging

$$\hat{\alpha}_{\mu,n} = \hat{\alpha}_{\mu,n-1} - \tilde{\gamma}_\mu(\hat{\rho}_{\mu,n} - \rho(y_n))\rho(U_n), \tag{6.28}$$

$$\tilde{\gamma}_{\mu} = \begin{cases} \gamma_{\mu}, \ if\, s(\mu)e_{\mu,n} > \delta_y, \\ 0, \ if\, s(\mu)e_{\mu,n} \le \delta_y, \end{cases} \quad (6.29)$$

where $e_{\mu,n} = \hat{\rho}_{\mu,n} - \rho(y_n)$.

Consider properties of algorithm (6.29) at $\mu = t$.

Theorem 3. *Let* $|\xi_n| \le \upsilon$, $\upsilon \ge 0$. *Then the algorithm (6.29) is a solution to target inequality* $|e_{t,n}| > \delta_y, \delta_y > 0$, *if*

$$0 < \tilde{\gamma}_t \le \frac{2|e_{t,n} + \upsilon_n|}{|e_{t,n}|\vartheta^2}$$

and number of errors for algorithm

$$m_{\xi} \le \frac{\delta\alpha_{t,0}^2}{c^2\delta_y^2}\vartheta^2$$

where $c \ge 1$, $\vartheta^2 = \max_n \rho(U_n)$, $\upsilon_n = |\xi_n|$.

The theorem 3 proof is like theorem 2 proof. Similarly, receive properties of estimations for the case $\mu = b$.

4.3. Example

Consider the example from section 3. $|\xi_n| \le 0.3$. ξ_n is random disturbance and $|\xi_n| \le 0.3$.

Figure 7. Informational portrait showing results of estimation in space $(\rho(y_n), (\rho(\hat{y}_n))$

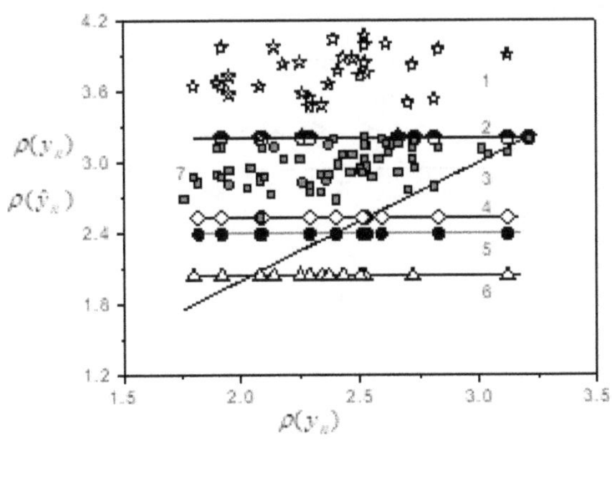

Receive estimations $\hat{\alpha}_t = 1.35$, $\hat{\alpha}_b = 1.07$ for vector A_n based on (6.20) and (6.21). It is the estimation for $a_{3,n} \in A_n$. They are overestimated, as $\rho(A_n) = 1$. Let $\hat{\alpha}_{t,0} = 1.35$, $\hat{\alpha}_{b,0} = 1.07$. Calculate average values $\rho(y_n), \rho(U_n)$ on interval $\boldsymbol{J}_N$. Apply procedure of a sliding average and receive sets $\{\rho_s(y_n)\}$, $\{\rho_s(U_n)\}$. Receive estimations $\hat{\alpha}_{t,s} = 1.14$, $\hat{\alpha}_{b,s} = 0.9$.

Apply procedure of a sliding average and receive arrays $\{\rho_s(y_n)\}$, $\{\rho_s(U_n)\}$. Then $\hat{\alpha}_{t,s} = 1.14$, $\hat{\alpha}_{b,s} = 0.9$.

Estimation results show in Figure 7, where following keys apply: 1 is the model output with $\hat{\alpha}_{t,0}$, 2 is the lower boundary of model output with $\hat{\alpha}_{t,0}$, 3 is $\rho(y_n)$, 4 is lower boundary of the model output with $\hat{\alpha}_{b,0}$, 5 is the lower boundary of the model output with $\hat{\alpha}_{t,s}$, 6 is lower boundary of model output with $\hat{\alpha}_{b,s}$. Simulation results confirm to apply the smoothing procedure for the system output.

Results of algorithm (6.28) work are shown in Figure 8. We use estimations received from (6.20) as initial values $\hat{\alpha}_{\mu,0}$. Apply exponential smoothing for estimating current average of model (6.26) output. The Figure 9 shows the domination condition for model (6.25) output. Received results confirm also values of indicators (6.23), (6.24) which, are equal $\delta_y = 0.2$, $P_{>\hat{\rho}_t} = 88\%$, $P_{>\hat{\rho}_b} = 63\%$. Parameter

$$c = \begin{cases} 2, \ \mu = t, \\ 1.5, \ \mu = b. \end{cases}$$

Figure 8. Tuning of model (6.25) parameters

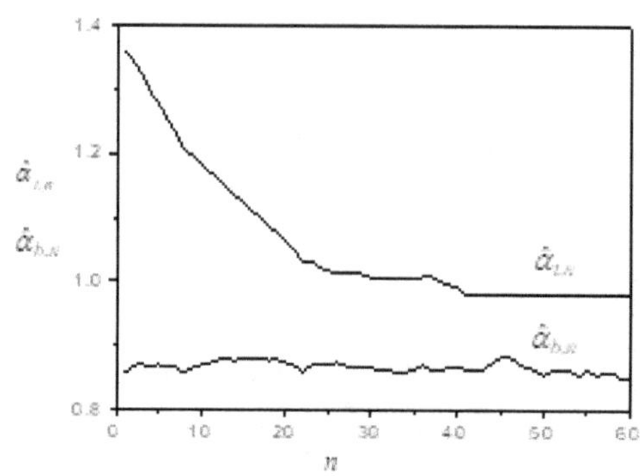

Figure 9. Domination of model (6.26) output after tuning of its parameters

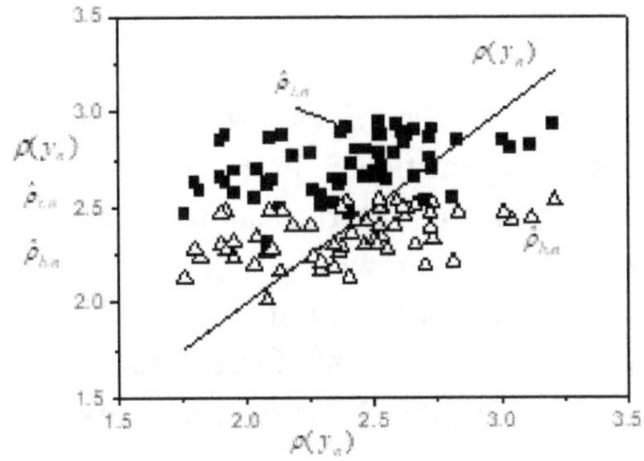

4.4. Area Ω_A Construction Based on Analysis Secant for $\rho(y_n)$

Consider informational portrait with mapping
$\Gamma_\rho \subseteq \{\rho(U_n)\} \times \{\rho(y_n)\}$.
Find the secant for Γ_ρ

$$\bar{\gamma}(\rho(y_n),\rho(U_n)) = \bar{\gamma}_\rho = a_0 + a_1\rho(U_n). \tag{6.30}$$

Value a_1 use as an initial estimation for $\hat{\alpha}_t$ in (6.20). We transform the CSP (Karabutov, 2009a) used $\bar{\gamma}_\rho$ and

$$k_{s,\rho,n} = \frac{\rho(y_n) - a_0}{\rho(U_n)}$$

$$\forall n \in \mathbb{J}_N \tag{6.31}$$

Calculate maximal value $k_{s,\rho,n}$

$$\kappa_t = \max_n k_{s,\rho,n} \tag{6.32}$$

If $\kappa_t < \hat{\alpha}_t$, where $\hat{\alpha}_t$ is the value received based on application (6.20), then κ_t we accept as new estimation $\hat{\alpha}_t$ in (6.20). κ_t, find the forecast for $\rho(y_n)$ based on the model (6.22).

If $\bar{\hat{\rho}}_{t,n} > \bar{\rho}(y_n)$ is fair, the upper bound is defined for area Ω_A. Here $\bar{\hat{\rho}}_{t,n}$, $\bar{\rho}(y_n)$ are average values of variables $\hat{\rho}_{t,n}$, $\rho(y_n)$.

Calculate

$$\kappa_b = \min_n k_{s,\rho,n} \qquad (6.33)$$

and checkup condition $\kappa_b > \hat{\alpha}_b$, where $\hat{\alpha}_b$ estimation $\hat{\alpha}_b$ in (6.20). If the inequality is fulfilled, κ_b consider as new estimation $\hat{\alpha}_b$ in (6.20).

To decide about κ_b, we use the rule described above, considering the specificity of the problem.

Statement 3. Let (i) set informational portrait $\Gamma_\rho \subseteq \{\rho(U_n)\} \times \{\rho(y_n)\}$, (ii) the secant (6.30) is received, (iii) the coefficient of structural properties (CSP) (6.31) is defined. Then values κ_t, κ_b, satisfying (6.32) and (6.33), use as estimations for $\hat{\alpha}_t$, $\hat{\alpha}_b$ in (6.20), if

$$\bar{\bar{\rho}}_{t,n} > \bar{\rho}(y_n), \bar{\bar{\rho}}_{b,n} < \bar{\rho}(y_n).$$

Application of the statement 3 gives the following estimation $\Omega_A = [0.78; 1.22]$ for the example from section 4.3.

4.5. About Influence of Vectorial Norms on Area Ω_A Estimation

Estimation Ω_A methods stated above are based on norm (6.21). Often norms (Lancaster, 1969)

$$\gamma(X) = \sum_{i=1}^{n} |x_i|,$$

$$h(X) = \sqrt{\sum_{i=1}^{n} x_i^2} \qquad (6.34)$$

are the base for shaping of restrictions.

Comparison $\rho(X)$ and (6.34) shows that

$$\rho(X) \le \gamma(X), \rho(X) \le h(X).$$

Hence, use of norms (6.34) gives the underestimated (pessimistic) estimations (6.20) for static system (6.18). To resolve this issue, follow these steps. Apply as the initial norm $\rho(X)$ and stated above approaches, and receive estimations (6.20). Let:

$\hat{\alpha}_b^\rho \le \|A_n\| \le \hat{\alpha}_t^\rho$.

Generate sets

$$G_v = \{k_v(X_n), n \in \mathbb{J}_N\}$$

$$v = \gamma, h \tag{6.35}$$

where:

$$k_v(X_n) = \frac{v(X_n)}{\rho(X_n)}.$$

Now define such $\hat{\alpha}_b^v$, $\hat{\alpha}_t^v$ covering some area corresponding to the average value $v(y_n)$. For this purpose, find average values $k_v(X_n)$.

$$\upsilon_v = \bar{k}_v(X_n) = \frac{1}{N}\sum_{n=1}^{N}\bar{k}_v(X_n).$$

Receive estimations for Ω_A used norm $v(X)$

$$\hat{\alpha}_b^v \leq v(A_n) \leq \hat{\alpha}_t^v, \hat{\alpha}_b^v = \upsilon_v \hat{\alpha}_b^\rho, \; \hat{\alpha}_t^v = \upsilon_v \hat{\alpha}_t^\rho.$$

Another approach to the estimation of APR based on the application of norms $\gamma(X)$ and $h(X)$ is described by Karabutov (2011).

5. CONCLUSION

We consider the area estimation problem of parametric restrictions for static object under uncertainty. Advantages and shortages of vectorial norms are noted. APR estimation algorithms are stated. We consider procedures of the data analysis and decision making. Working of procedures and methods is confirmed by results of modelling.

Most methods are based on the construction of the field of secants in some space. We show that the most adequate estimations of APR are based on the norm $\rho(X)$ use. Apply of other norms gives the estimations, not having practical value. Therefore, algorithms of the APR estimation are based on norm $\rho(X)$ if applied norms $\gamma(X)$, h (X). Decision-making criteria are offered for norms $\gamma(X)$, $h(X)$.

REFERENCES

Bardsley, J. M. (2004). A bound-constrained Levenberg-Marquardt algorithm for a parameter identification problem in electromagnetics. https://www.researchgate.net/publication/255577468 (2005).

Chia, T. L., Chow, P. C., & Chizeck, H. J. (1991). Recursive parameter identification of constrained systems: An application to electrically stimulated muscle. *IEEE Transactions on Biomedical Engineering*, *38*(5), 429–442. DOI: 10.1109/10.81562 PMID: 1874525

Corrêa, M. V., Aguirre, L. A., & Saldanha, R. R. (2002). Using steady-state prior knowledge to constrain parameter estimates in nonlinear system identification. *IEEE Transactions on Circuits and Systems. I, Fundamental Theory and Applications*, *49*(9), 1376–1381. DOI: 10.1109/TCSI.2002.802345

Gagliardini, P., Gourieroux, C., & Renault, E. (2011). Efficient derivative pricing by the extended method of moments. *Econometrica. Econometrica*, *79*(4), 1181–1232. DOI: 10.3982/ECTA7192

Giacomini, R., & Rossi, B. (2009). Model comparisons in unstable environments, *ERID Working Paper 30*, Duke.

Hametner, C., & Jakubek, S. (2011). Nonlinear identification with local model networks using GTLS tecniques and equality constraints. *IEEE Transactions on Neural Networks*, *22*(9), 1406–1418. DOI: 10.1109/TNN.2011.2159309 PMID: 21788188

Karabutov, N. (2013). Identification of parametrical restrictions in staic systems in conditions of uncertainty. *International Journal of Intelligent Systems and Applications*, *5*(4), 43–55. DOI: 10.5815/ijisa.2013.04.04

Karabutov, N. N. (2006). *Adaptive system identification: information synthesis*. URSS.

Karabutov, N. N. (2009a). Structural identification of systems: analysis of informational structures. Moscow: URSS: Book house Librokom.

Karabutov, N. N. (2009b). Structural identification of a static object by processing measurement data. *Measurement Techniques*, *52*(1), 7–15. DOI: 10.1007/s11018-009-9219-1

Karabutov, N. N. (2011). Structural identification of static objects: Fields, structures, methods. Moscow: URSS: Book house Librokom.

Lancaster, P. (1969). *Theory of Matrices*. Academic Press.

Mavroeidis, S., & Magnusson, L. (2014). Identification using stability restrictions. *Econometrica*, *82*(5), 1799–1851. DOI: 10.3982/ECTA9612

Mazunin, V. P., & Dvojnikov, D. A. (2010). Parametrical restrictions in nonlinear control systems of mechanisms with elasticity. *Russian Electrical Engineering*, *81*(5), 227–231. DOI: 10.3103/S1068371210050020

Mead, J. L., & Renaut, R. A. (2010). Least squares problems with inequality constraints as quadratic constraints. *Linear Algebra and Its Applications*, *432*(8), 1936–1949. DOI: 10.1016/j.laa.2009.04.017

Palanthandalam-Madapusia, H. J., Van Peltb, T. H., & Bernstein, D. S. (2010). Parameter consistency and quadratically constrained errors-in-variables least-squares identification. *International Journal of Control*, *83*(4), 862–877. DOI: 10.1080/00207170903470666

Rossi, B. (2005). Optimal tests for nested model selection with underlying parameter instability. *Econometric Theory*, *21*(05), 962–990. DOI: 10.1017/S0266466605050486

Van Pelt, T. H., & Bernstein, D. S. (2001). Quadratically constrained least squares identification. *Proceedings of the 2001 American Control Conference. (Cat. No.01CH37148)*, Arlington, VA, USA, *5*, 3684-3689. DOI: 10.1109/ACC.2001.946207

Vanli, O. A., & Del Castillo, E. (2007). Closed-Loop System Identification for Small Samples with Constraints. *Technometrics*, *49*(4), 382–394. DOI: 10.1198/004017007000000146

Wen-Ming, S. (2010). Parameter Estimation with Constraints Based on Variational Method. *Journal Marine. Sci.*, *9*, 105–108.

Section 2
Structural Identifiability

Chapter 7
Structural Identifiability of Nonlinear Systems

ABSTRACT

In this section, we consider the problem of structural identifiability (SID). Its solution is based on the application of the GF proposed in Part 1. We show the relationship between structural identification (SI) and SID for nonlinear systems, as they are based on the analysis of geometric frameworks. The basic concepts of parametric identifiability are considered. The system input influence is studied on the structural properties of a nonlinear system. The causes of insignificant GF are shown. A criterion for evaluating insignificant frameworks is presented. We evaluate the effect of the excitation constancy on SID. The conditions of a structurally unidentifiable system are obtained, and an indicator is presented for estimating the degree of non-identifiability. The approach to assessing SID is outlined.

1. INTRODUCTION. IDENTIFIABILITY

The identification problem of dynamic systems, despite the set of obtained results, is one of the important study directions. Fundamental results are got on parametrial identification. Approach to the identifiability estimation is based on R. Kalman idea (Kalman1961). Further development of these ideas is given by Lee (1964) and Elgerd (1967). R. Li gives the following identifiability definition.

Consider the system

$$\begin{aligned} X_{n+1} &= AX_n, \\ y_n &= C^T X_n, \end{aligned} \tag{7.1}$$

DOI: 10.4018/979-8-3373-0822-7.ch007

where $X_n \in \mathbb{R}^m$ is a state, $A \in \mathbb{R}^{m\times m}$, $y_n \in \mathbb{R}$ is an exit, $n = \mathbb{J}_n = [0,N]$ is the discrete time.

Problem: determine by conditions under what the system (7.1) is identified based on of the data analysis:

$$\mathbb{I}_o = \{y_n, n = \overline{0,N}, N < \infty\}. \tag{7.2}$$

The following sufficient and necessary conditions are obtained for $y_n \in \mathbb{R}^m$ (Lee (1964).

Definition 1. The system (7.1) is n-identified if the matrix A2 is possible to determine based on variable X measurement.

Definition 2. The system (7.1) is 1-identified if the matrix A is possible to determine based on measurement y.

The condition n-identifiability: the matrix $[X_0 \quad AX_0 \quad A^2X_0 \quad \ldots \quad A^{m-1}X_0]$ is nondegenerate.

1-identifiability conditions:

1. The system (7.1) is n-identified.
2. The pair (A, C) is observable.

The identifiability case when the dynamic system is order m is considered.

The analysis of considered results shows that the identifiability (ID) estimation reduces to identification of parameters. Name parametric identifiability as IP-identifiability (IPI). The publication set is devoted to the IPI research. The difference from the approach stated by Lee (1964) consists that ID results present in the form accepted in parametric estimation. Walter (1982) introduced the concept of the SID. SID is understood as the proximity of models, which is interpreted as the proximity of their parameters. It is shown that the local ID has conditions for the global ID. Different approaches and methods are applied for SID verification (Audoly et al., 1998; Avdeenko, 2013; Bodunov, 2012; Balonin, 2010).

The work majority devoted to IPI and SID not contain methods for identifiability estimation.

Remark 1. Most works, considered the IPI and available to the author, do not contain methods of the system structure estimation, accepted the identification theory. Therefore, the proposed SID concept does not reflect the essence of the considered problem. As this terminology is applied on ID estimation, in this section, we will hold onto this concept. Further, we will introduce a concept which is directly related to the SID of NDS in a structural space.

Criteria (Bodunov, 2012) are proposed for local ID estimation of a linearized NDS. The case is considered when the matrix rank is equal m. The local identifiability estimation method is proposed for an inhomogeneous linear system based on the Lyapunov exponent analysis. IPI criteria are introduced by Balonin (2010), and also generalization and development of results (Bodunov, 2012) are given.

The IPI problem of nonlinear systems was studied by various authors (see, e.g., (Chis. 2011; Krasovsky, 1987; Stigter, 2007; Saccomani, 2010). Models for nonlinear system ID estimation are proposed based on the decomposition in the Taylor series, tables of identifiability, algebra of differentials. Work (Saccomani, 2010) is devoted to the practical identifiability study. The estimation of practical ID is based on the analysis of experimental information and the differential algebra application. The identifiability of the static object, interpreted as IPI, is considered by Aivazyan et al.,1985.

So, the analysis of publications shows that the model ID is the estimation possibility of model parameters. The proposed methods are based on the non-degeneracy estimation of an informational matrix (IM). Similar results are got in parametric estimation, and non-degeneracy condition (matrix rank) of the IM is presented as the excitation constancy condition for the input. The model structure is specified a priori, and the sense of the structural local ID is not always understandable. The structure concept is widely applied to solve the identifiability problem. NDS identifiability also is transformed to the IPI problem based on model linearization methods. These researches do not consider the SID problem of NDS in the sense: is it possible to decide about the nonlinear system structure? In such a form, the problem is not set.

We propose an approach to the SID of NDS under uncertainty. This is the very complex problem as structure formalization methods of the system are not developed. The concept (Karabutov, 2016) of the structural identifiability (h-identifiability) is introduced for nonlinear systems. The h-identifiability (HI) is based on the GF analysis introduced in Part 1. Below, we give a generalization of results (Karabutov, (2016, 2021).

2. PROBLEM STATEMENT

Consider dynamic system

$$\begin{aligned} \dot{X} &= AX + B_{\varphi}\varphi(y) + B_{u}u, \\ y &= C^{T}X, \end{aligned} \tag{7.3}$$

where $u \in \mathbb{R}$, $y \in \mathbb{R}$ are an input and an output, $A \in \mathbb{R}^{q \times q}$; $B_u \in \mathbb{R}^q$, $B_\varphi \in \mathbb{R}^q$ $C \in \mathbb{R}^q$ are matrixes of corresponding dimensions; $\varphi(y)$ is a scalar nonlinear function. A is the Hurwitz matrix.

Assumption 1. The function $\varphi(y)$ belongs to the class:

$$\chi \in F_{\varphi} = \{\gamma_1\xi^2 \leq \varphi(\xi)\xi \leq \gamma_2\xi^2, \xi \neq 0, \varphi(0)=0, \gamma_1 \geq 0, \gamma_2 < \infty \} \quad (7.4)$$

where $\xi \in \mathbb{R}$ is the linear combination that is the vector X. We believe that the function $\varphi(y)$ is smooth.

Informational set for the system (7.3)

$$\mathbb{I}_o = \{u(t), y(t), t \in \mathbb{J} = [t_0, t_k]\}. \quad (7.5)$$

Problem: estimate the SID of the system (7.5) based on the analysis and processing $\mathbb{I}_o$.

3. DESIGN METHOD S_{ey}-FRAMEWORK

Design S_{ey}-framework demands the preliminary formation of the set $\mathbb{I}_{N,g}$ containing an information on the function $\varphi(y)$. Methods for obtaining $\mathbb{I}_{N,g}$ are described in Chapter 5.

4. ABOUT NEED HI ESTIMATION

The work review on the SID shows that the primary attention is given to the IPI problem. To understand the urgency of h-identifiability task, consider the raising problems on the example of the system (7.3) with:

$$A = \begin{bmatrix} 0 & 1 \\ -3 & -4 \end{bmatrix}, B_u = B_{\varphi} = \begin{bmatrix} 0 \\ 1 \end{bmatrix}, y(0) = 3, y'(0) = 2$$

$$\varphi(y) = \begin{cases} 2.2 & \text{if } (y - d > 2.2) \,\&\, (y' > 0), \\ y - d & \text{if } (y - d \leq 2.2) \,\&\, (y' > 0), \\ 1.5 & \text{if } (y - d \leq 1.5) \,\&\, (y' > 0), \\ 2.2 & \text{if } (y > 2.2) \,\&\, (y' < 0), \\ y & \text{if } (y \leq 2.2) \,\&\, (y' < 0), \\ 1.5 & \text{if } (y \leq 1.5) \,\&\, (y' < 0), \end{cases} \quad d = 1$$

Results given below are based on application of GF. Results show the input $u(t)$ influence on the system (7.3) nonlinear properties. System properties are estimated based on the framework S_{ey} analysis, and the recovered function $\varphi(y)$, corresponding to the input $u(t)$, is presented.

Figure 1 represents the phase portrait S and the framework S_{ey} for $u_{6,-4}(t) = 6 - 4 \sin(0.1\pi t)$, and the function $\varphi(y)$ recovered on data $\{y(t), y'(t)\}$. We consider the case of the established motion. Figure 1 shows that $u_{6,-4}(t)$ give to reference function $\varphi(y)$. The framework S_{ey} is almost symmetric and has features which are also in the framework S.

Figure 1. Structure estimation results for $u_{6,-4}(t)$

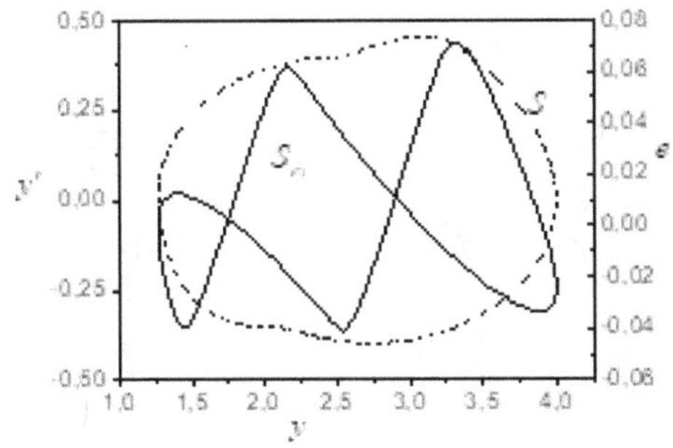

Figure 2. Structure estimation results for $u_{6,-2}(t)$

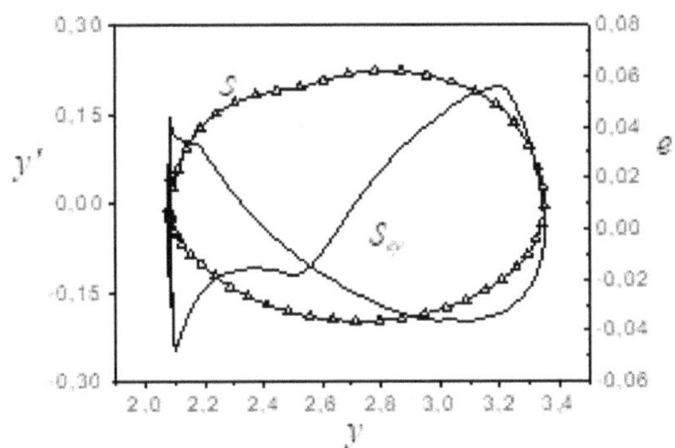

Further, a decrease in the sinusoidal amplitude makes framework S_{ey} asymmetry. We understand the restoration impossibility of the form function $\varphi(y)$ as an effect of the input property.

In Figure 2, this result is shown with $u_{6,-2}(t) = 6 - 2\sin(0.1\pi t)$. We see that the sinusoidal amplitude decrease gives to framework definition range compression, and the framework left part has more active changes. Left area is not recovered by identification methods. More cardinal changes in $\varphi(y)$ gives $u_{6,-0.5}(t) = 6 - 0.5\sin(0.1\pi t)$. Corresponding results are shown in Figure 3.

The modeling result analysis shows that there is some parameter set of the input $u(t)$ at which SID (structural identification) of the NDS is possible. These results for the system with $\omega = 0.1\pi$ are presented in Figure 4, where the following designations are used: D_y, D_e is the diameter of the variation domain y, e; a_u is the sinusoid amplitude.

Figure 3. Structure estimation results for $u_{6,-0.5}(t)$

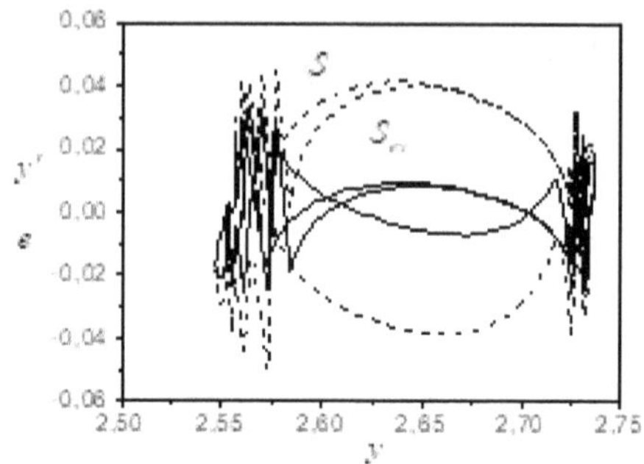

Figure 4. Input amplitude effect on the system (7.3) identifiability

(a) frameworks

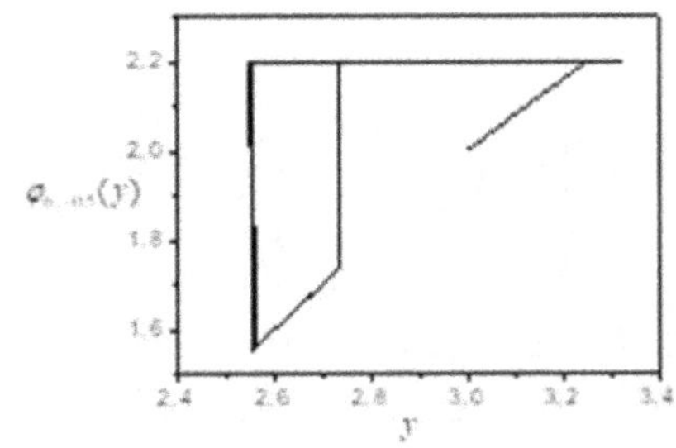

(b) nonlinearity

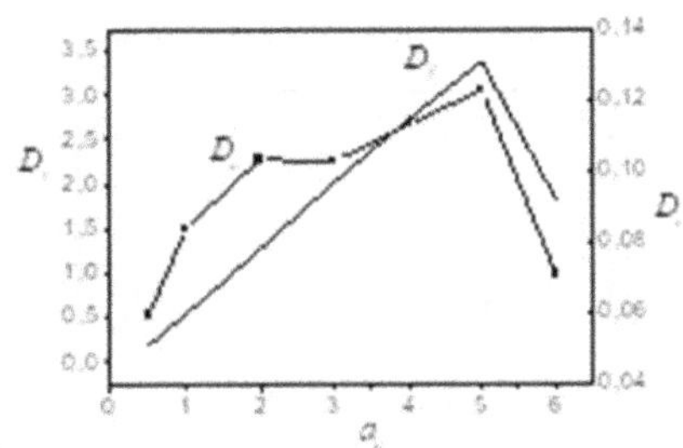

Figure 4 shows for the considered case, there is the input ($a_u = 5$), which solves the system SID problem. The system (7.3) (see Figure 1) is identified with the input having the amplitude $a_u = 4$.

We considered the input amplitude effect on the system features. Similarly, the effect gives to the frequency influence (Figure 5).

Remark 2. Modeling results show (Figure 5) that ensuring the excitation constancy (EC) condition for $u(t)$ can complicate the system HI estimation. Results, presented in Figures 5 show, that EC requirements have a distinct in structural and parametric identification. t should be considered solving the parametric identification task.

Figure 5. Structure estimation results for $u_{6-4.0.4}(t)$

a) frameworks

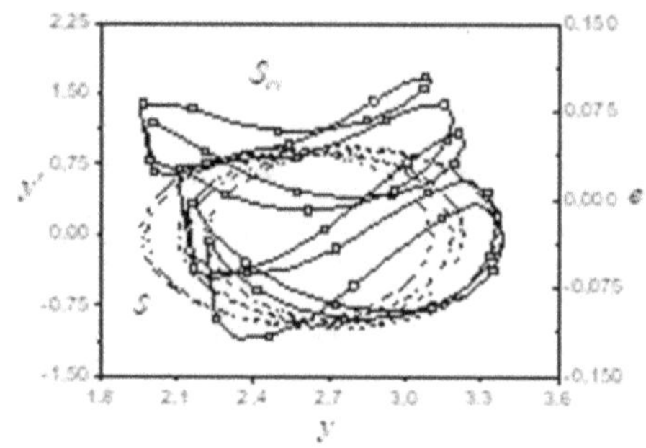

(b) nonlinearity

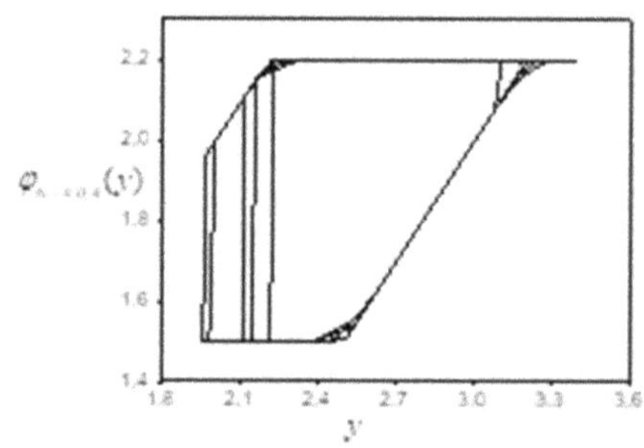

Modeling results allow giving to h-identifiability problem statement as the following task solution: find such input $u(t)$ for the system (7.3) which gives the definition range maximum for the output $y(t)$.

5. *h*-IDENTIFIABILITY

5.1. Properties *h* -Identifiability

Let fulfil to conditions:

1) the set $\mathbb{I}_o$ ensures the solution of the model (2.4) parametric identification problem.
2) the input $u(t)$ ensures obtaining informative framework $S_{ey}\left(\mathbb{I}_{N,g}\right)$ or $S_{ek}\left(\mathbb{I}_{N,g}\right)$.

If $u(t)$ has properties 1), 2), then input $u(t)$ $u(t)$ is *representative*.

Let the framework S_{ey} is closed and its area is not zero. Designate altitude S_{ey} as $h\left(S_{ey}\right)$, where the altitude is the distance between two points of the opposite sides of framework S_{ey}. Then the framework S_{ey} is identified on set $\mathbb{I}_{N,g}$.

Let $u \in PE_{\alpha}$, where PE_{α} is the EC property $PE_{\alpha}:u^2(t) \geq \alpha$, is fair for $\exists \alpha > 0$ and $\forall t \geq t_0$ on some interval $T > 0$.

Remark 3. We consider the relationship between S-synchronization and EC in the following sections.

Statement 1 (Karabutov, 2024). Let (i) the linear part of the system (7.3) is stable and nonlinearity satisfies the condition (11); (ii) the input $u(t)$ is piecewise continuous, limited and $u \in PE_{\alpha}$; (iii) exists $\delta_S > 0$ such that $h\left(S_{ey}\right) \geq \delta_S$. Then the framework S_{ey} is identified on set $\mathbb{I}_{N,g}$.

The framework S_{ey} which has referred properties, we will name h-identified. Further, we believe S_{ey} has the specified properties.

Features of the h-identifiability.

1. h-identifiability is a concept of no parametric identification, and SID.
2. The demand for parametric identifiability is the base for h-identifiability.
3. h-identifiability determines more stringent demands for the system input.

Feature 3 means that the bad input can satisfy the EC condition. Such input can give a so-called insignificant S_{ey}-framework (framework NS_{ey}) (Karabutov, 2024), which will have property of the HI.

5.2. Framework NS_{ey}

Consider the framework S_{ey}. Let $S_{ey} = F^l_{S_{ey}} \cup F^r_{S_{ey}}$, where $F^l_{S_{ey}}, F^r_{S_{ey}}$ are left and the right fragments S_{ey}. Determine for $F^l_{S_{ey}}, F^r_{S_{ey}}$ secants:

$$\gamma^l_S = a^l y$$

$$\gamma^r_S = a^r y, \tag{7.6}$$

where a^l, a^r are the numbers.

Theorem 1 (Karabutov, 2024). *Let (i) the framework S_{ey} is h-identifiable; (ii) the framework S_{ey} have the form $S_{ey} = F^l_{S_{ey}} \cup F^r_{S_{ey}}$, where $F^l_{S_{ey}}, F^r_{S_{ey}}$ are left and right fragments S_{ey} ; (iii) secants (7.6) are obtained for $F^l_{S_{ey}}, F^r_{S_{ey}}$. Then S_{ey} is NS_{ey}-framework, if*

$$|a^l - a^r| > \delta_h, \tag{7.7}$$

where $\delta_h > 0$ is some number.

6. SID

SID is a consequence of the structural identification. Since we use the GF method (GFM) and analyze the SID, the system (7.3) is represented as:

$$\begin{cases} S_y: \begin{cases} \dot{\tilde{X}} = A\tilde{X} + I\zeta, \\ \tilde{y} = C^T\tilde{X}, \end{cases} \\ S_\varphi: e = f(y, x_1), \end{cases} \tag{7.8}$$

where $\tilde{X} \in \mathbb{R}^n$ is the variable describing the general solution of the system (7.3), $\zeta \in \mathbb{R}$ is a bounded perturbation appearing as the analysis result of the variable e.

The main analysis of SID is related to the SI system.

6.1. System S_φ

Let conditions hold

B1. The input is constantly excited at the interval $\mathbb{J}$.

B2. The S_{ey} analysis gives the solution to the estimation problem of the nonlinear properties of the system $S_{y\varphi}$.

We state basic concepts and generalizations for SID.

Definition 3. If $u(t)$ satisfies B1 and B2 conditions, then the input $u(t)$ is representative.

Let the framework S_{ey} closed, and the area S_{ey} is not zero. Denote height S_{ey} as $h(S_{ey})$ where height is the distance between two points opposite sides of the framework S_{ey}.

Theorem 2. *Let (i) the linear part of the system (7.3) be stable; (ii) the nonlinearity $\varphi(\cdot)$ satisfies the condition (7.4); (iii) the input is bounded and constantly excited; (iv) $h(S_{ey}) \geq \delta_S$, where $\delta_S > 0$. Then the framework S_{ey} is identified on the set $\mathbb{I}_{N,g}$.*

Definition 4. The framework S_{ey} is called *h-identifiable* if theorem 2 holds for S_{ey}.

Let S_{ey} be h-identifiable. Designations: $\mathrm{D}_y = \mathrm{dom}(\mathrm{S}_{ey})$ is definition range of the framework S_{ey}, $D_y = D_y(\mathrm{D}_y) = \max_t y(t) - \min_t y(t)$ is diameter D_y. Let $u(t) \in \mathbb{U}$, where $\mathbb{U}$ is an acceptable set of inputs for the system (9). The set $\mathbb{U}$ contains representative inputs.

Definition 5. If D_y of the framework S_{ey} has the maximum diameter D_y, the input S-*synchronizes* the system (7.3).

Let S_{ey}^{ref} is a reference framework. S_{ey}^{ref} is the framework S_{ey} reflecting all properties of the function $\varphi(y)$. Designate by the diameter $D_y\left(S_{ey}^{ref}\right)$ as D_y^{ref}. D_y^{ref} exists if the input the system (7.3) is S-synchronizing.

Definitions 4 and 5 show if $S_{ey} \cong S_{ey}^{ref}$, then $\left|D_y - D_y^{ref}\right| \leq \varepsilon_y$, where $\varepsilon_y \geq 0$, $\cong$ is the proximity sign. Subset $\mathbb{U}_S$ have property:

$$\left|D_y\left(\mathrm{S}_{ey}\left(u(t)\big|_{u\in\mathrm{U}_S}\right)\right) - D_y^{ref}\right| \leq \varepsilon_y \tag{7.9}$$

Synchronization $u(t) \in \mathbb{U}$ is the choice of such input $u_h(t) \in \mathbb{U}$ that reflects all features $\varphi(y)$ in S_{ey}. It is true if $u(t)$ ensures $\max\limits_{u_h} D_y$ and $S_{ey} \neq NS_{ey}$. We interpret the choice $u_h(t) \in \mathbb{U}$ as ensuring synchronization between structures of the model and the system. $\mathrm{d}_{h,y} = \max\limits_{u_h} D_y$ is the condition of h-identifiability which can represent as

$$\left|D_y\left(\mathrm{S}_{ey}\left(u(t)\big|_{u\in\mathrm{U}_S}\right)\right) - \mathrm{d}_{h,y}\right| \leq \varepsilon_y \tag{7.10}$$

The condition for NS_{ey}:

$$\left|D_y\left(\mathrm{S}_{ey}\left(u(t)\big|_{u\in\mathrm{U}_S}\right)\right) - \mathrm{d}_{h,y}\right| \geq \varepsilon_y \tag{7.11}$$

(7.10) can be interpreted as proximity domain

$$\mathrm{Q}_D = \left|\mathrm{S}_{ey}\left(u(t)\big|_{u\in\mathrm{U}_S}\right) - S_{ey}^{ref}\right| \tag{7.12}$$

which is understood as $|\dot{y}(t) - \dot{y}^{ref}(t)| \leq \varepsilon_y$ for almost $\forall t \geq \tilde{t}$.

We will write $\mathrm{Q}_D \leq \varepsilon_y$ if the considered framework is closed.

Domain Q_D is the S-synchronizability area on $\{u_h(t)\}$ or the structural identifiability domain on $\{\mathrm{S}_h(u_h(t))\}$, where S_h is the phase portrait of the system (7.3), if the condition $\delta\mathrm{Q}_D \leq \varepsilon_y$ is true for Q_D almost $\forall t \geq t^*$.

Definition 6. If framework S_{ey} is h-identified and conditions $\|a^r| - |a^r\| \leq \delta_h$ satisfies, then the framework S_{ey} or the system (7.8) (system (7.3)) is structurally identified or h_{δ_h}-identifiable.

Remark 4. Conditions specified in Definition 6 are conditions for the structural identification of systems (7.3), (7.8).

Definition 7. The model (2.4) is *SM-identifying* if the framework S_{ey} is h_{δ_h}-identifiable.

We will write $u_h(t) \in \mathrm{S}$ if the input $u_h(t)$is S-synchronizing. Here $u_h(t)$ is the input that guarantees h_{δ_h}-identifiability.

The framework S_{ey} is defined on $u_h(t) \in S$ and $u_h(t)$ satisfies condition B1. Therefore, S_{ey} corresponds to the nonlinearity $\varphi(y)$ defined in the class:

$$\varphi(y) \in F_\varphi = \{\varphi(y) \in \mathbb{R} | \varphi(y, A), A \in \mathbb{R}^{n_A}, \alpha_i \in A, \alpha_i \in [\bar{\alpha}_i, \bar{\bar{\alpha}}_i]\} \tag{7.13}$$

where $\overline{\alpha}_i, \overline{\overline{\alpha}}_i$ are some numbers.

Theorem 3 (Karabutov, 2021). *Let (i) the input $u(t) \in S$ is excited constantly; (ii) the system (7.3) phase portrait has m features; (iii) S_{ey}-framework is h_{δ_h}-identified and contains fragments corresponding to features of the system (7.3). Then the model (2.4) is SM-identified.*

The theorem 3 shows if the model (2.4) is not *SM*-identifying, then model (7.3) structure or the informational set $\mathbb{I}_o$ need to change.

Let (a) c_S is the center of the framework S_{ey} on the set $\mathbb{J}_y = \{y(t)\}$, (b) c_{D_y} is the center of the area D_y, (c) $\mathbb{U}_S$ be the set of inputs, guaranteeing the system (7.3) S-synchronizability.

Theorem 4. *Let the set $\mathbb{U}_S$ given for the system $S_{y\varphi}$ and (i) $\varepsilon \geq 0$ exists such that $|c_S - c_{D_y}| \leq \varepsilon$; (ii) $||a^l| - |a^r|| \leq \delta_h$, where a^l, a^r are coefficients of secants (7.6) for $(F^l_{S_{ey}}, F^r_{S_{ey}}) \subset S_{ey}$. Then the system (7.8) is h_{δ_h}-identifiable, the input $u_h(t) \in S$, and the framework S_{ey} or $\varphi(t)$ define the class F_φ.*

Proof. Consider the input $u_h(t) \in \mathbb{U}_S$. Since condition $|a^l - a^r| \leq \delta_h$ is satisfied, the framework S_{ey} is symmetric concerning the point c_S plane (y, e). Definitional domains of diameters for fragments $F^l_{S_{ey}}, F^r_{S_{ey}}$ for the framework S_{ey} coincide up to a certain value $\varepsilon_F \geq 0$ on the set $\{y(t)\}$, i.e.

$$\left|D_{F^l_S}(D_{F^l_S}) - D_{F^r_S}(D_{F^r_S})\right| \leq \varepsilon_F \tag{7.14}$$

where $D_{F^l_S}, D_{F^r_S}$ are definitional domains $F^l_{S_{ey}}, F^r_{S_{ey}}$. Then the framework S_{ey} center is equal to $c_{D_y} = 0.5(D_{F^l_S} + D_{F^r_S})$. Since $D_{F^l_S} + D_{F^r_S} = D_y$, exist $\varepsilon \geq 0$ such that $|c_S - c_{D_y}| \leq \varepsilon$. Fulfillment of conditions (i), (ii) guarantees $u(t) = u_h(t)$ and $d_{h,y} = \max_{u_h} D_y$. Therefore, the framework S_{ey} contains all features characteristic of the function $\varphi(y)$ at $u_h(t)$. So $u_h(t) \in S$, and system (17) is h_{δ_h}-identifiable. ■

As $\varphi(y) \in F_\varphi$, then the area D_y have center $c_{D_y} \in \mathbb{J}_{c_{D_y}}$, $\mathbb{J}_{c_{D_y}}$ is some interval.

Definition 8. The framework $S_{ey,i}$ has $d_{h,\Sigma}$*-optimality property* on the set $\mathbb{U}_h$ if $\varepsilon_\Sigma > 0$ such that $|d_{h,\Sigma} - D_{y,i}| \leq \varepsilon_\Sigma \, \forall i = \overline{1, \#\mathbb{U}_h}$.

Len $\{u_{h,i}(t)\} \subset \mathbb{U}_S \subseteq \mathbb{U}$ $(i \geq 1)$, which elements have the property of S-synchronizability exists. The framework $S_{ey,i}(u_{h,i})$ has the diameter $D_{y,i}(D_{y,i})$ and corresponds to every $u_{h,i}(t)$. As $u_{h,i}(t) \in S$ the diameter $D_{y,i}$ has the property $d_{h,\Sigma}$-optimality.

Definition 9. If $\left(\left\{u_{h,i}(t)\right\} = \mathbb{U}_h \subset \mathbb{U}\right) \,\&\, \left(u_{h,i}(t) \in S\right), i \geq 1$ and frameworks $S_{ey,i}\left(u_{h,i}\right)$ have $\mathrm{d}_{h,\Sigma}$-optimality property, then frameworks $S_{ey,i}\left(u_{h,i}\right)$ are *structurally indiscernible* on sets $\left\{u_{h,i}(t)\right\}$.

Definition 10. If frameworks $S_{ey,i}\left(u_{h,i}\right)$ have $\mathrm{d}_{h,\Sigma}$-optimality property, then $S_{ey,i}\left(u_{h,i}\right)$ is locally structurally identifiable on the set $\mathbb{U}_h$.

Let the framework $S_{ey,i}\left(u_{h,i}\right)$ have $\mathrm{d}_{h,\Sigma}$-optimality property $S^{\Sigma}_{ey,i}$, and the locally structurally identified framework $S_{ey,i}\left(u_{h,i}\right)$ is $S^{LSI}_{ey,i}$.

The framework S_{ey} is locally structurally identifiable on the set $\mathbb{U}_h \subseteq \mathbb{U}_S$ if

$$\left(\exists u_h \in \mathrm{S}\right) \text{ what } \left(S_{ey} \cong S^{\Sigma}_{ey}\right) \to S_{ey} \cong S^{LSI}_{ey}$$

Remark 5. We consider nonlinearities satisfying condition (7.4). Therefore, notes made above are valid.

Definition 11. The framework $S_{ey,i}\left(u_i \notin \mathbb{U}_S\right)$ that does not have the $\mathrm{d}_{h,\Sigma}$-optimality property is locally structurally non-identifiable on the set $\mathbb{U}_h$.

The framework $S_{ey,i}\left(u_i \notin \mathbb{U}_S\right)$ that is structurally non-identifiable on the set $\mathbb{U}_h$ is defined in a class $F^N_{\varphi} \not\subset F_{\varphi}$.

Remark 6. The described approach is applied to the nonlinear system with a dynamic law of nonlinearity change. Here, we use the hierarchical immersion method for the structure estimation of the nonlinearity.

6.2. System S_y

The of system S_y identifiability is considered in Appendix A.

7. EXCITATION CONSTANCY EFFECT ON SID

Consider the input $u \in PE_{\alpha}$, where PE_{α} is the excitation constant property

$$PE_{\alpha}\colon u^2(t) \geq \alpha \tag{7.15}$$

holds for $\exists \alpha > 0$ and $\forall t \geq t_0$ on some interval $T > 0$.

Let input $u_k(t)$ of the system (7.8) have the property $u_k(t) \in PEF_{\alpha,\omega_k}$

$$u_k(t)\colon \left(u_k \in PE_{\alpha}\right) \,\&\, \left(u_k \in \mathrm{PF}_{\omega_k}\right) \,\&\, \left(\overline{u_h \in \mathrm{S}}\right),$$

$$PF_{\omega_h}\colon u_h(t) = PF_h\left(\Omega_h\right) \tag{7.16}$$

$PF_h(\Omega_k)$ is the model for $u_k(t)$ based on the Fourier series and given on the set of frequencies $\Omega_k = \{\omega_1, \omega_2, \ldots, \omega_k\}$.

Let $u_k \in \mathrm{U}_k$, $\mathbb{U}_k = \mathbb{U} \backslash \mathbb{U}_S$. Then $u_k \notin \mathrm{S}$. For $u_h \in \mathrm{S}$ is hold the condition:

$$u_h(t): (u_h \in PE_\alpha) \,\&\, \left(u_h \in \mathrm{PF}_{\omega_h}\right) \,\&\, (u_h \in \mathrm{S}),$$

$$PF_{\omega_h}: u_h(t) = PF_h(\Omega_h) \tag{7.17}$$

where $\Omega_h \neq \Omega_k$.

Compare conditions (7.16) and (7.17). Have:

$$\left(PF_h(\Omega_h) \neq PF_k(\Omega_k)\right) \Rightarrow S_{ey}^h \neq S_{ey}^k \Rightarrow S_{ey}^k = \mathrm{N}S_{ey} \tag{7.18}$$

Have

$$\left(\mathrm{D}_y\left(S_{ey}^h\right) \neq \mathrm{D}_y\left(S_{ey}^k\right)\right) \Rightarrow \left[D_y\left(S_{ey}^h\right) \geq D_y\left(S_{ey}^k\right)\right] \tag{7.19}$$

The definitional domain of frameworks S_{ey}^h, S_{ey}^k do not coincide, and S_{ey}^h is $d_{h,\Sigma}$-optimal on the set $\mathbb{U}_h$. The structure of the system (7.8) nonlinear part with u_k has indicators that do not coincide with structurally identifiable parameters of the system (7.8) with u_h.

So, the EC effects on SI of the S_φ-system and, therefore, the system (7.8).

The statement is genuine

Theorem 6. Let (i) the input u_k satisfies condition (7.16); (ii) the S_{ey}^k-framework corresponds to the input u_k; (iii) the input $u_h \in \mathrm{S}$ such that the condition (7.17) satisfied; (iv) conditions (7.18), (7.19) hold. Then (a) the S_φ-system is structurally non-identifiable by the input u_k; (b) structural parameters of the S_φ-system do not correspond to the system $S_{y\varphi}$ with the identifiable framework S_{ey}^h.

The input amplitude can influence on the SI of the nonlinear system. We transform conditions (7.16) and (7.17) as:

$$u_k(t): (u \in PE_\alpha) \,\&\, \left(u \in PF_{\omega_k}\right) \,\&\, (\overline{u_h \in \mathrm{S}}),$$

$$PF_{\omega_k}: u_k(t) = PF_k(G_k, \Omega_k) \tag{7.20}$$

$$u_h(t): (u_h \in PE_\alpha) \,\&\, \left(u_h \in PF_{\omega_h}\right) \,\&\, (u_h \in \mathrm{S}),$$

$$PF_{\omega_h}: u_h(t) = PF_h(G_h, \Omega_h) \tag{7.21}$$

where G_k, G_h are model PF_k, PF_h parameter vectors.

Present models PF_k, PF_h as

$$PF_h(G_h,\Omega_h) = g_h\widetilde{PF}_h(\widetilde{G}_h,\Omega_h),$$

$$PF_k(G_k,\Omega_k) = g_k\widetilde{P}F_k(\widetilde{G}_k,\Omega_k) \tag{7.22}$$

where $\widetilde{P}F_h(\widetilde{G}_h,\Omega_h)$, $\widetilde{P}F_k(\widetilde{G}_k,\Omega_k)$ are modifications of models (7.16), (7.17); $g_h = \max_i g_{h,i}$, $i = \overline{1,\#\Omega_h}$, $g_{h,i}$ is an element G_h; $g_k = \max_i g_{k,i}$, $i = \overline{1,\#\Omega_k}$. g_p ($p = k,h$) is the generalized amplitude of the input.

The condition (7.18) transforms as:

$$g_h\widetilde{P}F_h(\widetilde{G}_h,\Omega_h) \neq g_k\widetilde{P}F_k(\widetilde{G}_k,\Omega_k) \tag{7.23}$$

We see $u_h \in$ S then $g_h \geq g_k$. This conclusion follows from:

$$D_h(S(u_h)) \geq D_k(S(u_k)) \Rightarrow |\widetilde{P}F_h(\widetilde{G}_h,\Omega_h)| \geq |\widetilde{P}F_k(\widetilde{G}_k,\Omega_k)| \tag{7.24}$$

and the model $\widetilde{P}F_h(\widetilde{G}_h,\Omega_h)$ approximates the input, ensuring S-synchronization of the system $S_{y\varphi}$.

Obtain $d_{h,\Sigma}$-optimality of the diameter $D_h(S_{ey}^h)$ from $S(u_h) \Rightarrow S_{ey}^h$. The framework S_{ey}^k does not have this property (see (7.21)). Therefore, the input $u_k \notin$ S, which has a smaller generalized amplitude, gives the diameter $D_k(S_{ey}^k)$.

Theorem 7. *Let (i) the input u_k of the system (7.8) satisfy the condition (7.20); (ii) the framework S_{ey}^k corresponds to input u_k; (iii) the input $u_h \in$ S such that the condition (36) holds; (iv) conditions (7.18), (7.19) are held. Then (a) the S_φ-system is structurally non-identifiable by the input u_k; (b) structural parameters of the system S_φ do not correspond to the system (7.8) with the identifiable framework S_{ey}^h if $g_h \geq g_k$.*

We show the input property influence on SI and the SID of the system.

8. NON-IDENTIFIABILITY DEGREE

Consider the phase portrait S and framework S_{ey}. Get the non-identifiability degree (NID) estimate of the system S_y Definitional domains of S and S_{ey} are coincident. Therefore, the diameter $D(S_{ey})$ of the framework S_{ey} is known. Consider the set $\{u_i(t)\}$ having the property PE_α. Determine the framework $S_{ey,i}$ for each $u_i(t)$ and get $D_{y,i}(S_{ey,i})$. Suppose $d_{h,y} = \max_{u_i}|D_y(D(S_{ey,i}))|$ and denote the corresponding input as u_h. Determine diameters $d_{y,j} = |D_{y,j}(D[S_{ey,j}(u_j \in U)])|$ for all inputs $U = \{u_i(t)\}$

$\backslash\{u_h\}$. As $u_h \in$ S, therefore $d_{h,y} > D_{y,j}$ $\forall j \geq 1$. Evaluate the non-identifiability degree as:

$$SI_j = SI(S_{ey,j}) = \frac{d_{h,y} - d_{y,j}}{d_{h,y}} \tag{7.25}$$

If estimates know for the fragments F_S^l, F_S^r of the phase portrait S, then the identifiability degree is defined as:

$$SI = SI(S) = \frac{d_y^l(F_S^l)}{d_y^r(F_S^r)} \tag{7.26}$$

where $d_y^l(F_S^l)$, $d_y^r(F_S^r)$ are diameters of fragments F_S^l, F_S^r. The system S_φ is structurally identifiable if $SI(S) \leq O(1)$, where O(1) is neighborhood 1.

Example 1. Consider the system S_{BW} (Bouc-Wen model)

$$m\ddot{x} + c\dot{x} + F(x,z,t) = f(t),\ F(x,z,t) = \alpha kx(t) + (1-\alpha)kdz(t)$$

$$\dot{z} = d^{-1}(a\dot{x} - \beta|\dot{x}||z|^n sign(z) - \gamma\dot{x}|z|^n)$$

where $m > 0$ is mass, $c > 0$ is damp, $F(x,z,t)$ is the recovering force, $d > 0$, $n > 0$, $k > 0$, $\alpha \in (0,1)$, $f(t)$ is exciting force, a, β, γ are some numbers.

Parameters of S_{BW}-system: $n = 1.5, c = 2, m = 1, \beta = 0.5, \alpha = 0.7, k = 0.6$, $d = a = 1. f(t) = 2 - 2\sin(0.15\pi t)$. The system is modeled with initial conditions $x(0) = 1$, $\dot{x}(0) = 0$, $z(0) = 1$. Form the set $\mathbb{I}_o$. The phase portrait and output of the hysteresis are shown in Figure 6.

Figure 6. Phase portrait S_i for u_i

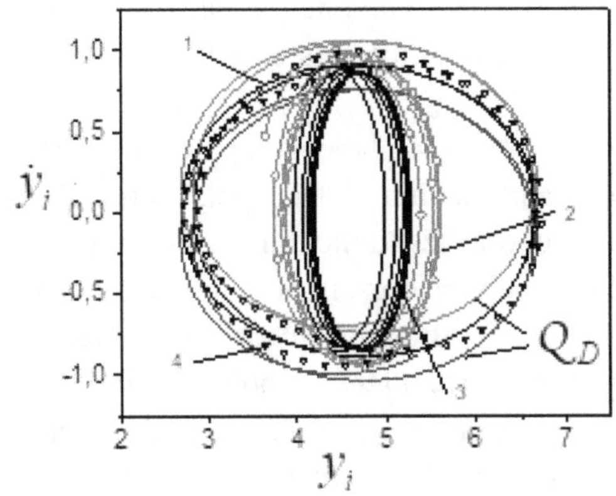

Results are obtained for the system S_{BW} steady state. We show the analysis results for $u_0(t) \in$ S. We assume that the system S_{BW} with S_0 is the standard and $d_{h,y} = D_{y,0}(S_0)$.

The non-identifiability degree of the system S_{BW} for various u_i:

$$SI_1 = 0.549, SI_2 = 0.718, SI_3 = -0.035.$$

Designations in Figure 1: 1 is S_0, 2 is S_2, 3 is S_1, 4 is S_3. Figure 6 confirms identifiability assessment results.

We see that the S_{BW}-system is structurally non-identifiable with u_1, u_2, and the S_{BW}-system with input u_3 is structurally indistinguishable from input u_0. So, frameworks $S_{ey,1}(u_1), S_{ey,2}(u_2)$ are frameworks of class NS_{ey}, and the framework $S_{ey,3}(u_3)$ belongs to class S_{ey}^{LSI}.

9. HIERARCHICAL IMMERSION METHOD

If nonlinear processes are complex, then the model (2.4) will be inadequate. Here, the hierarchical immersion method (HIM) (Karabutov, 2020) is used to design the S_{ey}-framework. HIM realizes the subsequent stages of synthesis S_{ey} if the model (2.4) is inadequate.

HIM is based on the application of the model (2.4) in a new structural space and the synthesis of the new framework $\widetilde{S}_{ey}$. If the new model (2.4) is significant, HIM stops. Otherwise, a new iteration is implemented.

10. APPROACH TO h_{δ_h}-IDENTIFIABILITY ESTIMATION

Consider the definition problem of an integral indicator, which allows deciding about the system (7.5) h_{δ_h}-identifiability. It is based on the analysis of framework S_{ey} properties.

In nonlinear dynamics and the fractal theory, approaches based on the principle of the covering (Bozhokin and Parshin, 2001) apply to the dimension estimation of a framework. Different approaches to the dimension are proposed. Topological dimension is one of the simplest indicators. It estimates the framework geometry and is it always reflects its internal features. Attractors and fractals often are heterogeneous. The heterogeneity reflects irregularity of point distribution on the framework. Heterogeneity estimations of frameworks are got based on parameters reflecting of system properties. It is estimated based on the probability analysis of the filling with certain objects of fractal geometrically identical elements. The heterogeneity characterizes discrepancy between probabilities of the fractal filling with the specified bodies and geometrical sizes of the respective areas. Such heterogeneous fractal objects call multifractals (Bozhokin and Parshin, 2001). S_{ey}-frameworks of dynamic systems with many-valued nonlinearities is the example of heterogeneous frameworks. Section 4 contains examples of heterogeneous frameworks.

Various indicators of the covering (correlation dimension, informational dimension, etc.) are approximate and labor-consuming (Bozhokin and Parshin, 2001). They give the assessment for framework fragment geometrical distinction, not always. Therefore, we introduce the integral characteristic of the framework, which is a distribution function of the variable e on the set $\{y(t)\}$ (Karabutov, 2017). Such an approach eliminates various a priori assumption concerning the framework covering local objects. We state to the proposed approach.

Let the framework S_{ey} be obtained for the system (7.5). Perform the fragmentation $S_{ey} = F^l_{S_{ey}} \cup F^r_{S_{ey}}$ where $F^l_{S_{ey}}$, $F^r_{S_{ey}}$ are left and right parts the framework S_{ey}. Fragments $F^l_{S_{ey}}$, $F^r_{S_{ey}}$ are described by functions $e^l(y)$, $e^r(y)$ where $\{e^l\} \subseteq \{e\}, \{e^r\} \subseteq \{e\}$.

Construct frequency distribution functions (histograms) H^l, H^r for $F^l_{S_{ey}}, F^r_{S_{ey}}$. Get cumulative frequency functions IH^l, IH^r based on H^l, H^r. Let $\mathrm{I}_H = \{i\Delta e,\ i = \overline{1,k}\}$ is the definition range of functions. We present the value range of functions H^l, H^r as:

$$L(IH^l) = \left[IH^l_1, IH^l_2, \ldots, IH^l_k\right]^T$$

$$R(IH^r) = \left[IH^r_1, IH^r_2, \ldots, IH^r_k\right]^T$$

Here, k is the quantity of pockets set on I_H, Δe is the pocket size on e.

Apply the model

$$\hat{R} = a_H L(IH^l) \tag{7.27}$$

also, determined the parameter a_H having applied the least-squares method.

The model is adequate if the parameter $a_H \in O(1)$ where $O(1)$ is the neighborhood 1. If the condition $a_H \in O(1)$ is fair, then the system (7.5) is h_{δ_h}- identifiable and $S_{ey} \neq NS_{ey}$. Otherwise, the framework S_{ey} is insignificant.

So, the following statement is fair.

Definition 12. If the system (7.5) is h_{δ_h}-identifiable, then the framework S_{ey} have dimension $DH_h = a_H$.

Definition 12 shows if $u(t) \in$ S, then structurally identified system dimension is approximate to 1. Value DH_h shows that the framework S_{ey} does not have complex areas and fragments $F^l_{S_{ey}}, F^r_{S_{ey}}$ are structurally identical or homothetic. If $DH_h \notin O(1)$, then it is a sign NS_{ey}-framework or a system with more complicated nonlinearity form. You can supplement results obtained based on the statement 2 with the histogram analysis for the framework S_{ey}. Get characteristics H^l, H^r and IH^l, IH^r functions and analyze their correlations, considering features S_{ey}. Some approaches are proposed by Karabutov (2016, 2017).

Examples. Consider the system from section 4 with the input $u_N(t) = 6 - 4\sin(0.5\pi t) + 0.4\sin(0.1\pi t)$. Frameworks S, S_{ey} are showed in Figure 5 for system steady state. We see that theorem 3 conditions are not satisfied. The sector which has to belong to the function $f_e = e(y)$ does not exist for S_{ey}. Therefore, the system is not S-synchronized and $S_{ey} = NS_{ey}$. So, the system is not h-identified.

Let $u(t) = 6 - 2\sin(0.1\pi t)$. The system has frameworks showed in Figure 2. Fulfill layout of the framework S_{ey} on segments $F^l_{S_{ey}}, F^r_{S_{ey}}$. It can be made of the basis of Figure 2. Secants for $F^l_{S_{ey}}, F^r_{S_{ey}}$ have the form:

$$\gamma^l_S = -0.0359y + 0.0792,$$

$$\gamma^r_S = 0.0211y - 0.0649 \tag{7.28}$$

Application of the theorem 1 will be showed that $S_{ey} = NS_{ey}$ i.e. system is it is not h-identifiable. This conclusion is confirmed with diameters $D_{F^l_S} = 0.478$, $D_{F^r_S} = 0.792$. Results h-identifiability estimation of the system based on statement 2 are showed in Figure 7. The model (7.27) has the form:

$$\hat{R} = 26 + 0.656L(IH^l) \tag{7.29}$$

Figure 7. Estimation h -identifiability of system

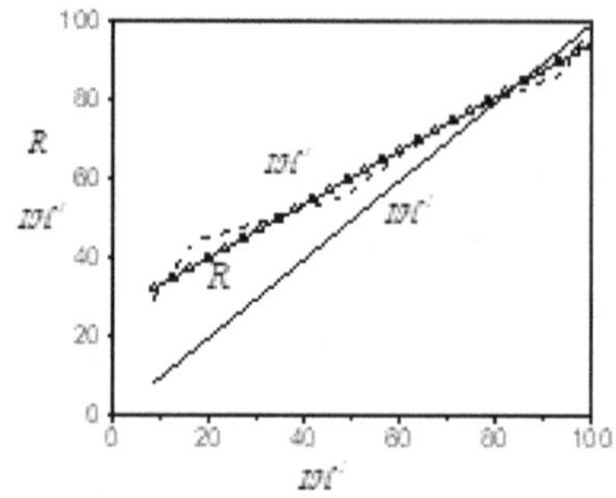

Adequacy of the model (7.15) is 97%. The framework S_{ey} dimension is 0.65. Analysis Results show that the system (7.5) with an input $u(t) = u_{6,-4}(t)$ is structurally non-identifiable.

Consider the system from section 4 with $u(t) = u_{6,-4}(t) = 6 - 4\sin(0.1\pi t)$. Corresponding frameworks are shown in Figure 1. We see the framework S_{ey} have some asymmetry that explains with nonlinear function characteristics (Figure 1b).

Figure 8. Functions IH^l, IH^r

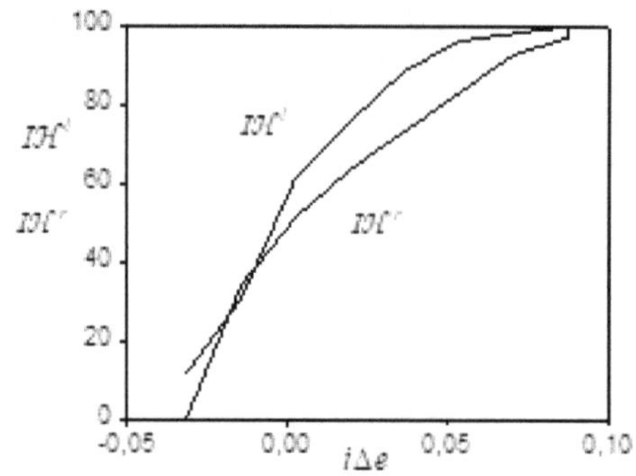

Following parameters of the secants (7.10) are obtained for fragments $F^l_{S_{ey}}, F^r_{S_{ey}}$: $a^l = -0.025, a^r = -0.027$. Let $\delta_h = 0.003$. The condition (7.11) will not be satisfied and, therefore $S_{ey} \neq NS_{ey}$. To confirm this inference, determine by functions IH^l, IH^r. They are shown in Figure 7, and results S_{ey}-framework dimension estimation is presented in Figure 9.

Figure 9. Estimation h- identifiability of system based on fragment cumulative frequency function with $u_{6,-4}(t)$

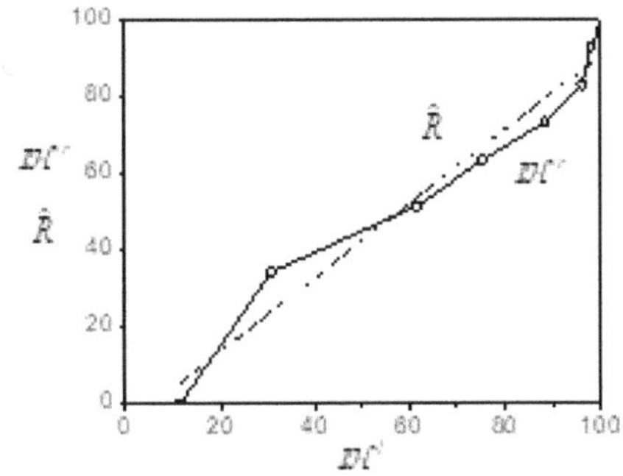

The model (7.27) has the form:

$$\hat{R} = -5.845 + 0.96L(IH^{l})$$

and the coefficient of determination is 96%. Conditions of statement 2 are satisfied, and the system is h_{δ_h}-identifiable. Framework S_{ey} dimension DH_h is 0.96. Framework fragment diameters are equal $D_{F^l_s} = 1.16$, $D_{F^r_s} = 1.43$. DiameterS_{ey} is equal to 2.59. This value coincides with $D_{F^l_s} + D_{F^r_s}$. If to choose $\varepsilon_F = 0.4$, then the condition (7.14) will be satisfied. The distinction between fragment $F^l_{S_{ey}}, F^r_{S_{ey}}$ definition ranges is depended on properties S_{ey}. The condition 2) theorem 4 is satisfied with $\varepsilon = 0$. Therefore, the system is the SI or h_{δ_h}-identifiable with $u_{6,-4}(t)$, and $u_{6,-4}(t) \in$ S.

11. CONCLUSION

The approach to structural identifiability analysis of nonlinear dynamic systems is proposed under uncertainty. This approach has a difference from methods applied to structural identifiability estimation of dynamic systems in the parametric space. We interpret the SI as structural identification possibility of a nonlinear part of the system. The input has to satisfy excitation constancy condition in SI problems. This condition differs from requirements to inputs in adaptive systems. We show that the input has to have synchronization property (S-synchronizability) for the SI problem solution. The SI estimation is based on special class framework S_{ey} analysis; therefore, synchronization has to give to the maximum value of the framework definition range. Non-synchronized input gives to an insignificant framework which does not allow solving structural identification problem; therefore, the system is not structurally identifiable. We get conditions under which it is possible to estimate system structural identifiability. Input subset is allocated have the S-synchronizability property and on which frameworks S_{ey} are indistinguishable.

REFERENCES

Aivazyan, S. A., Yenyukov, I. S., & Meshalkin, L. D. (1985). *Applied statistics. Study of relationships*. Moscow: Finansy i statistika.

Audoly, S., D'Angio, L., Saccomany, M. P., & Cobelli, C. (1998). Global identifiability of linear compartmental models – a computer algebra algorithm. *IEEE Transactions on Automatic Control, 45*, 36–47.

Avdeenko, T. V. (2013). Identification of linear dynamic systems with use parametrical space separators. *Automatics and program engineering*, 1(3), 16-23,

Balonin, N. A. (2010). *Theorems of identifiability*. Politekhnika publishing house.

Bodunov, N. A. (2012). Introduction to the theory of local parametrical identifiability. *Differential equations and management processes, 2*.

Bozhokin, S. V., & Parshin, D. A. (2001). *Fractals and Multifractals*. Scientific Publishing Centre Regular and Chaotic Dynamics.

Chis, O.-T., Banga, J. R., & Balsa-Canto, E. (2011). Structural identifiability of systems biology models: A critical comparison of methods. *PLoS One*, *6*(4), 1–16.

Elgerd, O. I. (1967). *Control Systems Theory*. McGraw-Hill.

Kalman, R. E. (1961). On the general theory of control systems. *Proceeding first IFAC Congress on Automatic Control,* Moscow, 1960; Butterworths, London, 1, 481-492.

Karabutov, N. (2015). Structural identification of nonlinear dynamic systems. *International Journal of Intelligent Systems and Applications*, *7*(9), 1–11.

Karabutov, N. (2016). Structural identification of dynamic systems with hysteresis. *International Journal of Intelligent Systems and Applications*, *8*(7), 1–13.

Karabutov, N. (2021). S-synchronization and excitation constancy in structural identifiability problem of nonlinear systems. *EPJ Web of Conferences*, *248*, 01004.

Karabutov, N. (2024). Bouc-Wen hysteresis identification: structural-parametric approach. In *Advances in Mathematics Research* (Vol. 36, pp. 99–144). Nova Science Publishers, Inc.

Karabutov, N. N. (2017). Structural methods of design identification systems. In Uvarova, L. A., Nadykto, A. B., & Latyshev, A. V. (Eds.), *Nonlinearity problems, solutions and applications* (Vol. 1, pp. 233–274). Nova Science Publishers Inc.

Karabutov, N. N. (2020). S-synchronization, structural identifiability and identification of nonlinear dynamic systems. *Mekhatronika, Avtomatizatsiya. Upravlenie, 21*(6), 323–336.

Krasovsky, A. A. (Ed.). (1987). *Reference book on automatic control theory*. Nauka.

Lee, R. C. K. (1964). *Optimal estimation, identification, and control.* MIT Press.

Saccomani, M. P., & Thomaseth, K. (2010). Structural vs practical identifiability of nonlinear differential equation models in systems biology. Bringing mathematics to life. In Rogato, A., Zazzu, V., & Guarracino, M. (Eds.), *Dynamics of mathematical models in biology* (pp. 31–42). Springer.

Stigter, J. D., & Peeters, R. L. M. (2007). On a geometric approach to the structural identifiability problem and its application in a water quality case study. *Proceedings of the European Control Conference 2007 Kos,* Greece, July 2-5, 3450-3456.

Walter, E. (1982). *Identifiability of state space models.* Springer-Verlag.

APPENDIX A

Present S_y as

$$y = \underbrace{C^T\left(sI_q - A\right)^{-1} I}_{W_y}\zeta \quad (A1)$$

where $I_q \in \mathbb{R}^{q\times q}$ is the unit matrix, $s = d/dt$. Let $A \in R^{q\times q}$ is the Frobenius matrix with the vector of parameters $A_S \in \mathbb{R}^q$, $A_S = \left[a_{s,1}, a_{s,2}, \ldots, a_{s,q}\right]^T$. Divide the left and right parts (A1) into a polynomial $v(s) = \prod_{i=1}^{q-1}(s+\mu_i)$, where $\mu_i > 0$. Then:

$$y(s) = s^{-1}\left\{b_1\zeta(s) - a_1 y(s) + \sum_{i=2}^{q}(b_i\zeta(s) + a_i y(s))\frac{1}{s+\mu_i}\right\} \quad (A2)$$

Denote $P = \left[y, p_{y2}, \ldots, p_{yq}; \zeta, p_{\zeta 2}, \ldots, p_{\zeta q}\right]^T$. Then:

$$\dot{y} = \bar{A}^T P, \bar{A}^T = \left[-a_1, a_2, \ldots, a_q; b_1, b_2, \ldots, b_q\right] \quad (A3)$$

where a_i и b_i depends on A_S.

We show global identifiability of system (A3). Consider the system (A4). Introduce the variable $e = \hat{y} - y$ and get $\dot{e} = \Delta A^T P$, where $\Delta A = \hat{A} - A$.

Consider the criterion $V(e) = 0.5e^2$ and get:

$$\dot{V} = -\hat{a}_1 e^2 + e\Delta A^T P \leq -\frac{\hat{a}_1}{2}e^2 + \frac{1}{2\hat{a}_1}\Delta A^T P P^T \Delta A \quad (A4)$$

As follows from (A4), $\Delta A = 0$ if $P(t) \in PE_\sigma$, where $\sigma > 0$.

Theorem 7A. *The system S_y is globally identifiable on output if: (i) the S_φ system is S-synchronized; (ii) S_y is controllable; (iii) $P(t) \in PE_\sigma$; (iv) the perturbation ζ is limited and extremely non-degenerate.*

Chapter 8
SID of Systems With Non-Symmetric Nonlinearities

ABSTRACT

In the next chapters, we apply the structural identifiability (SID) approach proposed in Chapter 7 to the SID evaluate of various nonlinear dynamical systems (NDS). The approach for SID generalization used in the analysis of systems with symmetric nonlinearity is presented. Homothety and identifiability conditions obtain for systems with non-symmetric nonlinearities (SNN). We present the conditions of-identifiability and S-synchronizability for NDS. We study the conditions of detectability recoverability of SNN based on GF analysis. These conditions guarantee the SID estimate under uncertainty. The role of excitation constancy in the analysis of SID systems with asymmetric non-linearities is noted.

1. INTRODUCTION

Non-symmetric nonlinear systems are used to control various objects (Hao et al., 2021; Shakiba et al., 2012; Song and Der Kiureghian (2006); Yang et al., 2018; Yang et al., 2020). A genetic procedure (Shakiba et al., 2012) is proposed for the non-symmetric hysteresis identification. An extended Prandtl-Ishlinskii model (Hao et al., 2021) based on the application of the asymmetric reproduction operator is proposed. Yang et al. (2020) proposes to consider the excitation frequency effect on hysteresis. The generalized Bouc-Wen model (Song and Der Kiureghian, 2006) is suggested to describe strongly asymmetric hysteresis. Inverse motion control scheme (Zhang et al., 2020) uses for estimates of dynamic surface. Other methods of using non-symmetric nonlinearities (NN) based on the use of parametric identification are discussed by Aljanaideh et al., 2014; Brandhofer et al., 2923; Do et al., 2023;

DOI: 10.4018/979-8-3373-0822-7.ch008

Hayashi, 1964; Popov, 1988; Wei and Wang, (2012); Wei et al., 2014; Xie et al., 2018; Zheng et al., 2018.

So, the analysis shows that NN applies to implement non-standard control laws. Standard synthesis methods use for the analysis and construction of systems with NN. The input effect on properties of nonlinear systems has not been studied. This conclusion is valid regarding the identifiability analysis of NDS. The identifiability estimation of nonlinear systems (NS) guarantees to implement control laws considering NS properties. Various approximation methods are used to estimate the NS identifiability (Karabutov, 2021). They level out the nonlinearity form and are applicable only under a priori certainty. The NS identifiability is related to SID (Karabutov, 2021), which bases on the analysis of geometric frameworks (GF) (Karabutov, 2020). NN has features and require of method GF modification. We give the development and generalization of the results got N. Karabutov (2020, 2024).

2. PROBLEM STATEMENT

Consider dynamic system

$$\begin{aligned} \dot{X} &= AX + B_{\varphi}\varphi(y) + B_{u}u, \\ y &= C^{T}X, \end{aligned} \tag{8.1}$$

where $u \in \mathbb{R}$, $y \in \mathbb{R}$ are input and output, $A \in \mathbb{R}^{q\times q}$; $B_u \in \mathbb{R}^q$, $B_\varphi \in \mathbb{R}^q$ $C \in \mathbb{R}^q$ are matrixes of corresponding dimensions; $\varphi(y,\tilde{A})$ is scalar nonlinear function; $\tilde{A} \in \mathbb{R}^k$ is vector of nonlinearity parameters; $B_\varphi = B_u = I = [0,0,\dots,0,1]^T$, $C = [1,0,\dots,0]^T$, $\varphi = \varphi_a = \varphi_a(y,A_a)$ is NN. A is the Hurwitz matrix.

Remark 1. Vector A_a elements may not coincide. It reflects the $\varphi_a(y,A_a)$ asymmetry. Symmetric nonlinearity is denoted as $\chi_s = \varphi_s(y,A_s)$. We believe:

$$\chi \in F_\varphi = \{\gamma_1\xi^2 \le \varphi(\xi)\xi \le \gamma_2\xi^2,\ \xi \ne 0,\ \varphi(0)=0,\ \gamma_1 \ge 0,\ \gamma_2 < \infty\} \tag{8.2}$$

where ξ is the linear combination X. We believe that the function $\varphi(y)$ is smooth. The informational set for the system (8.1)

$$\mathbb{I}_o = \{u(t), y(t), t \in \mathbb{J} = [t_0, t_k]\}. \tag{8.3}$$

Problem: estimate the SID of the system (7.3) based on the analysis and processing $\mathbb{I}_o$.

Apply the methods of the framework S_{ey} synthesis and the system (8.1) present as:

$$S_y:\begin{cases}\dot{\tilde{X}} = A\tilde{X} + I\zeta,\\ \tilde{y} = C^T\tilde{X},\end{cases} \tag{8.4}$$
$$S_\varphi: e = f(y, x_1),$$

where $\tilde{X} \in \mathbb{R}^q$ is a variable describing the general solution of the system (8.1); $\zeta \in \mathbb{R}$ is limited perturbation emerging as the result of the variable e define (see Chapter 2).

3. *h*-IDENTIFIABILITY AND S-SYNCHRONIZABILITY

We give a modification of the SID results described in Chapter 7. Input $u(t)$ plays an important role in nonlinear systems. $u(t)$ effects on S-synchronizability, h-identifiability (HI), and the appearance of insignificant S_{ey}-structures for the system (8.1).

Assumptions.

A1. The input is constantly excited (CE) at the interval $\mathbb{J}$

$$PE^S_{\alpha,\bar{\alpha}}: \ (\alpha \leq u(t) \leq \bar{\alpha}) \ \& \ (\Omega_u(\omega) \subseteq \Omega_S(\omega))$$

$$\forall t \geq t_0$$

where $(\alpha, \bar{\alpha}) > 0$ is some numbers, $\Omega_S(\omega)$ is the set of allowable input frequencies ensuring SID systems, $\Omega_u(\omega)$ is the set frequencies $u(t)$.

A2. Input $u(t)$ provides an informative structure $S_{ey}\left(\mathbb{I}_{N,g}\right)$.

Definition 1. An input $u(t)$ is presentative if it satisfies conditions A1 and A2.

Denote the framework S_{ey} with symmetric nonlinearity as S^s_{ey}.

Let the framework S^s_{ey} be closed, and its area is not zero. Denote the height S^s_{ey} as $h\left(S^s_{ey}\right)$ where height is the distance between two points on opposite sides of the structure S^s_{ey}.

Notations.

1. $\mathrm{D}_y = \mathrm{dom}\left(S^s_{ey}\right)$ is definitional domain S^s_{ey}.
2. $D_y = D_y\left(\mathrm{D}_y\right) = \max_t y(t) - \min_t y(t)$ is diameter D_y.
3. $u(t) \in \mathrm{U}$, where U is the allowed set of inputs for the system (8.1). The set $\mathbb{U}$ contains representative inputs.

Definition 2. If the definition domain D_y of the framework S^s_{ey} has a maximum diameter D_y on the set $\{y(t), t \in \mathbb{J}\}$, then the input $u(t) \in \mathbb{U}_S \subseteq \mathbb{U}$ is the *S-synchronizing* system (8.1).

Let $S_{ey} = F^l_{S_{ey}} \cup F^r_{S_{ey}}$, where $F^l_{S_{ey}}, F^r_{S_{ey}}$ are the left and right fragments S^s_{ey}. Define secant

$$\gamma^{r,s}_S = \alpha^{r,s} y,$$

$$\gamma^{l,s}_S = \alpha^{l,s} y \tag{8.5}$$

for $F^l_{S_{ey}}, F^r_{S_{ey}}$, where α^l, α^l are positive numbers.

Theorem 1 (Karabutov, 2024). *Let 1) the framework S^s_{ey} has the form $S^s_{ey} = \mathrm{F}^{l,s}_{S_{ey}} \cup \mathrm{F}^{r,s}_{S_{ey}}$; 2) secants described by equations (8.5) for $F^{l,s}_{S_{ey}}, F^{r,s}_{S_{ey}}$. Then S^s_{ey} is an NS^s_{ey}-framework if*

$$\left\| \alpha^{l,s} \right| - \left| \alpha^{r,s} \right\| > \delta_h \tag{8.6}$$

where $\delta_h > 0$ is some set value.

Consider the reference framework S^{ref}_{ey} reflecting all the properties of the function $\varphi(y)$. Let D^{ref}_y be the $D_y\left(S^{ref}_{ey}\right)$ diameter. D^{ref}_y exists for system (8.1) with S-synchronizing input.

Definitions 1, 2 show if $S^s_{ey} \cong S^{ref}_{ey}$, then $\left|D_y - D^{ref}_y\right| \le \varepsilon_y$, where $\varepsilon_y \ge 0$, $\cong$ is a sign of proximity. Subset U_S elements have the property:

$$S_{h_S}: \left| D_y\left(S^s_{ey}\left(u(t)\big|_{u \in \mathrm{U}_S}\right)\right) - D^{ref}_y \right| \le \varepsilon_y \tag{8.7}$$

As $\mathrm{d}_{h,y} = \max_{u_h} D_y$ is S-synchronizing input, the condition $\mathrm{d}_{h,y} = \max_{u_h} D_y$ fulfillment guarantees the system HI. *h*-identifiability condition

$$\left| D_y\left(S^s_{ey}\left(u(t)\big|_{u \in \mathrm{U}_S}\right)\right) - \mathrm{d}_{h,y} \right| \le \varepsilon_y \tag{8.8}$$

Remark 2. Interpret the condition (8.8) as an almost homothetic of the structure S^s_{ey}. The word almost emphasizes the computational aspects of the influence of structure obtaining. The homothety concept given by Hadamard (1948).

Let the input $u_h(t)$ synchronizes the set D_y. If $u(t)$ is S-synchronizing, then we write $u_h(t) \in \mathrm{S}$. A finite set $\{u_h(t)\} \in \mathrm{S}$ exists for the system (8.1). The optimal $u_h(t)$ choice depends on $\mathrm{d}_{h,y}$ and (8.8). The condition (8.8) fulfillment is the basis for the SID (structural identification (SI)) of the system (8.1).

Definition 3 (Karabutov, 2024). If condition (8.8) holds for $F^{l,s}_{S_{ey}}, F^{r,s}_{S_{ey}}$, then the S^s_{ey}-framework is *almost homothetic* (AH) or *ε-homothetic* (EH).

Theorem 2. *Let (i) $\{u_h(t)\} \in$ S; (ii) condition (8.8) is held; (iii) framework S^s_{ey} (system (8.1)) is HI; (iv) condition*

$$\left\|\alpha^{l,s}\right| - \left|\alpha^{r,s}\right\| \leq \delta_h \quad (8.9)$$

is satisfied. Then the system (8.1) (S^s_{ey}-framework) is structurally identifiable or h_{δ_h}-identifiable.

The Theorem 2 proof is based on the fulfillment of conditions AH or EH (conditions (8.8), (8.9)) for fragments $\left(F^{l,s}_{S_{ey}}, F^{r,s}_{S_{ey}}\right) \subset S^s_{ey}$. Condition (8.8) shows that lines (8.5) are almost parallel. Apply the Theorem 3 (Karabutov, 2020) proof and get that the condition $\left|c_S - c_{D_y}\right| \leq \varepsilon$, where $\varepsilon \geq 0$, holds for the framework S^s_{ey} with center c_S and the domain D_y with the center c_{D_y}. Hence,

(i) the S^s_{ey}-framework is almost symmetric or $S^s_{ey} \in S^H_{ey}(\chi)$ on the plane (y, e), where $S^H_{ey}(\chi)$ is a homothetic structures class for symmetric nonlinear systems.;

(ii) diameters of fragments $\left(F^{l,s}_{S_{ey}}, F^{r,s}_{S_{ey}}\right) \subset S^s_{ey}$ determination regions coincide accurately within value $\varepsilon_F \geq 0$ on $\{y(t)\}$:

$$\left|D_{F^l_s}\left(D_{F^l_s}\right) - D_{F^r_s}\left(D_{F^r_s}\right)\right| \leq \varepsilon_F \quad (8.10)$$

where $D_{F^l_s}, D_{F^r_s}$ are determination regions $F^{l,s}_{S_{ey}}, F^{r,s}_{S_{ey}}$. Obtain from (8.10)

$$u(t) = u_h(t) \Rightarrow \quad u(t) \in S \Rightarrow \quad d_{h,y} = \max_{u_h} D_y \quad (8.11)$$

Conditions (8.10), (8.11) guarantee SID (Karabutov, 2021) (h_{δ_h}-identifiability) for the system (8.1).

The nonlinearity χ_a asymmetry leads to $S^s_{ey}(\chi)$-framework deformation in comparison with nonlinearities χ_s. Denote the structure corresponding to χ_a as $S^a_{ey} = S_{ey}(\chi_a)$.

Let the input $u(t)$ satisfy A1 and A2 conditions. Hence, the subset $\{u_h(t)\} \in S$ that guarantees the system (8.1) h-identifiability exists. Structure $S^a_{ey}(\chi_a) \in S^{H_a}_{ey}(\chi_a)$ defines on $I_h = \left\{y(t), u(t)\big|_{u_h(t)\in S}\right\}$. Structural differences between χ_a and χ_s belonging a class $F(\chi)$ caused by the following condition.

Structural differences between χ_a and χ_s belonging to a class $F(\chi)$ we interpret the following condition:

$$\chi_a \neq \chi_s \Rightarrow \dim A_a \geq \dim A_s, A_a \in \mathbb{R}^k, A_s \in \mathbb{R}^m \quad (8.12)$$

We believe that $(\chi_a, \chi_s) \in F$ and χ_a, χ_s have different parametric content.

Theorem 3 (Karabutov, 2024). *Let (i) $\{u_h(t)\} \in$ S; (ii) $S^a_{ey} \in F$ for almost everyone y(t). Then the structure $S^a_{ey} \in F$ is $h^a_{\delta_h}$-identifiable or ε_a-homothetic if $F^{l,a}_{S_{ey}} \cong F^{r,a}_{S_{ey}}$*

hold for almost $\forall(y,u_h) \in \mathrm{I}_o$*, excluding the regions of fragments* $F^{l,a}_{S_{ey}}, F^{r,a}_{S_{ey}}$*, where elements of the vector for the function* χ_a *do not match for these regions.*

Proof. $u_h(t) \in \mathrm{S}$ and, therefore, the framework $\mathrm{S}^a_{ey} \in F$ has a maximum diameter. The condition $\left|c_{\mathrm{F}_{l,a}} - c_{\mathrm{F}_{r,a}}\right| \leq \varepsilon_a$ is satisfied for centers $F^{l,a}_{S_{ey}}, F^{r,a}_{S_{ey}}$, where $\varepsilon_a \geq 0$. Hence, fragments $F^{l,a}_{S_{ey}}, F^{r,a}_{S_{ey}}$ are ε_a-homothetic, and the system (8.1) is $h^a_{\delta_{\approx}}$ identifiable for almost all $y(t) \in \mathrm{I}_h$.■

Let (a) $(\chi_a, \chi_s) \in F$; (b) $\left|\mathrm{a}_{i,a} - a_{k,a}\right| \leq \upsilon$ is valid for some elements subset $\mathrm{A}_a \in A_a$, where$\upsilon \geq 0$, $\mathrm{a}_{i,a} \in \mathrm{A}_a$, $i \in \mathbb{Z}$ belongs to some interval, i can coincide with k, when $\dim A_a \rightarrow \dim A_s$, $a_{k,a} \in A_s$.

Consider a sub-domain $F^{q,a}_{S_{ey,i}}$ of the fragment $F^{q,a}_{S_{ey}}$, $q = l, r$. Select sub-domains $F^{l,a}_{S_{ey,i}}$ and$F^{l,a}_{S_{ey,i}}$, which are structurally similar (property $s_{\approx}$), as $\chi_a \in F$. Mark the $F^{l,a}_{S_{ey,i}}$ and $F^{l,a}_{S_{ey,i}}$ similarity as $\varepsilon_{\approx}$ is held. If the function $\chi_a \in F$ has the property $s_{\approx}$, then we write$\chi_a \in F_{s_{\approx}}$, and the corresponding property S^a_{ey}-framework denote as$\mathrm{S}^a_{ey} \in S_{\varepsilon_{\approx}}$.

Definition 4. The structure $\mathrm{S}^a_{ey} \in F$ is almost ε_a-homothetic or $h^a_{\delta_{\approx}}$-identifiable if S^a_{ey} is ε_a-homothetic, and the regions $F^{q,a}_{S_{ey,i}} \subset F^{q,a}_{S_{ey}}$ have the property$F^{q,a}_{S_{ey,i}} \in F_{s_{\approx}}$.

Theorem 4. *Let (i)* $\{u_h(t)\} \in \mathrm{S}$*; (ii) framework* $\chi_a \in F$*; (iii) condition* $\|a^{l,a}| - |a^{r,a}\| \leq \delta_{h,a}$ $(\delta_{h,a} \geq 0)$ *holds almost for all y(t) for secant fragments* $\left(F^{l,a}_{S_{ey}}, F^{r,a}_{S_{ey}}\right) \subset S^a_{ey}$*; (iv) condition* $\left|\mathrm{a}_{i,a} - a_{k,a}\right| \leq \vartheta$ *holds for the domain* $F^{q,a}_{S_{ey,i}} \subset F^{q,a}_{S_{ey}}$ $(q = i, r)$ *with parameters* $\mathrm{A}_a \in A_a$*, where*$\vartheta \geq 0$*,* $\mathrm{a}_i \in \mathrm{A}_a$*,* $a_k \in A_s$*. Then the framework* S^a_{ey} *is* ε_a*-homothetic or* $h^a_{\delta_{\approx}}$*-identifiable.*

Theorem 4 proof. Apply theorem 3 and get the ε_a-homotheticity for the S^a_{ey}-framework. If the condition of Theorem 4 (iv) is satisfied, then the region $F^{q,a}_{S_{ey,i}} \subset F^{q,a}_{S_{ey}}$ is almost ε_a-homothetic and is corresponded to the region $F^{q,s}_{S_{ey,i}}$ on the framework S^s_{ey} at point $\|\mathrm{A}_a - \mathrm{A}_s\| \leq \delta_{\mathrm{A}}$, where $\mathrm{A}_s \subset A_s$. Hence, S^a_{ey} corresponds to some$\mathrm{S}^s_{ey} \in F$.■

So, if theorem 4 conditions fulfill for S^a_{ey}, then we decide on the system identifiability with $\mathrm{S}^a_{ey} \in F$ based on the analysis $\mathrm{S}^s_{ey} \in F$.

As $u(t) \in PE^S_{\alpha,\overline{\alpha}}$, the $\zeta(t)$ perturbation in (8.4) is limited to according to the construction of the S_{φ}-system. Therefore, the S_y-system is stable and recoverable. The variable $y'(t)$ is detectable. The system S_y recoverability follows from the system S_{φ} detectability, as any recoverable state is stable. This conclusion shows that the framework S^a_{ey} is detectable and recoverable. The system input is representative and synchronizing on $u(t) \in \mathrm{S}$, since the system is recoverable. The system (8.1) is almost structural identifiable ($h^a_{\delta_{\approx}}$-identifiable), and structure S^a_{ey} is almost ε_a-homothetic. Its followers on theorem 4. So,

Theorem 5 *(Karabutov, 2024). Let (a)* $\left(u(t) \in PE^S_{\alpha,\overline{\alpha}}\right)$ & $(u(t) \in \mathrm{S})$*; (b) the* S_y *system is stable and recoverable. Then the system* S_{φ} *(framework* S^a_{ey}*) is detectable, recoverable and* $h^a_{\delta_{\approx}}$*-identifiable, and the structure* S^a_{ey} *is almost* ε_a*-homothetic.*

Let $(\chi_a, \chi_s) \in F$. The function χ_a structurally coincides with χ_s. They belong to the same class of structures, excluding regions$F^{q,a}_{S_{ey,i}}$. Therefore $\mathrm{S}^a_{ey} \approx \mathrm{S}^s_{ey}$, i.e., the frames are closed, excluding regions $F^{q,a}_{S_{ey,i}}$.

Introduce frameworks $\mathrm{S}^{a}_{ey}, \mathrm{S}^{s}_{ey}$ defined on sets $\mathbb{S}^{a}_{ey}, \mathbb{S}^{s}_{ey}$ and the proximity index $d\left(\mathbb{S}^{a}_{ey} \backslash \mathbb{S}^{s}_{ey}\right)$, understanding it as $d(t) = \left|e^{a}_{ey} - e^{s}_{ey}(t)\right| \leq \delta_e \forall y(t) \in \mathrm{D}_y\left(\mathrm{S}^{i}_{ey}\right)$, where $\delta_e > 0, i = a, s$. Get a set $\mathbb{D}\left(\mathbb{S}^{a}_{ey} \backslash \mathbb{S}^{s}_{ey}\right)$ whose elements correspond to the asymmetry areas of the framework S^{a}_{ey} with $d(t) > \delta_e$. Let define areas $\mathrm{D}_y\left(\mathrm{S}^{i}_{ey}\right)$ of structures coincide. The set $\mathbb{D}\left(\mathbb{S}^{a}_{ey} \backslash \mathbb{S}^{s}_{ey}\right)$ is finite in construction.

So, if the reference framework S^{s}_{ey} is known for$\chi_s \in F$, then the asymmetry structure of $\chi_a \in F$ is estimated by $d\left(\mathbb{S}^{a}_{ey} \backslash \mathbb{S}^{s}_{ey}\right)$. The filling of the set $\mathbb{D}\left(\mathbb{S}^{a}_{ey} \backslash \mathbb{S}^{s}_{ey}\right)$ depends on the condition (8.14) fulfillment.

Theorem 6. *Let (i) $u(t) \in$ S; (ii) the reference symmetric function $\chi_s \in F$ is given; (iii) the framework S^{s}_{ey} is obtained for χ_s; (iv) condition (8.12) is satisfied with the function $\chi_a \in F$. Then the structure S^{a}_{ey} is almost ε_a-homothetic for almost all $t \geq t_0$, and the system (8.4) is $h^{a}_{\delta_{\approx}}$-identifiable.*

Proof. Consider the system (8.4) with $\chi_s \in F$. As $u(t) \in$ S, there exists $u_h(t) \in$ Ssuch that the h_{δ_h}-identifiability of system (8.4) is guaranteed. The S^{s}_{ey}-framework is h_{δ_h}-identifiable. For system (8.4) with$\chi_a \in F_a$, according to theorem 5, there is an input $u_h(t) \in$ S such that $\mathrm{S}^{a}_{ey} \approx \mathrm{S}^{s}_{ey}$ is valid. So, the set $\mathbb{D}\left(\mathbb{S}^{a}_{ey} \backslash \mathbb{S}^{s}_{ey}\right)$, whose elements correspond to the asymmetry areas of the framework for $\forall d(t) > \delta_e$, exists. According to theorem 5, domains $F^{l,a}_{S_{ey,i}}$ and $F^{r,a}_{S_{ey,i}}$ are almost homothetic. So, the S^{a}_{ey}-framework is almost ε_a-homothetic, and the system is $h^{a}_{\delta_{\approx}}$-identifiable. ■

Consider the function $\chi_a \in F_a$, which is a combination of continuous functions $\chi_{a,i}\left(A_{a,i}\right)$. $\chi_{a,i}\left(A_{a,i}\right)$ (defines the framework χ_a), parameters $A_{a,i}$ of which may not coincide with$A_{a,i+1}$, where $i \in \mathbb{Z}_A$ is an integer. Class F_a contains functions with different structures.

Definition 5. If the system (8.4) input $u(t) \in$ S and $\left(F^{l,a}_{S_{ey}}, F^{r,a}_{S_{ey}}\right) \subset \mathrm{S}^{a}_{ey}$ has a full diameter, then the framework S^{a}_{ey} is structurally complete.

Theorem 7 *(Karabutov, 2024). Let (i) $u(t) \in$ S; (ii) $\chi_a \in F_a$; (iii) fragments $\chi_{a,i}\left(A_{a,i}\right) \in \chi_a$ are continuous functions; (iv) fragments $\left(F^{l,a}_{S_{ey}}, F^{r,a}_{S_{ey}}\right) \subset \mathrm{S}^{a}_{ey}$ are structurally complete. Then the S^{a}_{ey}-framework is detectable and recoverable.*

Theorem 7 proof. Consider the system (8.4) with $\chi_s \in F$. The input$\left(u(t) \in PE^{S}_{a,\overline{\alpha}}\right)$ & $(u(t) \in$ S). Hence, the input is an S-synchronizing. These properties of the input guarantee the closeness of the framework S^{a}_{ey}. Detectability and recoverability follow from Theorem 5. The input is representative and guarantees the system identifiability. The continuity $\chi_{a,i}\left(A_{a,i}\right) \in \chi_a$, $i > 1$ can lead to non-fulfillment of almost ε_a-homotheticityS^{a}_{ey}. Consequently, not all structural parameters χ_a can be identified. ■

Remark 3. The stated approach to SID assessment is general. Therefore, specifics of nonlinear systems must consider when analyzing SID. It demands the proposed approach modification.

4. EXAMPLES

1. Consider the engine operation stabilization system described by the equation:

$$\begin{cases} \dot{x}_1 = x_2, \\ \dot{x}_2 = -a_1x_1 - a_2x_2 + kf\big(g(t)-x_1\big), \end{cases} \tag{8.13}$$

$$f(z)= \begin{cases} d_1z^{\alpha}, \text{if } z \geq b, \\ d_2z, \quad \text{if } z \geq b, \end{cases} \tag{8.14}$$

where x_i is the state variable ($i = 1;2$);$g(t)$ is input; $a_1, a_2, k, d_1, d_2, b, \alpha$ are constant parameters.

The system (8.13), (8.14) is modeled with $a_1 = 3$, $a_2 = 4$, $k = 5$, $\alpha = 1.5$, $d_1 = 2$, $d_2 = 1$, $b = 0$,$g(t)= 2.5 \sin(0.05\pi t)$. Integration step on t is 2s. The set I_o and the phase portrait $S_{x_2x_1}$ of the system (Figure 1) are obtained for the steady-state mode. Relationship analysis in the system is showed that the variable x_2 does not depend on x_1,$g(t)$. We use x_1 for SID analysis, because $x_1$has close connections with $g(t)$ and $\upsilon = g - x_1$. Framework $S_{x_1\upsilon}$ reflecting the connection $x_1 = x_1(\upsilon)$ showing in Figure 1.

Figure 1. Phase portrait $S_{x_2x_1}$ and structure $S_{x_1\upsilon}$

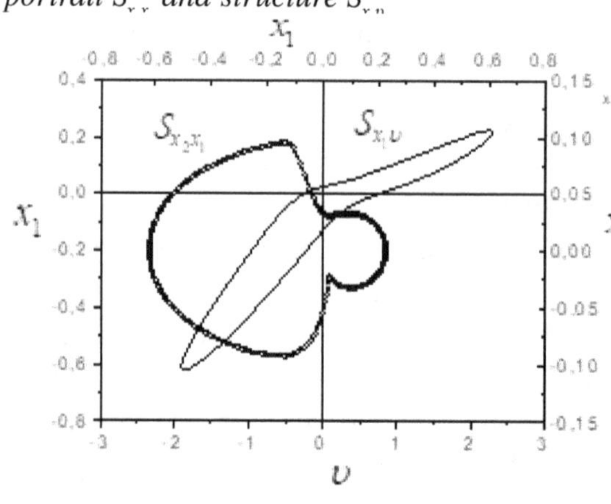

As follows from Figure1, the framework $S_{x_2x_1}$ is non-symmetric. This is a case of inadequate input selection. The framework $S_{x_2x_1}$ is not ε_a-homothetic. We cannot obtain an adequate S_{ex_2}-framework. The $S_{x_1\upsilon}$-framework gives a more complete performance on system nonlinear properties. We see (Figure 1) that the nonlinearity comprises two sections with a different slope of fragments. Fragment $F^{l,a}_{S_{ey}}$ is almost ε_a-homothetic to fragment $F^{r,a}_{S_{ey}}$.

Remark 4. The mutual influence redistribution of variables explains by the feedback influence in the system. This requires of the approach revision to the analysis of the system.

So, we decide about the system (8.13), (8.14) structural identifiability based on the analysis $S_{x_1 v}$. The mathematical structure choice for the nonlinear system SID analysis determines the result.

2. Control system with external and internal feedbacks

$$\begin{cases} \dot{x}_1 = x_2, \\ \dot{x}_2 = -a_2 x_2 + k(g(t) - x_1 - k_{os} f(x_1)), \end{cases} \tag{8.15}$$

$$\begin{cases} \dot{x}_1 = x_2, \\ \dot{x}_2 = -a_2 x_2 + k(g(t) - x_1 - k_{os} f(x_1)), \end{cases} \tag{8.16}$$

where $f(z)$ is a sign-function with different clipping levels; $a_2 > 0, k > 0, c > 0, c_1 > 0, k_{os} > 0$.

Parameters of systems: $a_2 = 2$, $k_{os} = 3.5$, $c = 2$, $c_1 = 0.25$, $g(t) = 4\sin(0.05\pi t)$. The system phase portrait $S_{x_2 x_1}$ (steady-state mode) showed in Figure 2. The analysis $S_{x_2 x_1}$ shows that a switching is in the region 0. $S_{x_2 x_1}$ is almost ε_a-homothetic. The function $f(z)$ is asymmetric with relation to zero.

Figure 2. Phase portrait of system (8.15), (8.16)

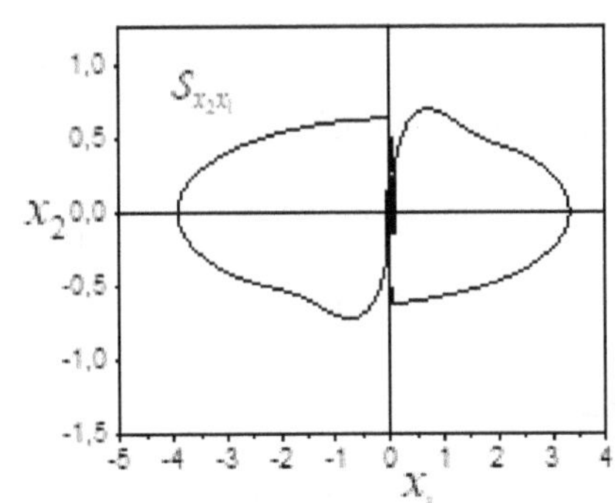

Figure 3. Framework $S_{x_1\varepsilon_1}$

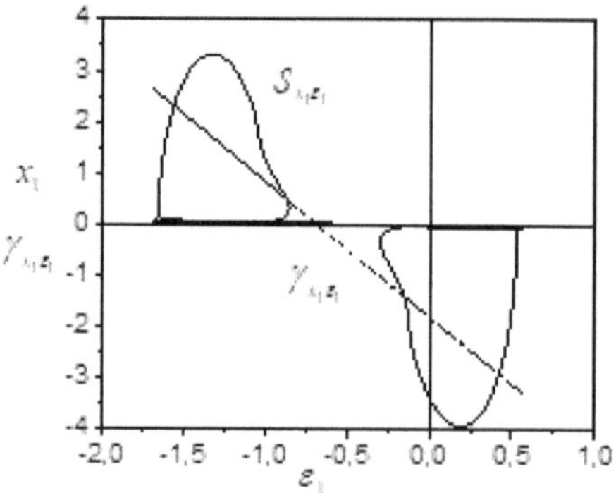

The approach described in the previous section is not applicable to $S_{x_2x_1}$, since the model (8.4) is inadequate. Therefore, consider the GF $S_{x_1\varepsilon_1}$ (Figure 3) described by the function $x_1 = x_1(\varepsilon_1)$, where $\varepsilon_1 = g(t) - x_1 - k_{os}f(x_1)$. We see that the framework $S_{x_2x_1}$ (Figure 3) is asymmetric. Apply the approach described in section 3, construct the secant $\gamma_{x_1\varepsilon_1}$ and calculate the residual $\upsilon_1 = x_1 - \gamma_{x_1\varepsilon_1}$. As a result, we have $S_{x_1\varepsilon_1}$.

The structure $S_{\upsilon_1\varepsilon_1}$ (Figure 4) describes by the function $\upsilon_1 = \upsilon_1(\varepsilon_1)$ and is almost ε_a-homothetic. Homothety process described the transformation of the left fragment $S_{\upsilon_1\varepsilon_1}$ into the right, also shown in Figure 4. The point (0.327; 0.012) is a marker for switching from the left fragment $S_{\upsilon_1\varepsilon_1}$-framework to the right. This result confirms conclusions based on Figure 2.

Figure 4. Checking $S_{\upsilon_1\varepsilon_1}$-structure ε_a-homothety

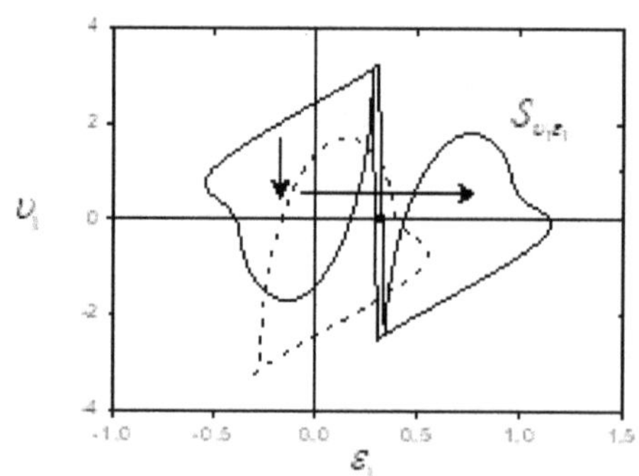

So, the system (8.15), (8.16) is ε_a-homothetic and $h^a_{\tilde{\delta}}$-identifiable, and the framework $S_{\upsilon_1\varepsilon_1}$ is detectable and recoverable. The conclusion is based on the analysis processing of the set $\mathbb{I}_o$ and theorems 4 and 7 applications.

5. CONCLUSION

We consider the structural identifiability problem of dynamical systems with NN. Requirements for system input guaranteed an assessment of the nonlinear system structural identifiability are suggested. The input must be constantly excited and S-synchronizing. We get the SID problem solution based on the construction and analysis of GF. ε_a-homothety and structural identifiability conditions are obtained for the system with NN. The detectability and recoverability of GF are proved under uncertainty. We get conditions under which the non-symmetric nonlinearity can be a hypothetical symmetric nonlinearity.

REFERENCES

Aljanaideh, O., Rakheja, S., & Chun-Yi, S. (2014). Experimental characterization and modeling of rate-dependent asymmetric hysteresis of magnetostrictive actuators. *Smart Materials and Structures*, *23*(3), 035002. DOI: 10.1088/0964-1726/23/3/035002

Brandhofer, S., Myers, C. R., Devitt, S. J., & Polian, I. (2023). Multiplexed pseudo-deterministic photon source with asymmetric switching elements. arXiv:2301.07258.

Du, C., Yang, C., Li, F., & Gui, W. (2019). A novel asynchronouss control for artificial delayed Markovian jump systems via output feedback sliding mode approach. *IEEE Transactions on Systems, Man, and Cybernetics. Systems*, *49*(2), 364–374. DOI: 10.1109/TSMC.2018.2815032

Hadamard, J. (1948). *Lecons de geometrie elementaire. T. 2: Geometrie dans l'espace*. Armand Colin & Cie.

Hao, L., Yang, H., Sun, Z., Xiang, C., & Xue, B. (2021). Modeling and compensation control of asymmetric hysteresis in a pneumatic artificial muscle. *Journal of Intelligent Material Systems and Structures*, *28*(19), 2769–2780. DOI: 10.1177/1045389X17698588

Hayashi, C. (1964). *Nonlinear oscillations in physical systems*. McGraw-hill Book Company.

Karabutov, N. N. (2020). S-synchronization structural identifiability and identification of nonlinear dynamic systems. *Mekhatronika, avtomatizatsiya, upravlenie, 21*(6), 323-336.

Karabutov, N. N. (2021). *Introduction to the structural identifiability of nonlinear systems*. URSS/LENAND.

Karabutov, N. N. (2024). Structural Identifiability Evaluation of System with Nonsymmetric Nonlinearities. *Mekhatronika, Avtomatizatsiya. Upravlenie*, *25*(2), 55–64.

Popov, E. P. (1988). *Theory of nonlinear automatic control and control systems*. Nauka.

Shakiba, S., Ourak, M., Poorten, E. V., Ayati, M., & Yousefi-Koma, A. (2021). Modeling and compensation of asymmetric rate-dependent hysteresis of a miniature pneumatic artificial muscle-based catheter. *Mechanical Systems and Signal Processing*, *154*(107532), 1–23. DOI: 10.1016/j.ymssp.2020.107532

Song, J., & Der Kiureghian, A. (2006). Generalized Bouc–Wen model for highly asymmetric hysteresis. *Journal of Engineering Mechanics*, *6*(6), 610–618. DOI: 10.1061/(ASCE)0733-9399(2006)132:6(610)

Wei, Z., & Wang, D. (2012). Non-symmetrical Bouc–Wen model for piezoelectric ceramic actuators. *Sensors and Actuators. A, Physical*, *181*, 51–60. DOI: 10.1016/j.sna.2012.03.048

Wei, Z., Xiang, B. L., & Ting, R. X. (2014). Online parameter identification of the asymmetrical Bouc-Wen model for piezoelectric actuators. *Precision Engineering*, *38*(4), 921–927. DOI: 10.1016/j.precisioneng.2014.06.002

Xie, S.-L., Liu, H.-T., Mei, J.-P., & Guo-Ying, G. (2018). Modeling and compensation of asymmetric hysteresis for pneumatic artificial muscles with a modified generalized Prandtl–Ishlinskii model. *Mechatronics*, *52*, 49–57. DOI: 10.1016/j.mechatronics.2018.04.001

Yang, H., Chen, Y., Sun, Y., & Hao, L. (2018). A novel kriging-median inverse compensator for modeling and compensating asymmetric hysteresis of pneumatic artificial muscle. *Smart Materials and Structures*, *27*(11), 115019. DOI: 10.1088/1361-665X/aad758

Yang, H., Zhang, Y., Chen, Y., Sun, Y., & Hao, L. (2020). A novel learning adaptive hysteresis inverse compensator for pneumatic artificial muscles. *Smart Materials and Structures*, *29*(1), 015035. DOI: 10.1088/1361-665X/ab4b84

Zhang, W., Zhang, Z., & Wang, H. (2019). Adaptive state feedback control f or nonlinear systems with multiple interval time-varying delays and non-symmetric dead-zone input: A Lyapunov-Krasovskii approach. *Transactions of the Institute of Measurement and Control*, *41*(14), 1–12. DOI: 10.1177/0142331219842284

Zhang, X., Chen, X., Zhu, G., & Su, C.-Y. (2020). Output feedback adaptive motion control and its experimental verification for time-delay nonlinear systems with asymmetric hysteresis. *IEEE Transactions on Industrial Electronics*, *67*(8), 6824–6834. DOI: 10.1109/TIE.2019.2938460

Chapter 9
SID of Feedback Systems With Nonlinear Adulterating

ABSTRACT

The structural identifiability (SID) problem of nonlinear feedback systems with nonlinear adulterating is considered. We study two cases of the adulterate: 1) additive effect of nonlinearity in feedback on the nonlinearity in the straight chain; 2) nonlinear adulterating of nonlinearity argument in the straight chain of the system. Identifiability conditions are received for these cases. The SID estimation is based on the application of geometric frameworks (GF) reflected properties of the nonlinear system, structural frequency diagrams, and the hierarchical immersion method. We obtain conditions of non-identifiability and local identifiability for the system. The nonlinear argument's influence on the system identifiability is analyzed.

1. INTRODUCTION

The identifiability problem of dynamic systems is a relevant area of study. Foundational results obtain on the parametric identification (Guillaume et al., 2019; Shin et al., 2019; Karabutov, 2015). The main direction is the a priori identifiability analysis in this area. Parametric identification methods are the basis for a priori identifiability (AI). The model structure sets a priori. In publications, AI is often interpreted as structural identifiability (SID), which is incorrect. A priori identifiability methods are transferred to the SIDE estimation problem for nonlinear systems. This approach eliminates the estimation problem of the nonlinearity structure.

Method for analyze identifiability of closed-loop systems with nonlinear feedback proposes Ko, S., and Bitmead, (2006). The method is based on the harmonic

DOI: 10.4018/979-8-3373-0822-7.ch009

balance and the double Fourier series (FS). FS uses to approximate the nonlinearity Parametric identifiability is the main direction in the analysis of linear closed-loop systems (see, for example, Wu and et al., 2021; Leboute et al., 2021; Chan-Zheng et al., 2021). Parametric identification (PI) guarantees the identifiability of the system. The SID problem does not solve by parametric methods under uncertainty. The parametric approach to nonlinearity applies only in a specific area. Changing of input properties give different identification results. This is the major problem of the parametric approach. Works analyze PI does not consider the estimation task of the structure of nonlinear systems.

We have noted that the type of nonlinear dynamic system (NDS) affects the SID. This remark is true for feedback NDS. Next, the SID problem solution is given for feedback systems (FS) with nonlinear adulterating (NA). We do not consider the identifiability of the system linear part. This problem has been well studied (see, for example, Karabutov, 2015).

2. PROBLEM STATEMENT

Consider the system

$$\begin{aligned} \dot{X} &= AX + B_{\varphi}\Phi(Y) + B_{u}U, \\ Y &= C^{T}X, \end{aligned} \tag{9.1}$$

where $u \in \mathbb{R}$, $Y \in \mathbb{R}^{g}$ are input and output, $X \in \mathbb{R}^{q}$ is the state vector, $A \in \mathbb{R}^{q\times q}$, $B_u \in \mathbb{R}^{q}$, $B_{\varphi} \in \mathbb{R}^{q\times k}$ $C \in \mathbb{R}^{g\times q}$ are matrices of corresponding dimensions, $\Phi(X) \in \mathbb{R}^{k}$ is vector of nonlinear functions. A is Hurwitz matrix.

The nonlinear function $\varphi_i(\xi) \in \Phi$ is smooth and satisfies the condition

$$\chi \in F_{\varphi} = \{\gamma_{1,i}\xi^2 \leq \varphi_i(\xi)\xi \leq \gamma_{2,i}\xi^2,\ \xi \neq 0,\ \varphi_i(0) = 0,\ \gamma_{1,i} \geq 0,\ \gamma_{2,i} < \infty\} \tag{9.2}$$

where $\xi \in \mathbb{R}$ is the input of a nonlinear element. ξ is a linear combination of elements X.

Information set

$$\mathbb{I}_o = \{U(t), Y(t), t \in \mathbb{J} = [t_0, t_k]\}$$

We assume that some state variable x_i may be a function depended on the difference in the output of nonlinearities.

Problem: analyze the set $\mathbb{I}_o$ and estimate the SID of the system (9.1).

3. PRELIMINARY CONCEPTS

Let $\varphi(\cdot) = \Phi(X)$, $\varphi(\cdot) \in \mathbb{R}$, $y = Y$, $y \in \mathbb{R}$. We apply the approach described in chapter 8 to the construction of GF (Karabutov, 2015, 2021) and synthesize a framework S_{ey}. S_{ey} describes by the function $f_{ey}: y \to e$, $e \in \mathbb{R}$. e reflects nonlinear processes in the system.

Assumption B1. The input $u(t)$ is constantly excited at the interval $\mathbb{J}$.

Assumption B2. The input $u(t)$ ensures an informative framework S_{ey}.

Definition 1. If $u(t)$ satisfies 1 and B2 assumptions, then the input $u(t)$ is *representative.*

Assumption 3. Let the framework S_{ey} closed, and the area S_{ey} is not zero.

Denote height S_{ey} as $h\left(S_{ey}\right)$, where height is the distance between two points opposite sides of the framework S_{ey}.

If the framework S_{ey} satisfies the assumption B3, then S_{ey} calls h-identifiable. Since the framework S_{ey} be h-identifiable. The properties of h-identifiability consider Karabutov, (2021).

Further results are based on the statements displayed in paragraph 6.1 of Chapter 8. Note, studying the mixing effect is the arduous task for NDS. Therefore, we consider at the NA effect and research the SID problem.

4. ADDITIVE ADULTERATE

Consider a case of the system (9.1) with feedback (self-oscillation system)

$$S_\varphi : \begin{cases} \dot{x}_1 = x_2, \\ \dot{x}_2 = x_3, \\ \dot{x}_3 = c_1(\varphi_1(u) - \varphi_2(x_4)) + c_2 x_4, \\ \dot{x}_4 = \varphi_1(u) - \varphi_2(x_4), \\ y = x_4, \end{cases} \tag{9.3}$$

where $u = -k_1 x_2 - k_2 x_1$ is input; $y \in \mathbb{R}$ is the output; c_1, c_2, k_1, k_2 are system parameters, $\varphi_1(x) = c\,\mathrm{sign}(x)$ is sign function, $c > 0$,

$$\varphi_2(x_4) = \begin{cases} c, \text{ if } x_4 > b, \\ 0, \text{if } x_4 \in [-b,b], \\ -c, \text{ if } x_4 < -b, \end{cases} \tag{9.4}$$

$b > 0$. Get the set $\mathbb{I}_o$

We see that the output of function $\varphi_1(u)$ adulterates by function $\varphi_2(x_4)$. System (9.3) is a nonlinear system of the second class (Besekersky and Popov, 1976). Harmonic linearization (Besekersky and Popov, 1976) does not apply to $\varphi_1(u)$, $\varphi_2(x_4)$. Find SID conditions for (9.3) using the results of chapter 8.

Determine the system S_φ solution under the following initial conditions and parameters: $x_1(0)= 2$, $x_2(0)= 1$, $x_3(0)= 0, x_4(0)= 1$, $c = 2$, $b = 1.5$, $c_1 = 2$, $c_2 = 2$, $k_1 = 1$, $k_2 = 1.5$. Form the information set $\mathbb{I}_o$. The system (9.3) phase portrait shows in Figure 1.

Figure 1. Phase portrait

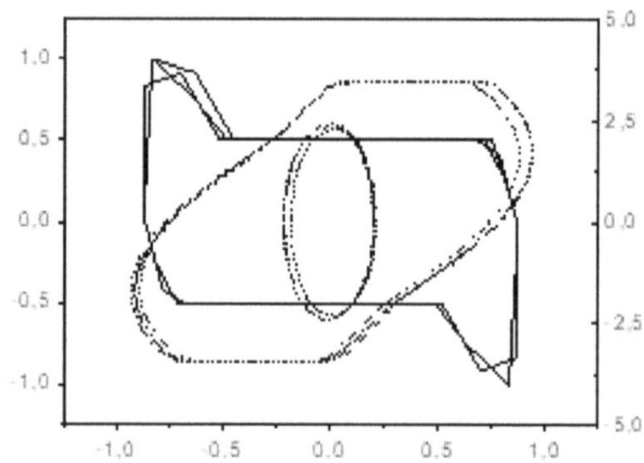

We see (Figure 1) that oscillations (variable x_2) emerge at the system input, and limited oscillations (variable x_4) occur at the output. They result from the action of nonlinearity in the system.

Presented frameworks show the processes do not have distortions (they are symmetrical). Therefore, the system input is S-synchronizing. The S-synchronizability condition is a necessary for the SID analysis. We see (Figure 1) that the conclusion about the SID of the system cannot make based on the mapping Apply the hierarchical immersion method (Karabutov, 2021). It gives mappings $\gamma_{x_1x_2}:x_1 \rightarrow x_2$ reflecting the nonlinear system (9.3) state.

Definition 2. System (9.3) is locally identifiable (LI) on the set $\{x_1,x_2\}$ if the framework $S_{x_2x_1}$ described by the mapping $\gamma_{x_1x_2}:x_1 \to x_2$ and has the maximum diameter of the definition domain.

Figure 1 shows that the framework $S_{x_2x_1}$ is LI. The analysis $S_{x_1\dot{x}_1}$ does not give the estimate for nonlinearity structural parameters (SP) from adulteration. Go into a specific structural space. The mathematical model design is based on the relationship between variables (the interaction graph (Karabutov, 2015)).

Synthesize models on sets $\{u,\dot{x}_4\}$ and $\{x_1,\dot{x}_4\}$

$$\hat{\dot{x}}_{4,u} = 2.234u - 0.004 \,,$$

$$\dot{x}_{4,x_1} = -0.0712x_1 + 0.003 \tag{9.5}$$

Determine residuals $\varepsilon_{4,x_1} = \dot{x}_4 - \hat{\dot{x}}_{4,x_1}$, $\varepsilon_{4,u} = \dot{x}_4 - \hat{\dot{x}}_{4,u}$ and introduce frameworks $S_{\dot{x}_4,\varepsilon_{4,x_1}}$ and $S_{\dot{x}_4,\varepsilon_{4,u}}$ described by functions $\gamma_{\varepsilon_{4,x_1}\dot{x}_4}:\varepsilon_{4,x_1} \to \dot{x}_4$ and $\gamma_{\varepsilon_{4,x_1}u}:\varepsilon_{4,u} \to \dot{x}_4$ (Figure 2).

Figure 2. System frameworks for estimating identifiability on analysis u, x_1

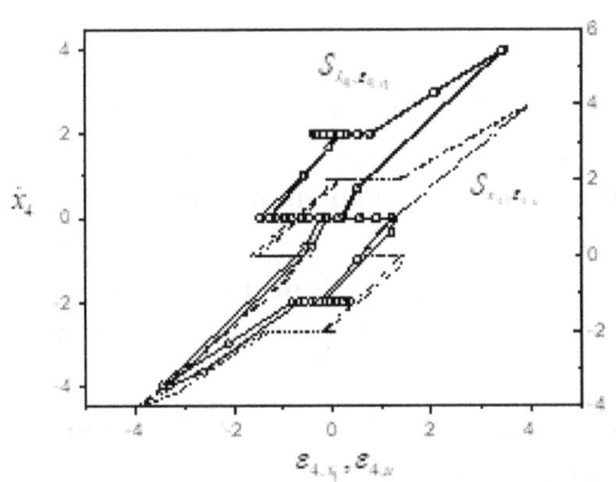

Frameworks $S_{\dot{x}_4,\varepsilon_{4,x_1}}$, $S_{\dot{x}_4,\varepsilon_{4,u}}$ reflect the influence of nonlinearities φ_1 and φ_2. We see (Figure 2) that frameworks have saturation equaling to 2 and -2, which coincides with the initial data of the system (9.3). Switching points of nonlinear functions shifted because of the adulterating effect.

Construct secants for $S_{\dot{x}_4,\varepsilon_{4,x_1}}$ and $S_{\dot{x}_4,\varepsilon_{4,u}}$

$$\gamma_{\dot{x}_4,\varepsilon_{4,x_1}} = a_{4,x_1}\varepsilon_{4,x_1},$$

$$\gamma_{\dot{x}_4,\varepsilon_{4,u}} = a_{4,u}\varepsilon_{4,u} + b_{4,u} \tag{9.6}$$

where $a_{4,x_1} = 1, a_{4,u} = 1.027, b_{4,u} = 0.22$.

Definition 3. The system (9.3) is *structurally unidentifiable* on the set $\{u(t), x_1(t)\}$ if the input is S-synchronizing, and the secant (9.6) coefficients satisfy the condition $|a_{4,x_1} - a_{4,u}| \leq \delta$, where $\delta \geq 0$.

Definition 3 applies to nonlinearities φ_1, φ_2 with $\delta = 0.005$. This conclusion confirms by Figure 3 (cantered diagram), which shows the variables ε_{4,x_1} (gray) and $\varepsilon_{4,u}$ (white).

Figure 3. Comparison of variables ε_{4,x_1} and $\varepsilon_{4,u}$

The nonlinearity φ_1 dominates over φ_2, but its witching occurs at $x_1 \approx \pm 1.5$, which corresponds to parameters φ_2. SID is the confirmation of structural identification. Therefore, we conclude that adulteration complicates the identifiability of the system (9.3) structure.

Go to space $\{\varepsilon_{4,x_1}, \varepsilon_{4,x_4}\}$, where ε_{4,x_4} forms as ε_{4,x_1}, and decide on φ_1. Obtained results show the structural identification possibility of nonlinearity $\chi(\varphi_1, \varphi_2)$ based on φ_1, φ_2.

Statement 1 (Karabutov, 2022). The system (9.3) nonlinear part $\chi(\varphi_1, \varphi_2)$ is locally structurally identified on the set $\{u(t), x_2(t), x_4(t)\}$ if the input $u(t)$ is S-synchronizing.

Proof. The statement follows from (9.3) and Figure 1 that the input is S-synchronizing. S-synchronizability is guaranteed the maximum definitional domain diameter of framework $S_{x_4,\varepsilon_{4,x_1}}$, $S_{x_4,\varepsilon_{4,u}}$. Therefore, $\chi(\varphi_1, \varphi_2)$ is *h*-identifiable (Karabutov, 2015). As the system output depends on the difference $\varphi_1(u)$ and $\varphi_2(x_4)$, the system (9.3) structure will depend on the structural parameters of the final function $\chi(\varphi_1, \varphi_2)$. ■

Corollary of statement 1. *Functions $\varphi_1(u)$ and $\varphi_2(x_4)$are structurally unidentifiable on set $\mathbb{I}_o$.*

Figure 4 gives indirect confirmation of statement 1, where the framework presents reflecting the function $\chi(\varphi_1, \varphi_2)$ influence, $\hat{\chi}(\varphi_1, \varphi_2)$ is function $\chi(\varphi_1, \varphi_2)$ estimation based on the analysis of the set $\mathbb{I}_o$.

Figure 4. Evaluation of effect $\chi(\varphi_1, \varphi_2)$ on system (9.3) structure

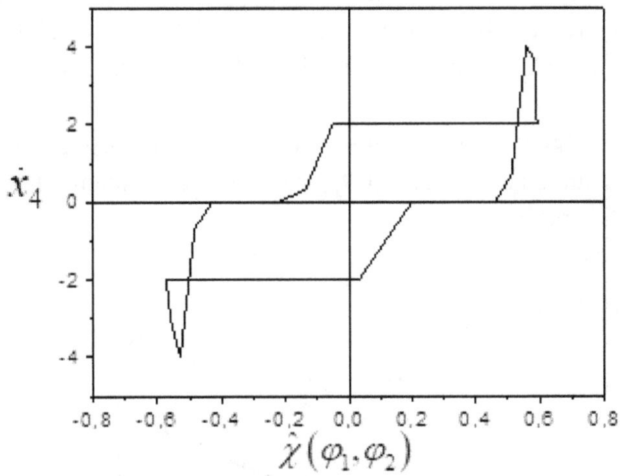

Introduce the variable $v = u - x_2$ and the model for $\dot{x}_4$:

$$\hat{\dot{x}}_4 = a_{4,0} + a_{4,1} v \tag{9.7}$$

where $a_{4,0} = 0.125, a_{4,1} = 1.45$. Find the residual $\vartheta_4 = \dot{x}_4 - \hat{\dot{x}}_4$ and construct the framework $S_{\dot{x}_4,\vartheta_4}$ described by function $\gamma_{,\dot{x}_4}:\vartheta_{4\,1} \rightarrow \dot{x}_4$. Determine for $S_{\dot{x}_4,\vartheta_4}$, $S_{\dot{x}_4,\varepsilon_{4,x_1}}$ secants

$$\gamma_{\vartheta_4,\dot{x}_4} = \alpha_{4,0}\vartheta_4 + \alpha_{4,1},$$

$$\gamma_{\varepsilon_{4,x_1},\dot{x}_4} = \mu_{4,0}\varepsilon_{4,x_1} + \mu_{4,1} \tag{9.8}$$

where $\alpha_{4,0} = 0.996$, $\alpha_{4,1} = -0.06$, $\mu_{4,0} = 1.028$, $\mu_{4,1} = -0.219$.

Theorem 1. *If 1) $u \in S$, 2) coefficients $\alpha_{4,0}, \mu_{4,0}$ of secants (9.8) for frameworks $S_{\dot{x}_4,\vartheta_4}, S_{\dot{x}_4,\varepsilon_{4,x_1}}$ satisfy condition:*

$$|\alpha_{4,0} - \mu_{4,0}| \leq \delta_{\alpha,\mu} \tag{9.9}$$

where $\delta_{\alpha,\mu} \geq 0$ is specified value, then the function $\chi(\varphi_1, \varphi_2)$ is structurally identifiable on the set $\{\vartheta_4(t), \dot{x}_4(t)\}$.

Proof. The input of the system (9.3) is S-synchronizing. According to statement 1, the framework $S_{\dot{x}_4,\varepsilon_{4,x_1}}$ has the maximum diameter of the definition domain. Therefore, the function $\chi(\varphi_1, \varphi_2)$ is locally structurally identifiable. The framework $S_{\dot{x}_4,\vartheta_4}$ has the maximum diameter of the definition domain as well. We construct secants (9.8) for $S_{\dot{x}_4,\vartheta_4}, S_{\dot{x}_4,\varepsilon_{4,x_1}}$. The condition (9.9) is satisfied with $\delta_{\alpha,\mu} = 0.04$. Hence, the function $\chi(\varphi_1, \varphi_2)$ is locally structurally identifiable on the set $\{\vartheta_4(t), \dot{x}_4(t)\}$. ■

The theorem 1 application simplifies the identifiability estimation of the system (9.3).

Figure 5 confirms the validity of theorem 1, where we present frameworks $S_{\dot{x}_4,\vartheta_4}$, $S_{\varepsilon_{4,x_1},\vartheta_4}$ and the GF adequacy in space $(\vartheta_{4\,1}, \varepsilon_{4,x_1})$. The coefficient of determination is 99% between $\vartheta_{4\,1}$ and ε_{4,x_1}.

Figure 5. Theorem 1 verification

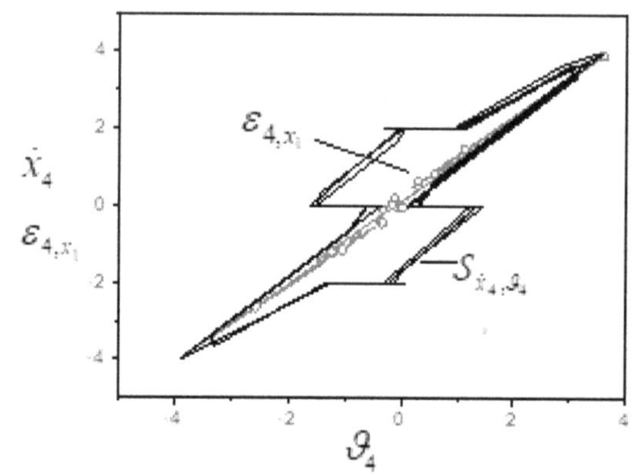

So, obtained results show that the system (9.3) is structurally unidentifiable on set $\mathbb{I}_o$. Go into space $(\vartheta_4(t), \dot{x}_4(t))$ and get estimates for φ_1, φ_2.

Apply the variable permutation procedure Karabutov, (2022). Let $\dot{x}_4$ be the input, and x_1, x_2 are outputs. This approach almost eliminates the adulteration influence. Construct frameworks $S_{x_1,\dot{x}_4}$, and $S_{x_2,\dot{x}_4}$. As the input is S-synchronizing, frameworks $S_{x_1,\dot{x}_4}$, $S_{x_2,\dot{x}_4}$ are significant (Karabutov, 2015, 2021) and have the maximum diameter of the definitional domain. We apple theorem 1 and confirm by h_{δ_h}-identifiability) of the system (9.3). Frameworks show in Figure 6. GF analysis gives to evaluate of the properties for nonlinearities on the set of function switching intervals. Figure 6 shows that the system has two nonlinearities. One nonlinearity (framework $S_{x_2,\dot{x}_4}$) is saturation, and the framework $S_{x_1,\dot{x}_4}$ is a saturation function with dead space (DS). DS does not coincide with the original area from the specifics of the system. The saturation level is 2.

Figure 6. Frameworks $S_{x_1,\dot{x}_4}$, $S_{x_2,\dot{x}_4}$

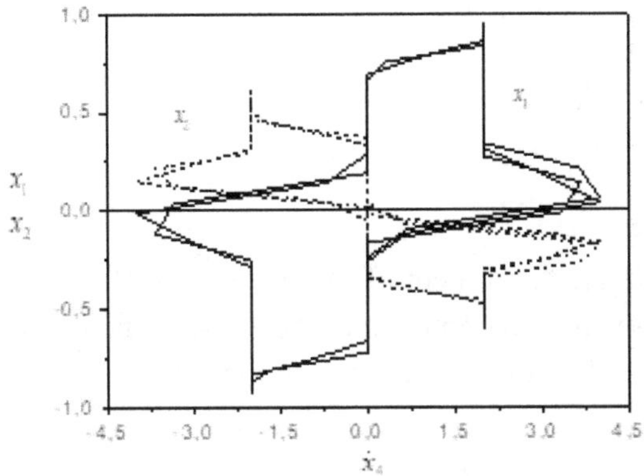

The relationship between u and x_2 (the determination coefficient is 0.98) affects the nonlinear properties of the system (9.3). Consequently, the system (9.3) and the nonlinearity $\chi(\varphi_1, \varphi_2)$ are SID. Parameters $\chi(\varphi_1, \varphi_2)$ depend on their connection to the system.

So, we have

Statement 2. *If: (i)* $u \in \mathrm{S}$ *and* $\dot{x}_4 \in \mathrm{S}$*; (ii) frameworks* S_{x_1,x_4}, S_{x_2,x_4} *described by functions* $\gamma_{\dot{x}_4,x_1}:\dot{x}_4 \to x_1, \gamma_{\dot{x}_4,x_2}:\dot{x}_4 \to x_2$ *have* $\mathrm{d}_{h,\Sigma}$*-optimal properties, then system (9.3) and function* $\chi(\varphi_1, \varphi_2)$ *are structurally identifiable.*

5. NA OF NONLINEARITY ARGUMENT

Consider the self-oscillation generation system with the nonlinear feedback (NF) (9.10). NF adulterates the nonlinear function argument in a straight chain. The system contains: 1) the engine with parameters k_4, T_2 and output y; 2) gain element with parameters k_3, T_1; 3) nonlinear feedback on x_2 with the function $f_2(x_2)$; 4) linear feedback on x_2 with parameters $k_1 k_5$:

$$S_f:\begin{cases} \dot{x}_1 = x_2, \\ \dot{x}_2 = x_3, \\ \dot{x}_3 = -\dfrac{1}{T_1 T_2}x_2 - \dfrac{T_1 + T_2}{T_1 T_2}x_3 + \dfrac{k_3 k_4}{T_1 T_2}f_1(u,b,c), \\ \dot{x}_4 = -k_1 k_5 x_2, \\ y = x_1, \end{cases} \tag{9.10}$$

where y is the rotation angle of the engine shaft, $u = x_4 - k_{os} f_2(x_2)$ is function f_1 argument, $f_2(x_2) = (x_2)^2 \operatorname{sign}(x_2)$,

$$f_1(x_4)\begin{cases} c, \text{ if } x > b \\ 0, \text{if } x \in [-b, b], \\ -c, \text{ if } x < -b \end{cases}$$

$$b > 0 \tag{9.11}$$

We have the adulterated argument of the function f_1. Hence, its argument $u(x_2, x_4)$ is the nonlinear function. It complicates the SI assessment of the system (9.10).

Consider the system S_f with two sets of parameters

$$P_i = \{k_{1(i)}, k_{3(i)}, k_{4(i)}, k_{5(i)}, k_{os(i)}, T_{1(i)}, T_{2(i)}, b_{(i)}, c_{(i)}\},$$

$$i = 1,2 \tag{9.12}$$

where:

$$\begin{aligned} P_1 &= \{0.35, 10, 1, 1.25, 0.005, 2.5, 0.4, 0.45, 0.75\}, \\ P_2 &= \{0.35, 10, 1, 0.55, 0.005, 2.5, 0.4, 0.25, 2\}. \end{aligned} \tag{9.13}$$

These sets guarantee self-oscillation in the system. Therefore, the input is constantly excited for P_i, and frameworks $S_{ey(i)}$ are h-identifiable. But the adulterating effect can give false h-identifiable frameworks $S_{ey(i)}$ (denote these frameworks and the system as $LS_{ey(i)}$, $LS_{f(i)}$). Extend the frameworks $S_{ey(i)}$ set and their derivatives (frameworks $KS_{ey(i)}$) to select a reference information set I_0^r for analysis of the SID.

Consider the variable $u(t)$ and construct the structurally-frequency diagram H_u for it. H_u reflects the distribution u according to the change in the system output.

Definition 4. If H_u contains areas conforming to the system (9.10) features, then the variable $u(t)$ is *structurally significant* or $u \in H(H_u)$.

Definition 5. $S_{ey(i)}$-framework based on $u \in H(H_u)$ is *structurally significant* or $S_{ey(i)} \in SS_{ey(i)}$. The system $S_{f(i)}$ with $u \in H(H_u)$ is called *a structurally significant* or $S_{f(i)} \in SS_{f(i)}(H(H_u))$. If $u \notin H(H_u)$, then $S_{f(i)} \in SIS_{f(i)}(H(H_u))$.

Consider frameworks $S_{x_{3(i)}, x_{1(i)}}$, $S_{u_{(i)}, x_{4(i)}}$ described by functions

$$\eta_{3,1(i)} : x_{1(i)} \rightarrow x_{3(i)},$$

$$\mu_{4,u(i)} : u_{(i)} \rightarrow x_{4(i)}. \tag{9.14}$$

In Figures 7 and 8, we present GF and its structural estimation used structural-frequency analysis (Karabutov, 2015) on sets (9.13).

Theorem 2. *Consider the system S_f with sets* P_1, P_2 *and (i) inputs $u_{(i)}$ are constantly excited; (ii) the system state reflects by the frameworks $S_{x_{3(i)},x_{1(i)}}$, $S_{u_{(i)},x_{4(i)}}$ and the structural-frequency diagrams* $H_{(i)}$ *$(i = 1,2)$. The frameworks defined on the set* P_1 *are structurally significant, i.e., they are applicable for SID analysis.*

Figure 7. System (9.10) phase portraits with parameters P_1

(a) framework $\mathscr{S}_{x_{3(1)},x_{1(1)}}$

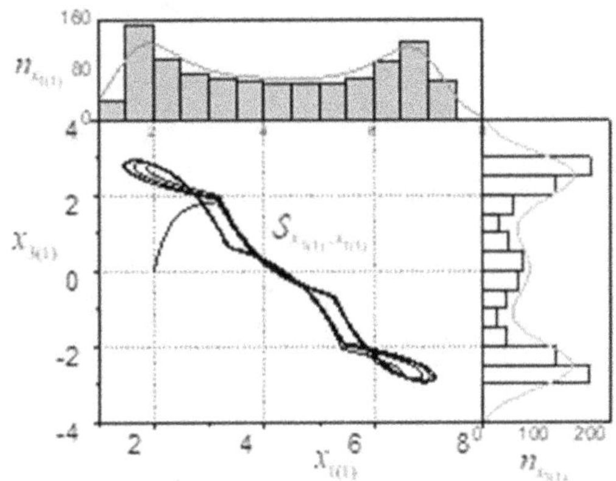

(b) framework $\mathscr{S}_{u_{(1)},x_{4(1)}}$

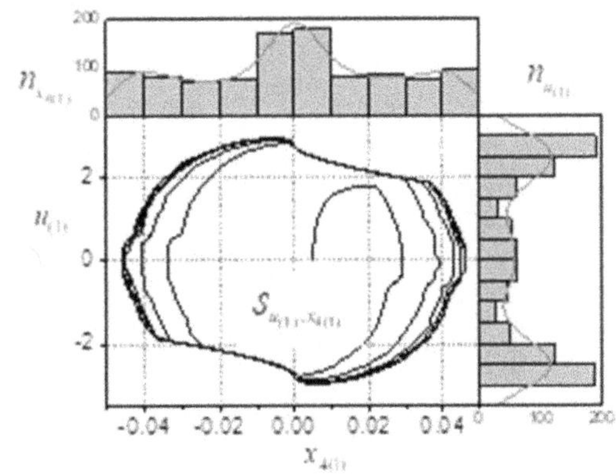

Proof. The system (9.10) generates self-oscillations. Therefore, the input $u_{(i)}$ satisfy the condition of constant excitation for the given P_i. Consider the structural-frequency diagrams (SDS) presented in Figures 7 and 8. The system S_f contains two nonlinearities. Therefore, SDS should reflect their features. Consider frameworks $S_{u_{(i)},x_{4(i)}}$. We see that the framework $S_{u_{(1)},x_{4(1)}}$ contains fragments (see the distribution of $n_{u_{1(i)}}$ in Figure 7), which are not represented in $S_{u_{(2)},x_{4(2)}}$. Compare distributions $n_{u_{1(i)}}$. We see that the input $u_{(1)}$ is more structurally significant than $u_{(2)}$. Then $u_{(1)} \in H\left(\mathrm{H}_{u_{(1)}}\right)$, and $u_{(2)} \notin H\left(\mathrm{H}_{u_{(2)}}\right)$. Thus, the system (9.10) with the set P_1 is a candidate for further analysis of the system S_f identifiability. Confirm these conclusions by the analysis of frameworks $S_{x_{3(i)},x_{1(i)}}$ (Figure 8). Further SID analysis bases on the use of the information set for the system $S_f(\mathrm{P}_1)$. Then $S_f(\mathrm{P}_1) \in SS_{f(1)}(H(\mathrm{H}_u))$, and $S_f(\mathrm{P}_2) \in SIS_{f(2)}(H(\mathrm{H}_u))$. ■

Figure 8. System (9.10) phase portraits with parameters P,

(a) framework $\mathscr{S}_{x_{3(2)},x_{1(2)}}$

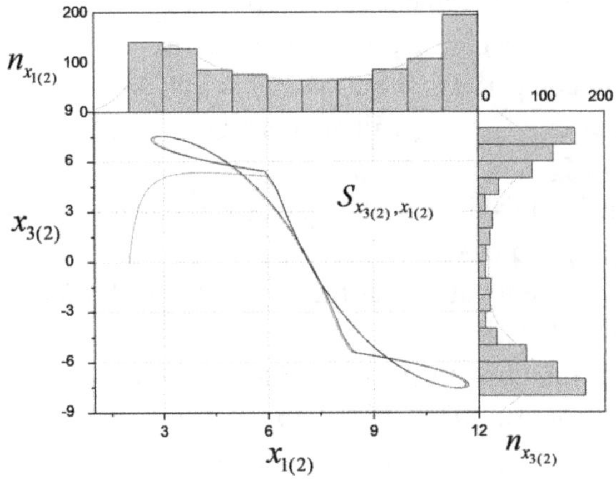

(b) framework $\mathscr{S}_{u_{(2)},x_{4(2)}}$

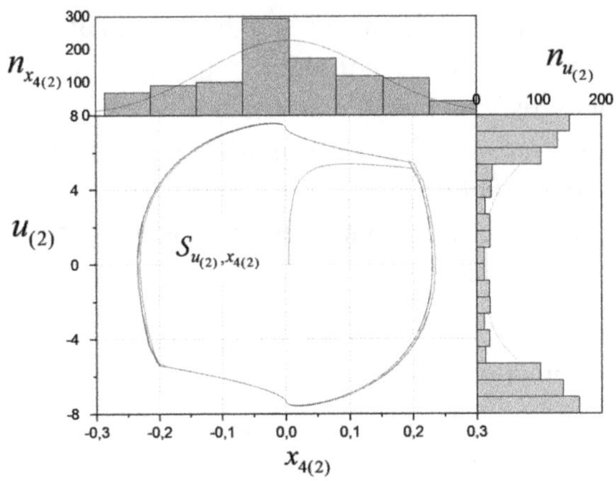

The framework $S_{x_{3(2)},x_{1(2)}}$ does not reflect the system (9.10) specifics. Explain it by the nonlinear feedback action.

Consider the system (9.10) defined on the set of parameters

$$\mathrm{P}_3 = \{0.35, 12, 1, 0.55, \{k_{os,i}\}, 3, 0.4, 0.5, 1.5\} \tag{9.15}$$

with different values k_{os}.

Statement 3. If the system (9.10) is defined on the set (9.15), then increasing the parameter k_{os} changes the function F2 domains in the coordinate origin.

Proof. Consider a neighborhood O_{f_1} of function f_1 coordinate origin. Let $k^*_{os} = 0.0025$ is the reference value of the coefficient k_{os}. Augmenting $k_{os,j} \in \{k_{os,i}\}$ increases function f_2 values. Increasing f_2 changes between the argument $u(k_{os,j})$ and the function f_1 form, compared to the values of $f_1(u(k^*_{os}))$ at the same t. Therefore, the domain of values $O_{f_1(u)}|_{u=u(\{k_{os,i}\})}$ will not coincide with the domain $O_{f_1(u)}|_{u=u(k^*_{os})}$ of the same function f_1. ■

Denote the domain of values and the diameter of neighborhood $O_{f_1(u)}|_{u=u(\{k_{os,i}\})}$ as $\mathrm{Im}\left(O_{f_1}(u \neq u(k^*_{os}))\right)$ and $D_{O_{f_1(u)}}$.

Definition 6. Call the input $u(t)$ of the system (9.10) U_h-*identifying* if it minimizes $\mathrm{Im}\left(O_{f_1}(u \neq u(k^*_{os}))\right)$.

Denote the set of inputs that have U_hG2-identifying as $\mathbb{U}_{U_h}$

Theorem 3 (Karabutov, 2022). *If the input $u(t)$ is constantly exciting and U_h-identifying, then the system (9.10) and the function f_1 are h_{δ_h}-identifiable.*

The proof of theorem 3 is based on the construction and analysis of S_{x_3,x_1}.

Domain $O_{f_1(u)}$ at $u \notin \mathbb{U}_{U_h}$ can be determined by the histogram or the framework S_{x_3,x_1}. The region $O_{f_1(u)}$ reflects the change in the interval $[-b, b]$ of the function f_1. It leads to the change in the definition area and the domain of values f_1. According to theorem 3, this violates the h_{δ_h}-identifiability property of the system (9.10).

Consider an input satisfying theorem 3 conditions.

Theorem 4. *If the input $u \in \mathbb{U}_{U_h}$ is constantly exciting, then the function $f_2(x_2)$ of the system S_f is h_{δ_h}-identifiable.*

The proof of theorem 4 follows from theorem 2 conditions fulfilment. Show the function $f_2(\cdot)$ estimation in Figure 9.

Figure 9. Change estimation of function f_2

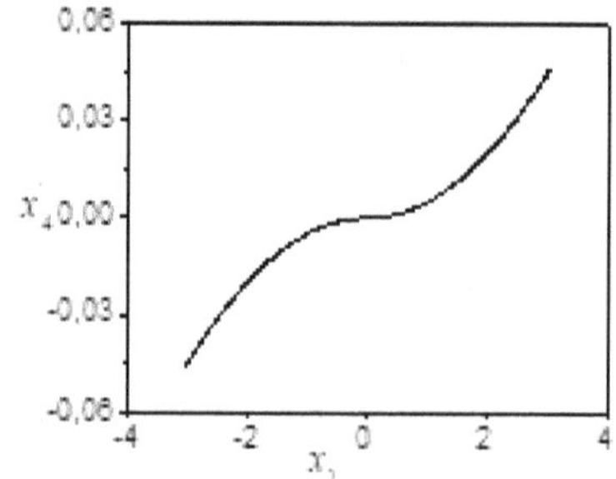

6. CONCLUSION

The structural identifiability problem of nonlinear feedback systems considers with adulterating nonlinearity. We apply the analysis of geometric frameworks and hierarchical immersion method for identifiability estimation study under uncertainty. We consider the nonlinear adulterating case of the nonlinear function argument. Conditions of structural identifiability obtain for this case. We study the input influence on the nonlinearity structural identifiability in the system straight chain. The influence of argumentative adulterating shows on the nonlinearity parameters estimation. Structurally-frequency diagrams are the basis for the analysis of these systems.

REFERENCES

Besekersky, V. A., & Popov, E. P. (1975). *Theory of automatic control systems*. Nauka.

Chan-Zheng, C., Borja, P., & Scherpen, J. (2021). Passivity-based control of mechanical systems with linear damping identification. arXiv: 2105.04324v2.

Guillaume, J. H. A., Jakeman, J. D., Marsili-Libelli, S., Asher, M., Brunner, P., Croke, B., Hill, M. C., Jakeman, A. J., Keesman, K. J., Razavi, S., & Stigter, J. D. (2019). Introductory overview of identifiability analysis: A guide to evaluating whether you have the right type of data for your modeling purpose. *Environmental Modelling & Software*, *119*, 418–432. DOI: 10.1016/j.envsoft.2019.07.007

Karabutov, N. (2015). Structural Identification of Nonlinear Dynamic Systems. *International Journal of Intelligent Systems and Applications*, *7*(9), 1–12. DOI: 10.5815/ijisa.2015.09.01

Karabutov, N. (2016). Structural identification of dynamic systems with hysteresis. *International Journal of Intelligent Systems and Applications*, *8*(7), 1–13. DOI: 10.5815/ijisa.2016.07.01

Karabutov, N. (2021). Geometrical Frameworks in Identification Problem. *Intelligent control and automation, 12*, 17-43. DOI: 10.4236/ica.2021.122002

Karabutov, N. (2022). Structural Identifiability of Feedback Systems with Nonlinear Adulterating. *Contemporary Mathematics*, *3*(2), 257–269. DOI: 10.37256/cm.3220221258

Ko, S., & Bitmead, R. R. (2007). Identifiability analysis for nonlinear feedback limit-cycle systems using harmonic balance methods. *Proceedings of the IEEE Conference on Decision and Control, 39*(1), 2506 – 2511. DOI: 10.1109/CDC.2006.377326

Leboutet, Q., Roux, J., Janot, A., Guadarrama-Olvera, J. R., & Cheng, G. (2021). Inertial parameter identification in robotics: A Survey. *Applied Sciences (Basel, Switzerland)*, *11*(9), 4303. DOI: 10.3390/app11094303

Shin, M.-J., Guillaume, J. H., Croke, A. B. F. W., & Jakeman, A. J. (2015). A review of foundational methods for checking the structural identifiability of models: Results for rainfall-runoff. *Journal of Hydrology (Amsterdam)*, *520*, 1–16. DOI: 10.1016/j.jhydrol.2014.11.040

Wu, B., Han, X., & Hui, N. (2021). System identification and controller design of a novel autonomous underwater vehicle. *Machines (Basel)*, *9*(109), 1–15. DOI: 10.3390/machines9060109

Chapter 10
Identifiability and Detectability of Lyapunov Exponents

ABSTRACT

We consider Lyapunov Exponent (LE) identifiability and detectability of dynamic systems (DS). Calculating the full spectrum of LE is a complex problem. Therefore, identification, recovery, and detection of LE is an important problem of control theory. The whole spectrum of LE cannot be obtained. The identifiability recoverability and detectability issues of Lyapunov exponents were not considered. This problem is actual. For the linear DS, we propose an approach to get and verify these characteristics. The approach bases on the analysis of geometric frameworks (GF) that are depend on the coefficient of structural properties (SPC). We got: (a) conditions of fully detectable LE; (b) -detectability conditions with level of -recoverability if the system contains no recoverable lineans.

1. PROBLEM STATEMENT

Consider the dynamic system

$$\begin{aligned} \dot{X} &= AX + BU, \\ Y &= C^T X, \end{aligned} \tag{10.1}$$

where $X \in \mathbb{R}^m$ is state vector, $U \in \mathbb{R}^k$, $Y \in \mathbb{R}^n$ are input and output, $A \in \mathbb{R}^{m\times m}$, $B \in \mathbb{R}^{m\times k}$, $C \in \mathbb{R}^{n\times m}$. Matrix A is the Hurwitz matrix.

The measurement information

DOI: 10.4018/979-8-3373-0822-7.ch010

$$\mathbb{I}_o = \{Y(t), U(t), t \in \mathbb{J} = [t_0, t_1]\} \tag{10.2}$$

The solution of the system (10.1) writes as:

$$X(t) = X(t_0, u, t) \tag{10.3}$$

where X is an operator determined by matrices A, B and initial conditions for $X(t_0)$.

The system (10.1) solution has the form

$$X(t) = X_g(t) + X_q(t) \tag{10.4}$$

where $X_q(t)$ is a particular solution with $U \in \mathbb{I}_o$, $X_g(t)$ is a general solution with $U = 0$.

Denote the general solution of with system (10.1) $X_0 = X_0(Y_0) \in \mathbb{I}_o$ as $X_g(X_0, t)$.

Problem: determine estimates for the $X_g(t) = X_g(X_q, X_0, t)$ on the set $\mathbb{I}_o$, get LE spectrum identifiability and detectability conditions.

The estimate method of the $X_g(t) = X_g(X_q, X_0, t)$ solution based on processing the set $\mathbb{I}_o$ is described in part 1. Therefore, we believe that the evaluation $X_g(t)$ is known.

2. RECOVERABILITY AND DETECTABILITY

Consider the basic concepts from the general theory of systems (Kwakernaak et al., 1972). Issues of recoverability and detectability are closely interrelated. The LE calculation is based on obtaining estimates for state variables (SV). The possibility of determining SV relates to the problem of system recoverability (10.1).

Definition 1. Let $y(t, t_0, X_0, u)$ describes the output change of the system:

$$\begin{aligned} \dot{X}(t) &= A(t)X(t) + Bu(t), \\ y(t) &= C^T X(t) \end{aligned} \tag{10.5}$$

with $X_0 = X(t_0)$. Then the system is fully recoverable if a moment t_1 exists for all $t_0 \in (-\infty, t_1)$ such that from

$$y(t, t_0, X_0, u) = y(t, t_0, X'_0, u) \tag{10.6}$$

$\forall u(t)$ and $t_0 \in [t_0, t_1]$ follow $X_0 = X'_0$.

As follows from definition 1, if the system is fully recoverable and the system output is observed until the moment t_1, then the moment $t_0 < t_1$ exists that the system state determines uniquely. Therefore, if $X(t_0)$ is known, then the value $X(t_1)$ can be determined.

Theorem 1. *System (10.5) is fully recoverable if and only if such a moment $t_0 \in (-\infty\ , t_1)$ exists for all t_1, that from equality*

$$y(t, t_0, X_0, 0) = 0 \tag{10.7}$$

fellow $X_0 = 0$.

Definition 2. Let A, C be matrices of dimension $n \times n$ and $l \times n$. Then the pair (A, C) is called fully recoverable if the system

$$\begin{aligned} \dot{X}(t) &= AX(t), \\ y(t) &= CX(t) \end{aligned} \tag{10.8}$$

is fully recoverable.

There may be a situation when the system state is not fully recoverable. This case complicates the estimation of the system current state.

Let $y = Y, u = U, y \in \mathbb{R}, u \in \mathbb{R}$ in (10.1).

Definition 3. The space of non-recoverable states PN_n for the system (10.1) is a linear subspace of states X_0 if $y(t, t_0, X_0, 0) = 0 \quad t \geq t_0$.

Theorem 2. *The non-recoverable states PN_n* subspace of the system (10.1) is the null space of the recoverability matrix

$$D = \begin{pmatrix} C \\ CA \\ CA^2 \\ \vdots \\ CA^{n-1} \end{pmatrix}$$

The state property belonged to PN_n: the system motion at $u = 0$ converges to zero. Therefore, the non-recoverable state component does not give a growth of error. The system with this property is detectable.

Definition 4. If the subspace of non-recoverable states belongs to the subspace of stable states, then the system (10.1) is detectable.

Definition 5. A pair (A, C) is detectable if the system

$$\begin{cases} \dot{X}(t) = A(t)X(t), \\ y(t) = CX(t) \end{cases} \tag{10.9}$$

is detectable.

Theorem 3. *Any asymptotically stable system (10.1) is detectable. Any fully recoverable system (10.1) is detectable.*

Consider the system (10.8) and a transition matrix $\Phi(t,t_0)$ corresponding to it. System (10.8) is fully recoverable if and only if

$$M(t,t_1) = \int_t^{t_1} \Phi(\tau,t)C^T(\tau)C(\tau)\Phi(\tau,t_0)d\tau > 0(\forall t_1) \;\&\; (-\infty < t_0 < t_1) \quad (10.10)$$

If

(a) $\alpha_0 I \leq M(t_1-\sigma,t_1) \leq \alpha_1 I \forall t_1,(\alpha_0,\alpha_1) > 0,$
(b) $\beta_0 I \leq \Phi^T(t_1-\sigma,t_1)M(t_1-\sigma,t_1)\Phi(t_1-\sigma,t_1) \leq \alpha_1 I \forall t_1, (\alpha_0,\alpha_1,\beta_0) > 0,$

then the system (10.8) is uniformly and fully recoverable.

Uniform recoverability guarantees the identification of the state at a time interval. Uniform full recoverability of systems (10.1) follows from full recoverability.

Next, using the approach from Chapter 5, we get the variable $X_g(t)$.

3. LE. COEFFICIENT OF STRUCTURAL PROPERTIES

LE and coefficient of structural properties are considered in the chapter 1 subsection 4.3. We will remind some definitions and notations. LE (Lyapunov, 1992) is

$$\chi[h] = \overline{\lim_{t\to\infty}} \frac{\ln|h(t)|}{t} \quad (10.11)$$

where $\overline{\lim_{t\to\infty}}$ is upper limit.

For the stationary system (10.1), LE χ_i $(i = \overline{1,m})$ coincide with the eigenvalues λ_i real part of the matrix A.

Let the estimate of the general solution $X_g(t)$ $\forall t \in \mathbb{J}_g$ known and the system (10.1) be stable. Therefore, the system (10.1) is recoverable and discoverable. Apply (10.11) to $\hat{y}_g(t)$:

$$\chi[\hat{y}_g] = \overline{\lim_{t\to\bar{t}}} \frac{\ln|\hat{y}_g(t)|}{t} \quad (10.12)$$

where $\bar{t} \in \mathbb{J}_g$ is the maximum value (upper bound) on $\mathbb{J}_g \subset \mathbb{J}$.

(10.12) is the largest (maximum) LE. If the limit (10.11) exists, then $\chi\left[\hat{y}_g\right]$ is the estimate of the matrix A maximum eigenvalue. Therefore, $\chi\left[\hat{y}_g\right]$ is an indicator of the stability degree for the system (10.1). If $m = 2$, then we obtain for $\dot{\hat{y}}_g$:

$$\chi\left[\dot{\hat{y}}_g\right] = \overline{\lim_{t \to \bar{t}}} \frac{\ln\left|\dot{\hat{y}}_g\right|}{t} \tag{10.13}$$

Perron bottom index (Izobov, 2006) is defined as:

$$\eta[h] = \underline{\lim}_{t \to \infty} \frac{\ln|h(t)|}{t} \tag{10.14}$$

where $\underline{\lim}_{t \to \infty}$ is a bottom limit.

Karabutov (2009, 2018) proposes using LE to solve identification problems. The approach is based on the analysis of the coefficient structural properties

$$k_s(t, \rho) = \frac{\rho\left(\hat{y}_g\right)}{t} \tag{10.15}$$

where $\rho\left(\hat{y}_g\right) = \rho_g = \ln\left|\hat{y}_g(t)\right| \; \forall t \in \mathbb{J}_g \subset \mathbb{J}_g$, $J_g = \left[t_0, \bar{t}\right]$ determine from (10.12).

4. FRAMEWORKS FOR ESTIMATE LE

Consider sets

$$\mathbb{I}_{k_s} = \left\{k_s\left(t, \rho\left(\hat{y}_g(t)\right)\right), t \in \mathbb{J}_g\right\},$$

$$\mathbb{I}_{k_s'} = \left\{k_s\left(t, \rho\left(\dot{\hat{y}}_g(t)\right)\right), t \in \mathbb{J}_g\right\}. \tag{10.16}$$

Define the mapping (framework) $S_{k_{s,\rho}} \to \mathbb{I}_{k_s} \times \mathbb{I}_{k_s'}$. Consider the function on $\mathbb{I}_{k_s'}$:

$$\Delta k_s'(t) = k_s\left(t, \rho\left(\dot{\hat{y}}_g(t+\tau)\right)\right) - k_s\left(t, \rho\left(\dot{\hat{y}}_g(t)\right)\right), \tag{10.17}$$

where $\tau > 0$.

Form the set

$$\mathbb{I}_{\Delta k_s'} = \left\{\Delta k_s\left(t, \rho\left(\hat{y}_g(t)\right)\right), t \in J_g\right\} \tag{10.18}$$

and introduce the frameworks

$$SK_{\Delta k'_{s,\rho}} \rightarrow \square_{k_{s,\rho}} \times \square_{\Delta k'_{s,\rho}},$$

$$LSK_{\Delta k'_{s,\rho}} \rightarrow \square_{k_{s,\rho}} \times \mathbb{B}\left(\square_{\Delta k'_{s,\rho}}\right) \tag{10.19}$$

where $\mathbb{B}\left(\square_{\Delta k'_{s,\rho}}\right) \subset \{-1;1\}$. Elements of the binary set $\mathbb{B}\left(\square_{\Delta k'_{s,\rho}}\right)B\left(\mathrm{I}_{\Delta k'_{s,\rho}}\right)$

$$b(t)=\begin{cases} 1, \ if \ \Delta k'_s(t)\geq 0, \\ -1, \ if \ \Delta k'_s(t)< 0, \end{cases} \tag{10.20}$$

where $t \in \mathbb{J}_g$.

Remark 1. Choose the upper limit boundary in (10.12) is based on changes in $S_{k_{s,\rho}}$.

Remark 2. The choice of the $b(t)$ value range is determined by its graphical representation.

Consider frameworks $SK_{k^i_{s,\rho}} \rightarrow \square_{k_s} \times \square_{k^i_s}$ and $LSK_{\Delta k'_{s,\rho}}$, where i is the ith derivative of $\hat{y}_g(t)$,

$$\square_{k^i_s} = \left\{k_s\left(t,\rho\left(\hat{y}^{(i)}_g(t)\right)\right), t \in \mathbb{J}_g\right\} \tag{10.21}$$

Largest LE (LLE) $\chi_l\left[\hat{y}_g\right]$ locate on the left side of the framework $SK_{k^i_{s,\rho}}$. All other $\chi_i\left[\hat{y}_g\right]$ coincide with Perron bottom indices $\eta_i\left[\hat{y}_g\right]$. They are located to the right of $\chi_l\left[\hat{y}_g\right]$. Indices $\chi_i\left[\hat{y}_g\right]$ correspond to local minima on $SK_{k^i_{s,\rho}}$. Minima $SK_{k_{s,\rho}}$ coincides with the change in the framework's sign $LSK_{\Delta k'_{s,\rho}}$. Hence, we get the set $M^m_{LE} = \{\hat{\lambda}_1,\hat{\lambda}_2,\ldots,\hat{\lambda}_m\}$ of eigenvalues for the system (10.1). If the system (10.1) is fully L-detectable (see below), then we obtain the full spectrum for LE, which corresponds to the full space of lineans $L^m(t)$.

5. IDENTIFIABILITY AND DETECTABILITY OF LE

5.1. Stationary System

Consider the input $u \in PE_\alpha$, where $u \in PE_\alpha$ is the constancy excitation (CE) property

$$PE_\alpha: u^2(t) \geq \alpha \tag{10.22}$$

fair for $\exists\alpha > 0$ and $\forall t \geq t_0$ at some interval $T > 0$. With such the input, the phase portrait of the system is closed, the system (10.1) is identifiable (Karabutov, 2022).

Definition 6 (Smirnov, 1968). A set $\mathbb{F}$ of elements from L_2 is linean. If functions $\varphi(p) \in \mathbb{F}$ and $\phi(p) \in \mathbb{F}$, then we have $c\varphi(p) \in \mathbb{F}$ $c\varphi(p) \in \mathbb{F}$ for any c and $(\varphi(p) + \varphi(p)) \in \mathbb{F}$.

Definition 7. If LE set detects based on lineans, then the system (10.1) is L-detectable.

We present the modification of Theorem 1 (Karabutov, 2015) for the system (10.1) with simple eigenvalues.

Theorem 4. *LE set is* LP-*detectable or fully detectable, and system (10.1) has the order m if the function b(t) changes sign m − 1 times on the interval* $[t_0, t^*] \subset \mathbb{J}_g$ $(t^* \leq \bar{t})$.

The proof of Theorem 4 coincides with the proof of Theorem 5 from Chapter 5.

Corollary of theorem 4. *If the changes number of function b(t) signs is less* $m - 1$, *than the pyramid* L^m *contains steps that are undetectable.*

Proof. Let the number of signs shift υ of the function $b(t)$ be $\upsilon < m - 1$. Therefore,

$$0 \equiv L^{l_0}(t) \subset L^{l_1}(t) \subset \ldots \subset L^{l_m}(t) \equiv L^m(t) \tag{10.23}$$

contains lineans $\mathrm{NL}_k(t)$, $k \geq 1$. These lineans $\mathrm{NL}_k(t)$ belong to L^m implicitly and are undetectable (non-computable). System (10.1) is stable. Therefore, these sets $\{\mathrm{NL}_k\}$ exist implicitly (in the sense of computation). Then the observed lineans pyramid L_υ has υ observed steps and $L_\upsilon = L^m \backslash \left\{ \mathrm{NL}_{k\,|k>0} \right\}$.■

Denote $\sigma = m - 1 - \upsilon$. From the corollary to theorem 4, we receive statement.

Definition 8. System (10.1) is σ-detectable with the υ-non-reproducibility level if υ lineans are undetectable.

Remark 3. We interpret LP-detectability and recoverability as the possibility of calculating LE.

Remark 4. If the system (10.1) is LP-detectable, then the set M_{LE}^m of Lyapunov exponents corresponds to the steps of the pyramid L^m.

Theorem 5 *(Karabutov, 2022). If the system (10.1) is stable and recoverable, and the framework* $LSK_{\Delta k'_{s,\rho}}$ changes sign $m - 1$ times, then (i) Lyapunov exponents are LP-detectable, (ii) the structure $SK_{k^i_{s,\rho}}$ contains the set M_{LE}^m.

The theorem 5 proof follows from the stability and recoverability of the system (10.1). It is analogous to the theorem 4proof.

Corollary of theorem 5. *If the change in signs of the framework* $LSK_{\Delta k'_{s,\rho}}$ is less $m - 1$, then pyramid L^m contains undetectable steps.

5.2. System With Periodic Coefficients

Consider the system

$$\begin{aligned} \dot{X} &= A(t)X + Bu, \\ y &= C^T X, \end{aligned} \qquad (10.24)$$

where $X \in \mathbb{R}^m$ is the state vector; $u \in \mathbb{R}, y \in \mathbb{R}$ are input and output; $A(t) \in \mathbb{R}^{m \times m}, B \in \mathbb{R}^m, C \in \mathbb{R}^m$.

We believe that assumptions and statements introduced in section 5 of Chapter 5 are fulfilled for the system (10.24).

Consider the framework $SK_{\Delta k'_{s,\rho}}$ defined on $\mathbb{I}_{k_{s,\rho}} \times \mathbb{I}_{\Delta k'_{s,\rho}}$ and described by the function $f_{sk}(t): k_s \to \Delta k_s . k_s(t)$ is a $\mathrm{BF}_{\alpha\pi}$-function. Therefore, $f_{sk}(t)$ contains areas D_{sk} that have sharply varying amplitude.

Definition 9 (Karabutov, 2022). The domain D_{sk} is $\alpha\pi$-domain of the function f_{sk} on the interval $\mathbb{J}_{sk} = [t, t+T]$ $(T > 0)$ if it is $\mathrm{BF}_{\alpha\pi}$-function on this interval.

The domain D_{sk} is $\alpha\pi$-domain of the function f_{sk} on the interval $\mathbb{J}_{sk} = [t, t+T]$ if it is $\mathrm{BF}_{\alpha\pi}$-function on this interval.

Theorem 6 *(Karabutov, 2022). If the system (10.24) is stable and contains prime eigenvalues, then frameworks* $SK_{k^i_{s,\rho}}, i = \overline{1,m}$ contain information about Lyapunov exponents.

Theorem 7 *(Karabutov, 2022). If the system (10.24) is stable and recoverable, and the function* $f_{sk}(t)$ contains at least m of domain D_{sk} on the interval $[t_0, t^*] \subset \mathbb{J}_g$ ($t^* \leq \bar{t}$), then the system (10.24) has the order m and is LP-identifiable (LP-detectable).

The theorem 7 proof follows from theorem 6.

The system (10.24) is stable and recoverable. According to theorem 7, the system (10.24) is identifiable on the set $\left\{ SK_{k^i_{s,\rho}} \right\}$. As shown Karabutov (2015), the local minima on the framework $SK_{k^i_{s,\rho}}$ coincides with domains D^i_{sk} of the structure $SK_{\Delta k'_{s,\rho}}$. As the result of the analysis D^i_{sk}, we obtain the set M_{LE} containing estimates for the LE system (10.28). The cardinal number M_{LE} does not match with the LE quantity. M_{LE} characterizes the available set of lineans for the system (10.24).

Denote the upper bound of the smallest $\eta_i[\hat{y}_g]$ as κ_m.

Theorem 8. *If (a) assumptions A1-A3 from section 5 of chapter 5 fulfill for the system (10.24), (b) the system (10.24) is recoverable, (c) set* $\left\{ k^i_{s,\rho}(t) \right\}$ elements are $\mathrm{BF}_{\alpha\pi}$-functions, (d) the framework $SK_{\Delta k^i_{s,\rho}}$ contains a point at which the sign of the derivative changes, then the indicator κ_m is detectable.

Proof. As assumptions A1-A3 fulfill, the system is stable and recoverable. According to theorem 6, LE is identifiable (detectable) on the framework $SK_{k^i_{s,\rho}}$. $k^i_{s,\rho,j}(t) \in \left\{ k^i_{s,\rho}(t) \right\}$ $(j > 0)$ is the $\mathrm{BF}_{\alpha\pi}$-function. Therefore, the framework $SK_{k^i_{s,\rho}}$ has local minima, the global minimum of which is the upper estimate κ_m for the least $\eta_i[\hat{y}_g]$. This confirms the detectability $\eta_i[\hat{y}_g]$. ■

Remark 5. We decide based on the analysis of various frameworks $SK_{k^i_{s,\rho}}$ $(i \geq 1)$. Therefore, κ_m is the upper bound of $\kappa_{m,i}$.

Remark 6. The approach based on the analysis $SK_{k^i_{s,\rho}}$ gives the estimate of the minimal $\eta_i[\hat{y}_g]$. It is the major advantage of this approach. If the structure $SK_{\Delta k^i_{s,\rho}}$ contains a single drastic in the value, then it is a sign $\eta_m[\cdot]$. As $|\eta_m[\cdot]| > |\chi_i[\cdot]|$, where $i = \overline{1, m-1}$, then $\eta_m[\cdot]$ is equal to κ_m, and it is an upper bound of the least LE. Therefore, the corresponding linear $L^m(t)$ has the minimal domain of definition. The function $\hat{y}_g(t)$ is not $\alpha\pi$-almost periodic on $L^m(t)$, as its parameters α, π decrease rapidly. Therefore, condition almost periodicity not fulfill. So, the linean $L^m(t)$ contains only one value that corresponds to κ_m.

Theorem 9. *Let (a) assumptions A1-A3 from section 5 of chapter 5 hold true for the system (10.24), (b) the system (10.24) is recoverable, (c) set* $\{k^i_{s,\rho}(t)\}$ elements are $BF_{\alpha\pi}$-functions, (d) the structure $SK_{\Delta k^i_{s,\rho}}$ contains not fewer m of regions D^v_{sk} $(v \leq m)$, corresponding to local minima on the structure $SK_{k^i_{s,\rho}}$. Then the set M_{LE} is LP-detectable or fully detectable.

Proof. The system (10.24) is stable and recoverable, and the structure $SK_{\Delta k^i_{s,\rho}}$ contains m regions D^v_{sk}. $SK_{\Delta k^i_{s,\rho}}$ is the framework obtained from $SK_{k^i_{s,\rho}}$. Therefore, local minima correspond to regions D^v_{sk} on $SK_{k^i_{s,\rho}}$ and reflects the LE position. It follows from theorems 7 and 8. Therefore, the set M_{LE} is identifiable or LP-detectable. ■

Remark 7. As the eigenvalues $\lambda_i(t)$ of the matrix A are periodic functions, lineans $L^i(t)$ and $L^{i+1}(t)$ can intersect. This case gives an infinite LE spectrum.

The system (10.24) is non-stationary. Therefore, estimate the area to which the set M_{LE} belongs. Explain this by the power of the set M_{LE} (see remark 7). Note that this area has a lower limit for the κ_m index.

Consider the criterion for evaluating the LP-detectability of the system (10.24). Theoretical indicators (Bylov et al., 1966) offer as strength, reachability complex solutions and do not work under uncertainty. Such indicators (Bylov et al., 1966) as strength, a reachability is not verifiable under uncertainty. The set M_{LE} and the mobility boundary (Karabutov, 2015) $\chi_1 \leq \sup J^1_{k^i_{s,\rho}}$ are the sign of LP-detectability. Therefore, we use analysis results of the structure properties. Apply the approach to confirm the detectability of the set M_{LE}. It bases on checking the χ-adequacy (Karabutov, 2018).

Consider the framework $S_{\hat{y}_g,\hat{\hat{y}}_g}$ described by the function $f_{\hat{y},\hat{\hat{y}}}: \hat{y}_g \rightarrow \hat{\hat{y}}_g$ in the space $R_y = (\hat{y}_g, \hat{\hat{y}}_g)$. The system (10.24) satisfies A1-A3 assumptions from section 5 of chapter 5. Therefore, $S_{\hat{y}_g,\hat{\hat{y}}_g}$ contains regions that reflect the $\alpha\pi$-almost periodic behaviors of the system. Next, consider frameworks $SL_{\Delta k^i_{s,\rho}}$ and $SL_{k^i_{s,\rho}}$ described by functions:

$$f_{SL_{k^i_{s,\rho}}}:\hat{y} \rightarrow k^i_{s,\rho},$$

$$f_{SL_{\Delta k^i_{s,\rho}}}:\hat{y} \rightarrow \Delta k^i_{s,\rho} \tag{10.25}$$

Definition 10. LE estimates are χ-adequate in the space $\mathbb{R}$ if their definition areas coincide with areas of the $\alpha\pi$-almost periodicity of the framework $S_{\hat{y}_g,\hat{\hat{y}}_g}$.

Introduce fragments $D^j_{sl} \subset SL_{\Delta k^i_{s,\rho}}$ $(j \geq 1)$ in space R_y. Denote the domain of definition D^j_{sl} as dom D^j_{sl}. The following modification of theorem 3 (Karabutov, (2018)) is valid.

Theorem 10 *(Karabutov, 2022). Let (i) system (10.24) be stable and recoverable, (ii) the set* M_{LE} is LP-detectable, (iii) definition areas D^j_{sl} of framework $SL_{\Delta k^i_{s,\rho}}$ coincide with areas of the $\alpha\pi$-almost periodicity of the framework $S_{\hat{y}_g,\hat{\hat{y}}_g}$. Then elements belonging to set M_{LE}, are χ-adequate to regions of $\alpha\pi$-almost periodicity $S_{\hat{y}_g,\hat{\hat{y}}_g}$.

So, Criterion $\chi_1 \leq \sup J^1_{k^i_{s,\rho}}$ gives an estimate of LP-detectability for the system (10.24).

6. EXAMPLES

Consider the model of the single-link manipulator robot with a flexible hinge (Veluvolu and Soh, 2011)

$$\begin{pmatrix}\dot{x}_1\\ \dot{x}_2\\ \dot{x}_3\\ \dot{x}_4\end{pmatrix} = \begin{pmatrix}0 & 1 & 0 & 0\\ -\frac{k}{J_m} & -\frac{b}{J_m} & \frac{k}{J_m} & 0\\ 0 & 0 & 0 & 1\\ -\frac{k}{J_l} & 0 & -\frac{k}{J_l} & 0\end{pmatrix}\begin{pmatrix}x_1\\ x_2\\ x_3\\ x_4\end{pmatrix} + \begin{pmatrix}0\\ \frac{k_\tau}{J_m}u\\ 0\\ -\frac{mgh}{J_l}\sin(x_3)\end{pmatrix}, \tag{10.26}$$

$$\begin{pmatrix}y_1\\ y_2\end{pmatrix} = \begin{pmatrix}x_1\\ x_3\end{pmatrix},$$

where x_1, x_3 are the angular rotations of the motor and the link; u is motor control torque; J_m, J_l are inertia moments of the motor and the link; k is the elastic constant; b is viscous friction coefficient; k_τ is the amplifier gain; m is the link mass; h is the center of mass and g is the gravity. The system (10.26) considers the influence of viscous friction and elastic deformation of the flexible joint. The measurement vector is $Y = [x_1, x_3]^T$. System parameters: $J_m = 0.037$ Kg m2, $J_l = 0.093$ Kg m2, $\kappa = 0.18$ Nm/rad, $B = 0.0083$ Nm/rad, $m = 0.21$ kg, $g = 9.81$ m/s2, $h = 0.15$ m, $k_\tau = 0.18$ Nm/V.

Phase portrait (PP) is shown on channel "$x_1 - x_2$" in Figure 1. Input $u(t)$ does not satisfy the CE condition. But, despite this, damping oscillations occur in the system. The system is stable and, therefore, recoverable. Structure (phase portrait) of the system is open-ended. But internal oscillations are a source for analyzing the system identifiability.

Figure 1. System (10.26) phase portrait on channel $x_1 - x_2$

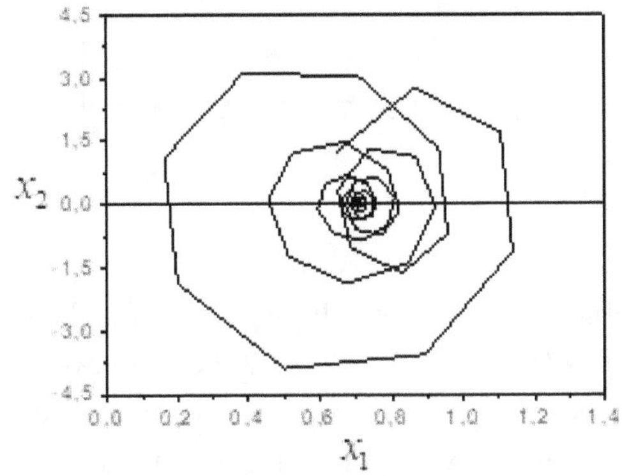

Frameworks $S_{k^1_{s,\rho},k^2_{s,\rho}}$ and $LSK_{\Delta k^2_{s,\rho}}$ for evaluating Lyapunov exponents present in Figure 2. Here $\hat{y}_1$ is the estimate x_2,

$$k_s^1\left(t,\rho\right) = \frac{\rho(y_1(t))}{t}, k_s^2\left(t,\rho\right) = \frac{\rho(\hat{y}_1(t))}{t}, LSK_{\Delta k^2_{s,\rho}} \rightarrow \mathbb{I}_{k^1_{s,\rho}} \times \mathbb{B}\left(\mathbb{I}_{\Delta k^2_{s,\rho}}\right).$$

Figure 2. Structures for evaluating LE

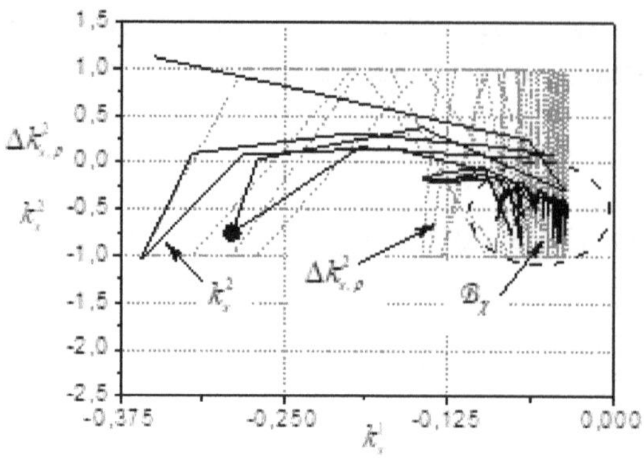

As the system is stable, we consider only negative LE. The presence of oscillations reflects by function *b*(*t*). *b*(*t*) represents by the framework $LSK_{\Delta k^2_{s,\varphi}}$ (Figure 2) described by the function $LSK_{\Delta k^2_{k,\varphi}} \to \square_{k^2_{k,\varphi}} \times \mathbb{B}\left(\square_{\Delta k^2_{k,\varphi}}\right)$. System state variables are B $F_{\alpha\pi}$-functions. Apply theorem 8. As Figure 2 follows, the order of the system is 4, and the evaluation of the Perron bottom index is known (it is equal to –0.7). The area to which the set LE $\chi_1[x_2]$ belongs, we denote as B_χ. B_χ corresponds to (Karabutov, 2018) the presence of a complex root. $\chi_3[x_2]$ locate between $\eta[x_2]$ and area B_χ. $\chi_3[x_2]$ belongs to the region approaching $k^2_s = 0$ from below. The linean corresponding to $\chi_3[x_2]$ is linearly dependent, and, therefore, the system is σ-detectable with level $\upsilon = 1$. Spectrum of roots is $s(\lambda) = [-0.65 \quad 0 \quad \pm 0.8 + 8.16i]$. The system (10.26) is sensitive to input influences.

Consider the linearized system (LS) for (10.26). We present estimation results of the LE identifiability and detectability in Figures 3 and 4. Show the PP system in the space $(x_{l,1}, x_{l,2})$ in Figure 3, where $x_{l,1}, x_{l,2}$ are state variables of the LS. PP has the similar form in space $(x_{l,3}, x_{l,4})$. The LS is stable, and damping oscillations occur in it. Therefore, LS is identifiable, and the state variables present asB $F_{\alpha\pi}$-functions (Figure 4). Figure 4 shows the frameworks

$$SL_{k^{1,l}_{s,\varphi}, k^{2,l}_{s,\varphi}} \to \square_{k^{1,l}_{s,\varphi}} \times \square_{k^{2,l}_{s,\varphi}}, \; LSK_{\Delta k^{2,l}_{s,\varphi}} \to \square_{k^1_{s,\varphi}} \times \mathbb{B}\left(\square_{\Delta k^{2l}_{s,\varphi}}\right).$$

confirm the LE detectability.

Figure 3. Phase portrait of linearized system (10.26)

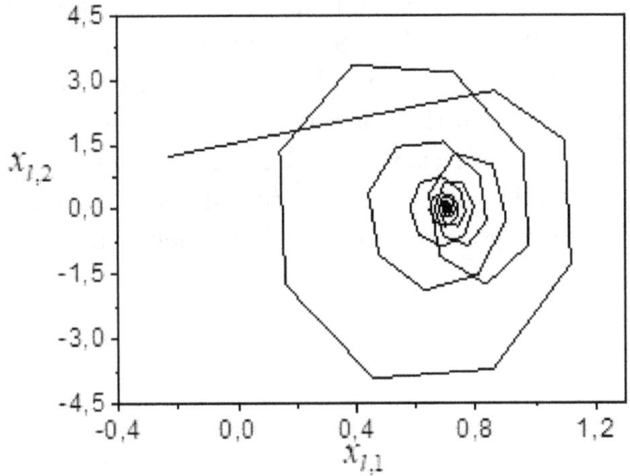

We see (Figure 4) that the linearized system (10.26) contains two complex indicators $\chi_i\left[x_{1,2}\right]$, and the system order is 4. So, the linearized system (10.26) is identifiable, and the LE set is detectable.

Remark 8. Linearization changes the LE spectrum location and effects on the stability degree. But linearization improves the system quality, as the system aways from the stability boundary. Note that the system is sensitive to parameter changes.

Figure 4. Frameworks for LE estimating of linearized system (10.26)

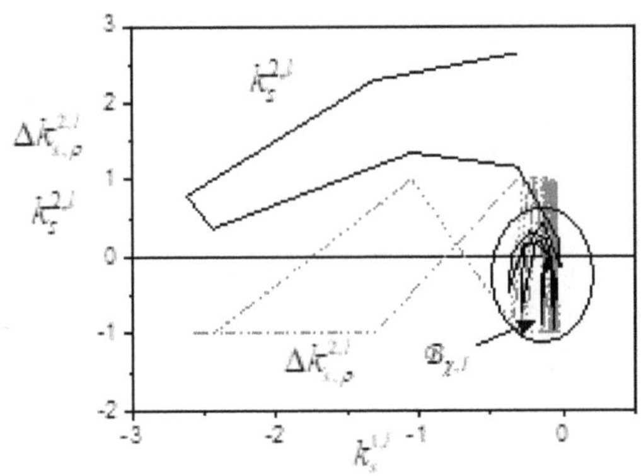

7. CONCLUSION

The problem of identifiability, recoverability and detectability of Lyapunov exponents is considered. We have obtained conditions guaranteeing the calculation (detection and identifiability) of the LE set. LE is based on the analysis of geometric structures. We obtain (i) σ-detectability conditions with the level of υ-non-recoverability and LP-detectability of LE for the stationary system; (ii) conditions for the detectability of Lyapunov exponents for the linear system with periodic coefficients. There shows that for nonlinear systems linearization gives to thy change of the LE set.

REFERENCES

Bylov, B. F., Vinograd, R. E., Grobman, D. M., & Nemytsky, V. V. (1966). *Theory of Lyapunov indexes and its application to stability problems*. Nauka.

Izobov, N. A. (2006). *Introduction in theory of Lyapunov indexes*. BGU.

Karabutov, N. (2015). Structural methods of estimation Lyapunov exponents linear dynamic system. *International Journal of Intelligent Systems and Applications*, *7*(10), 1–11. DOI: 10.5815/ijisa.2015.10.01

Karabutov, N. (2018). *Frameworks in identification problems: Construction and analysis*. Lenand.

Karabutov, N. N. (2009). *Structural identification of systems. analysis of information structures* [Russian language]. URSS/ Librokom.

Karabutov, N. N. (2022). Identifiability and Detectability of Lyapunov Exponents in Robotics. In Mellal, M. A. (Ed.), *Design and Control Advances in Robotics* (pp. 152–174). IGI Global. DOI: 10.4018/978-1-6684-5381-0.ch009

Kwakernaak, H., & Sivan, R. (1972). *Linear optimal control systems*. Wiley Interscience.

Lyapunov, A. M. (1992). *General Problem of the Stability of Motion*. Taylor & Francis, DC. DOI: 10.1080/00207179208934253

Smirnov, V. I. (1968). A course of higher mathematics.: Vol. 5. *Integration and functional analysis* [Russian language]. Nauka.

Veluvolu, K., & Soh, Y. (2011). Fault reconstruction and state estimation with sliding mode observers for Lipschitz non-linear systems. *IET Control Theory & Applications*, *5*(11), 1255–1263. DOI: 10.1049/iet-cta.2010.0171

Section 3
Adaptive Identification

Chapter 11
Method of f–Algorithms

ABSTRACT

In this part, we consider issues of adaptive identification (AI) of complex dynamic systems (DS). Such DS are specific and require novel approaches. The design of AI systems for DS is based on the development of the GF method and the application of AI-algorithms. The design of AI systems for DS is based on the development of the geometric framework (GF) method and the application off-algorithms. We present a scheme for the synthesis of -algorithms under constraints and uncertainty. Methods are proposed for robustness increasing of f-algorithms. Properties of adaptive algorithms are studied, and the quality estimates of the adaptive system are obtained.

1. INTRODUCTION

Now many algorithms of control considering existing restrictions are offered. They are obtained heuristically, since formalized methods of synthesis under constraints are not proposed. Such the problem status affects both a choice of criteria of quality, and on an estimation of performing a control system. The maximum principle is not applied to the synthesis of adaptive algorithms. The method of the dynamic programming (Bellman, 2003) allows synthesizing a class of control potential laws, which then are changed under restrictions. Therefore, the design of effective methods is still an actual problem.

We state the analytical approach to the synthesis of algorithms. This method applies to obtain as control laws, so and parametric dynamic algorithms.

The account of restrictions allows reasonable to suit to the choice (if the analysis of the a priori information on noise) or to synthesis (if the analysis of a priori and experimental data) algorithms of the parametric estimation. Distinctions between a choice and synthesis here are essential. The identification algorithm is configured based on a priori information about the environment, but the law of estimation is heuristic (Tsypkin, 1995). In the second case, the area of restrictions for the system

DOI: 10.4018/979-8-3373-0822-7.ch011

"plant + environment" is formed based on the analysis of the information, and then control algorithms are synthesized as optimization of a standard (square) functional. Such a distinction in approaches leads to various classes of nonlinear algorithms. The nonlinearity of identification and control procedures is a fee for considering the actual conditions reflecting the functioning of the system. The offered method of φ-algorithms (Karabutov, 2006) is based on Lyapunov functions (LF). It allows to consider on only restrictions, but also the requirements shown to quality of transients in system. This approach develops the LF method and the synthesis (Krasovsky et al., 1977) idea for the case restrictions.

2. CONSTRUCTION SCHEME OF φ–ALGORITHMS

Consider the dynamic system (DS)

$$\dot{X} = F(X, A, U, t) \tag{11.1}$$

where $X \in \mathbb{R}^m$ is a condition vector; $A(t)$ is a vector of parameters, inhering of some limited, but a priori unknown area $\mathbb{G}_A \subseteq \mathbb{R}^k$; $U \in \mathbb{R}^n$ is a vector of control (input); $t \in \mathbb{J} = [t_0, \infty)$; $F(\cdot)$ is continuously differentiated on X, A, U a vector function.

Restrictions

$$\Omega_s = \{(X, A, U) : L_s(X, A, U, t) \leq 0 \;\; \forall t \geq t_0\} \tag{11.2}$$

where L_s is some nonlinear operator, strategy of control influencing a choice.

We introduce functional $V(t) = V(X, t)$ characterizing the system work quality. The control purpose:

$$\lim_{t \to \infty} V(X, t) = 0 \tag{11.3}$$

The decision-making algorithm (adaptation, management) $U = U(X)$ is chosen from the condition

$$\dot{V}(X, U, t) \leq 0 \tag{11.4}$$

This classical condition of a stability expressed through the LF function. Synthesis of the control algorithm is hampered considering restriction (11.2).

The idea of the φ-algorithms method (FAM) (Karabutov, (2006, 2013)) is the transformation of the function $\dot{V}(t)$, considering Ω_s. By transformation over X, U, A, $\dot{V}(t)$ is presented as:

$$\dot{V}(t) = V\big(X, \Phi(X, U, A \mid (X, U, A) \in \Omega_s, t\big)$$

where $\Phi: \mathbb{R}^m \times \mathbb{R}^k \times \mathbb{R}^n \times \mathbb{J} \rightarrow \mathbb{R}^k$ is some nonlinear vector-function. Then the target condition (11.3) is reduced to a choice of vector $\Phi(X, U, A, t)$, which obtains as an Ω_s account result,

$$\Phi(t) = \Phi(\square, \Omega_s)$$

where $\square = \square(\square_a, \square_e)$ is the set containing the a priori and experimental information on the system.

The law of control presents as:

$$D_U U = -\Gamma \frac{\partial \Phi(X, A, U, t)}{\partial U^T} I \tag{11.5}$$

where D_U is a proportional or differential operator, $\Gamma = \Gamma^T > 0, I \in \mathbb{R}^n$ is the unit vector, T is a transposition sign.

(11.5) we will name φ-algorithm (FA) of control. Next, we consider the implementation of the described approach to the DS adaptive control problem.

3. ADAPTIVE φ-ALGORITHMS SYNTHESIS

3.1. Design of Adaptive FA

Consider the system DS (11.1) identification problem with considering the restrictions. Thus, the system is understood as the generalized adjusted system, including object and adjusted model.

We consider restrictions which are formed, based on a priori $\square_a$ and a posteriori information $\square_{ap}$. $\square_{ap}$ is the analysis result of the of experimental data

$$\Omega_s(\square_{ap}, \square_a) = \{(X, A, U, t): L_s(X, A, U, t) \leq 0 \ \ \forall t \geq t_0\} \tag{11.6}$$

When solving the problem, the area Ω_s can be set and transformed. Therefore, the time t reflects this in (11.6).

The problem is reduced to the adaptive algorithm choice parameters for the system:

$$\dot{A}(t) = A(\mathbb{I}_e, \Omega_S, t) \tag{11.7}$$

where A is the operator that is selecting from the fulfilment (11.3) or the weaker condition (11.4).

Consider $\dot{V}(t)$

$$\dot{V}(t) = \frac{\partial V(X,t)}{\partial t} + \text{grad}_X^T V(X,t)\ F(A,X,U,t) \tag{11.8}$$

Unlike known schemes of synthesis (Fradkov, 1979), based on the algorithm of a high-speed gradient

$$\dot{A}(t) = -\Gamma\ \text{grad}_A\ \dot{V}(X,t), \Gamma \in \mathbb{R}^{k\times k}, \Gamma = \Gamma^T \tag{11.9}$$

to consider $\Omega_S$gives to a more complicated algorithm of adaptation.

To consider restrictions, we will apply the following procedure. Realize decomposition of function $\chi(X,A,t) = \dot{V}(X,t)$ on two vectors: one vector does not depend on $A(t)$, and the second vector depends on $A(t)$. For the system, linear on $A(t)$, the first vector is present as:

$$\Psi_1(X,t) = \text{grad}_A \chi(X,A,t) \tag{11.10}$$

In the case of nonlinear dependence $\chi(\cdot)$ from A, such decomposition is fulfilled based on the analysis of function $\chi(\cdot)$form. The received component is designated as $\Psi_1(X,t)$. Then:

$$\chi(X,A,t) = \Psi_1^T(X,t)\ \Psi_2(X,A,t) \tag{11.11}$$

where $\Psi_2 \in \mathbb{R}^k$ is some vector.

The following stage is the account of restrictions. Here, vector Ψ_1 is transformed into Ψ, and we get the so-called F F-representation:

$$\chi(X,A,t) = \Psi^T(X,t)\ \Phi(X,A,t) \tag{11.12}$$

where $\Psi \in \mathbb{R}^k$ is some positive vector-function (i.e. Ψ is a vector, at which $\Psi > 0 \rightarrow \psi_i > 0, i = \overline{1,k}), \Phi(\cdot) \in \mathbb{R}^k$is vector continuously differentiated on A.

In the generalized aspect, vector $\Psi(X,t)$ reflects restrictions superimposed on the system. It is constructed based on vector $\Psi_1(X,t)$ considered restriction Ω_s (11.6)

$$\Psi(X,t)=\Psi\left(X,\ \Psi_1(X,t),\ \Omega_s\right)$$

Thus, the restrictions Ω_s account given such transformation R_+ for vector $\Psi_1(X,t)$ that vector $\Psi(X,t)$ was positive

$$R_+:\Psi_1(X,t)\ \rightarrow\ \Psi(X,t)\forall(X,U)\ \in\ \Omega_s \tag{11.13}$$

If area Ω_s is opened, then $R_+:\Psi_1(X,t)\ \rightarrow\ I$, where $I\ \in\ R^k$ is a unit vector.

As application R_+ to $\Psi_1(X,t)$, vector $\Psi_2(X,A,t)$ also is transformed to some vector $\Phi(X,A,t)$ which aspect is completely defined by set Ω_s,

As $\Psi(X,t)\ >\ 0$, then the adaptive FA has the form:

$$\dot{A}(t)=-\Gamma\,\frac{\partial\,\Phi(X,A,\,t)}{\partial\,A^T}\,I \tag{11.14}$$

The restrictions completely defined properties of adaptive φ-algorithm. We consider restrictions on perturbations $\xi(t)$ as $(X,U)\in\ \Omega_s$.

If on function $\chi(\cdot\,)$ to superimpose functional restriction

$$\chi(X,A,t)\ \leq\ -\,\omega(X,A,t) \tag{11.15}$$

where $\omega(\cdot\,)>\ 0$ is the quadratic form from $A(t)$, depending from Ω_s,then we will receive adaptive algorithm:

$$\dot{A}(t)=-\Gamma\left\{\lambda A(t)+\frac{\partial\,\Phi(X,A,\,t)}{\partial\,A^T}\,I\right\},\lambda\ >\ 0 \tag{11.16}$$

We name the (11.16) is proportional-integral φ-algorithm. The received laws describe a wide class of identification and control algorithms. If the function $\chi(X,A,t)$ is linear on the vector $A(t)$, then $\Phi(X,A,t)$ is depended on a pseudo-gradient [5] of function $\chi(X,A,t)$.

We will say, that vector $\Phi(X,A,t)\ \in\ H$, if:

1) $\Phi(X,A,t)$is continuously differentiated A

$$\Phi(X,A,t)=\Phi_1(X,t)A(t)$$

where $\Phi_1(X,t)$ is Jacobi matrix;

2) matrix $\Phi_1(X, t)$ is such that

$$\Phi_1(X, t)\Psi_1(X,A, t) \geq 0 \forall((X,A) \in \Omega_s) \& (t \geq t_0) \quad (11.17)$$

So, if function $\chi(X,A,t)$ is linear on A, then we receive the class of continuous pseudo gradient algorithms (see, for example, Fradkov, (1979); Karabutov, (2006); Polyak and Tsypkin (1980)]).

For the FA, the stability condition is

$$\Psi_1^T(X,A, t)\frac{\partial\Phi(X,A, t)}{\partial A^T}I \geq 0 \quad (11.18)$$

Other special classes, following from (11.14) and (11.16), are algorithms of the high-speed gradient (Kushner and Clark, 1978)

$$\dot{A}(t) = -\Gamma \,\mathrm{grad}_A\, \chi(X,A,t) \quad (11.19)$$

$$\dot{A}(t) = -\Gamma \left\{\lambda A(t) + \mathrm{grad}_A\, \chi(X,A,t)\right\} \quad (11.20)$$

The equation (11.20) is regularized the algorithm (11.19) variant.

The stated FAM applies in case of restrictions superimposed on an input, an exit (a perturbation), and the vector of parameters

$$\Omega_A = \left\{A \in R^k : L_A(A,t) \geq 0 \;\; \forall t \geq t_0\right\} \quad (11.21)$$

where L_A is a nonlinear operator.

In identification problems, L_A is the vector A norm in space $\mathbb{R}^k$. As with (11.6), the account of restrictions Ω_A leads to F-representation (11.12), but vector-function Ψ depends on vector $A(t)$. Despite it, demands presented to $\Psi(X,A,t)$ are coincident with the demands on the area Ω_s.

The set Ω_A account has the specificity and is reflected in a class of received algorithms. If $A \in \Omega_A$, then the adaptive φ-algorithm is presented as:

$$\dot{A}(t) = -\Gamma\frac{\partial\,\Phi(X, P(A,\Omega_A), A, t)}{\partial\, A^T}I \quad (11.22)$$

where matrix $P(A,\Omega_A)$ is the indicator belongs area Ω_A

$$P(A,\Omega_A)=\begin{cases}P_1, & A\in\Omega_A,\\ P_2, & A\notin\Omega_A,\end{cases} \tag{11.23}$$

P_1 and P_2 is some matrixes constructed based on set Ω_A. In particular $P_1 = I_k$, $P_2 = 0_k$, $I_k \in \mathbb{R}^{k\times k}$ and $0_k \in \mathbb{R}^{k\times k}$ are unit and zero matrixes accordingly. $P(A,\Omega_A)$ is the function of k-signed logic.

Integral functionals $V(X,t)$ can be used for FA design. In a local case, functional $Q(t)=Q(X,t)$ is had form $\dot{V}(X,t)=\dot{Q}(t)$.

3.2. FA Synthesis With the Set Quality

The scheme FA, stated in the previous section, is based on optimization degenerate concerning control (in this case, control is the tuning law of model parameters) function Lyapunov $V(X,t)$. The method of the dynamic programming is based on the optimization of the function of specific losses $\omega(\cdot)$ depending on control (vector $A(t)$). Most times, ω is the square form depends on restrictions and control, i.e. is $\omega = \omega(X,A)$ (Karabutov, 2006; Kuntsevich and Lichac, 1985). Synthesis problems of adaptive control systems with square function based on optimization of specific losses (SL) are considered by Krasovsky (1975) and Krasovsky et al., (1977). Specificity of the considered problem demands SL function introduction, which reflects the restrictions superimposed on the system. Control φ-algorithm of such type is proposed by Karabutov (2006). We apply results of these works and show as optimize non-singular criterion to receive the FA.

Consider the integral criterion of quality

$Q(A)=\int_{t_0}^{\infty}\omega(X,A)\,dt$ on $(X,A)\in\Omega_s$

We introduce the diagonal matrix $D[\Psi(X,A,U)]$ from vector $\Psi(X,A,U)$

$$D[\Psi(X,A,U)] = \operatorname{diag}\left(\phi_1(X,A,U),\ \ldots,\ \phi_k(X,A,U)\right)$$

where $\phi_i(X,A,U)\in\Psi(X,A,U)$, and criterion $Q(A)$ we present as

$$\tilde{Q}(A)=\int_{t_0}^{\infty}\varpi(X,A)\,dt = \int_{t_0}^{\infty}L_A A^T(t)\Gamma^{-1}D[\Psi(X,A,U,t)]A(t)\,dt$$

where L_A is a proportional or differential operator.

Consider LF $V(X,t)$. Then the Bellman equation for system (11.1):

$$-\dot{V} = \max_A\left\{\chi(X,A,t)+L_A A^T(t)\Gamma^{-1}D\left[\Psi(X,A,U,t)\right]A(t)\right\}(X,A)\in\Omega_s \tag{11.24}$$

or

$$\dot{V}(t) = \chi(X,A,t) + L_A A^T(t)\Gamma^{-1} D[\Psi(X,A,U,t)]A(t) \leq 0$$

Let $L_A = d/dt I_k$ and have an optimality condition $\partial \dot{V}(t)\,\partial \dot{V}(t)/\partial A^T = 0$. Solving this equation, we obtain the adaptive law. It has the form (11.14).

So, the basic difference from a class case of the dynamic programming consists in design of function of specific losses. In classical case, $\Psi(X,A,U,t) = I$ as area Ω_s is opened. In the non-classical case, the vector function $\Psi(X,A,U,t)$ is a component of the optimization criterion $\dot{Q}(A,\Omega_S)$. The proposed approach allows obtaining the proportional integral FA.

The synthesis of a discrete variant of AF is proposed by Karabutov (2006).

4. ABOUT ROBUSTNESS OF φ-ALGORITHMS

φ-algorithms are nonlinear and, therefore, robust to errors in the formulation Ω_s. But the system may be exposed to various uncontrolled disturbances. Therefore, the synthesis of FA should guarantee the system robustness.

Following approaches (Karabutov, 2006) are applied for robustness of φ-algorithms:

a) introducing additional signals;
b) application of nonlinear vector-functions

$$Z(X,\hat{A},U,t) = \frac{\partial \Phi(X,A,U,t)}{\partial A^T} I \; ; \tag{11.25}$$

c) modification of algorithm parameters.

Necessity of coarsening adaptive FA speaks that function $\chi(X,\hat{A},U,t)$ is not negatively defined in space $\{X,\hat{A}\}$. To ensure performance of condition $\chi < 0$ for $\forall(X,\hat{A}) \in \Omega_s$, additional signals are introduced in the system. In [6], the feedback signal was applied for coarsening adaptive algorithm and has been as a result received regularized algorithm (11.20).

The approach based on introducing additional signals is reduced in the problem (Karabutov, 2006): to find the adaptation algorithm of system parameters that the condition is fulfilled

$$\dot{V}(X,t) \leq -\omega(X,A,\Omega_s,t)$$

where $\omega(\cdot)$ is function of specific losses.

To solution of this problem gives the proportional-integral φ-algorithm (11.16). Unlike (11.14), the FA (11.16) works as the filter which properties depend on parameters λ and Γ.

The second coarsening mode is based on the choice of the vector Z (11.25), based on the a priori information about Ω_s and the perturbation $\xi(t)$. This average coarsening is more difficult to implement. To the dissipativity ensure of the closed system (11.1), (11.14), the vector $Z(X,A,U,t)$ must satisfy the condition:

$$\|Z(X,A,U,t)\| \leq c_1 V^{\pi_1}(X,t) - c_2 V^{\pi_2}(X,t) \tag{11.26}$$

where $c_1 > 0, c_2 = c_2(\xi) > 0, (\pi_1, \pi_2) \in (0, 1]$

As follows from (11.26), the vector Z should choose based on the requirements presented to the adaptive system to consider existing restrictions (the first item in (11.26)), and the analysis perturbations (an item in (11.26)). The third approach roughening is the most complicated in realization, as demands the definition $\Gamma(t) = \Gamma(X, \Omega_s, \xi, t)$. Here, matrix $\Gamma(t)$ in (11.14) is chosen so to ensure the dissipative of system. The general recommendations are not developed for choice matrixes Γ under perturbations $\xi(t)$. One of their approaches to the matrix Γ choice (Tsypkin, 1995) is based on the analysis of the a priori information about $\xi(t)$.

5. APPROACHES TO QUALITY ESTIMATION OF ADAPTIVE SYSTEMS

At design and the quality estimation of adaptive systems (AS), two approaches apply: statistical and is functional-plural. These approaches differ as analysis methods, so and received estimations. The first group of methods is based on statistical procedures and statistics. The convergence analysis of adaptive algorithms (AA) is formed the basis of AS quality. Establishing the AA convergence is of great importance, since the use of AA is possible only with guaranteed convergence. Therefore, the convergence research is an actual issue on AS realization.

Convergence of gradient iteration schemes at an action of stochastic perturbations present by Tsypkin (1995), Katkovnik (1976), Kushner and Clark (1978) and Poljak (1976). Convergence conditions are determined and convergence rate estimates are got. These results will give you exact estimates for system parameters. These estimations have an asymptotic character. Criteria-type estimates can be used in adaptive control systems, and they ensure the minimum of the optimization criterion (Polyak and Tsypkin (1983). To study the convergence of stochastic iterative procedures, the Lyapunov stochastic functions and the theory of super-martingales were mainly used. The received estimations as argument, and criteria type have an average character on probability space with the set measure, and convergence is understood in the probability sense. The received estimations as argument, and criteria type have an average character in probability space, and convergence is understood in the probability sense. Other approaches are described by Kulchitsky (1983).

The Lyapunov function method is the basis for the study of continuous adaptive algorithms. If there are no perturbations, then the convergence proof is based on the constant excitation condition (CE) of the data (Brian et al., 1986; Eikhoff, 1974).

Various approaches and methods are used in the non-statistical interpretation of the adaptive control (identification) problem. These are the principle of compressive maps; first and second Lyapunov methods (Evtushenko, 1982; Furasov, 2005); the approach based on the implicit function theorem (Katkovnik, 1976).

Study of the quality of continuous AS is based on the concepts of input and output stability, passivity, and averaging. The AS stability of adaptive systems and convergence of identification algorithms based on these approaches depend on the properties of experimental data.

Quality assessment of transients in AS has not been studied. For adaptive control tasks, integral quality estimations were obtained by Gromiko and Sankovsky (1974) and Kogan and Nejmark (1987). The analysis of transients in AS allows: (i) to get functional restrictions (FR); (ii) to identify the class of adaptive algorithms satisfied the FR. The mathematical theory of stability and integral inequalities (Martinyuk, 1980; Martinyuk et al., 1989) provided a direction for the research development in this area.

Next, we analyse the quality of continuous AS with φ-algorithms based on the performed analysis.

6. AS PROPERTIES WITH CONTINUOUS φ-ALGORITHMS

6.1. AS Properties

Let perturbations operate on the system (11.1)

$$\dot{X} = F(X,A,U,t)+\xi(t), \tag{11.27}$$

where $\xi \in \mathbb{R}^m$ is understood as both external and internal disturbances that arise as a discrepancy between the model and the system. We believe that $\xi \in \mathbb{G}_\xi$, where $\mathbb{G}_\xi$ is some closed set.

Assumptions.

1^0. Function $V(X,t) > 0$ at $X \neq 0$, $V(0,t) = 0$ and satisfies to a condition

$$\inf_{\|X\| \to \infty} V(X,t) \to \infty$$

2^0. the equation is fair for $\dot{V}(X,t)$

$$\dot{V}(X,t) = \chi(X,A,t) + \Psi_2(X,t)\,\xi(t)$$

Function $\chi(X,A,t)$ satisfies (11.12) and is convex on $\widehat{A}$

$$\chi\left(X,\widehat{A}_1,t\right) - \chi\left(X,\widehat{A}_2,t\right) \geq \Psi_1^T(X,t)\left(\widehat{A}_1 - \widehat{A}_2\right) \tag{11.28}$$

where $\Psi_1(X,t) = \text{grad}_{\widehat{A}}\chi(X,A,t)$, $\Psi_2(X,t) = \text{grad}_X\chi(X,A,t)$.

3^0. There is such a vector of parameters $A_* \in \mathbb{R}^k$, that:

$$\chi(X,A_*,t) \leq -V^\sigma(X,t), \alpha \geq 0,\ \sigma \in [0,\ 1]$$

4^0. Right parts of the system (11.1), (11.12) are locally bounded uniformly over $t \geq t_0$.
5^0. $\|\xi(t)\| \leq \kappa, 0 \leq \kappa < \infty$.
6^0. Matrix Γ is diagonal and $\beta_1 \leq \|\Gamma\| \leq \beta_2$, $\left(\beta_1,\beta_2\right) > 0$.

Target condition

$$\lim_{t \to \infty} V(X,t) \leq \varepsilon, \varepsilon \geq 0 \tag{11.29}$$

Designations: $\mathbb{J} = \left[t_0,t\right], t \to \infty$, $\mathbb{J}_\alpha = \left[t_0,t_\alpha\right]$, $\mathbb{J}_{0,\infty} = [0,\infty)$.

Show the limitation of trajectories in the system (11.14), (11.27). In assumption 30, the A_* is considered as a vector of unknown system parameters.

Theorem 1 (Karabutov, (1993, 2006). *Let assumptions 1^0, 2^0 with $\xi(t) = 0$, 3^0, 4^0 and 6^0 are satisfied. Then 1) the control goal (11.29) is achieved in the system (11.14) and (11.27) with $\varepsilon = 0$; 2) all trajectories are limited and fair*

$$\alpha\int_{t_0}^{t} V^\sigma(X,\tau)\,d\tau \leq V_1(t_0) - V_1(t)$$

$$V_1(t) = V(X,t) + \int_{t_0}^{t} \delta\dot{A}^T(\tau)\,\Gamma^{-1} D[\Psi(X,\tau)]\,\delta A(\tau)\,d\tau, \delta A(t) = \widehat{A}(t) - A_* \tag{11.30}$$

As follows from Theorem 1, the system (11.14), (11.27) is asymptotically stable in space $\{X\}$, and the vector $\delta A(t)$, and, therefore, $\hat{A}(t)$ are limited. For $\sigma = 1$, assumption 3^0 is the system exponential stability condition regarding the vector $X(t)$ at $\hat{A} = A_*$.

Properties of the system with FA (11.14)

Theorem 2 (Karabutov, (1993, 2006)). *Les theorem 1 conditions and the inequality*:

$$\bar{c}_D V^{\theta}(X,t) \le \left\|D\left[\Psi(X,\hat{A},t)\right]\right\|, 0 \le \theta < 1 \tag{11.31}$$

are satisfied. Then all trajectories are limited in the system (11.14) and (11.27), condition (11.29) are fulfilled and the estimate

$$\|\delta A(t)\|^2 \le \|\delta A(t_0)\|^2 + \frac{2\beta}{\bar{c}_D(1-\theta)}\left[V^{1-\theta}(t_0) - V^{1-\theta}(t)\right] - \frac{2\beta\alpha}{\bar{c}_D}\int_{t_0}^{t} V^{\sigma-\theta}(\tau)\,d\tau \tag{11.32}$$

is fair.

Corollary from the theorem 2. *If* $\bar{c}_D = 1, \theta = 0, \sigma = 1$, *then the estimate*

$$\|\delta A(t)\|^2 \le \|\delta A(t_0)\|^2 + 2\beta\left[V(t_0) - V(t)\right] - 2\beta\alpha\int_{t_0}^{t} V(\tau)\,d\tau \tag{11.33}$$

is valid for the algorithm (11.19)

We see that properties φ-algorithm (11.14) depends on function $V(X, t)$, the matrix Γ, system parameters (α) and the choice of initial conditions. The largest velocity of convergence has FA, for which set Ω_s is opened. Here, the estimation (11.32) is fair. If actual conditions to consider, then the convergence rate of the adaptive φ-algorithm (11.14) slows. This is true because the inequality $V^{\sigma}(t^*) \le V(t^*)$ will be satisfied by some $t^* > t \ge t_0$. If

$\int_{t_0}^{\infty} V^{\sigma}(\tau)\,d\tau \to \infty$ by $t \to \infty$

then the algorithm (11.14) is identifying on $\left\{X(t_0), \hat{A}(t_0)\right\}$.

Theorem 3 (Karabutov, 1993; Salyga et al., 1988). *Let assumptions 1^0-6^0, (11.31) and inequalities*

$$\left\|D\left[\Psi((X,\hat{A},t)\right]\right\| \le c_D V^{\theta}(X,t) \tag{11.34}$$

$$\|\mathrm{grad}_X V(X,t)\| \le c_{\phi} V^{\mu}(X,t) \tag{11.35}$$

$$\|\mathrm{grad}_{\hat{A}} \dot{V}(X,t)\| \le c_{\rho} V^{\rho}(X,t) \tag{11.36}$$

by fulfilled, where $c_\rho > 0, c_\phi > 0, 0 \le (\mu,\rho) < 1$. *Then system (11.14), (11.27) is dissipative and estimations*

$$V(t) \le G^{-1}\left(G(V(t_0)) + t - t_0\right), t_0 \le t \le t_* \tag{11.37}$$

$$\|\delta A(t)\|^2 \le \|\delta A(t_0)\|^2 + \frac{2\beta}{\overline{c}_D(1-\theta)} V^{1-\theta}(t_0) + \frac{2\kappa\beta\, c_\phi}{\overline{c}_D} \int_{t_0}^{t} V^{\mu-\theta}(\tau)\, d\tau \tag{11.38}$$

are fair, where G^{-1} *is function inverse to*

$$G(V) = \int_{V(t_0)}^{V(t)} \frac{dV}{g(V)}$$

$$V \in \mathbb{J}_{0,\infty} \tag{11.39}$$

$$g(V) = -\alpha V^\sigma + \kappa c_\phi V^\mu + c_\rho \max_t \|\delta A\| V^\rho, t_* = \min\left(\overline{t}_\gamma, \overline{t}_\beta\right) \tag{11.40}$$

$$\overline{t}_\gamma = \sup\left\{t_\gamma \in J; V(t_0) + \int_{t_0}^{t} g\left(V(s)\right) ds \in \mathbb{J}_{0,\infty}; t \in \mathbb{J}_\gamma\right\}$$

$$\overline{t}_\beta = \left\{\sup t_\beta \in \mathbb{J}; \left[G(V(t_0)) + t - t_0\right] \in G\left(\mathbb{J}\right); t \in \mathbb{J}_\beta\right\}$$

Corollary 1 of Theorem 3. *If the system (11.14), (11.27) no restrictions are imposed, then the estimates are fair*

$$V(t) \le G^{-1}\left[G\left(V(t_0)\, e^{-\alpha(t-t_0)}\right) + \frac{1}{\alpha}\left(1 - e^{-\alpha(t-t_0)}\right)\right], t_0 \le t \le t_\beta \tag{11.41}$$

$$\left\|\delta A(t)\right\|^2 \le \left\|\delta A(t_0)\right\|^2 + 2\beta V(t_0) + 2\kappa\beta\, c_\phi \int_{t_0}^{t} V^\mu(\tau)\, d\tau \tag{11.42}$$

where G has the form (11.39) and

$$g(V) = \kappa c_\phi V^\mu + c_\rho \max_t \|\delta A\| V^\rho$$

$$\overline{t}_\beta = \sup\left\{t \in J; \left[G\left(V(t_0)\, e^{-\alpha(t-t_0)}\right) + \frac{1}{\alpha}\left(1 - e^{-\alpha(t-t_0)}\right)\right] \in G\left(\mathbb{J}_{0,\infty}\right)\right\}$$

Corollary 2 of Theorem 3. *If* $\sigma = 1$ *and* $\vartheta = \min(\mu,\rho)$, *then for a system (11.27) with an* φ*-algorithm (11.19), estimates (11.38) and*

$$V(t) \leq \left(e^{-\tilde{\alpha}(t-t_0)} V^{1-\vartheta}(t_0) + \frac{c}{\tilde{\alpha}}(1 - e^{-\tilde{\alpha}(t-t_0)})\right)^{\frac{1}{1-\vartheta}} \tag{11.43}$$

are valid, where $c = \kappa c_\phi + c_\rho \max_t \left\| \delta A(t) \right\|$, $\tilde{\alpha} = \alpha(1 - \vartheta)$. $0 \leq \vartheta < 1$.

From corollary 2 of Theorem 3, we get the parameter εin the target condition (11.29):

$$V(\infty) \leq \varepsilon = \left(\frac{c}{\alpha}\right)^{\frac{1}{1-\vartheta}} = \left(\frac{\kappa c_\phi + c_\rho \max_t \|\delta A(t)\|}{\alpha}\right)^{\frac{1}{1-\vartheta}} \text{ by } t \to \infty \tag{11.44}$$

It follows from (11.44) that the limiting properties of the system (11.7), (11.14) depend on the level of disturbances $\xi(t)$ and FA properties. At $\xi(t) = 0$

$$V(t) = \left(e^{-\overline{\alpha}(t-t_0)} V^{1-\rho}(t_0) + \frac{c}{\overline{\alpha}}(1 - e^{-\overline{\alpha}(t-t_0)})\right)^{\frac{1}{1-\rho}} \tag{11.45}$$

where $\overline{\alpha} = \alpha(1 - \rho)$, $c = c_\rho \max_t \|\delta A(t)\|$

If the algorithm (11.19) is identifying for some $A(t_0)$, then $V(\infty) = 0$. Here, the assumption 3^0 with $\sigma = 1$ is fulfilled and the system (11.19), (11.27) is exponentially stable with indicator α. If $\xi(t) = 0$, then the properties of the algorithm (11.19) follow from the theorem 1. For $\xi(t) \neq 0$, the convergence of the φ-algorithm (11.14) (or algorithm (11.19) for $\Omega_s \subseteq \mathbb{R}^k$), as seen from (11.39) or (11.42), will be ensured if for $\left\|\hat{A}(t_0)\right\| < \infty$

$$\int_{t_0}^{t} V^{\mu-\theta}(\tau)\, d\tau < \infty$$

6.2. AS Properties With Robust FA

Consider AS with the algorithm (11.16)

$$\dot{A}(t) = -\Gamma \{\lambda A(t) + Z(X, U, t)\} \tag{11.45}$$

where:

$$Z(X, U, t) = \frac{\partial \Phi(X, A, t)}{\partial A^T} I$$

Theorem 4 (Karabutov, (1993, 2006)). *Let the conditions of Theorem 3 and the inequality*

$$\hat{A}^T(t) D\left[\Psi(X, \hat{A}, U, t)\right] \delta\hat{A}(t) \geq \delta_1 \overline{c}_D V^\theta(t) \left\|\delta\hat{A}(t)\right\|^2 \tag{11.46}$$

be satisfied. Then the system (11.27), (11.45) is dissipative and estimates are fair:

$$V(t) \leq G^{-1}\left[G(V(t_0))+t-t_0\right], t_0 \leq t \leq t_*, \tag{11.47}$$

$$\|\delta A(t)\|^2 \leq e^{-\nu(t-t_0)}\left(\|\delta A(t_0)\|^2 + \frac{2\beta}{\overline{c}_D(1-\theta)}V^{1-\theta}(t_0)\right) + \\ + \frac{2\beta}{\overline{c}_D}\int_{t_0}^{t} e^{-\nu(t-\tau)}\left(\kappa c_\phi V^{\mu-\theta}(\tau) + \frac{\nu}{1-\theta}V^{1-\theta}(\tau)\right)d\tau, \tag{11.48}$$

where $\nu = 2\beta\lambda\delta_1$,

$$G(V) = \int_{V(t_0)}^{V(t)} \frac{dV}{g(V)} \tag{11.49}$$

$$g(V) = -\alpha V^\sigma + \kappa c_\phi V^\mu + c_\rho \max_t \|\delta A(t)\| V^\rho - \lambda\delta_1 \min_t \|\delta A(t)\|^2 V^\theta, t_* = \min\left(\bar{t}_\alpha, \bar{t}_\beta\right)$$

$$\bar{t}_\alpha = \sup\left\{t_\alpha \in \mathbb{J}; \left(V\left(t_0\right) + \int_{t_0}^{t} g(V(s))ds\right) \in \mathbb{J}_{0,\infty}; t \in \mathbb{J}_\alpha\right\}$$

$$\bar{t}_\beta = \sup\left\{t_\beta \in \mathbb{J}; \left[G(V(t_0)) + t - t_0\right] \in G\left(\mathbb{J}\right); t \in \mathbb{J}_\beta\right\}$$

From Theorem 4 at $\lambda = 0$, we get estimates for the system (11.14) and (11.27).

Corollary 1 of Theorem 4. *If the system (11.27), (11.45) is not under restrictions, then*

$$V(t) \leq G_1^{-1}\left(G_1\left(V(t_0)e^{-\alpha(t-t_0)}\right) + \frac{1}{\alpha}\left(1 - e^{-\alpha(t-t_0)}\right)\right), t_0 \leq t \leq t_\beta \tag{11.50}$$

$$\|\delta A(t)\|^2| \leq e^{-\nu(t-t_0)}\left(\|\delta A(t_0)\|^2 + 2\beta V(t_0)\right) + 2\beta\int_{t_0}^{t} e^{-\nu(t-\tau)}\left(\kappa c_\phi V^\mu(\tau) + \nu V(\tau)\right)d\tau, \tag{11.51}$$

where:

$$G(V) = \int_{V(t_0)}^{V(t)} \frac{dV}{\kappa c_\phi V^\mu + c_\rho V^\rho \max_t \|\delta A(t)\| - \lambda\delta_1 \min_t \|\delta A(t)\|}$$

$$t_\beta = \sup\left\{t \in \mathbb{J}; G_1\left(\left(V(t_0)e^{-\alpha(t-t_0)}\right) + \frac{1}{\alpha}\left(1 - e^{-\alpha(t-t_0)}\right)\right) \in G_1\left(\mathbb{J}_{0,\infty}\right)\right\}$$

As inequality is true for $V(t)$

$$V(t) \leq \left\{V(t_0) + \lambda\delta_1 \min_t \|\delta A(t)\|^2 t_0\right\} - \alpha\int_{t_0}^{t} V(\tau)\,d\tau + \int_{t_0}^{t} g_2(V(\tau))\,d\tau \tag{11.52}$$

where $g_2(V) = \kappa c_\phi V^\mu + c_\rho V^\rho \max_t \|\delta A\|$, then applying Theorem 1.3.5 from [38] to (11.52), we get

$$V(t) \leq G_2^{-1}\left\{G_2\left(V(t_0) + \lambda\delta_1 \min_t \|\delta A(t)\|^2 t_0\right) e^{-\alpha(t-t_0)} + +\alpha^{-1}(1 - e^{-\alpha(t-t_0)})\right\},\ t_0 \leq t \leq t_\beta \tag{11.53}$$

,

where:

$$G(V) = \int_{V(t_0)}^{V(t)} \frac{dV}{g_2(V)}$$

$$t_\beta = \sup\left\{t \in \mathbb{J};\left(G_2\left[\left(V(t_0) + \lambda\delta_1 \min_t \|\delta A(t)\|^2 t_0\right) e^{-\alpha(t-t_0)}\right] + +\alpha^{-1}(1 - e^{-\alpha(t-t_0)}\right) \in G_2(\mathbb{J}_{0,\infty})\right\}$$

Corollary 2 of Theorem 4. *If the $\Omega_s = \mathbb{R}^k$ and $\xi(t) = 0$, then the following estimates are valid for the system (11.20) and (11.27)*

$$\|\delta A(t)\|^2 \leq e^{-\nu(t-t_0)}\left(\|\delta A(t_0)\|^2 + 2\beta V(t_0)\right) + 2\beta\int_{t_0}^{t} e^{-\nu(t-\tau)} V(\tau)\,d\tau, \tag{11.54}$$

$$V(t) \leq \left[\left\{V(t_0) + \lambda\delta_1 \min_t \|\delta A(t)\|^2 t_0\right\}^{1-\rho} e^{-\upsilon(t-t_0)} + \frac{c}{\alpha}(1 - e^{-\upsilon(t-t_0)})\right]^{\frac{1}{1-\rho}} \tag{11.55}$$

$$c = c_\rho \max_t \|\delta A(t)\|,\ \upsilon = \alpha(1-\rho)$$

We see that at $t \to \infty$

$$V(\infty) \leq \left(\frac{c_\rho \max_t \|\delta A(t)\|}{\alpha}\right)^{\frac{1}{1-\rho}} \tag{11.56}$$

that is the limiting properties of the system depend on the quality of the work FA (11.20) at $\Omega_s = \mathbb{R}^k$. If $\max_t \|\delta A(t)\| \to 0$, then the system is exponentially stable in the space $\{X\}$ with the exponent α.

From (11.49), we get the limiting value for the dissipation range of the system (11.27), (11.12) at $\vartheta = \min(\mu, \rho)$ and $t \to \infty$

$$\varepsilon \geq \left(\frac{\kappa c_\phi + c_\rho \max_t \|\delta A(t)\|}{\alpha}\right)^{\frac{1}{1-\vartheta}}$$

This estimate coincides with the value obtained in corollary 2 of Theorem 3.

Compare the properties of the φ-algorithms (11.14) and (11.16). As follows from estimates (11.39) and (11.48), the FA (11.14) coarsening, based on introducing feedback, guarantees its operability under any initial conditions. Adaptive φ-algorithm (11.16) is rough to perturbations, since it is a filtering system. The parameter λ, which depends on the algorithm (11.16) properties, sets the level of influence $\xi(t)$. The influence of disturbances in the algorithm (11.14) is compensated by choosing the vector $Z(X, U, t)$; in the algorithm (11.16), we use the parameters λ and Γ.

The φ-algorithm (11.16) provides restoration of the parameter vector A_* in the absence of disturbances. Show it. Let the goal of adaptation (11.12) be achieved in the system (11.16), (11.27). Put $\|\delta A(t)\| = \|\delta A\|$ at $t \to \infty$ and obtain from (11.48)

$$\|\delta A\| \leq \frac{2\beta\varepsilon^{1-\theta}}{\overline{c}_D}\left(\frac{\kappa c_\phi}{\nu} + \frac{1}{1-\theta}\right) \tag{11.57}$$

At $\xi(t) = 0$, we have $\dot{V}(t) < 0$, and the control goal with $\varepsilon = 0$ has been reached in the system (11.1), (11.16). Therefore, we get $\widehat{A}(t) \to A_*$ from (11.57). So, the system (11.1), (11.16) is asymptotically stable in the space $\{X, \widehat{A}\}$ at.

Asymptotic properties of the FA (11.16) for $\xi(t) \neq 0$ follow from (11.57). In the general case, convergence conditions of the φ-algorithm (11.16) for $\Gamma = \text{const}$ have the form:

$$\int_{t_0}^{\infty} e^{-\nu(t-\tau)} V^{\mu-\theta}(\tau)\, d\tau < \infty\ , \int_{t_0}^{\infty} e^{-\nu(t-\tau)} V^{1-\theta}(\tau)\, d\tau < \infty$$

If $\Gamma = \Gamma(t)$, then, for $\Gamma(t) = \gamma(t) I_k$ and $\gamma \in \mathbb{R}$ the algorithm (11.16) restores the vector of parameters A_* if

$$\int_{t_0}^{\infty} \gamma(\tau)\, d\tau < \infty\ , \int_{t_0}^{\infty} e^{-\nu(t-\tau)} \gamma(\tau) V^{\mu-\theta}(\tau)\, d\tau < \infty\ , \int_{t_0}^{\infty} e^{-\nu(t-\tau)} \gamma(\tau) V^{1-\theta}(\tau)\, d\tau < \infty$$

The algorithm (11.20) properties follow from corollaries 1, 2 of Theorem 4. If we compare the estimates (11.48) and (11.51), we get that algorithm (11.20) has a higher convergence rate than algorithm (11.45). If $\xi(t) \neq 0$, then algorithm (11.45) properties can be improved using additional information. This is the FA (11.45) advantage from (11.20). If there are no restrictions and perturbations, then the properties of the ψ-algorithm (11.20) follow from (11.54). It can be seen from (11.54) that if the function satisfies assumption 10, then the φ-algorithm (11.20) is identifying. Estimates (11.51) and (11.54) have of a criterion nature.

Apply Theorem 4 to evaluate the quality of transients in the system (11.1) and (11.45). The AS effectiveness is determined both by properties of the main contour (including the object and the model) and the characteristics of the adaptation contour. The AS effectiveness is determined both by properties of the main contour (including the object and the model) and the characteristics of the adaptation contour.

6.3. Robustness Based on Z Choice

Consider conditions imposed on the vector $Z(X, U, t)$ (11.25), which guarantee the dissipativity of the system (11.1), (11.14). As shown Karabutov (2006), the inequality

$$\dot{V} \leq -\alpha V^{\sigma} + \Psi_2^T \xi + Z^T D[\Psi]\delta A$$

holds for $\dot{V}$.

Apply conditions of Theorem 3 and get an estimate

$$\dot{V} \leq -\alpha V^{\sigma} + \kappa c_{\phi} V^{\mu} + c_D V^{\theta} \|Z\| \, \|\delta A\|$$

The system (11.14), (11.27) is dissipative if

$$-\alpha V^{\sigma} + \kappa c_{\phi} V^{\mu} + c_D V^{\theta} \|Z\| \, \|\delta A\| \leq 0 \tag{11.58}$$

Then

$$\|Z(X, U, t)\| \leq \frac{-\alpha V^{\sigma-\theta}(X, t)}{c_D \max_t \|\delta A(t)\|} - \frac{\kappa c_{\phi} V^{\mu-\theta}(X, t)}{c_D \max_t \|\delta A(t)\|} \tag{11.59}$$

In the presence of disturbances $\xi(t)$, the vector $Z(X, U, t)$ structure is based on the restriction Ω_s (the first term in the right part (11.59)). The structure should be adjusted to consider the information about $\xi(t)$ (the second term in the right part (11.59)).

As $\|Z(X, U, t)\| \geq 0$, then we get from (11.58)

$$\alpha V^{\sigma} \geq \kappa c_{\phi} V^{\mu} \tag{11.60}$$

Remark 1. The condition (11.60) is valid only for $\xi(t) \neq 0$, since otherwise parameters α, κ and c_{ϕ} are determined by the domain Ω_s.

In the absence of restrictions on X, U and$A$, the estimate for $Z = \text{grad}_{\hat{A}} \chi$ has the form:

$$\|Z\| \leq \frac{\alpha V^{\sigma} - \kappa c_{\phi} V^{\mu}}{c_A}, c_A > 0 \tag{11.61}$$

Parameters α, κ and c_{ϕ} are fixed, therefore, the system dissipativity is provided only for low-intensity disturbances, the magnitude of which satisfies the condition

$$\sup_t \|\xi(t)\| \leq \frac{\alpha}{c_{\phi}} V^{1-\mu}$$

At $0 \leq c_{\phi} \leq 1$, the function $V^{\mu}(t)$ decreases more slowly than the function $V(t)$. Therefore, the inequality (11.61) is violated for some $t \geq t' \geq t_0$. This causes a loss of stability in the system. This conclusion explains why the velocity gradient algorithms are not robust to $\xi(t)$.

Example. Consider of the φ-algorithm (11.14):

$$\hat{A} = \begin{cases} -\Gamma Z(X,t)+\zeta, & V(X,t) > \varepsilon, \\ 0, & V(X,t) \leq \varepsilon, \end{cases} \tag{11.62}$$

where $\varepsilon \geq 0$ is a threshold that depended on the level of active disturbances $\xi(t),\zeta(t)$; $\zeta(t)$ is a disturbance that considers the discrepancy between the algorithm and the operating conditions of the system.

Let $\|\xi(t)\| \leq \kappa_{\xi}$, $\|\zeta(t)\| \leq \kappa_{\zeta}$, where $\kappa_{\xi} \geq 0,\ \kappa_{\zeta} \geq 0$.

Theorem 5 [3]. *Let the assumptions 1^0-3^0 be fulfilled, the conditions (11.32), (11.33) and inequality $\|\delta A(t)\| \leq c_A$, $c_A \geq 0$. Then, all trajectories of the system (11.27), (11.62) are limited, and the condition $\overline{\lim_{t\to\infty}} V(X,t) \leq \varepsilon$ is fulfilled.*

Theorem 5 gives the choice of the threshold in the target condition (11.3) $\varepsilon \geq \varepsilon_*(\xi,\zeta)$, where ε_* is the positive root of the equation

$$\varepsilon^{\sigma-\mu} - \frac{c_{\phi}\kappa_{\xi}}{\alpha} - \frac{c_A \kappa_{\zeta} c_D}{\alpha}\left\|\Gamma^{-1}\right\|\varepsilon^{\theta-\mu} = 0$$

If $\theta = \mu$, then

$$\varepsilon \geq \left(\frac{c_{\phi}\kappa_{\xi} + c_A\kappa_{\zeta} c_D\left\|\Gamma^{-1}\right\|}{\alpha}\right)^{\frac{1}{\sigma-\mu}}$$

Let $\zeta(t) = 0$. Then:

$$\varepsilon \geq \left(\frac{c_{\phi}\kappa_{\xi}}{\alpha}\right)^{\frac{1}{\sigma-\mu}} \tag{11.63}$$

If $\sigma = 1$, then the estimate for ε coincides with the value obtained Fomin et al. (1981).

7. CONCLUSION

The approach is proposed for the synthesis of continuous adaptive control algorithms under restrictions. We have developed a way to construct a quality criterion under constraints. Methods of coarsening continuous adaptive φ-algorithms are offered. We analyse approaches to estimating the quality of adaptive systems. The properties of adaptive systems based on FA are studied under constraints. We estimate the impact of coarsening methods on the quality of adaptive systems.

REFERENCES

Bellman, R. E. (2003). *Dynamic programming*. Dover Publications.

Brian, D. O., & Anderson, R. R., Bitmead, C., Johnson R. Jr., Kokotovic, P. V. et al. (1986). *Stability of adaptive system: passivity and averaging analysis*. Cambridge: MIT Press.

Eikhoff, P. (1974). *System identification: parameter and state estimation*. Wiley-Interscience.

Evtushenko, Ju. G. (1982). Methods of solution of extremum problems and their application in optimisation problems. Moscow: Nauka.

Fomin, V. N., Fradkov, I. L., & Yakubovich, V. A. (1981). *Adaptive control in dynamic plants*. Nauka, Ru.

Fradkov, A. L. (1979). Shema of a high-speed gradient and its application in adaptive control problems. *Automatics and telemechanics, 9*, 90-102.

Furasov, V. D. (2005). *Tasks of the guaranteed identification. Digital systems*. Moscow: BINOMIAL. Laboratory of knowledge.

Gromiko, D., & Sankovsky, I. (1974). *Self-customized systems with model*. Energy.

Karabutov, N. N. (1993). Integral estimations of quality of transients in adaptive systems in the presence of restrictions and perturbations. *Reports of Academy of sciences, 331*(2), 158-160.

Karabutov, N. N. (2006). *Adaptive identification of systems: informational synthesis*. URSS.

Karabutov, N. N. (2013). Method of φ-algorithms of control. In Nadykto, A. B., Uvarova, L., & Latyshev, A. V. (Eds.), *Mathematical models of non-linear phenomena, processes, and systems: from molecular scale to planetary atmosphere* (pp. 259–396). Published by Nova Science Publishers.

Katkovnik, V. J. (1976). *Linear estimations and stochastic problems of optimization*. Nauka, Ru.

Kogan, M. M., & Nejmark, J. I. (1987). Adaptive locally-optimum control. *Automatics and telemechanics, 8*, 126-136.

Krasovsky, A.A. (1975). Optimal algorithms in problems of identification with adaptive model. *Automatics and telemechanics, 8*, 75-82.

Krasovsky, A. A., Bukov, V. N., & Shendrik, V. S. (1977). *Universal algorithms of optimum control of continuous processes*. Nauka.

Kulchitsky, O. J. (1983). Algorithms of type of stochastic approximation in a contour of adaptive discrete stochastic linear dynamic system. I. *Automatics and telemechanics, 9*, 102-119.

Kuntsevich, V. M., & Lichac, M. M. (1985). *Synthesis of optimum and adaptive control systems*. Naukova dumka, UA.

Kushner, H. J., & Clark, D. S. (1978). *Stochastic approximation methods for constrained and unconstrained systems*. Springer. DOI: 10.1007/978-1-4684-9352-8

Martinyuk, A. A. (1980). Method of integral inequalities in the theory of stability of motion. *Soviet Applied Mechanics, 16*(4), 267–281. DOI: 10.1007/BF00884242

Martinyuk, A. A., Lanshmikantam, V., & Lila, S. (1989). *Stability of movement: a method of integral inequalities*. Naukova dumka, UA.

Ortega, J., & Rheinboldt, W. (1970). *Iterative solution of nonlinear equations in several variables*. Academic Press.

Poljak, B. T. (1976). Convergence and a velocity of convergence iterative rocedures I. *Automatics and telemechanics, 4*, 101-107.

Poljak, B. T. (1983). *Introduction in optimization*. Nauka, Ru.

Polyak, B. T., & Tsypkin, J. Z. (1980). Optimum pseudogradients algorithms of adaptation. *Automatics and telemechanics, 8*, 74-84.

Polyak, B. T., & Tsypkin, J. Z. (1983). Optimum algorithms of criteria optimisation in the conditions of uncertainty. *Reports AS USSR, 273*(2), 315–318.

Salyga, V. I., Karabutov, N. N., & Kusmina, O. I. (1988). Identification of nonlinear plants by the method of φ-algorithms. *Identification and system parameter estimation. 8th IFAC/IFOS symposium,* Beijing, China, *2*. 661–665.

Tsypkin, J. Z. (1995). *Bases of identification informational theory*. Nauka.

Chapter 12
Adaptive Identification of Systems With Multiple Nonlinearities

ABSTRACT

The solution of identification problem off with multiple nonlinearities depends on such factors as feedback, ways of connecting nonlinear links, and signal properties. We consider the constant excitation (CE) condition and analyze it as the link with S-synchronizability of the nonlinear system. The system decomposition ensures identification by measured outputs. We apply a structural-parametric approach to identification. It is based on the S-synchronizability of the system and CE. Adaptive algorithms are synthesized for each of the subsystems. The limitation of trajectories is proved in an adaptive system in parametric and coordinate spaces. Conditions guaranteeing exponential stability of the trajectories are got. Systems of self-oscillation generation and nonlinear correction of a nonlinear system are considered. The conditions of the limited trajectories in the adaptive system are proved. The influence of nonlinearity and feedback on the quality of assessment is studied.

1. INTRODUCTION

Studies (Aykan and Özgüven, 2012; Binczak et al.,2000; Li, 2021; Noël and Schoukens, 2018; Pavlov1 et al., 2015; Zhao et al., 2016; Pai and Palazotto, 2008; Pintelon, 2002; Prawin et al., 2020; Safari and Monsalve, 2021; Singh and Moore, 2021) are devoted to the identification of systems with several nonlinearities (SN). The case when a system comprises several nonlinearities in series (Singh and Moore, 2012) is considered and an identification method is proposed. The detection of nonlinearity is performed using sinusoidal tests. A similar approach based

DOI: 10.4018/979-8-3373-0822-7.ch012

on the functional description method is applied (Zhao et al., 2016) for parametric identification of the system having two nonlinear elements in different locations. An approach to estimating parameters (Pavlov et al., 2015) of the transfer function for the second-order nonlinear system containing two nonlinearities is proposed. Here, the harmonic linearization of nonlinearities is performed beforehand. Various methods based on the nonlinearity approximation are considered by Noël and Schoukens (2018), Pai and Palazottoin (2005), Safari and Monsalve (2021). Li (2021) studies approach to identifying discrete feedback systems. Identifiability conditions are obtained by applying the least squares method (Li, 2021).

Pintelon (2002) considers the identification of the system having nonlinear mechanical oscillations. The proposed model, which has a gray box appearance, is based on the application of nonlinear basis functions using a limited number of measured output variables. Other approaches to identification are considered by Andersson et al. (2019), Gedon et al. (2021), Singh and Moore (2021), Van den Hof (1997). These are based on considering physical laws when selecting the model (Singh and Moore, 2021), applying frequency methods for the feedback system (Van den Hof, 1997), along with learning theory (Andersson et al., 2019.

The review (Kerschen et al., 2006) is devoted to the analysis of methods for identifying nonlinear processes in structural dynamics. Here, modifications to the frequency approach are mainly applied. Disadvantages of approaches based on linearization, harmonic balance, and the restoring force surface method are noted. Holmes (1982) applies regression analysis and Hilbert transform are applied. Azeez and Vakakis (1999) use chaos theory methods to identify bifurcation processes.

Various approaches (Aljamaan, 2016; Forssell and Ljung, 1999; Mejari, et al., 2018; Van den Hof, 1998) based on estimating parameters of the transfer function for a feedback system is presented. The identification of the feedback system is often reduced to identifying an open system. Piga and Tyth (2014), Song et al. (2019) note difficulties in identifying the feedback system.

Thus, it follows from the presented review that frequency methods are used for identifying a nonlinear system. In some cases, approaches to estimating the nonlinearity structure are proposed. Typically, different procedures of nonlinearity linearization are applied. However, the identification of systems having multiple nonlinearities has received less attention: this is possible because of the inherent complexity of such systems. In these cases, different approaches and identification methods based on the localization of nonlinearities are used.

The problem of SN identification is complex and insufficiently studied. It requires solving of problems whose consideration is given below.

2. PROBLEM STATEMENT

Consider system S_F

$$\dot{X}(t) = AX(t) + DF_1(X,t) + BU(t), \tag{12.1}$$

$$\mathrm{L}Y(t) = CX(t) + F_2(X,t), \tag{12.2}$$

where $X \in \mathbb{R}^m$ is the state vector; $A \in \mathbb{R}^{m\times m}$ is the state matrix; $D \in \mathbb{R}^{m\times q}$; $F_1(X,t): \mathbb{R}^m \to \mathbb{R}^q$ is the nonlinear vector function; $U \in \mathbb{R}^k$ is the input (control) vector; $B \in \mathbb{R}^{m\times k}$; $Y \in \mathbb{R}^n$ is the output vector; $C \in \mathbb{R}^{n\times m}$; $F_2(X,t): \mathbb{R}^m \to \mathbb{R}^n$ is the disturbance (measurement errors) vector; L is the operator defining the way of forming vector Y; t is time. Sometimes, L may be the differential operator, shown dynamic properties of the measurement system.

We shall consider the following data set:

$$\mathbb{I}_o = \{Y(t), U(t), t \in [t_0, t_N]\}, t_N < \infty \tag{12.3}$$

Assumption 1. Elements $\varphi_{1,i}(x_j) \in F_1(X,t)$, $\varphi_{2,i}(x_j) \in F_2(X,t)$ (non-linear functions belonging to F) are smooth single-valued functions

$$F_\varphi = \{\gamma_1 \xi^2 \le \varphi(\xi)\xi \le \gamma_2 \xi^2,\ \xi \ne 0,\ \varphi(0) = 0,\ \gamma_1 \ge 0,\ \gamma_2 < \infty\}$$

Sometimes, condition $\varphi_i^1(x_j) = \varphi_i^1(\varphi_k^1(x_j))$ may be satisfied. For estimating the parameters of matrixes A, D, B and C, the following model is applied:

$$M_F: \begin{cases} \dot{\hat{X}}(t) = \hat{A}(t)\hat{X}(t) + \hat{D}(t)F_1(X,t) + \hat{B}(t)U(t), \\ \mathrm{L}\hat{Y}(t) = CX(t) + F_2(X,t), \end{cases} \tag{12.4}$$

where $\hat{A}(t)$, $\hat{D}(t)$, $\hat{B}(t)$ are matrices with adjustable parameters.

Problem: for system (12.1) satisfying assumption 1, construct model (12.4) based on the analysis of $\mathbb{I}_o$, and find such regularities of adjusting parameters of matrices $\hat{A}(t)$, $\hat{D}(t)$ and $\hat{B}(t)$, that

$$\lim_{t\to\infty} \|\hat{Y}(t) - Y(t)\| \le \delta_y,$$

$$\delta_y \ge 0$$

where $\| \cdot \|$ is the Euclidean norm.

3. EXCITATION CONSTANCY CONDITION

The excitation constancy (EC) condition plays an important role in parametric estimation problems. If the system is nonlinear, then the fulfillment of this condition may not be sufficient. As shown Karabutov (2018, 2021), the system should have the property of S-synchronizability to consider the nonlinear properties of the system.

Let: (1) bounded vector $P \in \mathbb{R}^m$ and its corresponding set of frequencies $\Omega_P(\omega)$; (2) a set of allowable input frequencies $\Omega_S(\omega)$ providing S-synchronizability (see chapter 5) of the system. Then the EC condition for matrix $B_P(t) = P(t) P^T(t)$ has the following form:

$$EC^S_{\alpha,\bar{\alpha}}: \ \left(\alpha I_l \leq B_P(t) \leq \bar{\alpha} I_l\right) \ \& \ \left(\Omega_P(\omega) \subseteq \Omega_S(\omega)\right)$$

$$\forall t \geq t_0 \qquad (12.5)$$

for $\alpha > 0$ and $\forall t \geq t_0$ on some interval $T > 0$, where $\bar{\alpha} > 0$ is some number, while $I_l \in \mathbb{R}^{m \times m}$ is the identity matrix. Usually, $P(t)$ is the vector of measurements and state variables.

4. STRUCTURAL-PARAMETRIC APPROACH TO IDENTIFICATION

The following displays the procedure for identifying system SF based on the structural-parametric approach (SPA) (Karabutov, 2018). Depending on the available a priori information, several stages implementing SPA can be applied. The system (12.1), (12.2) has a complex form; synthesis of adaptive algorithms requires a priori information about its structure. The composition of the subsystems included in system S_F is assumed to be known. Hence, based on the dimensionality of the system output vector, matrix A can be divided into n blocks (subsystems $S_{F,n} \subseteq S_F$, $\left\{s_{F,j}\right\} \in S_{F,n}, j < n$). Analyzing the subsystems (blocks), we shall distinguish those containing nonlinearities, i.e., $S_{F,nel} \subseteq S_{F,n}$. Then we shall apply SPA to each $s_{F,nel,k} \in S_{F,nel}$ element. If subsystem $\left\{s_{F,j}\right\} \in S_{F,n} \backslash S_{F,nel}$ does not contain nonlinearities, then the adaptive identification procedure is applied to it.

Remark 1. The structural-parametric approach is based on the S-synchronizability of the system and the fulfillment of condition (12.5).

Under uncertainty, SPA can be divided into two procedures: (1) structural S_F analysis and (2) parametric estimation (adaptive identification). These stages are proposed by Karabutov (2018).

Remark 2. Subsystems and the mutual influence of variables have a significant impact on the structural identifiability (S-synchronizability) of the system. Therefore, estimating the identifiability of the system nonlinear elements requires the construction of a diagram of the mutual influence in the system (Karabutov, 2017). By analyzing the interrelations, the effect of influencing variables can be excluded to determine the structural identifiability of the system (nonlinearity). Constructing the mutual influence diagram is possible only if condition $EC^{S}_{\alpha,\overline{\alpha}}$ is satisfied.

Remark 3. Estimating the S-synchronizability of the system is based on the analysis of dynamic structures S_{ey} reflecting the nonlinearity structure of the corresponding subsystem $s_{F,nel,k}$. Karabutov (2018, 2021) proposes the method of their construction.

Remark 4. Got estimates of the nonlinearity structure in (12.1), (12.2) comprise the basis for implementing the adaptive parametric identification of the system S_F.

Remark 5. If a priori information about the nonlinear properties is known, then the structural analysis stage of system S_F may be skipped.

5. ADAPTIVE IDENTIFICATION OF S_F SYSTEM

We shall consider the subsystem $s_{F,i} \in S_F$, $\dim s_{F,i} = n_i$, L being the linear operator in (12.2). Let set $\mathbb{I}_{o,i} \subset \mathbb{I}_o$ be known for $s_{F,i}$. Subsystem $s_{F,i}$ is described by the following equations:

$$\begin{cases} \dot{X}_{s_{F,i}} = A_{s_{F,i}} X_{s_{F,i}} + D_{1,s_{F,i}} F_{1,s_{F,i}}(X) + B_{s_{F,i}} U_{s_{F,i}}, \\ y_{s_{F,i}} = x^1_{s_{F,i}}. \end{cases} \tag{12.6}$$

We will transform (12.6) to the form:

$$W_{s_{F,i}}(\upsilon) y_{s_{F,i}} = \sum_{k=1}^{n_i} \sum_{j=1}^{n_i} \left(d^{k,h}_{1,s_{F,i}} f^{h,j}_{s_{F,i}} + b^{k,j}_{s_{F,i}} u^j_{s_{F,i}} \right), \tag{12.7}$$

where $X_{s_{F,i}} \in \mathbb{R}^{n_i}$ is the state vector of the subsystem $s_{F,i}$, $x^1_{s_{F,i}}$ being the first $X_{s_{F,i}}$ element; $D_{1,s_{F,i}}$ $F_{1,s_{F,i}}$ and $B_{s_{F,i}}$ are matrices of fitted dimensions; $d^{k,h}_{1,s_{F,i}} \in D_{1,s_{F,i}}$, $f^j_{s_{F,i}} \in F_{1,s_{F,i}}$, $b^{k,j}_{s_{F,i}} \in B_{s_{F,i}}$, $u^j_{s_{F,i}} \in U_{s_{F,i}}$, $\upsilon = d/dt$, $W_{s_{F,i}}(\upsilon)$ is the polynomial of n_i degree. Matrix $A_{s_{F,i}} \in \mathbb{R}^{n_i \times n_i}$ is the Hurwitz matrix. We shall divide left and right parts of (12.7) by polynomials of $n_i - 1$ degree:

$$H_{s_{F,i}}(\upsilon)=\prod_{k=1}^{n_i-1}(\upsilon+\mu_k) \tag{12.8}$$

Then:

$$\dot{y}_{s_{F,i}} = -\eta_{s_{F,i}} y_{s_{F,i}} + \sum_{k=1}^{n_i-1}\sum_{j=1}^{n_i-1}\left(\tilde{d}^{k,h}_{s_{F,i}} p_{f^{h,j}_{s_{F,i}}} + \tilde{b}^{k,j}_{s_{F,i}} p_{u^j_{s_{F,i}}}\right) \tag{12.9}$$

where υ, $\upsilon > 0, \mu_k > 0$ are positive numbers; $\eta_{s_{F,i}} > 0$, $\tilde{d}^{k,h}_{s_{F,i}}$ and $\tilde{b}^{k,j}_{s_{F,i}}$depend on parameters of subsystem $s_{F,i}$ and μ_k, and variables$p_{z^j_{s_{F,i}}}$ $(z = f, u)$ satisfy the following equation:

$$\dot{p}_{z^j_{s_{F,i}}} = -\mu_j p_{z^j_{s_{F,i}}} + z^j_{s_{F,i}} \tag{12.10}$$

Remark 6. The structure of the right part of (9) is determined by the type of matrix $A_{s_{F,i}}$ and polynomial $H_{s_{F,i}}(\upsilon)$.

The adaptive model for estimate parameters (12.9), (12.10):

$$\dot{\hat{y}}_{s_{F,i}} = -k_{s_{F,i}}\left(\hat{y}_{s_{F,i}} - y_{s_{F,i}}\right) + \hat{\kappa}_{s_{F,i}} y_{s_{F,i}} + \sum_{k=1}^{n_i-1}\sum_{j=1}^{n_i-1}\left(\hat{\tilde{d}}^{k,h}_{s_{F,i}} p_{f^{h,j}_{s_{F,i}}} + \hat{\tilde{b}}^{k,j}_{s_{F,i}} p_{u^j_{s_{F,i}}}\right) \tag{12.11}$$

where $k_{s_{F,i}} > 0$; $\hat{\kappa}_{s_{F,i}}$, $\hat{\tilde{d}}^{k,h}_{s_{F,i}}$, and $\hat{\tilde{b}}^{k,j}_{s_{F,i}}$ are adjustable parameters.

The equation for the identification error:

$$\dot{e}_{s_{F,i}} = -k_{s_{F,i}} e_{s_{F,i}} + \Delta\kappa_{s_{F,i}} y_{s_{F,i}} + \sum_{k=1}^{n_i-1}\sum_{j=1}^{n_i-1}\left(\Delta\tilde{d}^{k,h}_{s_{F,i}} p_{f^{h,j}_{s_{F,i}}} + \Delta\tilde{b}^{k,j}_{s_{F,i}} p_{u^j_{s_{F,i}}}\right), \tag{12.12}$$

where

$$\Delta\tilde{d}^{k,h}_{s_{F,i}} = \hat{\tilde{d}}^{k,h}_{s_{F,i}} - \tilde{d}^{k,h}_{s_{F,i}}.$$

$$\Delta\tilde{b}^{k,j}_{s_{F,i}} = \hat{\tilde{b}}^{k,j}_{s_{F,i}} - \tilde{b}^{k,j}_{s_{F,i}},$$

$$e_{s_{F,i}} = \hat{y}_{s_{F,i}} - y_{s_{F,i}},$$

$$\Delta\kappa_{s_{F,i}} = \hat{\kappa}_{s_{F,i}} - \eta_{s_{F,i}}.$$

We shall consider the Lyapunov function $V_{e,i}\left(e_{s_{F,i}}\right) = 0.5\,e^2_{s_{F,i}}$ Then the following is obtained for $\dot{V}_{e,i}\left(e_{s_{F,i}}\right)$

$$\dot{V}_{e,i}\left(e_{s_{F,i}}\right) = e_{s_{F,i}}\left[-k_{s_{F,i}} e_{s_{F,i}} + \Delta\kappa_{s_{F,i}} y_{s_{F,i}} + \sum_{k=1}^{n_i-1}\sum_{j=1}^{n_i-1}\left(\Delta\tilde{d}^{k,h}_{s_{F,i}} p_{f^{h,j}_{s_{F,i}}} + \Delta\tilde{b}^{k,j}_{s_{F,i}} p_{u^j_{s_{F,i}}}\right)\right]$$

If variables $p_{f^{h,j}_{s_{F,i}}}, p_{u^j_{s_{F,i}}}$ have property $EC^S_{\alpha,\bar{\alpha}}$ then the following is obtained from the condition $\dot{V}_i\left(e_{s_{F,i}}\right) < 0$:

$$\begin{aligned}
\Delta\dot{\kappa}_{s_{F,i}} &= -\gamma_{\kappa,s_{F,i}} e_{s_{F,i}} y_{s_{F,i}}, \\
\Delta\dot{\tilde{d}}^{k,h}_{1,s_{F,i}} &= -\gamma_{k,h,s_{F,i}} e_{s_{F,i}} p_{f^{h,j}_{s_{F,i}}}, \\
\Delta\dot{\tilde{b}}^{k,j}_{s_{F,i}} p_{u^j_{s_{F,i}}} &= -\gamma_{k,j,s_{F,i}} e_{s_{F,i}} p_{u^j_{s_{F,i}}},
\end{aligned} \tag{12.13}$$

where $\gamma_{k,j,s_{F,i}} > 0$ is the amplification factor.

It is not difficult to obtain algorithms for the parameters of the model (12.11) from (12.13).

Thus, the adaptive identification system of the subsystem $s_{F,i}$ is described by (12.12) and (12.13). We shall denote it as $AS_{F,i}$.

Consider the Lyapunov function (LF)

$$V_{\Delta,i} = 0.5\gamma^{-1}_{\kappa,s_{F,i}}\left(\Delta\kappa_{s_{F,i}}\right)^2 + 0.5\mathrm{Sp}\left(\Delta D_{1,s_{F,i}} \Gamma^{-1}_{k,h,s_{F,i}} \Delta D^T_{1,s_{F,i}}\right) + 0.5\mathrm{Sp}\left(\Delta B_{s_{F,i}} \Gamma^{-1}_{k,j,s_{F,i}} \Delta B^T_{s_{F,i}}\right) \tag{12.14}$$

where $\Gamma_{k,h,s_{F,i}} = \mathrm{diag}(\gamma_{k,h,s_{F,i}})$, $\mathrm{Sp}(\cdot)$ is the spur of matrix, $\Gamma^{-1}_{k,j,s_{F,i}} = \mathrm{diag}\left(\gamma_{k,j,s_{F,i}}\right)$, $\Delta D_{1,s_{F,i}}$ and $\Delta B_{s_{F,i}}$ contain elements $\Delta\tilde{d}^{k,h}_{s_{F,i}}$ and $\Delta\tilde{b}^{k,j}_{s_{F,i}}$, respectively. Let $\Delta K_{s_{F,i}} \triangleq \left[\Delta D_{1,s_{F,i}}, \Delta B_{s_{F,i}}\right]$ and $V_{s_{F,i}}(t) = V_e(t) + V_{\Delta,i}(t)$.

Theorem 1. (Karabutov, 2023). *Let: (i) functions $V_{e,i}(t)$, $V_{\Delta,i}(t)$ are positively definite and admit the infinitesimal higher limit at $|e| \to \infty$ and $\left\|\Delta K_{s_{F,i}}\right\| \to \infty$, $\left|\Delta\kappa_{s_{F,i}}\right| \to \infty$, (ii) matrix $A_{s_{F,i}} \in \mathbb{R}^{n_i \times n_i}$ is the Hurwitz matrix; (iii) $p_{f^{h,j}_{s_{F,i}}}$ and $p_{u^j_{s_{F,i}}}$ have property $EC^S_{\alpha,\bar{\alpha}}$. Then all trajectories of system $AS_{F,i}$ are bounded, lie in the following domain:*

$$\mathrm{G}_t = \left\{\left(e_{s_{F,i}}, \Delta\kappa_{s_{F,i}}, \Delta K_{s_{F,i}}\right) : V_{s_{F,i}}(t) \le V_{s_{F,i}}(t_0)\right\}$$

and the following estimation is valid:

$$2k_{s_{F,i}}\int_{t_0}^{t} V_{e,i}(\tau)\,d\tau \le V_{s_{F,i}}(t_0) - V_{s_{F,i}}(t)$$

The proof of Theorem 1 is given in Appendix A.

Let estimation A1 be true for $V_{\Delta,i}$:

$$0.5\,\underline{\vartheta}\left\{\mathrm{Sp}\left(\Delta K_{s_{F,i}}(t)\Delta K_{s_{F,i}}^{T}(t)\right)+\left(\Delta\kappa_{s_{F,i}}\right)^{2}\right\}\leq V_{\Delta,i}(t)\leq 0.5\,\overline{\vartheta}\left\{\mathrm{Sp}\left(\Delta K_{s_{F,i}}(t)\Delta K_{s_{F,i}}^{T}(t)\right)+\left(\Delta\kappa_{s_{F,i}}\right)^{2}\right\},$$

where $\Gamma_{s_{F,i}} = \Gamma_{k,h,s_{F,i}} \dotplus \Gamma_{k,j,s_{F,i}}$, $\dotplus$ is the sign of the direct sum of matrices, $\beta_1\left(\Gamma_{s_{F,i}}\right)$, and $\beta_l\left(\Gamma_{s_{F,i}}\right)$ are the minimal and maximal eigenvalues of matrix $\Gamma_{s_{F,i}}$, $\underline{\vartheta} = \min\left(\beta_l^{-1}\left(\Gamma_{s_{F,i}}\right), \gamma_{\kappa,s_{F,i}}^{-1}\right)$ and $\overline{\vartheta} = \max\left(\beta_1^{-1}\left(\Gamma_{s_{F,i}}\right), \gamma_{\kappa,s_{F,i}}^{-1}\right)$.

Definition 1 (Karabutov, 2017). A non-positive quadratic form has the $W(Y,X)$ property M⁺ or $W(Y,X) \in M^{+}$ if

$$W(Y,X) = -c_y \|Y\|^2 + c_x \|X\|^2$$

is true for any $Y \in \mathbb{R}^m$ and $X \in \mathbb{R}^n$ in some bounded region

$$\Omega_D = \{Y \in \mathbf{R}^m, X \in \mathbf{R}^n: \ \|Y\|^2 + \|X\|^2 \leq \chi,\ \chi \geq 0\}$$

The M⁺ property is a sign of the constructive completeness for the quadratic form $W(Y,X)$. It allows you to reduce the analysis of $W(Y,X)$ properties to the assessment of the characteristics for the M-matrix.

Lemma 1. $\dot{V}_{e,i}(t) \in M^{+}$ *or*

$$\dot{V}_{e,i} \leq -k_{s_{F,i}} V_{e,i} + \frac{\overline{\alpha}\overline{\vartheta}}{2k_{s_{F,i}}} V_{\Delta,i}$$

if

$$P_{f_{s_{F,i}}^{h,j},u_{s_{F,i}}^{j}}^{T}(t)\, P_{f_{s_{F,i}}^{h,j},u_{s_{F,i}}^{j}}(t) \in PE_{\alpha_P,\overline{\alpha}_P}^{S}$$

and

$$y_{s_{F,i}}(t) \in PE_{\alpha_{y_i},\overline{\alpha}_{y_i}}^{S},\ \textit{where}$$

$$\alpha_{P_i}, \overline{\alpha}_{P_i}, \alpha_{y_i}, \overline{\alpha}_{y_i}$$

are positive numbers,

$$\overline{\alpha} = \max\left(\overline{\alpha}_{P_i}, \overline{\alpha}_{y_i}\right).$$

Lemma 2. If the conditions of Lemma 1 are satisfied, then $\dot{V}_{\Delta,i}(t) \in M^{+}$ or

$$\dot{V}_{\Delta,i} \leq -\tfrac{3}{8}\pi\alpha_i V_{\Delta,i} + \tfrac{8}{3}\pi V_{e,i}$$

where $\pi \geq 0$.

Theorem 2 (Karabutov, 2023). *Let: 1) positive definite Lyapunov functions* $V_{e,i}(t) = 0.5 e_{s_{F,i}}(t)$,

$$V_{\Delta,i}(t) = 0.5 \gamma^{-1}_{\kappa,s_{F,i}} \left(\Delta \kappa_{s_{F,i}}\right)^2 + 0.5\mathrm{Sp}\left(\Delta D_{1,s_{F,i}} \Gamma^{-1}_{k,h,s_{F,i}} \Delta D^T_{1,s_{F,i}}\right) + 0.5\mathrm{Sp}\left(\Delta B_{s_{F,i}} \Gamma^{-1}_{k,j,s_{F,i}} \Delta B^T_{s_{F,i}}\right)$$

admit the infinitesimal higher limit at $\left|e_{s_{F,i}}(t)\right| \to 0$, $\left|\Delta \kappa_{s_{F,i}}\right| \to 0$, $\left\|\Delta K_{s_{F,i}}\right\| \to 0$ *2) matrix* $W_{s_{F,i}}(t) = P^T_{f^{h,j}_{s_{F,i}}, u^j_{s_{F,i}}}(t) P_{f^{h,j}_{s_{F,i}}, u^j_{s_{F,i}}}(t)$ *and* $y_{s_{F,i}}$ *are piecewise continuous bounded and* $W_{s_{F,i}}(t) \in EC^S_{\alpha,\overline{\alpha}}$, $y_{s_{F,i}}(t) \in EC^S_{\alpha_{y_i},\overline{\alpha}_{y_i}}$ *3) the following equality is valid*

$$e_{s_{F,i}} \sum_{k=1}^{n_i-1}\sum_{j=1}^{n_i-1}\left(\Delta \tilde{d}^{k,h}_{s_{F,i}} p_{f^{h,j}_{s_{F,i}}} + \Delta \tilde{b}^{k,j}_{s_{F,i}} p_{u^j_{s_{F,i}}}\right) = \pi\left\{ \mathrm{Sp}\left(\Delta K_{s_{F,i}}(t) P^T_{f^{h,j}_{s_{F,i}}, u^j_{s_{F,i}}} P_{f^{h,j}_{s_{F,i}}, u^j_{s_{F,i}}} \Delta K^T_{s_{F,i}}(t)\right) + \left(\Delta \kappa_{s_{F,i}} y_{s_{F,i}}\right)^2 + e^2_{s_{F,i}} \right\}$$

in domain $O_v(O)$, *where* $\pi > 0$, $O = \left\{0, 0^{n_i \times n_i}\right\} \subset \mathbb{R} \times \mathbb{R}^{n_i \times n_i} \times \mathbb{J}_{0,\infty}$, , $0^{n_i \times n_i} \in {}^{n_i \times n_i}$ *is a zero matrix,* O_v *is some neighborhood of point O,* $t \in \left[0, \infty\right] = \mathbb{J}_{0,\infty}$ *is time interval; 4) estimation A1 is valid for function* $V_{\Delta,i}(t)$*; 5) the following system of inequalities is fulfilled:*

$$\begin{bmatrix} \dot{V}_{e,i} \\ \dot{V}_{\Delta,i} \end{bmatrix} \leq \underbrace{\begin{bmatrix} -k_{s_{F,i}} & \dfrac{\overline{\alpha}\overline{\vartheta}}{2k_{s_{F,i}}} \\ \dfrac{8}{3}\pi & -\dfrac{3}{8}\pi\alpha_i \end{bmatrix}}_{A_V} \begin{bmatrix} V_{e,i} \\ V_{\Delta,i} \end{bmatrix}$$

for $\dot{V}_{e,i}$, $\dot{V}_{\Delta,i}$ *;6) the upper solution for* $V_{e,\Delta_i}(t) = \left[V_{e,i}(t) \;\; V_{\Delta,i}(t)\right]^T$ *satisfies equation* $\dot{S} = A_V S$ *if the following inequality is valid:*

$$V_\rho(t) \leq s_\rho(t)$$

$$\forall\left(t \geq t_0\right) \;\&\; \left(V_\rho(t_0) \leq s_\rho(t_0)\right)$$

$\rho = e,i;\Delta,i$ *for elements* $V_{e,i}$, $V_{\Delta,i}$. *Then the adaptive system* $AS_{F,i}$ *is exponentially stable with the following estimate:*

$$V_{e,\Delta_i}(t) \leq e^{A_V(t-t_0)} S(t_0)$$

if

$$k_{s_{F,i}} > 0,$$

$$k_{s_{F,i}} \geq \frac{4}{3}\sqrt{\frac{2\bar{\alpha}\vartheta}{\alpha_i}} \tag{12.15}$$

From Theorem 2, we see if the information matrix $W_{s_{F,i}}(t) = P^T_{f^{N,J}_{s_{F,i}},u^J_{s_{F,i}}}(t)P_{f^{N,J}_{s_{F,i}},u^J_{s_{F,i}}}(t) \in EC$, then the adaptive system $AS_{F,i}$ allows obtaining accurate estimates of the system $s_{F,i}$ parameters. Here, the system parameters should satisfy conditions (12.15).

6. EXAMPLES

1. We shall consider the system with nonlinear correction of the nonlinear system. It contains the amplifier with the electric motor and the relay control described by function $f_1(u)$. The nonlinear feedback on speed with parabolic characteristic $f_2(x_2)$ is used as a correction device, as follows:

$$\begin{cases} \dot{x}_1 = x_2, \\ \dot{x}_2 = x_3, \\ \dot{x}_3 = -a_1x_2 - a_2x_3 + bf_1(u), \\ y = x_1 \end{cases} \tag{12.16}$$

$$f_1(u) = \begin{cases} c, & \text{if } u \geq d, \\ 0, & \text{if } -d \leq u \leq d, \\ -c, & \text{if } u < -d, \end{cases}$$

where $u = g - x_1 - f_2(x_2)$ is control; $f_2(x_2) = k_{fb}x_2^2\,sign(x_2)$;g is the system input; $c > 0, d > 0; k_{fb} > 0$, x_i is a state variable; a_i and b_i are system parameters.

Equation (13) and polynomials in (7), (8) have the following form:

$$\left(\upsilon + \eta - \frac{v_1}{\upsilon + \mu_1} - \frac{v_2}{\upsilon + \mu_2}\right)y = \left(\frac{b_1}{\upsilon + \mu_1} + \frac{b_2}{\upsilon + \mu_2}\right)f_1(u),$$

$$\upsilon = d/dt \tag{12.17}$$

or

$$\dot{y} = -\eta y + v_1 p_{y,\mu_1} + v_2 p_{y,\mu_2} + b_1 p_{f_1,\mu_1} + b_2 p_{f_1,\mu_2} \tag{12.18}$$

$$\begin{aligned} \dot{p}_{y,\mu_1} &= -\mu_1 p_{y,\mu_1} + y,\ \dot{p}_{y,\mu_2} = -\mu_2 p_{y,\mu_2} + y, \\ \dot{p}_{f_1,\mu_1} &= -\mu_1 p_{f_1,\mu_1} + f_1,\ \dot{p}_{f_1,\mu_2} = -\mu_2 p_{f_1,\mu_2} + f_1, \end{aligned} \tag{12.19}$$

where $b_1 = 1.4$, $b_2 = 0.4, \eta = 0.35$, $\nu_1 = -1$, $\nu_2 = 1.35$, $\mu_1 = 2/05$, $\mu_2 = 2.25$. The parameters of nonlinearity f_1: $c_1 = 2$, $d = 0.5$. Input $g(t) = \sin(0.2t)$.

The adaptive model, algorithms, and error equation are present as:

$$\dot{\hat{y}} = -k_y e - \hat{\eta} y + \hat{\nu}_1 p_{y,\mu_1} + \hat{\nu}_2 p_{y,\mu_2} + \hat{b}_1 p_{f_1,\mu_1} + \hat{b}_2 p_{f_1,\mu_2}, \tag{12.20}$$

$$\dot{\hat{\eta}} = -\gamma_\eta e y, \dot{\hat{\nu}}_1 = -\gamma_{\nu_1} e p_{y,\mu_1},$$

$$\dot{\hat{\nu}}_2 = -\gamma_{\nu_2} e p_{y,\mu_2}, \dot{\hat{b}}_1 = -\gamma_{b_1} p_{f_1,\mu_1}, \tag{12.21}$$

$$\dot{\hat{b}}_2 = -\gamma_{b_2} p_{f_1,\mu_2}$$

$$\dot{e} = -k e - \Delta\eta y + \Delta \nu_1 p_{y,\mu_1} + \Delta \nu_2 p_{y,\mu_2} + \Delta b_1 p_{f_1,\mu_1} + \Delta b_2 p_{f_1,\mu_2} \tag{12.22}$$

The adaptive law for ΔK follows from (12.21):

$$\Delta \dot{K} = -\Gamma_K e P_y \tag{12.23}$$

where $\Gamma_K = \mathrm{diag}\left(\gamma_\eta, \gamma_{\nu_1}, \gamma_{\nu_2}, \gamma_{b_1}, \gamma_{b_2}\right)$, $P_y = \left[y, p_{y,\mu_1}, p_{y,\mu_2}, p_{f_1,\mu_1}, p_{f_1,\mu_2}\right]^T$.

The boundedness of the system (12.22), (12.23) trajectories follow from Theorem 1. Adaptive system results are shown in Figures 1–4. The tuning of the model (12.20) parameters is shown in Figures 1 and 2. The change in estimation error $e(t)$ is shown in Figure 3. This change in the error is related to the change in the system output.

Figure 1. Tuning parameters of model (12.20)

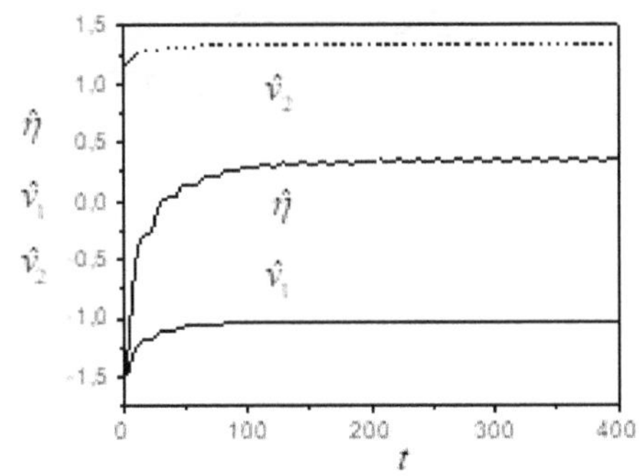

Figure 2. Tuning parameters $\hat{b}_i$ of model (12.20)

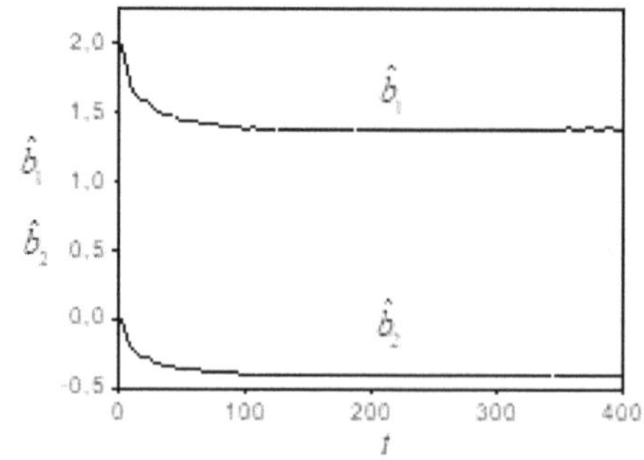

Karabutov (2021) notes that by function $f_2(x_2)$ is unidentifiable on the set of measurements. Here, the indirect information about the dependence of $u_1 = \omega - x_1$ on x_2 may be used. This is true since there is a relationship between u_1 and u.

2. Consider the self-oscillation generation system contains an object (variables y_1, y_2), nonlinear (variable y_3) and linear (variable y_4) converters, and the linear amplifier-converter with nonlinear actuator (variable y_5).

Figure 3. Change of error

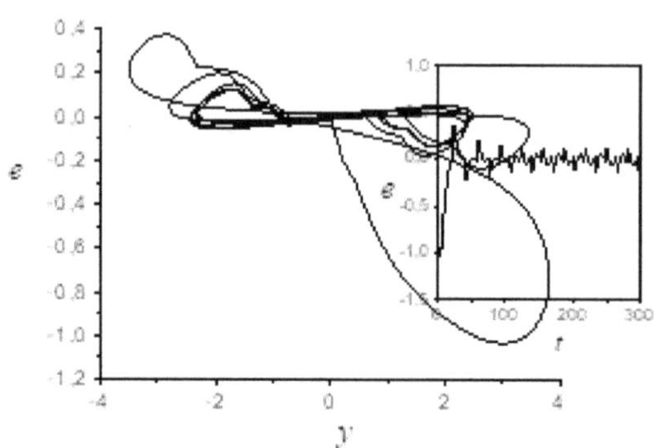

Function $f_i(x)$ $(i = 1,3)$is the saturation function with a dead zone

$$f_i(x)=\begin{cases} c, & \text{if } x \geq d_{2,i}, \\ 2(x-d_{1,i}), & \text{if } d_{1,i} < x < d_{2,i}, \\ 0, & \text{if } -d_{1,i} \leq x \leq d_{1,i}, \\ 2(x+d_{1,i}), & \text{if } -d_{1,i} < x, \\ -c, & \text{if } x < -d_{2,i}, \end{cases}$$

where $c > 0$, $d_{1,i} > 0$, and $d_{2,i} > 0$ are some numbers,

$$\dot{Y} = AY + DF(Y)$$

$$Y = \begin{bmatrix} y_1 \\ y_2 \\ y_3 \\ y_4 \\ y_5 \end{bmatrix}, A = \begin{bmatrix} 0 & 1 & 0 & 0 & 0 \\ 0 & -g & 0 & 0 & k_0 \\ 0 & 0 & -\frac{1}{T_1} & 0 & 0 \\ 0 & \frac{k_2}{T_2} & 0 & -\frac{1}{T_2} & 0 \\ 0 & 0 & 0 & 0 & -\frac{1}{T_3} \end{bmatrix}, D = \begin{bmatrix} 0 & 0 \\ 0 & 0 \\ \frac{1}{T_1} & 0 \\ 0 & 0 \\ 0 & -\frac{1}{T_3} \end{bmatrix}, F(Y) = \begin{bmatrix} f_1(y_1) \\ f_3(y_3 + y_4) \end{bmatrix} \tag{12.24}$$

where $T_i > 0$ is a time constant, $g \geq 0$. Variable y_5 is used as the input.

The phase portrait of the object is shown in Figure 4. It shows that self-oscillations arise in the system. For identifying parameters of the system (12.24), we shall use the ideas of the adaptive observer for the object described above. It is necessary to transform only the first two equations in (12.24). For this, we represent them in the form (12.8) and divide the resulting equation by $\upsilon + \mu$. Then, variables y_1, y_5 are transformed as:

$$\begin{aligned} \dot{p}_{y_1,\mu} &= -\mu p_{y_1,\mu} + y_1, \\ \dot{p}_{y_5,\mu} &= -\mu p_{y_5,\mu} + y_5, \end{aligned} \tag{12.25}$$

Figure 4. Phase portrait of object (12.24)

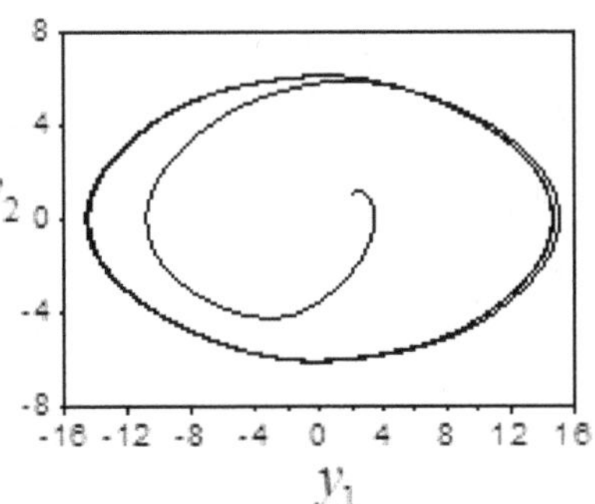

and the following identification representation is obtained:

$$
\begin{aligned}
\dot{y}_1 &= a_{11}y_1 + a_{12}p_{y_1} + a_{15}p_{y_5},\\
\dot{y}_3 &= -a_3 y_3 + a_{31}f_1(y_1),\\
\dot{\hat{y}}_4 &= -a_4 y_4 + a_{42}y_2,\\
\dot{\hat{y}}_5 &= -a_5 y_5 + a_{53}f_3(y_3 + y_4).
\end{aligned}
\tag{12.26}
$$

The adaptive system for estimating parameters of the system (12.26) has the following form:

$$
\begin{aligned}
\dot{\hat{y}}_1 &= -k_1 e_1 + \hat{a}_{11}y_1 + \hat{a}_{12}p_{y_1} + \hat{a}_{15}p_{y_5},\\
\hat{y}_2 &= +\hat{a}_{11}y_1 + \hat{a}_{12}p_{y_1} + \hat{a}_{15}p_{y_5}\\
\dot{\hat{y}}_3 &= -k_3 e_3 + \hat{a}_{31}(f_1(y_1) - y_3),\\
\dot{\hat{y}}_4 &= -k_4 e_4 + \hat{a}_{41}y_4 + \hat{a}_{42}\hat{y}_2,\\
\dot{\hat{y}}_5 &= -k_5 e_5 + \hat{a}_{51}(y_5 + f_3(y_3 + y_4)),
\end{aligned}
\tag{12.27}
$$

where $\gamma_{ij} > 0$

Adaptive algorithms for adjusting parameters of system (12.27)

$$
\begin{aligned}
\dot{\hat{a}}_{11} &= -\gamma_{11}e_1 y_1, \quad \dot{\hat{a}}_{12} = -\gamma_{12}e_1 p_{y_1},\\
\dot{\hat{a}}_{15} &= -\gamma_{15}e_1 p_{y_5}, \quad \dot{\hat{a}}_{31} = -\gamma_{31}e_3(f_1(y_1) - y_3),\\
\dot{\hat{a}}_{41} &= -\gamma_{41}e_4 y_4, \quad \dot{\hat{a}}_{42} = -\gamma_{42}e_4\hat{y}_2,\\
\dot{\hat{a}}_{51} &= -\gamma_{51}e_5(y_5 + f_3(y_3 + y_4)),
\end{aligned}
\tag{12.28}
$$

The adaptive system has the following form:

$$
\begin{aligned}
\dot{e}_1 &= -k_1 e_1 + \Delta a_{11}y_1 + \Delta a_{12}p_{y_1} + \Delta a_{15}p_{y_5},\\
e_2 &= +\Delta a_{11}y_1 + \Delta a_{12}p_{y_1} + \Delta a_{15}p_{y_5},\\
\dot{e}_3 &= -k_3 e_3 + \Delta a_{31}(f_1(y_1) - y_3),\\
\dot{e}_4 &= -k_4 e_4 - \Delta a_{41}y_4 + \Delta a_{42}\hat{y}_2,\\
\dot{e}_5 &= -k_5 e_5 - \Delta a_{51}(y_5 + f_3(y_3 + y_4)),
\end{aligned}
\tag{12.29}
$$

$$
\begin{aligned}
\Delta\dot{a}_{11} &= -\gamma_{11}e_1 y_1, \quad \Delta\dot{a}_{12} = -\gamma_{12}e_1 p_{y_1},\\
\Delta\dot{a}_{15} &= -\gamma_{15}e_1 p_{y_5}, \quad \Delta\dot{a}_{31} = -\gamma_{31}e_3(f_1(y_1) - y_3),\\
\Delta\dot{a}_{41} &= \gamma_{41}e_4 y_4, \quad \Delta\dot{a}_{42} = -\gamma_{42}e_4\hat{y}_2,\\
\Delta\dot{a}_{51} &= \gamma_{51}e_5(y_5 + f_3(y_3 + y_4)),
\end{aligned}
\tag{12.30}
$$

where $\Delta a_{ij}(t) = \hat{a}_{ij}(t) - a_{ij}$.

The tuning results of the system (12.29), (12.30) are shown in Figures 5–7. In Figures 5 and 6 present tuning parameters of the system (12.27).

The process of tuning parameters of the system (12.27) is presented in Figures 5 and 6, while the change in the output discrepancies of the system (12.27) is shown in Figure 7. The results confirm the boundedness of the adaptive system trajectories.

Figure 5. Adjusting parameters of the models for estimating y_1

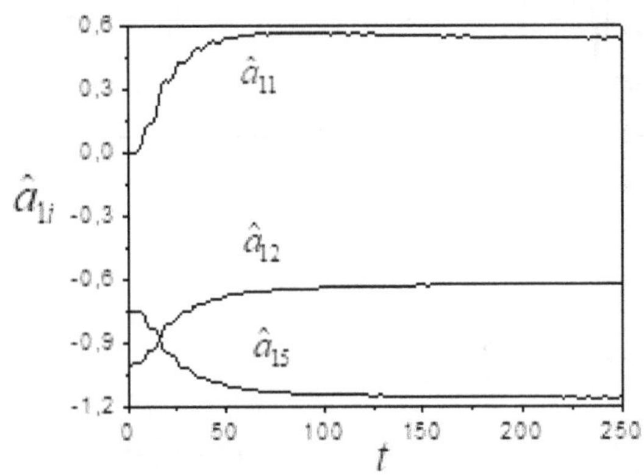

Despite fulfill condition $y_5 \in EC^S_{\alpha,\bar{\alpha}}$ for the input and the system S-synchronizability, it is not possible to ensure the condition of asymptotic stability.

Figure 6. Adjusting parameters of the models for estimating y_3, y_4 and y_5.

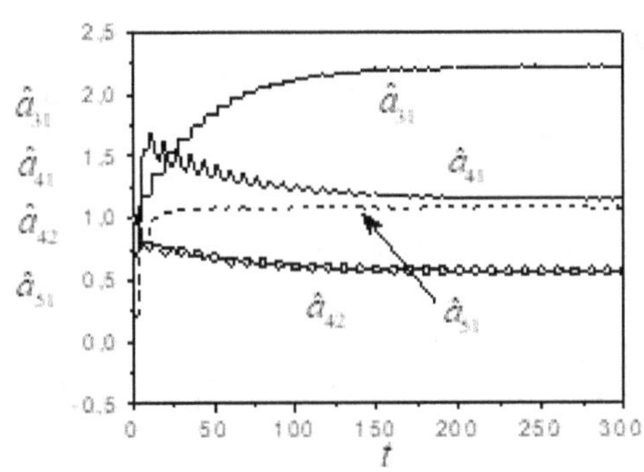

Figure 7. Change in the discrepancy of outputs of models (12.27)

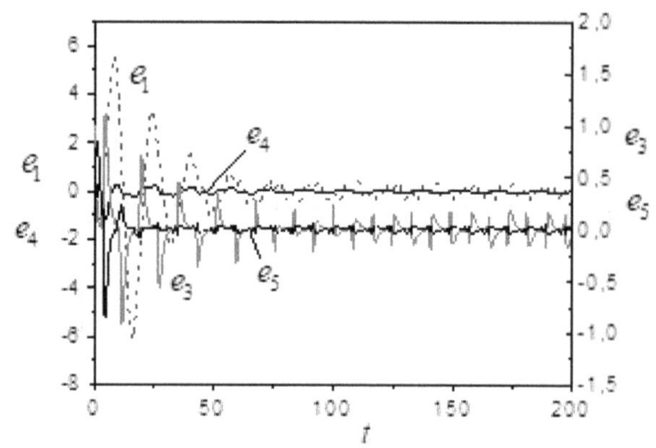

Consider LF: $V_{e,1-5}(t) = 0.5\sum_{i=1,i\neq 2}^{5} e_i^2(t)$

$$V_{\Delta,1-5}(t) = 0.5\underbrace{\left(\gamma_{11}^{-1}\Delta a_{11}^2(t) + \gamma_{12}^{-1}\Delta a_{12}^2(t) + \gamma_{15}^{-1}\Delta a_{15}^2(t)\right)}_{V_{\Delta,1}} + \underbrace{0.5\gamma_{31}^{-1}\Delta a_{31}^2(t)}_{V_{\Delta,3}} + \\ + \underbrace{0.5\left(\gamma_{41}^{-1}\Delta a_{41}^2(t) + \gamma_{42}^{-1}\Delta a_{42}^2(t)\right)}_{V_{\Delta,4}} + \underbrace{0.5\gamma_{51}^{-1}\Delta a_{51}^2(t)}_{V_{\Delta,5}}. \quad (12.31)$$

$$V_{e,\Delta} = \left[V_{e,1}, V_{\Delta,1}, V_{e,3}, V_{\Delta,3}, V_{e,4}, V_{\Delta,4}, V_{e,5}, V_{\Delta,5}\right]^T \quad (12.32)$$

Theorem 3. *Let the following conditions be satisfied: (i) positive definite Lyapunov functions* $V_{e,i}(t)$, $V_{\Delta,i}(t) = 0.5\gamma_{ij}^{-1}\left(\Delta a_{ij}\right)^2$ $V_{\Delta,1}(t) = 0.5\left(\gamma_{11}^{-1}\Delta a_{11}^2(t) + \gamma_{12}^{-1}\Delta a_{12}^2(t) + \gamma_{15}^{-1}\Delta a_{15}^2(t)\right)$ *admit the infinitesimal higher limit at* $\left|e_i(t)\right| \to 0$, $\left|\Delta a_{ij}\right| \to 0$; *(ii)* y_i *are piecewise continuous bounded while* $y_i(t) \in EC^S_{\alpha_{y_i},\bar{\alpha}_{y_i}}$, $p_{y_1} \in EC^S_{\alpha_{p_{y_1}},\bar{\alpha}_{p_{y_1}}}$, $p_{y_5} \in EC^S_{\alpha_{p_{y_5}},\bar{\alpha}_{p_{y_5}}}$, $\alpha = \max\left(\bar{\alpha}_{y_1}, \bar{\alpha}_{p_{y_1}}, \bar{\alpha}_{p_{y_5}}\right)$, *and* $i = 1,3,4,5$; *(iii) for* $V_{\Delta,1}(t) = 0.5\Delta G_1^T(t)\Gamma_1^{-1}\Delta G_1(t)$, *the following inequation is valid:*

$$0.5\underline{\vartheta}_1\Delta G_1^T(t)\Delta G_1(t) \leq V_{\Delta,1}(t) \leq 0.5\overline{\vartheta}_1\Delta G_1^T(t)\Delta G_1(t), \quad (12.33)$$

where $\Delta G_1(t) = \left[\Delta a_{11}(t), \Delta a_{12}(t), \Delta a_{15}(t)\right]^T$, $\Gamma_1 = \mathrm{diag}\left(\gamma_{11}, \gamma_{11}, \gamma_{15}\right)$, $\underline{\vartheta}_1 = \beta_l^{-1}(\Gamma_1)$, $\overline{\vartheta}_1 = \beta_1^{-1}(\Gamma_1)$, $\beta_1(\Gamma_1)$, $\beta_l(\Gamma_1)$ *are minimal and maximal eigenvalues of matrix* Γ_1; *(iv)* $\alpha_{f_1,y_3} - \underline{\upsilon_3} \leq \left(f_1(y_1) - y_3\right)^2 \leq \overline{\alpha}_{f_1,y_3} +$, *where* $\upsilon_3 \geq 0, \upsilon_3 \geq 0$; *(v)* $\gamma_4 = \max\left(\gamma_{41}, \gamma_{42}\right)$; *(vi) equality* $e_i\Delta a_{ij}\omega_i = \pi_i\left(e_i^2 + \left(\Delta a_{ij}\right)^2\omega_i^2\right)$ *is satisfied in domain* $O_\nu(O)$, *where* $\pi > 0$, $O = \left\{0^4, 0^n\right\} \subset \mathbb{R}^4 \times \mathbb{R}^n \times \mathbb{J}_{0,\infty}$, $0^4, 0^n$ *are zero vectors, n is the number of adjustable parameters,* O_ν *is some neighborhood of point O,* $t \in \left[0,\infty\right] = \mathbb{J}_{0,\infty}$; *(vii) for* $V_{e,\Delta}(t)$ *matrix system of inequalities* $\dot{V}_{e,\Delta} \leq A_{e,\Delta}V_{e,\Delta} + B_{e,\Delta}$ *is valid, where* $A_{e,\Delta}$ *is the following block diagonal matrix:*

$$A_{e,\Delta} = \operatorname{diag}\left(A_{e,\Delta,1}, A_{e,\Delta,3} A_{e,\Delta,4}, A_{e,\Delta,5}\right), B_{e,\Delta} = [\begin{matrix} 0 & 0 & \tilde{\upsilon}_3 \vartheta_3 & \pi_3 \chi & 0 & 0 & 0 & 0 \end{matrix}]^T$$

$$A_{e,\Delta,1} = \begin{bmatrix} -k_1 & \dfrac{\alpha \beta_l(\Gamma_1)}{k_1} \\ \dfrac{\pi_1}{3} & -\frac{3}{4}\pi_1 \alpha_{P_1} \beta_1(\Gamma_1) \end{bmatrix}, A_{e,\Delta,3} = \begin{bmatrix} -k_3 & \dfrac{\alpha_{f_1,y_3} \gamma_{31}}{k_3} \\ \dfrac{\pi_3 \overline{\alpha}_{f_1,y_3}}{3(\alpha_{f_1,y_3} - \underline{\upsilon_3})} & -\frac{3}{4}\pi_3 \alpha_{f_1,y_3} \gamma_{31} \end{bmatrix},$$

$$A_{e,\Delta,4} = \begin{bmatrix} -k_4 & \dfrac{\alpha_{y_1,\hat{y}_2}}{k_4} \gamma_4 \\ \dfrac{2\pi_{41}}{3} & -\frac{3}{4}\pi_{41} \overline{\alpha}_{y_1,\hat{y}_2} \gamma_4 \end{bmatrix}, A_{e,\Delta,5} = \begin{bmatrix} -k_5 & \dfrac{\overline{\alpha}_{y_5} + c^2}{k_5} \gamma_{51} \\ \dfrac{2\pi_5}{3} & -\dfrac{3\pi_5}{4} \alpha_{y_5,f_3} \gamma_{51} \end{bmatrix};$$

(viii) the upper solution for $V_{e,\Delta}(t)$ satisfies equation $\dot{S}_{e,\Delta} = A_{e,\Delta} S_{e,\Delta} + B_{e,\Delta}$ if exist functions $s_i(t) \geq 0$ such that $V_{e,\Delta,i}(t) \leq s_i(t)\ \forall (t \geq t_0)\ \&\ \left(V_{e,\Delta,i}(t_0) \leq s_i(t_0)\right)$, $i = 1,3;4;5$, where $S_{e,\Delta} \in \mathbb{R}^6$, s_i are elements of vector $S_{e,\Delta}$. Then the adaptive system (12.29), (12.30) is exponentially dissipative with the estimation:

$$V_{e,\Delta}(t) \leq e^{A_{e,\Delta}(t-t_0)} S(t_0) + \int_{t_0}^{t} e^{A_{e,\Delta}(t-\tau)} B_{e,\Delta} d\tau,$$

if $k_1 \geq \frac{2}{3}\sqrt{\dfrac{\alpha \beta_l(\Gamma_1)}{\alpha_{P_1} k_1 \beta_1(\Gamma_1) + e_1^2}}$, $k_3 \geq \frac{2}{3}\sqrt{\dfrac{\overline{\alpha}_{f_1,y_3}}{\left(\alpha_{f_1,y_3} - \underline{\upsilon_3}\right)}}$, $k_4 \geq \frac{2}{3}\sqrt{2\dfrac{\alpha_{y_1,\hat{y}_2}}{\overline{\alpha}_{y_1,\hat{y}_2}}}$, $k_5 \geq \frac{2}{3}\sqrt{2\dfrac{\overline{\alpha}_{y_5}}{\alpha_{y_5,f_3}}}$.

The proof of Theorem 3 is given in Appendix B.

As follows from Theorem 3, limiting properties of system (12.29), (12.30) depend on nonlinear properties, feedback, and compliance with the excitation constancy condition.

7. CONCLUSION

In the paper, the approach to adaptive identification of systems with several nonlinearities is proposed. The approach is based on transforming the system to eliminate immeasurable state variables. The synthesis of an adaptive identification system is presented. In order to simplify the adaptation process, we suggest the approach based on the system decomposition into several subsystems. Boundedness of trajectories in the adaptive system is proven. The problem of S-synchronizability and its influence on the choice of the CE condition is considered. Lyapunov vector functions apply to prove the exponential stability of the identification system.

REFERENCES

Aljamaan, I. (2016). *Nonlinear closed-loop system identification in the presence of non-stationary noise source (Unpublished doctoral thesis)*. University of Calgary., http://hdl.handle.net/11023/3070

Andersson, C., Ribeiro, A. H., Tiels, K., Wahlström, N., & Schön, T. B. (2019). Deep Convolutional Networks in System Identification. *2019 IEEE 58th Conference on Decision and Control (CDC), Nice, France*, 3670-3676.

Aykan, M., & Özgüven, H. N. (2012). Parametric Identification of Nonlinearity from Incomplete FRF Data Using Describing Function Inversion. In D. Adams, G. Kerschen, A. Carrella (Eds). *Topics in Nonlinear Dynamics,* Volume 3. *Conference Proceedings of the Society for Experimental Mechanics Series.*, New York: Springer. DOI: 10.1007/978-1-4614-2416-1_25

Azeez, M. A. F., & Vakakis, A. F. (1999). Numerical and experimental analysis of a continuously overhung rotor undergoing vibro-impacts. *International Journal of Non-linear Mechanics, 34*(3), 415–435. DOI: 10.1016/S0020-7462(98)00022-5

Binczak, S., Busvelle, E., Gauthier, J.-P., & Jacquir, S. (2000). Identification of Unknown Functions in Dynamic Systems. https://pageperso.lis-lab.fr/eric.busvelle/papers/bioid.pdf

Forssell, U., & Ljung, L. (1999). Closed-loop identification revisited. *Automatica, 35*(7), 1215–1241. DOI: 10.1016/S0005-1098(99)00022-9

Gantmakher, F. R. (2000). *The Theory of Matrices*. Chelsea Pub. Co.

Gedon, D., Wahlström, N., Schön, T. B., & Ljung, L. (2021). Deep State Space Models for Nonlinear System Identification, *arXiv:2003.14162v3* [eess.SY].

Holmes, P. (1982). The dynamics of repeated impacts with a sinusoidally vibrating table. *Journal of Sound and Vibration, 84*(2), 173–189. DOI: 10.1016/S0022-460X(82)80002-3

Karabutov, N. (2017). Adaptive observers with uncertainty in loop tuning for linear time-varying dynamical systems. *International Journal of Intelligent Systems and Applications, 9*(4), 1–13. DOI: 10.5815/ijisa.2017.04.01

Karabutov, N. (2018). Structural-parametrical design method of adaptive observers for nonlinear systems. *International Journal of Intelligent Systems and Applications, 10*(2), 1–16. DOI: 10.5815/ijisa.2018.02.01

Karabutov, N. (2021). Structural Identifiability of Systems with Multiple Nonlinearities. *Contemporary Mathematics*, *2*(2), 103–172. DOI: 10.37256/cm.222021763

Karabutov, N. (2023). On adaptive identification of systems having multiple nonlinearities. *Russian technological journal, 11*(5), 94–105.

Karabutov, N. N. (2021). *Introduction to structural identifiability of nonlinear systems*. URSS/Lenand.

Kerschen, G., Worden, K., Vakakis, A., & Golinval, J.-C. (2006). Past, present and future of nonlinear system identification in structural dynamics. *Mechanical systems and signal processing*, Elsevier, *20*(3), 505-592.

Li, C. (2021). Closed-loop identification for a class of nonlinearly parameterized discrete-time systems. *Automatica, 131*, 109747. DOI: 10.1016/j.automatica.2021.109747

Mejari, M., Piga, D., & Bemporad, A. (2018). A bias-correction method for closed-loop identification of linear parameter-varying systems. *Automatica*, *87*, 128–141. DOI: 10.1016/j.automatica.2017.09.014

Noël, J., & Schoukens, J. (2018). Grey-box state-space identification of nonlinear mechanical vibrations. *International Journal of Control*, *91*(5), 1118–1139. DOI: 10.1080/00207179.2017.1308557

Pai, P. F., & Palazotto, A. N. (2008). Detection and identification of nonlinearities by amplitude and frequency modulation analysis. *mechanical systems and signal processing, 22*(5), 1107-1132.

Pavlov, Y. N., Nedashkovskii1, V. M., & Tihomirova, E. A. (2015). Identification of nonlinear dynamic systems possessing some nonlinearities. *Science and education of the Bauman MSTU, 07*, 217–234.

Piga, D., & Tyth, R. (2014). A bias-corrected estimator for nonlinear systems with output-error type model structures. *Automatica*, *50*(9), 2373–2380. DOI: 10.1016/j.automatica.2014.07.021

Pintelon, R. (2002). Frequency-domain subspace system identification using nonparametric noise models. *Automatica*, *38*(8), 1295–1311. DOI: 10.1016/S0005-1098(02)00036-5

Prawin, J., Rao, A. R. M., & Sethi, A. (2020). Parameter identification of systems with multiple disproportional local nonlinearities. *Nonlinear Dynamics*, *100*(1), 289–314. DOI: 10.1007/s11071-020-05538-1

Safari, S., & Monsalve, J. M. L. (2021). Direct optimization-based model selection and parameter estimation using time-domain data for identifying localised nonlinearities. *Journal of Sound and Vibration, 501*, 116056. DOI: 10.1016/j.jsv.2021.116056

Singh, A., & Moore, K. J. (2021). Identification of multiple local nonlinear attachments using a single measurement case. *Journal of Sound and Vibration, 513*, 116410. DOI: 10.1016/j.jsv.2021.116410

Song, G., Xu, L., & Ding, F. (2019). Recursive least squares algorithm and stochastic gradient algorithm for feedback nonlinear equation-error systems. international journal of modelling, identification and control, *32*(3-4), 251-257.

Van den Hof, P. (1997). Closed-loop issues in system identification. In Y. Sawaragi & S. Sagara (Eds.), System Identification (SYSID'97) (1547-1560*) Proc. 11th IFAC Symp. System Identification, Kitakyushu, Japan, 8-11 July 1997. IFAC Proceedings Series, 3,* Van den Hof, P. (1998). Closed-loop issues in system identification. *Annual reviews in control, 22*(0), 173 – 186.

Zhao, Z., Li, C., Ahlin, K., & Du, H. (2016). Onlinear system identification with the use of describing functions – a case study. *VibroenginIeering PROCEDIA*, *8*, 33–38. DOI: 10.1016/j.proeng.2016.01.076

APPENDIX A

Consider the $V_{s_{F,i}}(t) = V_e(t) + V_{\Delta,i}(t)$, where $V_{\Delta,i}(t)$has the form (12.14). Then

$$\dot{V}_{s_{F,i}} = -k_{s_{F,i}} e^2 + \dot{V}_{i,\Delta} - \dot{V}_{i,\Delta} \leq -2k_{s_{F,i}} V_{e,i} \text{ (A1)}$$

Apply condition 1) of Theorem 1. Since $\dot{V}_{s_{F,i}}(t) < 0$, the $s_{F,i}$-subsystem is stable. Integrate $\dot{V}(t)$ in time and get:

$$V_{s_{F,i}}(t_0) - 2k_{s_{F,i}} \int_{t_0}^{t} V_e(\tau) d\tau \geq V_{s_{F,i}}(t) \text{ (A2)}$$

$V_{s_{F,i}}\left(e, \Delta\kappa_{s_{F,i}}, \Delta K_{s_{F,i}}\right)$ satisfies condition 1) of Theorem 1. All trajectories of the $AS_{F,i}$sub-system belong to the $G_t = \left\{\left(e, \Delta\kappa_{s_{F,i}}, \Delta K_{s_{F,i}}\right) : V_{s_{F,i}}(t) \leq V_{s_{F,i}}(t_0)\right\}$ domain. From (1.2), we obtain the desired estimate. ■

APPENDIX B

Consider the Lyapunov vector function F2 (12.32), the elements of which satisfy condition 1) of Theorem 3. Provide the M^+ property to each $\dot{V}_{e,\Delta,i}$ component. Consider $V_{e,1}$:

$$\dot{V}_{e,1} = -k_1 e_1^2 + e_1\left(\Delta a_{11} y_1 + \Delta a_{12} p_{y_1} + \Delta a_{15} p_{y_5}\right) \leq \\ \leq k_1 e_1^2 + |e_1|\left|\Delta a_{11} y_1 + \Delta a_{12} p_{y_1} + \Delta a_{15} p_{y_5}\right|. \quad \text{(B1)}$$

Apply the inequality

$$-az^2 + bz \leq -\frac{-az^2}{2} + \frac{b^2}{2a}, \quad a > 0, \; b \geq 0, \; z \geq 0$$

and obtain

$$\dot{V}_{e,1} \leq -\frac{k_1}{2} e_1^2 + \frac{1}{2k_1}\left((\Delta a_{11})^2 y_1^2 + (\Delta a_{12})^2 p_{y_1}^2 + (\Delta a_{15})^2 p_{y_5}^2\right) \quad \text{(B2)}$$

Let the condition (12.33) be fulfilled. Then

$$\dot{V}_{e,1} \leq -\frac{k_1}{2} e_1^2 + \frac{1}{2k_1}\left((\Delta a_{11})^2 y_1^2 + (\Delta a_{12})^2 p_{y_1}^2 + (\Delta a_{15})^2 p_{y_5}^2\right) \leq -k_1 V_{e,1} + \frac{\alpha\beta_l(\Gamma_1)}{k_1} V_{\Delta,1}. \quad \text{(B3)}$$

So, $\dot{V}_{e,1} \in M^+$. Similarly, we have for other $\dot{V}_{e,\Delta,i}$ components:

$$\dot{V}_{e,3} \leq -k_3 V_{e,3} + \frac{\overline{\alpha}_{f_1 y_3}\gamma_{31}}{k_3} V_{\Delta,3} + \overline{\upsilon}_3 \vartheta_3, \quad \text{(B4)}$$

where $\alpha_{f_1 y_3} - \upsilon_3 \leq (f_1(y_1) - y_3)^2 \leq \overline{\alpha}_{f_1 y_3} + \upsilon_3$, $\vartheta_3(t) \geq 0$ is the uncertainty arising from $(f_1(y_1) - y_3)^2 \notin \overline{E}C^S_{\alpha_{f_1 y_3}, \overline{\alpha}_{f_1 y_3}}$. M^+ property for other $\dot{V}_{e,i}$:

$$\begin{aligned}
\dot{V}_{e,1} &\leq -\frac{k_1}{2} e_1^2 + \frac{\alpha\beta_l(\Gamma_1)}{k_1} V_{\Delta,1} = -k_1 V_{e,1} + \frac{\alpha\beta_l(\Gamma_1)}{k_1} V_{\Delta,1}, \\
\dot{V}_{e,3} &\leq -\frac{k_3}{2} e_3^2 + \frac{\alpha_{f_1 y_3}}{2k_3}(\Delta a_{31})^2 \leq -k_3 V_{e,3} + \frac{\alpha_{f_1 y_3}\gamma_{31}}{k_3} V_{\Delta,3} + \overline{\upsilon}_3 \vartheta_3, \\
\dot{V}_{e,4} &\leq -\frac{k_4}{2} e_4^2 + \frac{\alpha_{y_1 \mathcal{G}}}{2k_4}\left((\Delta a_{41})^2 + (\Delta a_{42})^2\right) \leq -k_4 V_{e,4} + \frac{\overline{\alpha}_{y_1 \mathcal{G}}}{k_4}\gamma_4 V_{\Delta,4}, \\
\dot{V}_{e,5} &\leq -\frac{k_5}{2} e_5^2 + \frac{\overline{\alpha}_{y_5} + c^2}{2k_5}(\Delta a_{51})^2 \leq -k_5 V_{e,5} + \frac{\overline{\alpha}_{y_5} + c^2}{k_5}\gamma_{51} V_{\Delta,5},
\end{aligned} \quad \text{(B5)}$$

where $\gamma_4 = \max(\gamma_{41}, \gamma_{42})$.

If we apply the approach stated above and the results of Karabutov (2017), then we get for $\dot{V}_{\Delta,i}$:

$$\begin{cases} \dot{V}_{\Delta,1} \leq -\frac{3}{4}\pi_1 \alpha_{P_1} \beta_1(\Gamma_1) V_{\Delta,1} + \frac{\pi_1}{3} V_{e,1}, \\ \dot{V}_{\Delta,3} \leq -\frac{3\pi_3}{4} \alpha_{f_1,y_3} \gamma_{31} V_{\Delta,3} + \frac{\pi_3 \bar{\alpha}_{f_1,y_3}}{3(\alpha_{f_1,y_3} - \underline{\upsilon_3})} V_{e,3} + \pi_3 \chi, \\ \dot{V}_{\Delta,4} \leq -\frac{3}{4}\pi_{41} \tilde{\alpha}_{y_1,\hat{y}_2} \gamma_4 V_{\Delta,4} + \frac{2\pi_{41}}{3} V_{e,4}, \\ \dot{V}_{\Delta,5} \leq -\frac{3\pi_5}{4} \alpha_{y_5,f_3} \gamma_{51} V_{\Delta,5} + \frac{2\pi_5}{3} V_{e,5}, \end{cases} \quad \text{(B6)}$$

where χ considers the uncertainty because of non-fulfillment of the CE condition. The limitation follows from Theorem 1.

We obtain a vector system of inequalities from (B5) and (B6)

$$\dot{V}_{e,\Delta} \leq A_{e,\Delta} V_{e,\Delta} + B_{e,\Delta}. \text{ (B7)}$$

It follows from matrix $A_{e,\Delta}$ that the channels of the system (12.24) are isolated and stability analysis can be performed for each channel. As $A_{e,\Delta}$ is an M matrix (Gantmakher, 2000), the asymptotic stability of the adaptive system follows from the positivity of all $A_{e,\Delta}$ minors. We have

$$k_1 \geq \frac{2}{3}\sqrt{\frac{\alpha\beta_l(\Gamma_1)}{\alpha_{P_1} k_1 \beta_1(\Gamma_1)}}, k_3 \geq \frac{2}{3}\sqrt{\frac{\bar{\alpha}_{f_1,y_3}}{(\alpha_{f_1,y_3} - \underline{\upsilon_3})}}, k_4 \geq \frac{2}{3}\sqrt{2\frac{\alpha_{y_1,\hat{y}_2}}{\bar{\alpha}_{y_1,\hat{y}_2}}}, k_5 \geq \frac{2}{3}\sqrt{2\frac{\bar{\alpha}_{y_5} + c^2}{\alpha_{y_5,f_3}}}.$$

Limiting properties of the system (12.24) depend on the nonlinear properties, feedback, and fulfillment of the CE condition. In particular, this applies to Channel 3 in (B7). ■

Chapter 13
Adaptive Identification of Interconnected Systems

ABSTRACT

In this chapter, an approach is proposed to design models for interconnected systems (IS) under incomplete a priori information. As an example, we consider synthesis an adaptive model for two-channel systems (TCS) having cross-connections and identical channels. The case of asymmetric cross-links is considered, and estimates of their impact on the adaptive identification quality are obtained. We generalize the proposed approach to the case of the interconnected system. An approach is proposed to identifiability estimate of cross-linked TCS. The excitation constancy effect is shown on estimates of TCS parameters. Convergence estimates of the parameters are obtained.

1. INTRODUCTION

Interconnected systems (ICS) applied (Bukov et al., 1998; Morozovsky, 1970; Zyryanov, 2010; Meerov and Litvak,1972; Voronov, 1985) to control actuators for robots and manipulators (Egorov and Umnov, 2022; Egorov, 2010), and various technical systems (Gupta and Chopra, 2016; Kholmatov, 2023). ICS identification methods are based on an adaptive approach (Aliyeva, 2019), a black box concept for non-stationary multi-connected systems with uncertainty (Hua et al., 2004). Adaptive algorithms of decentralized robust control with a reference model are proposed for ICS with a time delay (Glentis and Slump, 1995). The identifiability issues are considered for a closed interconnected stochastic system by Ali et al. (2011). System decomposition proposes into subsystems. The identifiability considers as a system of individual elements and individual closed circuits. Convergence sufficient conditions

DOI: 10.4018/979-8-3373-0822-7.ch013

got almost certainly for parameter estimates. Convergence sufficient conditions got almost certainly for parameter estimates.

Different approaches are used in TCS identification tasks: an estimation transfer function model (Yang et al., 2016) for 2-D spatially interconnected systems based on neural network; adaptive nonlinear damping and a robust adaptive observer are the base for controlling laws (Ali et al., 2010). Krasovsky (1957) presents the identification method for a spatially interconnected system in a rational form.

Transfer functions and the state space models are the basis for analysing processes in ICS. Adaptive procedures are used to synthesize control algorithms. Adaptive control algorithms apply under the perturbations and unmodeled dynamics action. Explain this the identification problem (IP) status (i) incomplete information about the state and parameters of the system and (ii) the accounting complexity of relationships in the system. The identification of ICS parameters is based on statistical procedures, frequency methods, and neural network technologies. Apply of adaptive identification methods (AIM) study in some works. AIM is used to the tuning (identification) control parameters. Parametric uncertainties that arise in this case compensate for the choice of a control algorithm. The Identification of two-channel systems has not been studied.

We propose the approach to the ICS adaptive identification based on the use of measurement results. First, two-channel systems (TCS) with cross-links (LC) are considered. The approach proposes to the TCS identifiability under various assumptions about LC. The adaptive system stability is proven. We propose the approach to ICS identification.

2. TWO-CHANNEL SYSTEMS

2.1. Problem Setting

Consider TCS with CL. We believe that elements in channels are identical. As shown Karabutov (2024) transfer functions are the basis of TCS analysis. The TCS description in the state space is a more adequate representation in identification problems. Let TCS have n vertical layers:

(i) the first layer

$$\begin{cases} \dot{X}_{11} = A_1 X_{11} + B_1 e_{11} = A_1 X_{11} + B_1 v_{11}, \\ y_{11} = C_1^T X_{11}, \end{cases}$$
$$\begin{cases} \dot{X}_{21} = A_1 X_{21} + B_1 e_{21} = A_1 X_{21} + B_1 v_{21}, \\ y_{21} = C_1^T X_{21}, \end{cases} \tag{13.1}$$

(ii) kth layer ($13.1 < k \leq n$)

$$\begin{cases} \dot{X}_{1,k} = A_k X_{1,k} + B_k u_{1,k}, \\ y_{1,k} = C_k^T X_{1,k}, \\ u_{1,k} = y_{1,k-1} + v_{1,k-1}, \\ v_{1,k-1} = d_{k-1}\left(y_{2,k-1} + d_{k-2} v_{2,k-1}\right), \end{cases}$$
$$\begin{cases} \dot{X}_{2,k} = A_k X_{2,k} + B_k u_{2,k}, \\ y_{2,k} = C_k^T X_{1,k}, \\ u_{2,k} = y_{2,k-1} + v_{2,k-1}, \\ v_{2,k-1} = d_{k-1}\left(y_{1,k-1} + d_{k-2} v_{1,k-1}\right), \end{cases} \tag{13.2}$$

where $v_{11} = g_1 - y_{1,n}$, $v_{21} = g_2 - y_{2,n}$, $v_{i,k-1}$ is i-channel CL output, $i = 1, 2$; $X_{ik} \in \mathbb{R}^q$ is the $k$layer state vector of a i channel, $A_k \in \mathbb{R}^{q_k \times q_k}$, $C_k \in \mathbb{R}^{q_k}$, $B_k \in \mathbb{R}^{q_k}$, is output k layer of the i channel; $v_{i,k-1}$ is CL output; $g_i(t) \in \mathbb{R}$ is input (.upsetting control) the system. A_k is Hurwitz matrix. d_{k-1} is an operator. Tasks solved by TCS determine the d_{k-1} type. d_{k-1} can be a constant, a nonlinear function, or a differential operator.

d_{k-1} is an operator. Tasks solved by TCS determine the d_{k-1} type. d_{k-1} can be a constant, a nonlinear function, or a differential operator.

Information set

$$\mathbb{I}_o = \left\{ g_i(t), y_{i,k}(t) \forall (i = 1;2) \,\&\, (k = \overline{1,n}), t \in \mathbb{J} = [t_0, t_e] \right\} \tag{13.3}$$

where $\mathbb{J}$ time interval

The structure of the TCS model coincides with (13.1), (13.2). Let $\hat{y}_{i,k} \in \mathbb{R}$, $i = 1, 2$; $k = \overline{1,n}$ is the output model.

Task: design algorithms for tuning model parameters to

$$\lim_{t \to \infty} \left| \hat{y}_{i,k} - y_{i,k} \right| \leq \delta_{i,k} \tag{13.4}$$

where $\delta_{i,k} \geq 0$ is a set value.

2.2. About TCS Structural Aspects

The system (13.1), (13.2) identification depends on the evaluating possibility of its parameters. Introduce the model for (13.1)

$$\begin{cases} \dot{\hat{X}}_{11} = K_1(\hat{X}_{11} - X_{11}) + \hat{A}_1 X_{11} + \hat{B}_1 v_{11}, \\ \hat{y}_{11} = C_1^T \hat{X}_{11}, \end{cases}$$
$$\begin{cases} \dot{\hat{X}}_{21} = K_1(\hat{X}_{21} - X_{21}) + \hat{A}_1 \hat{X}_{21} + \hat{B}_1 v_{21}, \\ \hat{y}_{21} = C_1^T X_{21}, \end{cases} \quad (13.5)$$

where $K_1 \in \mathbb{R}^{q_1 \times q_1}$ is stable matrix (reference model); $\hat{A}_1 \in \mathbb{R}^{q_1 \times q_1}$, $\hat{B}_1 \in \mathbb{R}^{q_1}$ are matrixes (13.5), $\hat{X}_{i,1}$ is model state vector.

Let

$$E_{11} \triangleq \hat{X}_{11} - X_{11},$$

$E_{21} \triangleq \hat{X}_{21} - X_{21}$. Then for the first layer

$$\begin{aligned} \dot{E}_{11} &= K_1 E_{11} + \Delta A_1 X_{11} + \Delta B_1 v_{11}, \\ \dot{E}_{21} &= K_1 E_{21} + \Delta \hat{A}_1 \hat{X}_{21} + \Delta \hat{B}_1 v_{21}, \end{aligned} \quad (13.6)$$

where $\Delta A_1 \triangleq \hat{A}_1 - A_1$, $\Delta B_1 \triangleq \hat{B}_1 - B_1$ are matrices of parametric residuals. Similarly, we get the error equations for the remaining layers TCS

$$\begin{aligned} \dot{E}_{1,k} &= K_k E_{1,k} + \Delta A_k X_{1,k} + \Delta B_k u_{1,k}, \\ \dot{E}_{2k} &= K_k E_{2,k} + \Delta \hat{A}_k \hat{X}_{2,k} + \Delta \hat{B}_k u_{2,k}. \end{aligned} \quad (13.7)$$

Let the input $g_i(t)$, $i = 1;\ 2$ satisfy the constant excitation (CE) condition

$$EC_{\underline{\alpha}_i, \overline{\alpha}_i} : \underline{\alpha}_i \le g_i^2(t) \le \overline{\alpha}_i \quad \forall t \in [t_0, t_0 + T] \quad (13.8)$$

where $\underline{\alpha}_i, \overline{\alpha}_i$ are positive numbers, $T > 0$.
Designations:

1) if condition (13.8) is true, then we write $g_i(t) \in EC_{\underline{\alpha}_i, \overline{\alpha}_i}$;
2) if condition (13.8) is not satisfied, then $g_i(t) \notin EC_{\underline{\alpha}_i, \overline{\alpha}_i}$.

Theorem 1 (Karabutov, 2023). *Let 1) $g_i(t) \in EC_{\underline{\alpha}_i, \overline{\alpha}_i}$, where $(\underline{\alpha}_i, \overline{\alpha}_i) > 0$; 2) the system (13.5) is stable and detectable; 3) $K_1 \in \mathbb{R}^{q_1 \times q_1}$ is Hurwitz matrix; 4) $v_{i1}(t) \in EC_{\underline{\sigma}_{i1}, \overline{\sigma}_{i1}}$, where $\underline{\sigma}_{i1} > 0$, $\overline{\sigma}_{i1} > 0$ $i = 1, 2$; 5) $X_{i1}(t) \in EC_{\underline{\alpha}_{X_{i1}}, \overline{\alpha}_{X_{i1}}}$, where $(\underline{\alpha}_{X_{i1}}, \overline{\alpha}_{X_{i1}}) > 0$. Then the system (13.5) is identifiable if:*

$$\pi_{12}\,\|\Delta A_1\|^2 + 0.5\,\upsilon_{12}\,\|\Delta B_1\|^2 \le (\lambda - 0.5)\,V_1 \tag{13.9}$$

where $\lambda_1 > 0.5$, $\pi_{12} = 2\max\left(\overline{\alpha}_{X_{11}}, \overline{\alpha}_{X_{12}}\right)$, $\upsilon_{12} = \overline{\sigma}_{11} + \overline{\sigma}_{21}$, $\|\Delta A_1\|^2 = \mathrm{tr}(\Delta A_1^T \Delta A_1)$, $\mathrm{tr}(\cdot)$ *is spur of matrix,* $V_1(t) = 0.5\,E_{11}^T(t)\,R_1\,E_{11}(t) + 0.5\,E_{21}^T(t)\,R_1\,E_{21}(t)$, $R_1 = R_1^T > 0$ *is symmetric matrix.*

The theorem 1 proof we give in the appendix A.

Definition 1. If condition (13.9) satisfies, then the system (13.5) is $PI_{1,X}$-identifiable on a set of state variables.

The subsystem layer (13.1) identifiability depends on properties of the TCS output.

Consider the system (13.7) and the Lyapunov function (LF)

$$V_k(t) = 0.5\,E_{1,k}^T(t)\,R_k\,E_{1,k}(t) + 0.5\,E_{2,k}^T(t)\,R_k\,E_{2,k}(t) \tag{13.10}$$

Theorem 2. *Let (i)* $K_k \in \mathbb{R}^{q_k \times q_k}$ *is Hurwitz matrix; (ii)* $\|X_{1,k}\|^2 \in EC_{\underline{\alpha}_X, \overline{\alpha}_X}$, $\|X_{2,k}\|^2 \in EC_{\underline{\alpha}_X, \overline{\alpha}_X}$, $\pi_k = \max\left(\overline{\alpha}_{X_{1,k}}, \overline{\alpha}_{X_{2,k}}\right)$, $y_{2k-1} \in EC_{\underline{\alpha}_y, \overline{\alpha}_y}$*; (iii) the system (13.5) is* $PI_{1,X}$*- identifiable; (iv) the system (13.7) is stable; (v)* $v_{2,k-1}^2 \in EC_{\underline{\alpha}_v, \overline{\alpha}_v}$*; (vi) the* d_{k-1} *operator is constant:* $d_k \le \omega_k \le \omega$*, where* ω *is some number; (vi)* $\lambda_k \ge 0.5$*. Then the system (13.7) is* $PI_{k,X}$*-identifiable if*

$$0.5\,\pi_{k,i}\,\|\Delta A_k\|^2 + 0.5\,\|\Delta B_k\|^2\left(\tilde{\overline{\alpha}}_k + 2\omega^2(\tilde{\overline{\alpha}}_k + 2\omega^2\beta_k)\right) \le \left(\lambda_k - 0.5\right)V_k \tag{13.11}$$

where $\pi_{k,i} = 2\max\left(\overline{\alpha}_{X_{i,k}}, \overline{\alpha}_{X_{2,k}}\right)$, $\tilde{\overline{\alpha}}_k = 2\max\left(\overline{\alpha}_{y_{2,k-1}}, \overline{\alpha}_{y_{1,k-1}}\right)$, $\beta_k = \max_i\left(\overline{\alpha}_{y_{i,k-1}} + \omega^2\overline{\alpha}_{v_{i,k-1}}\right)$.

The theorem 2 proof gives in the appendix B.

Remark 1. The conditions $\|X_{i,k}\|^2 \in EC_{\underline{\alpha}_{X_{i,k}}, \overline{\alpha}_{X_{i,k}}}$, $i = 1; 2$ followed from $v_{i1}(t) \in EC_{\underline{\alpha}_v, \overline{\alpha}_v}$.

Remark 2. The k-layer (13.7) identifiability depends on properties of TCS previous layers and CL. Select CL parameters so that condition (13.11) is fulfilled.

Let d_k is differentiating.

Theorem 3. Let Theorem 2 conditions be satisfied and (i) operator d_k is differentiating, i.e. $v_{1,k-1} = d\left(y_{2,k-1} + d_{k-2}v_{2,k-1}\right)/dt$; (ii) the system (13.1), (13.2) is stable, detectable and recoverable. Then the system (13.5) $PI_{k,X}$- identifiable if:

$$0.5\,\pi_{k,i}\,\|\Delta A_k\|^2 + 0.5\,\|\Delta B_k\|^2\left(\tilde{\overline{\alpha}}_{k,\dot{y}} + 2\left(\overline{\alpha}_{\dot{y}} + \tilde{\overline{\alpha}}_{\dot{v}}\right)\right) \le \left(\lambda_k - 0.5\right)V_k \tag{13.12}$$

where $\tilde{\overline{\alpha}}_{k,\dot{y}} = 2\max_i \overline{\alpha}_{\dot{y}_{2k-1}}$ $\tilde{\overline{\alpha}}_{k,\dot{v}} = 2\max_i \overline{\alpha}_{\dot{v}_{i,k-1}}$, $\pi_{k,i} = 2\max\left(\overline{\alpha}_{X_{i,k}}, \overline{\alpha}_{X_{2,k}}\right)$, $V_k(t)$ *has the form (13.10).*

The theorem 3 proof of gives in the appendix C.

Obtained results give estimates of the system (13.1), (13.2) identifiability based on the analysis of the TCS state vector.

Let the set (13.3) be measured. Convert TCS to the form based on the set (Karabutov, 2023). Consider the system (13.1). Let A_1 be the Frobenius matrix with a parameter vector $A_{1,s} \in R^{q_1}, A_{1,s} = \left[a_{1,s,1}, a_{1,s,2}, \ldots, a_{1,s,q_1}\right]^T, C_1 = [1,0,\ldots,0]^T, B_1 = \left[0,\ldots,0,b_{1,s}\right]^T$. In the space (v_{11}, y_{11}), the system (13.1) has a representation (see appendix D)

$$\dot{y}_{1,1} = \bar{A}^T_{1,1} P_{1,1} \tag{13.13}$$

where $\bar{A}^T_{1,1} = \left[-a_{1,1,1}, a_{1,1,2}, \ldots, a_{1,1,q_1}; b_{1,s}, b_{1,2}, \ldots, b_{1,q_1}\right], \bar{A}_{1,1} \in \mathbb{R}^{2q_1}, P_{1,1} \in \mathbb{R}^{2q_1}$. Considering (13.13), the system (13.1) has the form

$$\begin{cases} \dot{y}_{1,1} = \bar{A}^T_{1,1} P_{1,1}, \\ \dot{y}_{2,1} = \bar{A}^T_{1,1} P_{2,1}. \end{cases} \tag{13.14}$$

Evaluate the system (13.14) identifiability by output ($PI_{1,y}$-identifiability). Consider the model:

$$\begin{cases} \dot{\hat{y}}_{1,1} = -\chi e_{1,1} + \hat{\bar{A}}^T_{1,1} P_{1,1}, \\ \dot{\hat{y}}_{2,1} = -\chi e_{2,1} + \hat{\bar{A}}^T_{1,1} P_{2,1}, \end{cases} \tag{13.15}$$

where $\chi_1 > 0$, $e_{i,1} = \hat{y}_{i,1} - y_{i,1}$ is output $y_{i,1}$ prediction error, $i = 1; 2$. The equation for $e_{i,1}$:

$$\dot{e}_{i,1} = -\chi_1 e_{i,1} + \Delta\bar{A}^T_{1,1} P_{i,1},$$

$$\Delta\bar{A}_{1,1} = \hat{\bar{A}}_{1,1} - \bar{A}_{1,1} \tag{13.16}$$

Consider LF $V_{1,2}(e_{1,1}, e_{2,1}) = 0.5\left(e^2_{1,1} + e^2_{2,1}\right)$, and $\dot{V}_{1,2}$ has the form

$$\dot{V}_{1,2} = -2\chi_1 V_{1,2} + \Delta\bar{A}^T_{1,1}\left(P_{1,1} e_{1,1} + P_{2,1} e_{2,1}\right) \leq -\tfrac{\chi_1}{2} V_{1,2} + \tfrac{1}{2\chi_1}\Delta\bar{A}^T_{1,1} P_1 P_1^T \Delta\bar{A}_{1,1}$$

where $P_1 P_1^T = P_{1,1} P^T_{1,1} + P_{2,1} P^T_{2,1}$.

It follows from (13.16) that $\Delta\bar{A}_{1,1} = 0$, if the vector $P_i(t) \in EC_{\underline{\alpha}_{P_i}, \overline{\alpha}_{P_i}}$ $(i = 1; 2)$ and subsystem (13.1) is parametrically $PI_{1,y}$-identifiable by output. The $PI_{1,y}$-identifiability of subsystems (13.2) is justified similarly.

Representation for the k-layer (system (13.2))

$$\begin{cases} \dot{y}_{1,k} = \bar{A}^T_{1,k} P_{1,k}, \\ \dot{y}_{2,k} = \bar{A}^T_{1,k} P_{2,k}. \end{cases} \tag{13.17}$$

where $P_{i,k} \in \mathbb{R}^{2q_k}$ is generalized input, $\bar{A}_{i,k} \in \mathbb{R}^{2q_k}$ is parameter vector.

Determining the CL type is one of the TCS identification tasks under uncertainty. Apply the following approach. Consider the k-layer of the system (13.2) with identical cross-links and $d_{k-1} = \text{cons}$. Construct the $S_{u_{i,k}, y_{i,k}}$-structure described by the function $f_{u_{i,k}, y_{i,k}} : u_{i,k} \to y_{i,k}$, $i = 1; 2$ for both channels of the k-layer. $S_{u_{i,k}, y_{i,k}}$ reflects the input-output state of the TCS. Determine the secants for $S_{u_{i,k}, y_{i,k}}$

$$\xi_{u_{i,k}, y_{i,k}} = a_{\xi_{0,i,k}} + a_{\xi_{1,i,k}} u_{i,k} \tag{13.18}$$

where $a_{\gamma_{0,i,k}}$, $a_{\gamma_{1,i,k}}$ are parameters determined using the least squares method.

Since CL is rigid, the angle between secants $\xi_{u_{i,k}, y_{i,k}}$ does not exceed a certain value $\delta_{\xi_{i,k}} : \left| a_{\xi_{1,1,k}} - a_{\xi_{1,2,k}} \right| \le \delta_{\xi_{i,k}}$. Therefore, CLs are positive. If the cross-links are asymmetric, then $\left| a_{\xi_{1,1,k}} - a_{\xi_{1,2,k}} \right| > \delta_{\xi_{i,k}}$. The signal $v_{1,k-1}$ is acted in the anti-phase with the output $y_{1,k-1}$, the previous layer of the first channel.

Remark 3. Proved theorems apply if each layer channels are non-identical.

2.3. TCS Adaptive Identification

Consider the representation (13.13) - (13.15), (13.17). Introduce the model for (13.17)

$$\begin{cases} \dot{\hat{y}}_{1,k} = -\chi_k e_{1,k} + \hat{\bar{A}}^T_{1,k} P_{1,k}, \\ \dot{\hat{y}}_{2,k} = -\chi_k e_{2,k} + \hat{\bar{A}}^T_{1,k} P_{2,k}, \end{cases} \tag{13.19}$$

where $\chi_k > 0$, $e_{i,k} = \hat{y}_{i,k} - y_{i,k}$ is i-output prediction error of the i-element for the k-layer.

The equation for $e_{i,k}$

$$\dot{e}_{i,k} = -\chi_k e_{i,k} + \Delta \bar{A}^T_{i,k} P_{i,k},$$

$$\Delta \bar{A}_{1,k} = \hat{\bar{A}}_{1,k} - \bar{A}_{1,k} \tag{13.20}$$

Adaptive parameter tuning algorithm for (13.19)

$$\Delta \dot{\bar{A}}_{1,k} = \dot{\hat{\bar{A}}}_{1,k} = -\Gamma_k e_{i,k} P_{1,k} \tag{13.21}$$

where $\Gamma_k \in \mathbb{R}^{q_k \times q_k}$ is a diagonal matrix with $\gamma_{k,j} > 0$.

Consider the subsystem (13.14) - (13.16) and LF $\tilde{V}_1(e_{1,1}) = 0.5 e_{1,1}^2$. From condition $\dot{\tilde{V}}_1 \leq 0$ we obtain then adaptive algorithm for adjusting vector $\hat{\bar{A}}_{1,1}$:

$$\Delta \dot{\bar{A}}_{1,1} = \dot{\hat{\bar{A}}}_{1,1} = -\Gamma_1 e_{1,1} P_{1,1} \tag{13.22}$$

where $\Gamma_1 \in \mathbb{R}^{q_1 \times q_1}$ is diagonal matrix with $\gamma_{1,j} > 0$, j is the diagonal element number.

Remark 4. As TCS has identical channels, we adjust only one channel.

The feedback effect leads to unidentifiable of the system (13.14) - (13.16) and (13.22) parameters. The equation (13.16) writes as:

$$\dot{e}_{1,1} = -\chi_1 e_{1,1} + \Delta \bar{A}_{1,1}^T P_{1,1} + f(\bar{A}_{1,1}, P_{1,1}) \tag{13.23}$$

where $f(\cdot)$ is the uncertainty that is the result from $P_{1,1} \notin EC_{\underline{\alpha}_p, \bar{\alpha}_p}$.

Consider LF $\tilde{V}_{1,e}(e_{1,1}) = 0.5 e_{1,1}^2$, $\tilde{V}_{1,\Delta}(\Delta \bar{A}_{1,1}) = 0.5 \Delta \bar{A}_{1,1}^T \Gamma_1^{-1} \Delta \bar{A}_{1,1}$.

Theorem 4. *Let 1)* $g_i(t) \in EC_{\underline{\alpha}_i, \bar{\alpha}_i}$, $P_{1,1} \notin EC_{\underline{\alpha}_p, \bar{\alpha}_p}$; *2)* $y_{i,1} \notin EC_{\underline{\alpha}_{y_{i,1}}, \bar{\alpha}_{y_{i,1}}}$, $i = 1, 2$; *3)* $|f|^2 \leq \alpha_f$, *where* $\alpha_f \geq 0$; *4) there is a* $\upsilon > 0$ *such that for* $t >> t_0$ *it is*

$$e_{1,1} \Delta \bar{A}_{1,1}^T P_{1,1} = \upsilon \left(e_{1,1}^2 + \Delta \bar{A}_{1,1}^T P_{1,1} P_{1,1}^T \Delta \bar{A}_{1,1} \right)$$

5) $\lambda_{\max}^{-1}(\Gamma_1) \|\Delta \bar{A}_{1,1}\|^2 \leq \tilde{V}_{1,\Delta} \leq \lambda_{\min}^{-1}(\Gamma_1) \|\Delta \bar{A}_{1,1}\|^2$; *6) the system of differential inequalities is valid for the system (13.16), (13.23)*

$$\begin{bmatrix} \dot{\tilde{V}}_{1,e} \\ \dot{\tilde{V}}_{1,\Delta} \end{bmatrix} \leq \underbrace{\begin{bmatrix} -\frac{\chi_1}{2} & 2\frac{\pi_{P_{1,1}}}{\chi_1} \\ 4\upsilon & -\frac{\upsilon\eta}{2} \end{bmatrix}}_{A_{S_1}} \begin{bmatrix} \tilde{V}_{1,e} \\ \tilde{V}_{1,\Delta} \end{bmatrix} + \underbrace{\begin{bmatrix} \frac{\alpha_f}{\chi_1} \\ 0 \end{bmatrix}}_{B_{S_1}} \tag{13.24}$$

and the comparison system $\dot{S}_1(t) = A_{S_1} S_1(t) + B_{S_1}$ *is for (13.24), where* $S_1(t) = [s_{1,e}(t), s_{1,\Delta}(t)]^T$, $s_{1,w}(t)$ $(w = e, \Delta)$ *is a majority factor for* $\tilde{V}_{1,w}(t)$ *and* $s_{1,w}(t_0) \geq \tilde{V}_{1,w}(t_0)$. *Then the system (13.14) - (13.16) and (13.23) is exponentially dissipative with the estimate*

$$\left[\tilde{V}_{1,e}(t)\ \tilde{V}_{1,\Delta}(t) \right]^T \leq e^{A_{S_1}(t-t_0)} S_1(t_0) + \int_{t_0}^{t} e^{A_{S_1}(t-\tau)} B_{S_1} d\tau$$

where $\chi_1^2\eta > 8\vartheta_{P_{1,1}}$, *then* $\eta = \underline{\pi}_{P_{1,1}}\lambda^2_{\min}(\Gamma_1), \underline{\pi}_{P_{1,1}} \geq 0$, $\|P_{1,1}P^T_{1,1}\| \leq \vartheta_{P_{1,1}}(\overline{\alpha}_{P_{1,1}})$, $\lambda_{\min}(\Gamma_1)$ *is the minimum eigenvalue of the matrix* Γ_1.

The theorem 4 proof gives in the appendix E.

As follows from Theorem 4, the adaptive identification system guarantees offset estimates for system (13.14) - (13.16) parameters.

Consider the system (13.17), (13.19), (13.21). Let d_{k-1} in (13.2) is constant.

Present $P_{1,k}$ and $\Delta\overline{A}_{1,k}$ as: $P_{1,k} = \left[\widetilde{P}^T_{1,k}, p_{j,v_{1,2,k-1}}\right]^T, \Delta\overline{A}_{1,k} = \left[\Delta\widetilde{\overline{A}}^T_{1,k}, \Delta d_{j,k}\right]^T$, where $p_{j,v_{i,2,k-1}}$ is transformation $v_{1,2,k-1}$, j is jth element of vector $\Delta\overline{A}_{1,k}$. Then (13.20)

$$\dot{e}_{1,k} = -\chi_k e_{1,k} + \Delta\overline{A}^T_{1,k}P_{1,k} + f_{1,k}(\overline{A}_{1,k}P_{1,k}) \tag{13.25}$$

where $f_{1,k}(\cdot) \in \mathbb{R}$ is uncertainty, which is $P_{1,k} \notin EC_{\underline{\alpha}_P,\overline{\alpha}_P}$.

Consider the system (13.19), (13.20) and (13.25) and LF

$$\widetilde{V}_{k,\Delta}(\Delta\overline{A}_{1,k}) = 0.5\Delta\overline{A}^T_{1,k}\Gamma_k^{-1}\Delta\overline{A}_{1,k}, \widetilde{V}_{k,e}(e_{1,k}) = 0.5e^2_{1,k}.$$

Theorem 5. *Let the Theorem 4 conditions be satisfied and (i)* $y_{1,k}(t) \notin EC_{\underline{\alpha}_{y_{1k}},\overline{\alpha}_{y_{1k}}}$, $P_{1,k} \notin EC_{\underline{\alpha}_{P_{1k}},\overline{\alpha}_{P_{1k}}}$; *(ii)* $u_{1,k} = \upsilon_{u_{1,k}}(y_{1,k-1}, v_{1,k-1}), u_{1,k} \notin EC$; *(iii)* $f^2_{1,k} \leq \alpha_{f_{1,k}}$, *where* $\alpha_{f_{1,k}} \geq 0$; *(iv)*

$$\lambda^{-1}_{\max}(\Gamma_k)\|\Delta\overline{A}_{1,k}\|^2 \leq \widetilde{V}_{k,\Delta} \leq \lambda^{-1}_{\min}(\Gamma_k)\|\Delta\overline{A}_{1,k}\|^2$$

where $\lambda_{\min}(\Gamma_k)$ *is the minimum eigenvalue of the matrix* Γ_k; *(v)* $\|P_{1,k}P^T_{1,k}\| \leq \pi_{P_{1,k}} < \overline{\alpha}_{P_{1,k}}$; *(vi) there is a* $\upsilon > 0$ *such that for* $t >> t_0$ *is fair*

$$e_{1,k}\Delta\overline{A}^T_{1,k}P_{1,k} = \upsilon\left(e^2_{1,k} + \Delta\overline{A}^T_{1,k}P_{1,k}P^T_{1,k}\Delta\overline{A}_{1,k}\right) \tag{13.26}$$

(vii) the system of differential inequalities

$$\underbrace{\begin{bmatrix}\dot{\widetilde{V}}_{k,e}\\ \dot{\widetilde{V}}_{k,\Delta}\end{bmatrix}}_{\dot{\widetilde{V}}_k} \leq \underbrace{\begin{bmatrix}-\frac{\chi_k}{2} & 2\frac{\vartheta_{P_{1,k}}}{\chi_k}\\ 4\upsilon & -\frac{\upsilon\eta}{2}\end{bmatrix}}_{A_{S_k}}\underbrace{\begin{bmatrix}\widetilde{V}_{k,e}\\ \widetilde{V}_{k,\Delta}\end{bmatrix}}_{\widetilde{V}_k} + \underbrace{\begin{bmatrix}\frac{\alpha_{f_{1,k}}}{\chi_k}\\ 0\end{bmatrix}}_{B_{S_k}} \tag{13.27}$$

is valid for the system (13.21), (13.25) and the comparison system $\dot{S}_k(t) = A_{S_k}S_k(t) + B_{S_k}$ *for (13.27), where* $S_k(t) = \left[s_{k,e}(t), s_{k,\Delta}(t)\right]^T$, $s_{k,w}(t)$ $(w = e, \Delta)$ *is the majority factor for* $\widetilde{V}_{k,w}(t)$, $s_{k,w}(t_0) \geq \widetilde{V}_{k,w}(t_0)$. *Then the system (13.19), (13.20), (13.25) is exponentially dissipative with the estimate:*

$$\tilde{V}_k(t) \leq e^{A_{S_k}(t-t_0)} S_k(t_0) + \int_{t_0}^{t} e^{A_{S_k}(t-\tau)} B_{S_k} d\tau$$

if

$$\chi_k^2 \eta > 32 \vartheta_{P_{1,k}},$$

$$\underline{\pi}_{P_{1,k}} \geq 0,$$

$$\eta = \underline{\pi}_{P_{1,k}} \lambda_{\min}(\Gamma_k)$$

The theorem 5 proof scheme coincides with the theorem 4 proof.

As follows from (13.27), adaptive identification system properties depend on the k-layer on cross-links.

So, we have proved the TCS identification by the state and output of the k-layer. Offered results confirming the convergence of estimates for the system parameters. Adaptive identification system properties depend on the CL parameters and properties of TCS signals.

Remark 5. If the TCS contains non-identical channels, then we apply the models (13.17) for each k-layer.

3. INTERCONNECTED SYSTEMS

Consider a system S_{MS}

$$\dot{X}(t) = AX(t) + DF_1(X,t) + BU(t), \tag{13.27}$$

$$\mathrm{L}Y(t) = CX(t) + F_2(X,t), \tag{13.28}$$

where $X \in \mathbb{R}^m$ is state vector, $A \in \mathbb{R}^{m \times m}$ is state matrix, $D \in \mathbb{R}^{m \times q}$, $F_1(X,t)$: $\mathbb{R}^m \to \mathbb{R}^q$ is nonlinear vector function, $U \in \mathbb{R}^k$ is input vector (control), $B \in \mathbb{R}^{m \times k}$, $Y \in \mathbb{R}^n$ is output vector, $C \in \mathbb{R}^{n \times m}$, $F_2(X,t) : \mathbb{R}^m \to \mathbb{R}^n$ is perturbation vector (measurement errors), L is an operator that forms the vector Y. L can be: 1) a differential operator reflecting the dynamic properties of the measurement system; 2) as the operator describing the method of interaction between subsystems. Matrices A, D, B are block-based and reflect the state of independent subsystems. The F_2 (X, t) vector can be a perturbation (measurement error) or a variable reflecting the influence of independent subsystems.

The observed data set

$_o = \{Y(t), U(t), t \in [t_0, t_N]\},$

$$t_N < \infty \tag{13.29}$$

Assumption 1. Elements $\varphi_{1,i}(x_j) \in F_1$, $\varphi_{2,i}(x_j) \in F_2$ are smooth, one-valued functions.

Apply the model

$$\begin{cases} \dot{\hat{X}}(t) = \hat{A}(t)\hat{X}(t) + \hat{D}(t)F_1(X,t) + \hat{B}(t)U(t) \\ \mathrm{L}\hat{Y}(t) = C\hat{X}(t) + F_2(\hat{X},t) \end{cases} \tag{13.30}$$

to estimate parameters of matrices A, D, B, C, where $\hat{A}(t)$ and $\hat{D}(t)$ are matrices with tuning parameters.

Problem: synthesize the model (13.30) for the system (13.27) satisfying the assumption 1, and find parameter tuning laws of matrices $\hat{A}(t)$, $\hat{D}(t)$ и $\hat{B}(t)$ to

$$\lim_{t\to\infty} \|\hat{Y}(t) - Y(t)\| \leq \delta_y, \delta_y \geq 0$$

where $\| \cdot \|$ is the Euclidean norm.

Remark 6. Consider (13.28) as the equation describing the connections between subsystems for some class of ICS. The vector $F_2(X,t)$ form (see the section 2) must be evaluated in this case.

The synthesis of adaptive identification algorithms bases on the approach described in section 2.

Remark 7. If the TCS contains nonlinear subsystems, then the decision on nonlinearity form bases on the construction of an interconnection graph (Karabutov, 2021) under uncertainty.

Models and adaptation algorithms match the equations obtained in the previous section. Apply theorems from section 2 to the AS quality estimate.

4. EXAMPLES

1. Consider The TCS for target angular tracking with identical azimuth and elevation channels and asymmetric CL

$$\dot{x} = -a_x x + b_x(g - y), \tag{13.31}$$

$$\ddot{y} = -a_y \dot{y} + b_y\left(x - (g_1 - y_1) - k(g_1 - y_1)'\right), \tag{13.32}$$

$$\dot{x}_1 = -a_x x_1 + b_x(g_1 - y_1), \tag{13.33}$$

$$\ddot{y}_1 = -a_y \dot{y}_1 + b_y\left(x_1 + (g - y) + k(g - y)'\right), \tag{13.34}$$

where $a_x = T_x^{-1}, b_x = T_x^{-1}k_x, T_x, k_x$ are time constant and amplifier gain; k is the cross-linking parameter; $a_y = T_y^{-1}, b_y = T_y^{-1}k_y$ are servomotor parameters; $g,\ g_1$ are inputs. CL represents in parentheses as a differentiating link with the parameter k.

Outputs $x_i(t), y_i(t)$ of links and the inputs $g_i(t)$are measured. Evaluate the system parameters (13.31), (13.32).

Apply the transformation described in Appendix D to obtain the model for $y(t)$. Consider the system of filters (transformations)

$$\begin{aligned} \dot{p}_y &= -\mu_1 p_y + y, \\ \dot{p}_\vartheta &= -\mu_1 p_\vartheta + \vartheta, \\ \dot{p}_\upsilon &= -\mu_1 p_\upsilon + \upsilon, \end{aligned} \tag{13.35}$$

where $\vartheta \triangleq x - g_1 + y_1$, $\upsilon \triangleq d(g_1 - y_1)/dt$, $p_i(0) = 0$, $i = y, \vartheta, \upsilon$, $\mu_1 > 0$ is a number that does not coincide with roots of the characteristic equation for the second equation in (13.31). The model for (13.31)

$$\dot{\hat{x}} = -\chi_x e_x + \hat{a}_x x + \hat{b}_x(g - y), \tag{13.36.1}$$

$$\dot{\hat{y}} = -\chi_y e_y + \hat{a}_y y + \hat{a}_{p_y} p_y + \hat{a}_\vartheta p_\vartheta + \hat{a}_\upsilon p_\upsilon, \tag{13.36.2}$$

where $e_x = \hat{x} - x, e_y = \hat{y} - y$, χ_x, χ_y is positive numbers (reference model). Adaptive algorithms for tuning system parameters (13.36.1), (13.36.2) have form:

$$\begin{aligned} \dot{\hat{a}}_x &= -\gamma_{a_x} e_x x, \quad \dot{\hat{b}}_x = -\gamma_{b_x} e_x(g - y), \\ \dot{\hat{a}}_y &= -\gamma_{a_y} e_y y, \quad \dot{\hat{a}}_{p_y} = -\gamma_{a_{p_y}} e_y p_y, \\ \dot{\hat{a}}_\vartheta &= -\gamma_{a_\vartheta} e_y p_\vartheta, \quad \dot{\hat{a}}_\upsilon = -\gamma_{a_\upsilon} e_y p_\upsilon \end{aligned} \tag{13.37}$$

where γ_{a_i}, γ_{b_i} $(i = x, y, a_y, p_y, \vartheta, \upsilon)$ are positive numbers guaranteeing convergence (13.37).

Figure 1. Evaluation of structure for cross-links

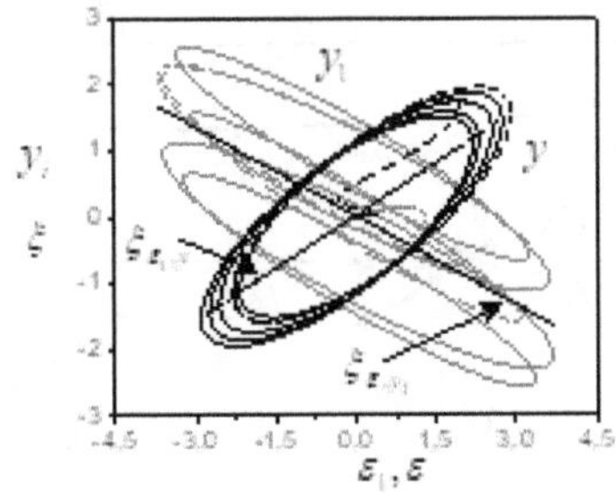

The system (13.31), (13.32) is modelled with the parameters: $a_x = 1.2, b_x = 2$, $a_y = 5.95, b_y = 1, k = 5$. Inputs $g_2(t) = 1.5\sin(0.1\pi t)$, $g_2(t) = 1.5\sin(0.1\pi t)$.

Figure 1 shows the structures confirming asymmetric connections in the system. The analysis of parameters $a_{\xi_{0,\varepsilon,y}} = 0.53, a_{\xi_{0,\varepsilon,y_1}} = -0.45$ for secants $\xi_{\varepsilon_1,y}, \xi_{\varepsilon,y_1}$ shows that the condition $\left|a_{\xi_{0,\varepsilon,y}} - a_{\xi_{0,\varepsilon,y_1}}\right| > \delta$ fulfils with $\delta_{\xi_0} = 0.2$. Therefore, CL is antisymmetric.

We consider remark 4 when evaluating the parameters of the system (13.31), (13.32). The identification system parameters: $\mu_1 = 1.5, \chi_x = 1.5, \chi_y = 2$.

Adaptive identification results are shown in Figure 2-6. Adaptive identification of model parameters (13.36.1), (13.36.2) presented in Figures 2 and 3.

Figure 2. Tuning model (13.36.1) parameters

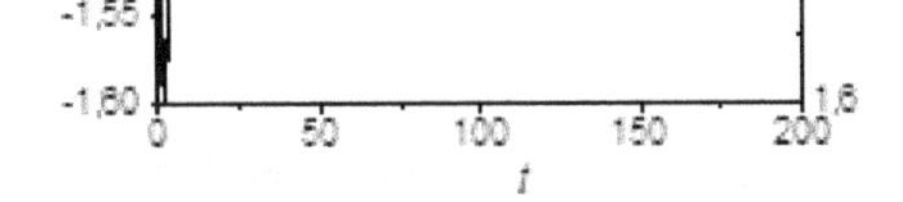

Figure 3. Tuning model (13.36.2) parameters

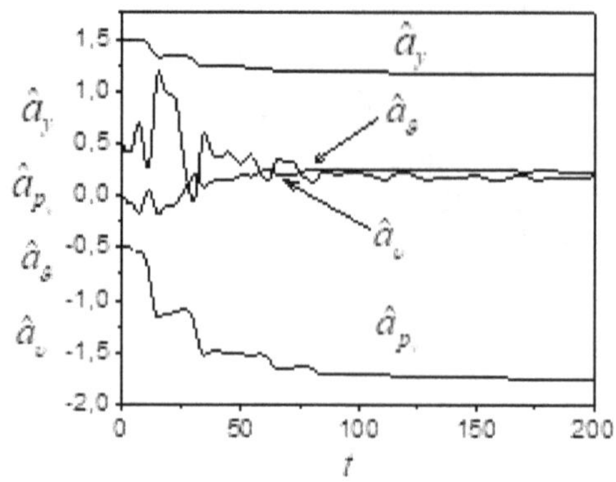

Identification errors show in Figure 4.

Figure 4. Changing identification error

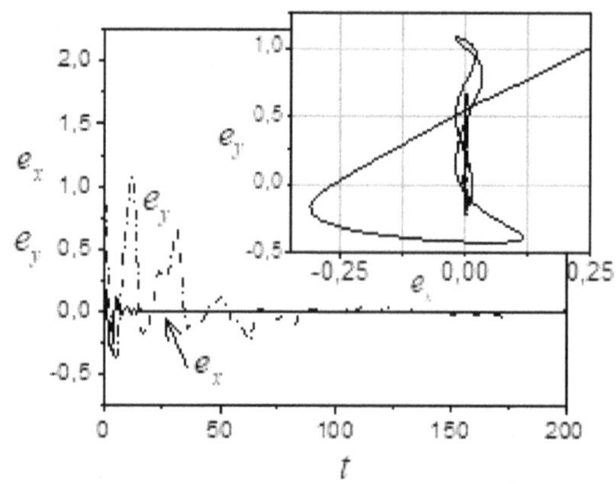

We see that the tuning of layer 2 parameters depends on the layer 1 output (model (13.36.1)) and the CL properties. The quality of the system parameter tuning (13.36.2) depends on the correlation between outputs of elements (13.31) and (13.33). This influence is transmitted through CL. This conclusion confirms by structures shown in Figure 5. These results confirm statements of Theorem 5. The tuning process of model (13.36.1) parameters has a smooth character (Figure 6).

Figure 5. Tuning model (13.36.2) parameters

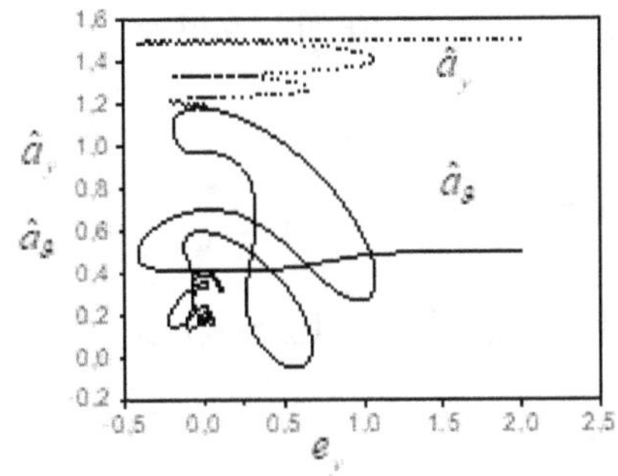

Figure 6. Tuning model (13.36.2) parameters

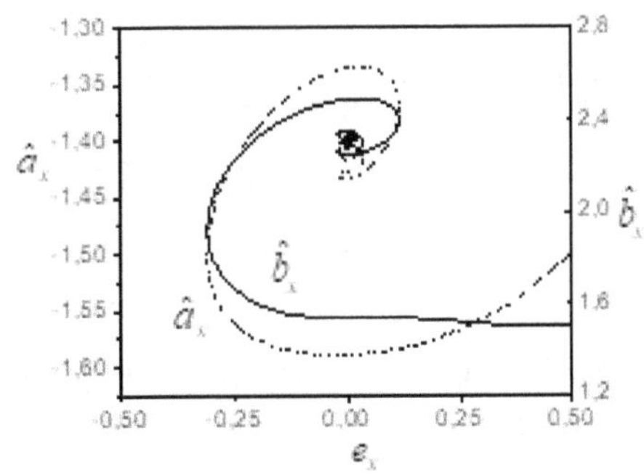

2. Consider a pseudo-linear two-channel corrector (PLTCC) (Skorospeshkin, 2008)

$$\begin{cases} \dot{x}_1 = -\alpha_1 x_1 + \beta_1(g - x_6) \\ \dot{x}_2 = -\alpha_2 x_2 + \mu_2(g - x_6)' + \beta_2(g - x_6) \\ \dot{x}_6 = x_7 \\ \dot{x}_7 = -\alpha_{01} x_7 - \alpha_{02} x_6 + \beta_0\left(x_1 sign(x_1)\right)\left(sign(x_2)\right) \end{cases} \quad (13.38)$$

where g is the setting effect, x_6 is output; $g - x_6$ is misalignment error; x_1 is amplitude of channel output; x_2 is a phase of channel output; $u = \left(x_1 sign(x_1)\right)\left(sign(x_2)\right)$ is regulator output; $\left(g - x_6\right)' = d\left(g - x_6\right)/dt$; $\alpha_1, \beta_1, \alpha_2, \mu_2, \beta_2, \alpha_{01}, \alpha_{02}, \beta_0$ are corrector parameters.

Remark 8. Skorospeshkin (2008) uses a preliminary selection of PLTCC parameters.

The set $_o = \{x_1(t), x_2(t), x_6(t), t \in [0, t_k]\}$, where t_k is known number. Apply models

$$\dot{\hat{x}}_1 = -k_1 e_1 + \hat{\alpha}_1 x_1 + \hat{\beta}_1 (g - x_6) \tag{13.39}$$

$$\dot{\hat{x}}_2 = -k_2 e_2 + \hat{\alpha}_2 x_2 + \hat{\mu}_2 d(g - x_6)/dt + \hat{\beta}_2 (g - x_6) \tag{13.40}$$

$$\dot{\hat{x}}_6 = -k_6 e_6 + \hat{\alpha}_{01} x_6 + \hat{\alpha}_{p_{x_6}} p_{x_6} + \hat{\alpha}_{p_u} p_u \tag{13.41}$$

where k_1, k_2, k_6 are known numbers; $e_i = \hat{x}_i - x_i$, $i = 1, 2, 6$; $\hat{\alpha}_i, \hat{\beta}_i, i = 1, 2$, $\hat{\alpha}_{p_{x_6}}$, $\hat{\alpha}_{p_{x_6}}$ are tuning parameters; $\hat{x}_i \in \mathbb{R}$ (13.$i = 1; 2; 6$) are model outputs; p_{x_6}, p_u obtain as (13.33).

Adaptation algorithms

$$\begin{aligned}
&\dot{\hat{\alpha}}_1 = -\gamma_{\alpha_1} e_1 x_1,\ \dot{\hat{\beta}}_1 = -\gamma_{\beta_1} e_1 (g - x_6),\\
&\dot{\hat{\alpha}}_2 = -\gamma_{\alpha_2} e_2 x_2,\ \dot{\hat{\beta}}_2 = -\gamma_{\beta_2} e_2 (g - x_6),\ \dot{\hat{\mu}}_2 = -\gamma_{\mu_2} e_2 (g - x_6)',\\
&\dot{\hat{\alpha}}_{01} = -\gamma_{\alpha_{01}} e_6 x_6,\ \dot{\hat{\alpha}}_{p_{x_6}} = -\gamma_{p_{x_6}} e_6 p_{x_6},\ \dot{\hat{\alpha}}_{p_u} = -\gamma_{p_u} e_6 p_u,
\end{aligned} \tag{13.42}$$

where $\gamma_i > 0$ is the gain factor in the corresponding parameter tuning circuit.

System (13.38) parameters: $\alpha_1 = 1.05$, $\beta_1 = 3.5$, $\alpha_2 = 2.2$, $\beta_2 = 2.2$, $\mu_2 = 0.4$, $\alpha_{01} = 3$, $\alpha_{02} = 5.03$, $\beta_0 = 5.2$ $g(t) = \sin(0.05\pi t)$. Simulation results show in Figure 7-12.

Figure 7 represents structures (the transition process is excluded) reflecting the phase processes in the system (13.38). We see that the system is nonlinear. The application of results (Karabutov, 2024) shows that the system is structurally identifiable. Therefore, the input is S-synchronizing and considers the nonlinear properties of the PLTC.

Figure 7. Phase portraits of system (13.38) (13.1 is $x_6(x_2)$, 2 is $x_6(x_1)$)

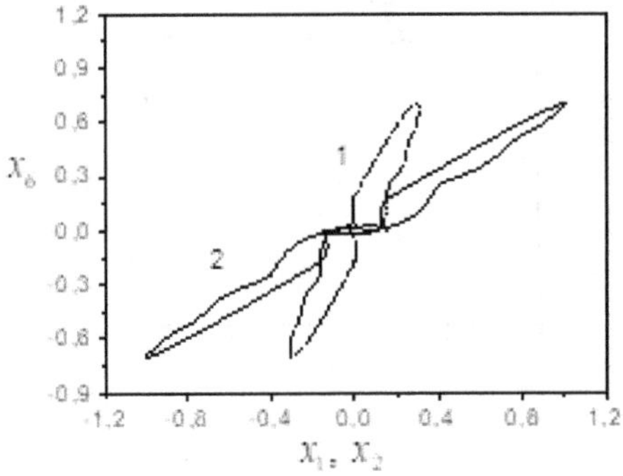

The analysis shows the relationship between x_1 and x_2 (the correlation coefficient is 0.94). The density diagram (Figure 8) confirms the influence x_1 and x_2 on x_6. We see clear influence boundaries of variables on x_6. These relationships affect the convergence of adaptive algorithms.

Figure 8. Effect evaluation of variables x_1, x_2 on x_6

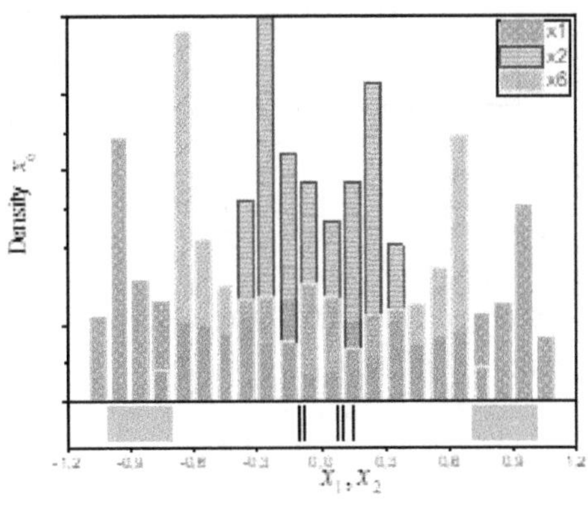

Figure 9. Tuning parameters of model (13.39)

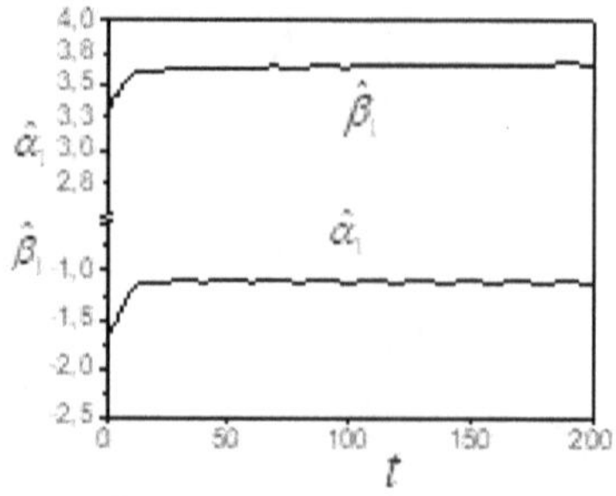

Figure 10. Tuning parameters of model (13.40)

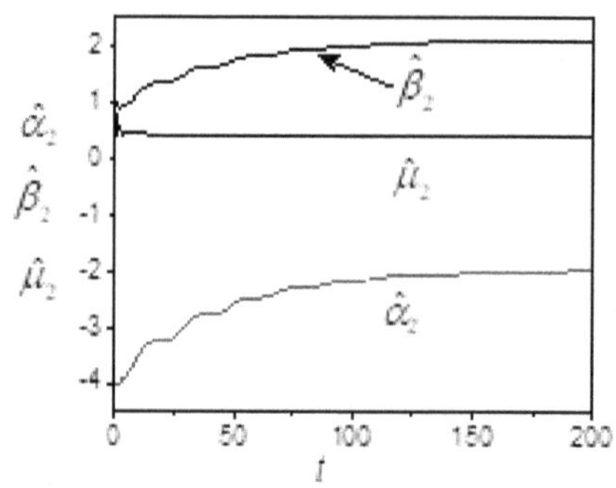

The tuning of model parameters for the system shows in Figure 11. The tuning process depends on the control action.

Figure 11. Tuning parameters of model (13.41)

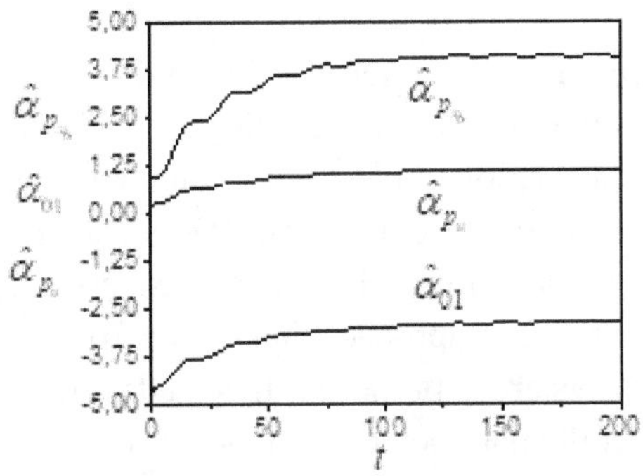

Figure 12. Dynamics of contour for tuning model (13.40)

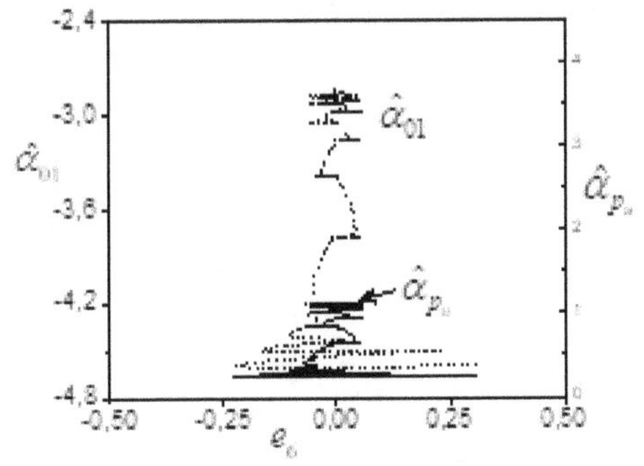

The tuning dynamics of model (13.41) has the more complex form. The input influence plays an important role. Theorem 5 statement holds for the PTLCC system.

5. CONCLUSION

The approach to the identification of interconnected systems is proposed. Adaptive identification features of TCS with cross-links considered. TCS identifiability conditions the in the state space and the output space are got. Structural aspects of the identification for two-channel systems are considered. We study the influence of input properties on the estimation of TCS parameters. Adaptive algorithms are designed for TCS identification with identical channels. Estimates and conditions confirming convergence estimates of TCS parameters are obtained. Properties of the adaptive identification system depend on the CL parameters and signal properties in the TCS.

We consider the interconnected systems and note the a priori information role in the analysis existing relationships for identification. It is shown that approaches used for the TCS identification apply to the interconnected system. Adaptive identification examples systems are considered. We consider TCS with nonlinear control and complex interconnections. The paper considers only certain aspects of this multifaceted subject area. The approach that allows to evaluate the dynamics of processes in the adaptive system is proposed. It considers the influence of system elements on the quality of the parameter tuning process.

REFERENCES

Ali, M., Abbas, H., Chughtai, S. S., & Werner, H. (2011). Identification of spatially interconnected systems using neural network. *49th IEEE Conference on decision and control (CDC)*. DOI: 10.1109/CDC.2010.5717080

Ali, M., Chughtai, S. S., & Werner, H. (2010). Identification of spatially interconnected systems. *Proceedings of the 48h IEEE Conference on Decision and Control (CDC) held jointly with 2009 28th Chinese Control Conference*. DOI: 10.1109/CDC.2009.5399748

Aliyeva, A. S. (2019). Identification of multiconnected dynamic objects with uncertainty based on neural technology and reference converters. *Informatics and control problems, 39*(2), 93-102.

Bukov, V. N., Maksimenko, I. M., & Ryabchenko, V. N. (1998). Control of multivariable systems. *Automation and Remote Control, 6*, 97–110.

Egorov, I. N. (2010). *Positional force control of robotic and mechatronic devices*. Publishing House of Vladimir State University.

Egorov, I. N., & Umnov, V. P. (2022). *Control systems for electric drives of technological robots and manipulators: textbook*. Vladimir: Publishing house of Vladimir State University.

Gantmacher, F. R. (2000). *The theory of matrices*. AMS Chelsea Publishing.

Glentis, G.-O., & Slump, C. H. (1995). A highly modular normalized adaptive lattice algorithm for multichannel least squares filtering. *1995 International Conference on Acoustics, Speech, and Signal Processing*. DOI: 10.1109/ICASSP.1995.480508

Gupta, N., & Chopra, N. (2016). Stability analysis of a two-channel feedback networked control system. *Conference: 2016 Indian Control Conference (ICC)*. DOI: 10.1109/INDIANCC.2016.7441129

Hua, C., Guan, X., & Shi, P. (2004). Decentralized robust model reference adaptive control for interconnected time-delay systems. *Proceeding of the 2004 American Control Conference Boston,* Massachusetts June 30 - July 2, 2004, 4285-4289.

Karabutov, N. (2021). Structural identifiability of systems with multiple nonlinearities. *Contemporary Mathematics*, 2(2), 140–161. DOI: 10.37256/cm.222021763

Karabutov, N. N. (2023). On adaptive identification of systems having multiple nonlinearities. *Russian technological journal*, 11(5), 94–105. doi: org/DOI: 10.37256/cm.222021763

Karabutov, N. N. (2024). On identification of interconnected systems. *Russian technological journal, 12*(5), 63-76.

Kholmatov, U. (2023). The possibility of applying the theory of adaptive identification to automate multi-connected objects. *The American journal of engineering and technology*, 31-38.

Krasovsky, A. A. (1957). On two-channel automatic control systems with antisymmetric connections. *Automation and telemechanics, 18*(2), 126–136.

Meerov, M. V., & Litvak, B. L. (1972). *Optimization of multicommunicated control systems*. Nauka.

Morozovsky, V. E. (1970). *Multi-connected automatic control systems*. Energiya.

Skorospeshkin, M. V. (2008). Adaptive two-channel correction device for automatic control systems. *Bulletin of the Tomsk Polytechnic University, 312*(5), 52–57.

Voronov, A. A. (1985). *Introduction in dynamics of complex controlled systems*. Nauka.

Yang, Q., Zhu, M., Jiang, T., He, J., Yuan, J., & Han, J. (2016). Decentralized Robust Adaptive output feedback stabilization for interconnected nonlinear systems with uncertainties. *Journal of Control Science and Engineering, 2016*, 1–12. DOI: 10.1155/2016/3656578

Zyryanov, G. V. (2010). *Control systems for multi-connected objects: a textbook*. SUSU Publishing Center.

APPENDIX A

Theorem 1 proof. Consider the Lyapunov function for layer (13.6):
$V_1(t) = 0.5E_{11}^T(t)R_1E_{11}(t) + 0.5E_{21}^T(t)R_1E_{21}(t)$. (A1)
The derivative of $V_1(t)$

$$\dot{V}_1 = E_{11}^T Q_1 E_{11} + E_{11}^T R_1(\Delta A_1 X_{11} + \Delta B_1 v_{11}) + E_{11}^T Q_1 E_{11} + E_{21}^T R_1(\Delta A_1 X_{21} + \Delta B_1 v_{21}), \quad (A2)$$

where $R_1 K_1 + K_1^T R_1 = Q$, Q is symmetric negatively defined matrix.

Since the matrix K_1 is stable, then $E_{11}^T Q E_{11} + E_{21\,1}^T Q E_{21} \leq -\lambda_1 V_1$, where $\lambda_1 > 0$. Apply the inequality $ab \leq 0.5(a^2 + b^2)$ and get

$$\begin{aligned}\dot{V}_1 &\leq -\lambda_1 V_1 + |E_{11}^T R_1(\Delta A_1 X_{11} + \Delta B_1 v_{11})| + |E_{21}^T R_1(\Delta A_1 X_{21} + \Delta B_1 v_{21})| \leq \\ &\leq -\lambda_1 V_1 + 0.5(\|R_1 E_{11}\|^2 + \|R_1 E_{21}\|^2) + 0.5(X_{11}^T \Delta A_1^T \Delta A_1 X_{11} + X_{21}^T \Delta A_1^T \Delta A_1 X_{21}) + \quad (A3) \\ &+0.5\,\|\Delta B_1\|^2(v_{11}^2 + v_{21}^2).\end{aligned}$$

$g_i(t) \in EC_{\underline{\alpha}_i,\overline{\alpha}_i}(g_i)$, and the system (13.5) is stable and detectable. Therefore, the residuals ΔA_1, ΔB_1 K2 are limited. Then (A3)

$$\begin{aligned}\dot{V}_1 &\leq -(\lambda_1 - 0.5)V_1 + 0.5(X_{11}^T \Delta A_1^T \Delta A_1 X_{11} + X_{21}^T \Delta A_1^T \Delta A_1 X_{21}) + \\ &+0.5\,\|\Delta B_1\|^2(v_{12}^2 + v_{21}^2) \leq -(\lambda_1 - 0.5)V_1 + \pi_{12}\|\Delta A_1\|^2 + 0.5\upsilon_{12}\|\Delta B_1\|^2, \quad (A4)\end{aligned}$$

where $\pi_{12} = 2\max\left(\overline{\alpha}_{X_{1,1}}, \overline{\alpha}_{X_{2,1}}\right)$, $X_{i,1}(t) \in EC_{\underline{\alpha}_{X_i},\overline{\alpha}_{X_i}}$, $v_{i1}(t) \in EC_{\underline{\alpha}_{i1},\overline{\sigma}_{i1}}$, $\upsilon_{12} = \overline{\sigma}_{11} + \overline{\sigma}_{21}$.
The system (13.6) will be stable and identifiable by $\Delta A_1, \Delta B_1$ if

$$\pi_{12}\|\Delta A_1\|^2 + 0.5\upsilon_{12}\|\Delta B_1\|^2 \leq (\lambda_1 - 0.5)V_1, \quad (A5)$$

and system variables satisfy conditions 1), 3), 4) of Theorem 1.

APPENDIX B

Theorem 2 proof. Consider the Lyapunov function (13.10) for layer $k \geq 2$ (13.7). The derivative of $V_k(t)$

$$\dot{V}_k = E_{1,k}^T Q_k E_{1,k} + E_{1,k}^T R_k \left(\Delta A_k X_{1,k} + \Delta B_k u_{1,k}\right) + E_{2,k}^T Q_k E_{2,k} E_{2,k} + \\ + E_{2,k}^T R_k \left(\Delta A_k X_{2,k} + \Delta B_k u_{2,k}\right), \tag{B1}$$

where Q_k is symmetric negatively defined matrix.
Apply the approach described in Appendix A and get

$$\dot{V}_k \leq -(\lambda_k - 0.5) V_k + 0.5 \|\Delta A_k\|^2 \left(\|X_{1,k}\|^2 + \|X_{2,k}\|^2\right) + 0.5 \|\Delta B_k\|^2 \left(u_{1,k}^2 + u_{2,k}^2\right). \text{(B2)}$$

Let

$$\|X_{1,k}\|^2 \in EC_{\underline{\alpha}_{X_{1,k}}, \overline{\alpha}_{X_{1,k}}}, \quad \|X_{2,k}\|^2 \in EC_{\underline{\alpha}_{X_{2,k}}, \overline{\alpha}_{X_{2,k}}}; \quad u_{1,k}^2 \leq \overline{\alpha}_{y_{1,k}} + \left|d_{k-1}(y_{2,k-1} + d_{k-2} v_{2,k-1})\right|^2,$$
$$u_{2,k}^2 \leq \overline{\alpha}_{y_{2,k}} + \left|d_{k-1}(y_{1,k-1} + d_{k-2} v_{1,k-1})\right|^2, \pi_{k,i} = 2 \max\left(\overline{\alpha}_{X_{1,k}}, \overline{\alpha}_{X_{2,k}}\right), y_{2k-1} \in EC_{\underline{\alpha}_{y_{i,k-1}}, \overline{\alpha}_{y_{i,k-1}}},$$

$$\left(d_{k-1}(y_{1,k-1} + d_{k-2} v_{1,k-1})\right)^2 \leq 2 d_{k-1}^2 \left(y_{1,k-1}^2 + (d_{k-2} v_{1,k-1})^2\right), \\ \left(d_{k-1}(y_{2,k-1} + d_{k-2} v_{2,k-1})\right)^2 \leq 2 d_{k-1}^2 \left(y_{2,k-1}^2 + (d_{k-2} v_{2,k-1})^2\right). \tag{B3}$$

Let $d_k(\cdot)$ is a constant, i.e. $d_k \leq \omega_k \leq \omega$, where ω is some number. Then (B3)

$$\left(d_{k-1}(y_{i,k-1} + d_{k-2} v_{i,k-1})\right)^2 \leq 2\omega^2 \left(\overline{\alpha}_{y_{i,k-1}} + \omega^2 v_{i,k-1}^2\right), \text{(B4)}$$

Matrices $A_j \, \forall (j \geq 1)$ are stable. Therefore, the j-element of each channel is stable and detectable. Applying assumptions made above, we have $v_{i,k-1}^2 \in EC_{\underline{\alpha}_{v_{i,k-1}}, \overline{\alpha}_{v_{i,k-1}}}$, and

$$\left(d_{k-1}(y_{i,k-1} + d_{k-2} v_{i,k-1})\right)^2 \leq 2\omega^2 \left(\overline{\alpha}_{y_{i,k-1}} + \omega^2 \overline{\alpha}_{v_{i,k-1}}\right). \text{(B5)}$$

Then (B2)

$$\dot{V}_k \leq -(\lambda_k - 0.5) V_k + 0.5 \pi_{k,i} \|\Delta A_k\|^2 + 0.5 \|\Delta B_k\|^2 (\overline{\overline{\alpha}}_k + 2\omega^2 \beta_k). \text{(B6)}$$

where $\overline{\overline{\alpha}}_k = 2 \max\left(\overline{\alpha}_{y_{2,k-1}}, \overline{\alpha}_{y_{1,k-1}}\right), \beta_k = \max_i \left(\overline{\alpha}_{y_{i,k-1}} + \omega^2 \overline{\alpha}_{v_{i,k-1}}\right).$
If variables of the system have the property CE and

$$0.5 \pi_{k,i} \|\Delta A_k\|^2 + 0.5 \|\Delta B_k\|^2 \left(\overline{\overline{\alpha}} + 2\omega^2 (\overline{\overline{\alpha}}_k + 2\omega^2 \beta_k)\right) \leq (\lambda_k - 0.5) V_k, \text{(B7)}$$

then the system (13.7) is stable, and, therefore, identifiable by $\Delta A_k, \Delta B_k$.

APPENDIX C

Theorem 3 proof. $\dot{V}_k(t)$has the form (B1), and $v_{i,k-1}$

$$v_{i,k-1} = \begin{cases} \dot{y}_{2,k-1} + \ddot{v}_{2,k-1}, & i = 1, \\ \dot{y}_{1,k-1} + \ddot{v}_{1,k-1}, & i = 2. \end{cases} \quad \text{(C1)}$$

As Theorem 2 condition 4) fulfills, the system (13.2) is recoverable. Hence, the derivatives in (C1) exist and are bounded, $\dot{y}_{i,k-1}(t) \in EC_{\underline{\alpha}_{\dot{y}_{i,k-1}},\overline{\alpha}_{\dot{y}_{i,k-1}}}$, $\ddot{v}_{2,k-1} \in EC_{\underline{\alpha}_{\ddot{v}_{i,k-1}},\overline{\alpha}_{\ddot{v}_{i,k-1}}}$. Then $u_{i,k}^2 \le \overline{\alpha}_{\dot{y}_{i,k-1}} + \overline{\alpha}_{\ddot{v}_{i,k-1}}$ and (B2):

$$\dot{V}_k \le -(\lambda_k - 0.5)V_k + 0.5\pi_{k,i}\|\Delta A_k\|^2 + 0.5\|\Delta B_k\|^2\left(\overline{\overline{\alpha}}_{k,\dot{y}} + 2(\overline{\alpha}_{\dot{y}} + \overline{\overline{\alpha}}_{\ddot{v}})\right), \quad \text{(C2)}$$

where $\overline{\overline{\alpha}}_{k,\dot{y}} = 2\max_i \overline{\alpha}_{\dot{y}_{i,k-1}}$ $\overline{\overline{\alpha}}_{k,\ddot{v}} = 2\max_i \overline{\alpha}_{\ddot{v}_{i,k-1}}$.
If variables of the system have the property CE and

$$0.5\pi_{k,i}\|\Delta A_k\|^2 + 0.5\|\Delta B_k\|^2\left(\overline{\overline{\alpha}}_{k,\dot{y}} + 2(\overline{\alpha}_{\dot{y}} + \overline{\overline{\alpha}}_{\ddot{v}})\right) \le (\lambda_k - 0.5)V_k,$$

then the system (13.7) is stable, and, therefore, identifiable by $\Delta A_k, \Delta B_k$.

APPENDIX D

Getting representation (13.13). Transfer function for the first equation of the system (13.1)

$$y_{1,1} = C_1^T\left(sI_{q_1} - A_1\right)^{-1} B_1 v_{11}, \text{ (D1)}$$

where $I_{q_1} \in R^{q_1 \times q_1}$ is the unit matrix, $s = d/dt$.

Let A_1 be the Frobenius matrix with a vector of parameters $A_{1,s} = \left[a_{1,s,1}, a_{1,s,2}, \ldots, a_{1,s,q_1}\right]^T$.

Divide the left and right parts (B1) by the polynomial $v(s) = \prod_{i=1}^{q_1-1}(s+\mu_i)$, where $\mu_i > 0$, and obtain the identification representation for $y_{1,1}$.

$$y_{1,1}(s) = s^{-1}\left\{ b_1 v_{1,1}(s) - a_{1,1} y_{1,1}(s) + \sum_{i=2}^{q}(b_{i,i} v_{1,1}(s) + a_{1,i} y_{1,1}(s))\frac{1}{s+\mu_i} \right\} \text{ (D2)}$$

or

$$\dot{y}_{1,1} = -a_{1,1} y_{1,1} + \sum_{m=2}^{q_1}\left(a_{1,m} p_{y_{1,1,m}} + b_{1,m} p_{v_{i,1,m}}\right) + b_{1,s} v_{1,1}, \text{ (D3)}$$

$$\dot{p}_{y_{1,1,m}} = -\mu_i p_{y_{1,1,m}} + y_{1,1},\ \dot{p}_{v_{i,1,m}} = -\mu_i p_{v_{i,1,m}} + v_{1,1},$$

Let $P_{1,1} \triangleq \left[y_{1,1}, p_{y_{1,1,1}}, \ldots, p_{y_{1,1,q_1}}; _{1,1}, v_{1,1}, p_{v_{1,1,1}}, \ldots, p_{v_{1,1,q}}\right]^T$. Then

$$\dot{y}_{1,1} = \bar{A}_{1,1}^T P_{1,1},\ \bar{A}_{1,1}^T = \left[-a_{1,1}, a_{1,1,2}, \ldots, a_{1,1,q_1}; b_{1,s}, b_{1,2}, \ldots, b_{1,q_1}\right]. \text{ (D4)}$$

APPENDIX E

Theorem 4 proof. Consider the LF

$$\tilde{V}_{1,e}(e_{1,1}) = 0.5e_{1,1}^2, \tilde{V}_{1,\Delta}(\Delta\bar{A}_{1,1}) = 0.5\Delta\bar{A}_{1,1}^T\Gamma_1^{-1}\Delta\bar{A}_{1,1} \text{ (E1)}$$

for the system (13.16), (13.21).

Select the element corresponding to v_{11} in $P_{1,1}$. Let $v_{11} \in EC_{\underline{\alpha}_g,\bar{\alpha}_g} \backslash \overline{EC}_{\underline{\alpha}_y,\bar{\alpha}_y}$, i.e. the variable v_{11} is not the constant excitable (the property $EC_{\underline{\alpha}_g,\bar{\alpha}_g}$ depends on the variable $y_{1,n}$). Therefore, $v_{11} \notin EC_{\underline{\alpha}_v,\bar{\alpha}_v}$. Then, (13.16) is represented as:

$$\dot{e}_{1,1} = -\chi_1 e_{1,1} + \Delta\bar{A}_{1,1}^T P_{1,1} + f(\Delta\bar{A}_{1,1}, p_{v_{11}}), \text{ (E2)}$$

where $f(\cdot)$ is uncertainty caused by CE non-fulfilment for v_{11}.

Remark 9. $v_{11} \in EC_{\underline{\alpha}_g,\bar{\alpha}_g} \backslash \overline{EC}_{\underline{\alpha}_y,\bar{\alpha}_y}$ is not a set operation. This is the operation of properties. Then

$$\begin{aligned}\dot{\tilde{V}}_{1,e} &= -\chi_1 e_{1,1}^2 + e_{1,1}(\Delta\bar{A}_{1,1}^T P_{1,1} + f) \le -\chi_1 e_{1,1}^2 + |e_{1,1}\Delta\bar{A}_{1,1}^T P_{1,1}| + |e_{1,1}f| \le \\ &\le -\frac{\chi_1}{2}e_{1,1}^2 + \frac{1}{\chi_1}\Delta\bar{A}_{1,1}^T P_{1,1}P_{1,1}^T\Delta\bar{A}_{1,1} + \frac{1}{\chi_1}f^2 \le -\frac{\chi_1}{2}e_{1,1}^2 + \frac{1}{\chi_1}\bar{\pi}_{P_{1,1}}\|\Delta\bar{A}_{1,1}\|^2 + \frac{1}{\chi_1}f^2,\end{aligned} \text{ (E3)}$$

where $\|P_{1,1}P_{1,1}^T\| \le \bar{\pi}_{P_{1,1}} < \bar{\alpha}_{P_{1,1}}$. f is limited as f contains a component that is uncompensated by the adaptive algorithm. Therefore $|f|^2 \le \alpha_f$, where $\alpha_f \ge 0$ and $\alpha_f \ne \bar{\alpha}_{v_{1,1}}$. So:

$$\dot{\tilde{V}}_{1,e} \le -\frac{\chi_1}{2}\tilde{V}_{1,e} + \frac{2}{\chi_1}\bar{\pi}_{P_{1,1}}\tilde{V}_{1,\Delta} + \frac{\alpha_f}{\chi_1}. \text{ (E4)}$$

Consider $\dot{\tilde{V}}_{1,\Delta} = -e_{1,1}\Delta\bar{A}_{1,1}^T P_{1,1}$. Let $\exists\upsilon \le 1$ exist for $t >> t_0$, which is true:

$$e_{1,1}\Delta\bar{A}_{1,1}^T P_{1,1} = \upsilon(e_{1,1}^2 + \Delta\bar{A}_{1,1}^T P_{1,1}P_{1,1}^T\Delta\bar{A}_{1,1}),$$

and $\lambda_{\max}^{-1}(\Gamma_1)\|\Delta\bar{A}_{1,1}\|^2 \le \tilde{V}_{1,\Delta} \le \lambda_{\min}^{-1}(\Gamma_1)\|\Delta\bar{A}_{1,1}\|^2$, where $\lambda_{\min}(\Gamma_1)$, $\lambda_{\max}(\Gamma_1)$ are the minimum and maximum eigenvalues of the matrix Γ_1. Then

$$\Delta\bar{A}_{1,1}^T P_{1,1}P_{1,1}^T\Delta\bar{A}_{1,1} \ge \underline{\pi}_{P_{1,1}}\|\Delta\bar{A}_{1,1}\|^2 \ge \underline{\pi}_{P_{1,1}}\lambda_{\min}(\Gamma_1)\tilde{V}_{1,\Delta} \text{ (E5)}$$

and

$$\dot{\tilde{V}}_{1,\Delta} = -e_{1,1}\Delta\bar{A}_{1,1}^T\Gamma_1 P_{1,1} \le -2\upsilon\tilde{V}_{1,e} - 2\upsilon\eta\tilde{V}_{1,\Delta}, \text{ (E6)}$$

where $\underline{\pi}_{P_{1,1}} \ge 0$, $\eta = \underline{\pi}_{P_{1,1}}\lambda_{\min}(\Gamma_1)$.

Transform the right side (E6)

$$\begin{aligned}-2\upsilon\tilde{V}_{1,e} - 2\upsilon\eta\tilde{V}_{1,\Delta} &= -\upsilon(\eta\tilde{V}_{1,\Delta} \pm 2\sqrt{2\eta\tilde{V}_{1,\Delta}\tilde{V}_{1,e}} + 2\tilde{V}_{1,e}) - \upsilon\eta\tilde{V}_{1,\Delta} = \\ &= -\upsilon\left(\sqrt{\tilde{V}_{1,e}} + \sqrt{2\eta\tilde{V}_{1,\Delta}}\right)^2 + 2\upsilon\sqrt{2\eta\tilde{V}_{1,\Delta}\tilde{V}_{1,e}} - \upsilon\eta\tilde{V}_{1,\Delta} \le \\ &\le -\upsilon\eta\tilde{V}_{1,\Delta} + 2\upsilon\sqrt{2\eta\tilde{V}_{1,\Delta}\tilde{V}_{1,e}}.\end{aligned}$$

As $-2\upsilon\eta\tilde{V}_{1,\Delta} - 2\upsilon\tilde{V}_{1,e} \leq -\upsilon\eta\tilde{V}_{1,\Delta} + 2\upsilon\sqrt{2\eta\tilde{V}_{1,\Delta}\tilde{V}_{1,e}}$, then (E6)

$\dot{\tilde{V}}_{1,\Delta} \leq -\upsilon\eta\tilde{V}_{1,\Delta} + 2\upsilon\sqrt{2\eta\tilde{V}_{1,\Delta}\tilde{V}_{1,e}}$. (E7)

Apply the inequality $-az^2 + bz \leq -\frac{-az^2}{2} + \frac{b^2}{2a}$, $a > 0,\ b \geq 0,\ z \geq 0$ and obtain

$\dot{\tilde{V}}_{1,\Delta} \leq -\upsilon\eta\tilde{V}_{1,\Delta} + 2\upsilon\sqrt{2\eta\tilde{V}_{1,\Delta}\tilde{V}_{1,e}} \leq -\frac{\upsilon\eta}{2}\tilde{V}_{1,\Delta} + 4\upsilon\tilde{V}_{1,e}$. (E8)

Functions $\dot{\tilde{V}}_{1,e} \in \mathrm{M}^+$, $\dot{\tilde{V}}_{1,\Delta} \in \mathrm{M}^+$ and we have the system of inequalities for the system (13.16), (13.21)

$$\underbrace{\begin{bmatrix} \dot{\tilde{V}}_{1,e} \\ \dot{\tilde{V}}_{1,\Delta} \end{bmatrix}}_{} \leq \underbrace{\begin{bmatrix} -\frac{\chi_1}{2} & 2\frac{\pi_{P_{1,1}}}{\chi_1} \\ 4\upsilon & -\frac{\upsilon\eta}{2} \end{bmatrix}}_{A_{S_1}} \underbrace{\begin{bmatrix} \tilde{V}_{1,e} \\ \tilde{V}_{1,\Delta} \end{bmatrix}}_{\tilde{V}_{1,e,\Delta}} + \underbrace{\begin{bmatrix} \frac{\alpha_f}{\chi_1} \\ 0 \end{bmatrix}}_{B_{S_1}}. \text{ (E9)}$$

Comparison system (CS) for (E9): $\dot{S}_1(t) = A_{S_1} S_1(t)$, where $S_1(t) = \left[s_{1,e}(t), s_{1,\Delta}(t)\right]^T$, $s_{1,w}(t)$ ($w = e, \Delta$) is the majorant for $V_{1,w}(t)$, and $s_{1,w}(t_0) \geq V_{1,w}(t_0)$. CS is stable if (Gantmacher, 2000) $(-1)^i \Delta_i\left(A_{S_1}\right) > 0$, where $\Delta_i\left(A_{S_1}\right)$ is main i-minor of the matrix A_{S_1}.

The condition for stability has the form $\chi_1^2\eta > 32\pi_{P_{1,1}}$. We get the estimate from (E9):

$\tilde{V}_{1,e,\Delta}(t) \leq e^{A_{S_1}(t-t_0)} S_1(t_0) + \int_{t_0}^{t} e^{A_{S_1}(t-\tau)} B_{S_1}\, d\tau$. (E10)

Exponential dissipation of the system (13.14) - (13.16) and (13.22) follows from (E10).

Chapter 14
Adaptive Identification of Decentralized Systems

ABSTRACT

In this chapter, we consider the identification problem of decentralized control systems (DS). Analysis shows that this problem has not been given sufficient attention. The complexity of systems and a priori uncertainty require the development of approaches and methods. The primary attention is paid to the problem of developing control algorithms. Assumptions about the system are considered. Adaptive algorithms are used to compensate for uncertainties. We consider the adaptive identification problem of DS with nonlinearities (NDS) for which the quadratic condition is satisfied. Next, a solution to the parametric identifiability (PI) issue is given for NDS. We study the influence of the information space on PI. The approach to the synthesis of adaptive identification algorithms based on the second Lyapunov method is proposed. Properties of parametric algorithms and signaling algorithms are analyzed. We proof the exponential dissipativity of the adaptive identification system

1. PROBLEM STATEMENT

The system comprising m interconnected subsystems consider

$$S_i:\begin{cases} \dot{X}_i = A_i X_i + B_i u_i + \sum_{j=1, j\neq i}^{m} \bar{A}_{ij} X_j + F_i(X_i), \\ Y_i = C_i X_i, \end{cases} \tag{14.1}$$

where $X_i \in \mathbb{R}^{n_i}$, $Y_i \in \mathbb{R}^{q_i}$ are vectors of the state and output of S_i-subsystem, $u_i \in \mathbb{R}$ is control, $i = \overline{1,m}$, $\sum_{i=1}^{m} n_i = n$. Parameters of the matrices $A_i \in \mathbb{R}^{n_i \times n_i}$, $B_i \in \mathbb{R}^{n_i}$, $\bar{A}_{ij} \in \mathbb{R}^{n_i \times n_j}$ are unknown, $C_i \in \mathbb{R}^{q_i \times n_i}$. The matrix $\bar{A}_{ij}$ reflects the mutual

DOI: 10.4018/979-8-3373-0822-7.ch014

influence of the S_j subsystem. $F_i(X_i) \in \mathbb{R}^{n_i}$ consider the nonlinear state of the S_i-subsystem, and A_i is Hurwitz matrix (stable).

Assumption 1. $F_i(X_i)$ belongs to the class

$$N_F(\pi_1,\pi_2) = \{F(X)\in \mathbb{R}^n:\pi_1 X \leq F(X)\leq \pi_2 X, F(0)=0\} \tag{14.2}$$

and satisfies the quadratic condition

$$(\pi_2 X - F(X))^T(F(X) - \pi_1 X) \geq 0 \tag{14.3}$$

where $\pi_1 > 0,\ \pi_2 > 0.$

Information set of measurements for subsystems

$$\mathbb{I}_{o,i} = \{X_i(t),u_i(t),X_j(t),t \in \mathbb{J} = [t_0,t_k]\}.$$

Mathematical model for (14.1)

$$\dot{\hat{X}}_i = K_i(\hat{X}_i - X_i) + \hat{A}_i X_i + \hat{B}_i u_i + \sum_{j=1,j\neq i}^{m} \hat{A}_{ij} X_j + \hat{F}_i(X_i) \tag{14.4}$$

where $K_i \in H$ is Hurwitz matrix with known parameters (reference model); $\hat{A}_i$, $\hat{B}_i$, $\hat{A}_{ij}$ are tuning matrices of corresponding dimensions, $\hat{F}_i$ is a priori given nonlinear vector function.

Problem: find algorithms for estimating model (14.4) parameters based on the analysis of the set $\mathbb{I}_{o,i}$ and the assumption 1 fulfilment, so

$$\lim_{t\to\infty}\|\hat{X}_i(t)-X_i(t)\| \leq \delta_i$$

where $\delta_i \geq 0.$

2. ON IDENTIFIABILITY S_i-SUBSYSTEM

Identifiability is the basis for estimating S_i-subsystem parameters. It knows that fulfilment is the constant excitation (CE) condition for $\mathbb{I}_{o,i}$ guarantees identifying DS parameters.

The CE conditions

$$EC_{\underline{\alpha}_{u_i},\overline{\alpha}_{u_i}}:\underline{\alpha}_{u_i} \leq u_i^2(t)\leq \overline{\alpha}_{u_i}\ \forall t \in [t_0,t_0 + T] \tag{14.5}$$

where $\underline{\alpha}_{u_i}, \overline{\alpha}_{u_i}$ are positive numbers, $T > 0$. Next, condition (14.5) will be written as $u_i(t) \in EC_{\underline{\alpha}_{u_i}, \overline{\alpha}_{u_i}}$. If $u_i(t)$ does not have the CE property, then we will write $u_i(t) \notin EC_{\underline{\alpha}_{u_i}, \overline{\alpha}_{u_i}}$ or $u_i(t) \notin EC$.

Remark 1. Condition (14.5) requires modification, considering S-synchronization for NDS (Karabutov, 2021). We write property (14.5) as:

$$EC^S_{\underline{\alpha}_{u_i}, \overline{\alpha}_{u_i}}: \ \left(\underline{\alpha}_{u_i} \le u_i^2(t) \le \overline{\alpha}_{u_i}\right) \ \& \ \left(\Omega_{u_i}(\omega) \subseteq \Omega_S(\omega)\right)$$

where $\Omega_{u_i}(\omega)$ is the set of frequencies for u_i; $\Omega_S(\omega)$ is the set of acceptable frequencies for u_i, guaranteeing S-synchronizability. Next, we will denote the CE property as $EC_{\underline{\alpha}_{u_i}, \overline{\alpha}_{u_i}}$, assuming that it guarantees $EC^S_{\underline{\alpha}_{u_i}, \overline{\alpha}_{u_i}}$.

To obtain conditions of identifiability, consider the model (14.4). The error equation:

$$\dot{E}_i = K_i E_i + \Delta A_i X_i + \Delta B_i u_i + \sum_{j=1, j\neq i}^{M} \Delta \overline{A}_{ij} X_j + \Delta F_i(X_i) \tag{14.6}$$

where $E_i = \hat{X}_i - X_i$; $\Delta A_i = \hat{A}_i - A_i$, $\Delta B_i = \hat{B}_i - B_i$, $\Delta \overline{A}_{ij} = \hat{\overline{A}}_{ij} - \overline{A}_{ij}$, $\Delta F_i = \hat{F}_i - F_i$ are parametric residuals.

Lemma 1. *If the nonlinearity* $F(X) \in \mathbb{R}^n$, $X \in \mathbb{R}^n$ belongs to the class $N_F(\pi_1, \pi_2)$ and

$$\pi_1 X \le F(X) \le \pi_2 X, \tag{14.7}$$

then

$$\|F(X)\|^2 \le \eta \overline{\alpha}_X \tag{14.8}$$

where $\pi_1 > 0$, $\pi_2 > 0$, $\eta = \eta(\pi_1, \pi_2) > 0$, $\overline{\alpha}_X = \overline{\alpha}_X(X) > 0$.

Lemma 1 proof gives in Appendix A.

Lemma 2. *If Lemma 1 conditions are satisfied, then the estimate*

$$\Delta F^T \Delta F \le 2\eta \overline{\alpha}_X + \delta_F$$

is valid for $\Delta F(X)$, where $\eta = 2\overline{\pi} + \pi^2$, $\overline{\pi} = \pi_1 \pi_2$, $\pi = \pi_1 + \pi_2$, $\delta_F > 0$.

Consider the system (14.6) and Lyapunov function (LF) $V_i(E_i) = 0.5E_i^T R_i E_i$, where $R_i = R_i^T > 0$ is a positive symmetric matrix. Let $\|\Delta A_i\| = \sqrt{\mathrm{Sp}(\Delta A_i^T \Delta A_i)}$, $\|\Delta \bar{A}_{ij}\| = \sqrt{\mathrm{Sp}(\Delta A_{ij}^T \Delta A_{ij})}$ are matrix norms ΔA_i, $\Delta \bar{A}_{ij}$.

Lemma 2 proof gives in Appendix B.

Theorem 1. *Let 1)* $A_i \in H$; 2) $X_i(t) \in PE_{\underline{\alpha}_{X_i}, \bar{\alpha}_{X_i}}$, $X_j(t) \in PE_{\underline{\alpha}_{X_j}, \bar{\alpha}_{X_j}}$ $u_i(t) \in PE_{\underline{\alpha}_{u_i}, \bar{\alpha}_{u_i}}$; 3) conditions of Lemma 1, 2 are satisfied for $F_i(X_i)$. Then the subsystem (14.1) is identifiable on the set $\mathbb{I}_{o,i}$ if

$$2\left(\bar{\alpha}_{X_i}\|\Delta A_i\|^2 + \bar{\alpha}_{u_i}\|\Delta B_i\|^2 + \sum_{j=1, j\neq i}^{m} \bar{\alpha}_{X_j}\|\Delta \bar{A}_{ij}\|^2 + 2\eta\bar{\alpha}_{X_i} + \delta_{F_i}\right) \leq \bar{\lambda}_i V_i \tag{14.9}$$

where $\bar{\lambda}_i = \lambda_i - k_i, \lambda_i > 0$ *is the minimum eigenvalue of the matrix* Q_i, $k_i > 0$, $K_i R_i + K_i^T R_i = -Q_i$, Q_i *is positive symmetric matrix,* $\eta = 2\bar{\pi} + \pi^2, \pi = \pi_1 + \pi_2, \bar{\pi} = \pi_1 \pi_2$, $\delta_{F_i} \geq 0$.

Theorem 1 proof gives in Appendix C.

If Theorem 1 conditions are fulfilled, then the subsystem S_i is identifiable on the set $\mathbb{I}_{o,i}$ or PI_{X_i}-identifiable. Consider the identifiability of the S_i-subsystem on the set:

$$\mathbb{I}_{o,Y_i} = \left\{Y_i(t), u_i(t), Y_j(t), t \in \mathbb{J} = [t_0, t_k]\right\}.$$

Representation of the S_i-subsystem on $\mathbb{I}_{o,Y_i}$

$$\dot{Y}_i = A_{\#,i}Y_i + B_{\#,i}u_i + \sum_{j=1, j\neq i}^{m} \bar{A}_{\#,ij}Y_j + \tilde{C}_i^{\#} F_i(\tilde{C}_i Y_i), \tag{14.10}$$

where $\tilde{C}_i = (C_i^T C_i)^{\#} C_i^T$, $A_{\#,i} = \tilde{C}_i^{\#} A_i \tilde{C}_i$, $B_{\#,i} = \tilde{C}_i^{\#} B_i$, $\bar{A}_{\#,ij} = \tilde{C}_i^{\#} \bar{A}_{ij} \tilde{C}_j$, # is sign of a pseudo-inreversal of the matrix. The model for (14.10) has a similar structure.

Introduce the error $E_{\#,i} = \hat{Y}_i - Y_i$ and LF $V_{\#,i}(E_{\#,i}) = 0.5E_{\#,i}^T R_i E_{\#,i}$.

Theorem 2. *Let 1)* $A_{\#,i} \in H$; 2) $Y_i(t) \in EC_{\underline{\alpha}_{Y_i}, \bar{\alpha}_{Y_i}}$, $Y_j(t) \in EC_{\underline{\alpha}_{Y_j}, \bar{\alpha}_{Y_j}}$ $u_i(t) \in EC_{\underline{\alpha}_{u_i}, \bar{\alpha}_{u_i}}$; 3) $F_i(X_i)$ satisfies conditions of Lemmas 1 and 2; 4) the S_i subsystem is observable. Then the subsystem (14.1) is identifiable on the set $\mathbb{I}_{o,Y_i}$ if

$$2\left(\bar{\alpha}_{Y_i}\|\Delta A_{\#,i}\|^2 + \bar{\alpha}_{u_i}\|\Delta B_{\#,i}\|^2 + \sum_{j=1, j\neq i}^{m} \bar{\alpha}_{Y_j}\|\Delta \bar{A}_{\#,ij}\|^2 + 2\eta\bar{\alpha}_{Y_i} + \delta_{F_i}\right) \leq \bar{\lambda}_{\#,i} V_{\#,i}$$

where $\Delta A_{\#,i} = \hat{A}_{\#,i} - A_{\#,i}$, $\Delta B_{\#,i} = \hat{B}_{\#,i} - B_{\#,i}$, $\Delta \bar{A}_{\#,ij} = \hat{\bar{A}}_{\#,ij} - \bar{A}_{\#,ij}$, $\Delta F_i = \hat{F}_i - F_i$, $\bar{\lambda}_{\#,i} > 0$.

The Theorem 2 proof is analogically to the Theorem 1 proof.

3. SYNTHESIS OF ADAPTATION ALGORITHMS

Consider LF $V_i(E_i) = 0.5 E_i^T R_i E_i$ and

$$\dot{V}_i = -E_i^T Q_i E_i + E_i^T R_i \left(\Delta A_i X_i + \Delta B_i u_i + \sum_{j=1, j \neq i}^{m} \Delta \bar{A}_{ij} X_j + \Delta F_i(X_i) \right)$$

We require that the functional constraint be satisfied for all $\forall t \geq t_0$:

$$\dot{V}_i \leq -\chi(\Delta A_i, \Delta \tilde{A}_i, \Delta F_i)$$

Where:

$$\chi(\Delta A_i, \Delta \bar{A}_i, \Delta B_i, \Delta F_i) = 0.5\Big(\varphi_{A_i}(t) \, \|\Delta A_i(t)\|^2 + \varphi_{\bar{A}_i}(t) \, \|\Delta \bar{A}_i(t)\|^2 +$$
$$\varphi_{B_i}(t) \, \|\Delta B_i\|^2 + \varphi_{F_i}(t) \, \|\Delta F_i(t)\|^2 \Big)$$

$\varphi_{A_i}(t)$, $\varphi_{\bar{A}_i}(t)$, $\varphi_{B_i}(t)$, $\varphi_{F_i}(t)$ are limited non-negative functions. Then:

$$\dot{V}_i$$

$$= -E_i^T Q_i E_i + \chi(\Delta A_i, \Delta \bar{A}_i, \Delta F_i) + E_i^T R_i \left(\Delta A_i X_i + \Delta B_i u_i + \sum_{j=1, j \neq i}^{m} \Delta \bar{A}_{ij} X_j + \Delta F_i(X_i) \right) \quad (14.11)$$

From (14.11), we obtain adaptive algorithms:

$$\begin{aligned} \Delta \dot{A}_i &= -\Gamma_{A_i}\left(\varphi_{A_i}(t) \Delta A_i + E_i^T R_i X_i \right), \\ \Delta \dot{\bar{A}}_{ij} &= -\Gamma_{\bar{A}_{ij}}\left(\varphi_{\bar{A}_{ij}}(t) \Delta \bar{A}_{ij} + E_i^T R_i X_j \right), \\ \Delta \dot{B}_i &= -\Gamma_{B_i}\left(\varphi_{B_{ij}}(t) \Delta B_i + E_i^T R_i u_i \right), \end{aligned} \quad (14.12)$$

where $\Gamma_{A_i}, \Gamma_{\bar{A}_{ij}}, \Gamma_{B_i}$ are diagonal matrices of corresponding dimensions with positive diagonal elements, ensuring the stability of adaptation processes.

3.1. Parametric Algorithm for F_i

Parametric and signal algorithms can be used to evaluating F_i. Consider the parametric approach (Karabutov, 2018).

Assumption 2. The function $F_i(X_i)$ is given on the set

$$F_i \in \mathbb{F}_{F_i} = \{F_i \in N_F(\pi_1,\pi_2):F_i(X_i) = \tilde{F}_i^T(X_i,N_{i,1})N_{i,2}, N_i = [N_{i,1}^T, N_{i,2}^T]^T, N_i \in \mathbb{N}_a\} \tag{14.13}$$

where $N_{(i,a)} = \{N_i \in R^n:\underline{N}_i \leq N_i \leq \overline{N}_i\}$is a posteriori generated parametric domain for F_i; $\underline{N},\overline{N}$ are vectors boundaries for N, understood as $\underline{n}_i \in \underline{N}_i,\overline{n}_i \in \overline{N}_i$; $N_{i,1}$ is a priori set vector of nonlinearity parameters, $N_{i,2}$ is a priori an unknown vector of parameters, which we consider as a vector to be evaluated. Some elements of $\underline{N}_i$,$\overline{N}_i$ may be unknown. The structure $\tilde{F}_i(X_i,N_{i,1})$ is formed a priori considering the known vector $N_{i,1}$.

As follows from (14.13), the estimation for function $F_i(X_i)$ is defined as:

$$\hat{F}_i(X_i)=\tilde{F}_i^T(X_i,\hat{N}_{i,1})\hat{N}_{i,2} \tag{14.14}$$

where $\hat{N}_{i,1} \in \mathbb{R}^{n_{i,1}}$ is a priori estimation of known parameters, $\hat{N}_{i,2} \in \mathbb{R}^{n_{i,2}}$ is vector of tuning parameters.

We believe $\mathbb{N}_i = \mathbb{N}_{i,1} \cup \mathbb{N}_{i,2}$. The set $\mathbb{N}_{i,1} \subset \mathbb{R}^{n_{i,1}} (N_{i,1} \in \mathbb{N}_{i,1})$ contains elements that are not available for adjusting. Estimate elements $N_{i,2} \in \mathbb{N}_{i,2} \subset \mathbb{R}^{n_{i,2}}$ at the identification stage. The matrix $\tilde{F}_i(y,\hat{N}_{i,1})$ is formed at the stage of structural synthesis (analysis) of the system. Representation (14.14) is a consequence of the proposed parametric concept for $F_i(X_i)$.

Remark 2. The vector $\hat{N}_{i,1}$ can be adjusted iteratively based on the coercion algorithm (Karabutov, 2018).

As $\Delta F = \tilde{F}_i^T(y,\hat{N}_{i,1})\hat{N}_{i,2} - F_i(X_i)$, then we get an adaptive algorithm for $\hat{N}_{i,2}$ from the condition $\dot{V}_i \leq 0$:

$$\dot{\hat{N}}_{i,2} = -\Gamma_{F_i}\left(\varphi_{F_i}\dot{\hat{N}}_{i,2} + \tilde{F}_i^T R_i E_i(X_i,\hat{N}_{i,1})\right) \tag{14.15}$$

where Γ_{F_i} is a diagonal matrix with positive diagonal elements. Designate the system (14.6), (14.11), and (14.15) as AS_{AF_i}.

3.2. Signal Algorithm

Consider the model

$$\dot{\hat{X}}_i = K_i(\hat{X}_i - X_i) + \hat{A}_i X_i + \hat{B}_i u_i + \sum_{j=1, j\neq i}^{m} \hat{\bar{A}}_{ij} X_j + U_i \quad (14.16)$$

and the equation for the error

$$\dot{E}_i = K_i E_i + \Delta A_i X_i + \Delta B_i u_i + \sum_{j=1, j\neq i}^{m} \Delta \bar{A}_{ij} X_j + U_i(X_i) - F_i(X_i) \quad (14.17)$$

Then

$$\dot{V}_i = -E_i^T Q_i E_i + E_i^T R_i \left(\Delta A_i X_i + \Delta B_i u_i + \sum_{j=1, j\neq i}^{m} \Delta \bar{A}_{ij} X_j + U_i - F_i \right) \quad (14.18)$$

Choose the algorithm for U_i in the form:

$$U_i = -D_i R_i E_i \quad (14.19)$$

where $D_i \in \mathbb{R}^{n_i \times n_i}$ is a diagonal matrix with positive diagonal elements. As $F_i(X_i) \in N_F(\pi_1, \pi_2)$, then Lemma 1 is valid for $F_i(X_i)$.

Apply the approach, proposing by Gromyko and Sankovsky (1974). Select matrix D_i elements from the condition $\|D_i\| \geq d_i \geq \eta \bar{\alpha}_{X_i}$. Then (14.18):

$$\dot{V}_i = -E_i^T Q_i E_i + E_i^T R_i \left(\Delta A_i X_i + \Delta B_i u_i + \sum_{j=1, j\neq i}^{m} \Delta \bar{A}_{ij} X_j - D_i R_i E_i - F_i \right) \quad (14.20)$$

As $E_i^T R_i D_i R_i E_i \geq 2\eta \underline{\alpha}_{X_i} \underline{\lambda}_{R_i} V_i$, then

$$\dot{V}_i \leq -\sigma V_i + E_i^T R_i \left(\Delta A_i X_i + \Delta B_i u_i + \sum_{j=1, j\neq i}^{m} \Delta \tilde{A}_{ij} X_j \right)$$

where $\underline{\lambda}_{R_i}$ is the smallest eigenvalue of the matrix R_i, $\sigma = \lambda_{Q_i} + 2\eta \bar{\alpha}_{X_i} \underline{\lambda}_{R_i}$. Apply the approach described in the appendix and get:

$$\dot{V}_i \leq -\sigma V_i + 2\left(\overline{\alpha}_{X_i} \|\Delta A_i\|^2 + \overline{\alpha}_{u_i} \|\Delta B_i\|^2 + \sum_{j=1, j\neq i}^{m} \overline{\alpha}_{X_j} \|\Delta \overline{A}_{ij}\|^2\right)$$

If

$$2\left(\overline{\alpha}_{X_i} \|\Delta A_i\|^2 + \overline{\alpha}_{u_i} \|\Delta B_i\|^2 + \sum_{j=1, j\neq i}^{m} \overline{\alpha}_{X_j} \|\Delta \widetilde{A}_{ij}\|^2 + \eta \overline{\alpha}_{X_i i}\right) \leq \sigma V_i$$

then the system (14.16) is parametrically identifiable on the set $\{u_i(t), X_i(t), X_j(t)\}$ on the algorithms (14.12), (14.19) class, if $(u_i(t), X_i(t), X_j(t)) \in EC$.

Designate the system (14.6), (14.12), (14.19) as AS_{AS_i}.

4. ADAPTIVE ALGORITHM AS DYNAMIC SYSTEM

The aforementioned synthesis method for adaptive algorithms (AA) is typical for adaptive identification systems. Attempts to develop more complex AA are associated with considering the limitations on ASI. This approach requires some knowledge and does not always provide workable algorithms. We propose a method based on consideration of requirements for ASI.

Consider of a functional restriction

$$V_i^{\Delta}(E_i, \Delta A_i, \Delta \dot{A}_i) = 0.5 E_i^T R_i E_i + 0.5 \mathrm{Sp}(\Delta A_i^T \Delta \dot{A}_i)$$

Let

$$\dot{V}_i^{\Delta} \leq -\chi_{\Delta} = -\alpha_{\Delta} \mathrm{Sp}(\Delta A_i^T \Delta A_i)$$

Denote $\eta_{\Delta} = \dot{V}_i^{\Delta} + \chi_{\Delta}$ and obtain

$$\eta_{\Delta} = -E_i^T Q_i E_i + E_i^T R_i \Delta A_i X_i + \mathrm{Sp}(\Delta \dot{A}_i^T \Delta \dot{A}_i) + \mathrm{Sp}(\Delta A_i^T \Delta \ddot{A}_i) + \alpha_{\Delta} \mathrm{Sp}(\Delta A_i^T \Delta A_i).$$

Adaptive algorithm for ΔA_i

$$\Delta \ddot{A}_i = -\Delta \dot{A}_i - \alpha_\Delta \Delta A_i - \Gamma_{A_i} E_i R_i X_i^T$$

Let $\Delta A_i = Z_1$. Then

$$S_{AA} : \begin{cases} \dot{Z}_1 = Z_2, \\ \dot{Z}_2 = -\alpha_\Delta Z_1 - Z_2 - \Gamma_{A_i} E_i R_i X_i^T. \end{cases}$$

So, if the functional restriction $\chi_\Delta \geq 0$ is imposed on ASI, then AA is described by the system S_{AA}.

5. PROPERTIES OF ASI

5.1. System AS_{AF_i}

Consider systems AS_{AF_i}, AS_{AF_j} and LF $V_i(E_i) = 0.5 E_i^T R_i E_i$,

$$V_{\Delta,i} = 0.5 \cdot \mathrm{Sp}\left(\Delta A_i^T \Gamma_i^{-1} \Delta A_i\right) + 0.5 \sum_{j=1}^{m} \mathrm{Sp}\left(\Delta \overline{A}_{ij}^T \Gamma_{ij}^{-1} \Delta \overline{A}_{ij}\right) + 0.5 \Delta B_i^T \Gamma_i^{-1} \Delta B_i + 0.5 \Delta N_{i,2}^T \Gamma_{N_{i,2}}^{-1} \Delta N_{N_{i,2}} \quad (14.21)$$

where Sp(·) is matrix spur. We believe that the interference matrix $\overline{A}_{ij}$ ensures the stability of the S_i-subsystem.

Theorem 3. *Let (i) Lyapunov functions $V_i(t)$, $V_{\Delta,i}(t)$, admit an infinitesimal upper limit; (ii) $A_i \in H$; (iii) $\overline{A}_{ij}$ ensures the stability of the subsystem S_i; (iv) $X_i(t) \in EC_{\underline{\alpha}_{X_i}, \overline{\alpha}_{X_i}}$, $X_j(t) \in EC_{\underline{\alpha}_{X_j}, \overline{\alpha}_{X_j}}$ $u_i(t) \in EC_{\underline{\alpha}_{u_i}, \overline{\alpha}_{u_i}}$; (v) $F_i \in \mathbb{F}_{F_i}$; (vi) the system of inequalities*

$$\begin{bmatrix} \dot{V}_i \\ \dot{V}_{\Delta,i} \end{bmatrix} \leq \underbrace{\begin{bmatrix} -\mu_i & \frac{2}{\mu_i}\kappa_i \\ \vartheta_{\chi\alpha,i}\rho_i & -\beta_{\lambda\chi,i} \end{bmatrix}}_{A_{W_i}} \begin{bmatrix} V \\ V_{\Delta,i} \end{bmatrix} + \underbrace{\begin{bmatrix} \frac{2}{\mu_i}\varpi_{F_i} \\ 0.5\,\delta_{N,i} \end{bmatrix}}_{L_i} \quad (14.22)$$

is valid for the Lyapunov vector-function $W_i = \left[V_i, V_{\Delta,i}\right]^T$, where μ_i, κ_i, $\vartheta_{\chi\alpha,i}\rho_i$, $\beta_{\lambda\chi,i}$, ϖ_{F_i}, $\delta_{N,i}$ are positive numbers depending on the subsystem S_i parameters and set $\mathbb{I}_{o,i}$ properties; (vi) the upper solution for W_i satisfies the system of equation $\dot{S}_{W_i} = A_{W_i} S_{W_i} + L_i$ if

$$w_\rho(t) \leq s_\rho(t)$$

$$\forall (t \geq t_0) \ \& \ (w_\rho(t_0) \leq s_\rho(t_0))$$

$w_\rho \in W_i, s_\rho \in S_{W_i}, \rho = e, i; \Delta, i$. *Then the* $AS_{F,i}$*-system is exponentially dissipative with the estimate:*

$$W_i(t) \leq e^{A_{W_i}(t-t_0)} S_{W_i}(t_0) + \int_{t_0}^{t} e^{A_{W_i}(t-\tau)} L_i d\tau \tag{14.23}$$

if

$$\mu_i^2 \beta_{\lambda\chi,i} \geq 2\kappa_i \vartheta_{\chi\alpha,i} \rho_i \tag{14.24}$$

As follows from (14.23), the system $AS_{F,i}$ limiting properties are determined by vector L_i elements. If the vector $\widehat{N}_{i,1}$ structure and parameters are known, then the $AS_{F,i}$-subsystem is exponentially stable if $\widetilde{F}_i \in EC$.

Consider the system $AS_{F,i,j}$ with subsystems $AS_{F,i}$ and $AS_{F,j}$. We have the system of inequalities for $AS_{F,i,j}$:

$$\begin{bmatrix} \dot{W}_i \\ \dot{W}_j \end{bmatrix} \leq \begin{bmatrix} A_{W_i} & 0 \\ 0 & A_{W_j} \end{bmatrix} \begin{bmatrix} W_i \\ W_j \end{bmatrix} + \begin{bmatrix} L_i \\ L_j \end{bmatrix} \tag{14.25}$$

where A_{W_i} and L_i have the form (14.22). Exponential dissipative conditions:

$$\mu_i^2 \beta_{\lambda\chi,i} \geq 2\kappa_i \vartheta_{\chi\alpha,i} \rho_i, \mu_i^2 \beta_{\lambda\chi,j} \geq 2\kappa_j \vartheta_{\chi\alpha,j} \rho_j.$$

5.2. System AS_{AS_i}

Consider LF $V_i(t)$ and

$V_{\Delta_s,i} = 0.5 \cdot \mathrm{Sp}\left(\Delta A_i^T \Gamma_i^{-1} \Delta A_i\right) + 0.5\sum_{j=1}^{m} \mathrm{Sp}\left(\Delta \overline{A}_{ij}^T \Gamma_{ij}^{-1} \Delta \overline{A}_{ij}\right) + 0.5\Delta B_i^T \Gamma_i^{-1} \Delta B_i + 0.5$
$\int_{t_0} \Delta_{U_i} F_i^T(\tau) \Delta_{U_i} F_i(\tau) d\tau,$

where $\Delta_{U_i} F_i = U_i - F_i$.

Theorem 4. Let (i) Lyapunov functions $V_i(t)$ and $V_{\Delta_s,i}(t)$ have an infinitesimal upper limit; (ii) $A_i \in H$; (iii) $\overline{A}_{ij}$ ensures the stability of the subsystem S_i; (iv) $F_i(X_i) \in N_F(\pi_1,\pi_2)$;(v)$u_i(t) \in EC_{\underline{\alpha}_u,\overline{\alpha}_u}$,$X_i(t) \in EC_{\underline{\alpha}_X,\overline{\alpha}_X}$,$X_j(t) \in EC_{\underline{\alpha}_X,\overline{\alpha}_X}$;(v)exist$\nu_i > 0$ such that the condition $-F_i^T \Delta_{U_i} F_i = -\nu_i\left(\|F_i\|^2 + \|\Delta_{U_i} F_i\|^2\right)$ satisfy at $t >> t_0$ in some area of the origin; (vi) the system of inequalities:

$$\underbrace{\begin{bmatrix} \dot{V}_i \\ \dot{V}_{\Delta_{S,i}} \end{bmatrix}}_{W_{S_i}} \leq \underbrace{\begin{bmatrix} -\mu_i & \frac{2}{\mu_i}\kappa_{\Delta_{S,i}} \\ \vartheta_{\Delta_{S,i}} \rho_i & -\beta_{\Delta_{S,i}} \end{bmatrix}}_{A_{W_{S_i}}} \underbrace{\begin{bmatrix} V_i \\ V_{\Delta_{S,i}} \end{bmatrix}}_{W_{S_i}} + \underbrace{\begin{bmatrix} 0 \\ \sqrt{0.125}\,\nu_i \eta_i \overline{\alpha}_{X_i} \end{bmatrix}}_{L_{S,i}} \tag{14.26}$$

is valid for the Lyapunov vector function $W_{S,i} = \left[V_i, V_{\Delta_s,i}\right]^T$, where $\mu_i, \kappa_{\Delta_s,i}, \vartheta_{\Delta.i}, \rho_i, \beta_{\Delta,i}, \nu_i, \eta_i$ are positive numbers depending on the subsystemS_i parameters and set properties; (vii) the upper solution for W_i satisfies the system of equation $\dot{S}_{W_i} = A_{W_i} S_{W_i} + L_i$ if

$$w_\rho(t) \leq s_\rho(t) \forall (t \geq t_0) \;\&\; \left(w_\rho(t_0) \leq s_\rho(t_0)\right)$$

$\rho = e, i; \Delta_{S_i,i}$ *for elements* W_{S_i}, $w_\rho \in W_{S_i}$, $s_\rho \in S_{W_{S,i}}$. *Then the* AS_{AS_i}*-system is exponentially dissipative with an estimate*

$$W_{S_i}(t) \leq e^{A_{W_{S_i}}(t-t_0)} S_{W_{S_i}}(t_0) + \int_{t_0}^{t} e^{A_{W_{S_i}}(t-\sigma)} L_{S,i} d\sigma \tag{14.27}$$

if $\mu_i^2 \beta_{\Delta,i} \geq 2\vartheta_{\Delta.i} \rho_i \kappa_{\Delta_s,i}$.

As follows from Theorem 4, the AS_{AS_i}-system application gives displaced estimates for the parameters of the S_i-subsystem.

Remark 3. Signaling algorithms (SA) are widely used in adaptive control systems (Podval'ny and Vasil'ev, 2022). The rationale SA is based on ensuring on non-positivity derivative LF. This is a feature of using quadratic LF, which does not fully reflect the specifics of the processes in the system AS_{AS_i}. The Lyapunov function $V_{\Delta_s,i}$ proposed in the paper allows to prove the properties of the adaptive system.

Remark 4. Algorithm (14.19) is a compensating control. Therefore, the term "signal adaptation" reflects only the gain factor in (14.19). In identification systems, the SA use depends on the quality requirements of the identification system.

Remark 5. The analysis of algorithms (14.12) with $\varphi_i = 0$ is based on the results of N. Karabutov (2023).

6. EXAMPLE

Consider the system

$$
\begin{aligned}
&S_1: \begin{cases} \begin{bmatrix} \dot{x}_{11} \\ \dot{x}_{12} \end{bmatrix} = \begin{bmatrix} 0 & 1 \\ -a_{21} & -a_{22} \end{bmatrix} \begin{bmatrix} x_{11} \\ x_{12} \end{bmatrix} + \begin{bmatrix} 0 \\ \bar{a}_1 \end{bmatrix} x_2 + \begin{bmatrix} 0 \\ b_1 \end{bmatrix} u_1 + \begin{bmatrix} 0 \\ c_1 \end{bmatrix} f_1(x_{11}), \\ y_1 = x_{11}, \end{cases} \\
&S_2: \begin{cases} \dot{x}_2 = -a_2 x_2 + \bar{a}_2 x_{11} + b_2 u_2 + c_2 f_1(x_{11}), \\ y_2 = x_2, \end{cases}
\end{aligned}
\tag{14.28}
$$

where $X_1 = \begin{bmatrix} x_{11} & x_{12} \end{bmatrix}^T$ and y_1 is the state vector and output of the subsystem S_1; u_1 is input (control)); $f_1(x_{11}) = \text{sat}(x_{11})$ is saturation function; $f_2(x_2) = \text{sign}(x_2)$ is sign function; y_2 is subsystem S_2 output. System parameters (14.28): $b_1 = 1$, $a_{21} = 2$, $a_{22} = 3$, $\bar{a}_1 = 1.5$, $b_1 = 1$, $c_1 = 1$, $a_2 = 1.25$, $\bar{a}_2 = 0.2$, $b_2 = 1$, $c_2 = 0.25$. Inputs $u_i(t)$ are sinusoidal.

Since the variable x_{12} is not measured, the subsystem S_1 is converted to a form where only observable variables (Karabutov, 2023) are used. Representation in the input-output space for S_1:

$$
\dot{y}_1 = -\alpha_1 y_1 + \alpha_2 p_{y_1} + \beta_{12} p_{x_2} + b_1 p_{u_1} + c_1 p_{f_1} \tag{14.29}
$$

where α_1, α_2, β_{12}, b_1, c_1 are unknown coefficients; $\mu > 0$,

$$
\begin{aligned}
\dot{p}_{y_1} &= -\mu p_{y_1} + y_1, \quad \dot{p}_{x_2} = -\mu p_{x_2} + x_2, \\
\dot{p}_{u_1} &= -\mu p_{u_1} + u_1, \quad \dot{p}_{f_1} = -\mu p_{f_1} + f_1.
\end{aligned}
\tag{14.30}
$$

We present the phase portrait for S_1 in Figure 1. Processes in S_1 are nonlinear. There is a relationship between y_1 and y_2 (the determination coefficient is 75%). This reflects in properties of the subsystem S_1 (see Figure 1). In particular, y_2 is affected by S-synchronizability and parameter estimations of parameters. Apply the approach (Karabutov, 2021) and get that the S_1 subsystem is structurally identifiable.

Figure 1. Phase portrait of subsystem S_1

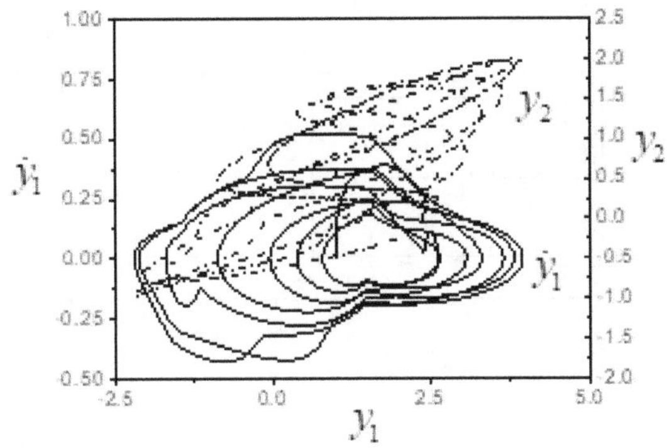

Models for subsystems S_1 and S_2:

$$\dot{\hat{y}}_1 = -k_1 e_1 + \hat{a}_{11} y_1 + \hat{a}_{12} p_{y_1} + \hat{\beta}_{12} p_{x_2} + \hat{b}_1 p_{u_1} + \hat{c}_1 p_{f_1} \tag{14.31}$$

$$\dot{\hat{y}}_2 = -k_2 e_2 + \hat{a}_2 y_2 + \hat{\bar{a}}_2 y_1 + \hat{b}_2 u_2 + \hat{c}_1 f_2 \tag{14.32}$$

where k_1, k_2 are a priori set positive numbers (reference model); $e_1 = \hat{y}_1 - y_1$, $e_2 = \hat{y}_2 - y_2$ are identification errors; $\hat{a}_i$, $\hat{\bar{a}}_i$, $\hat{b}_i$, $\hat{c}_i$ are tuning parameters.

Apply algorithms (14.12) with $\varphi_i = 0$:

$$\begin{aligned} &\dot{\hat{a}}_{11} = -\gamma_{a_{11}} e_1 y_1,\ \dot{\hat{a}}_{12} = -\gamma_{a_{12}} e_1 p_{y_1},\ \dot{\hat{\beta}}_{12} = -\gamma_{\beta_{12}} e_1 p_{x_2}, \\ &\hat{b}_1 = -\gamma_{b_1} e_1 p_{u_1},\ \hat{c}_1 = -\gamma_{c_1} e_1 p_{f_1}, \end{aligned} \tag{14.33}$$

$$\begin{aligned} &\dot{\hat{a}}_2 = -\gamma_{a_2} e_2 y_2,\ \dot{\hat{\bar{a}}}_2 = -\gamma_{\bar{a}_2} e_2 y_1, \\ &\hat{b}_2 = -\gamma_{b_2} e_2 u_1,\ \hat{c}_2 = -\gamma_{c_2} e_2 f_2, \end{aligned} \tag{14.34}$$

where $\gamma_{a_{ij}} > 0, \gamma_{\bar{a}_{ij}} > 0,\ \gamma_{\beta_{ij}} > 0,\ \gamma_{b_i} > 0$, and $\gamma_{c_i} > 0$ are gain factors of the adaptation subsystem.

Figure 2 shows tuning parameters of the model (14.31) for S_1.

Figure 2. Tuning model (14.31) parameters

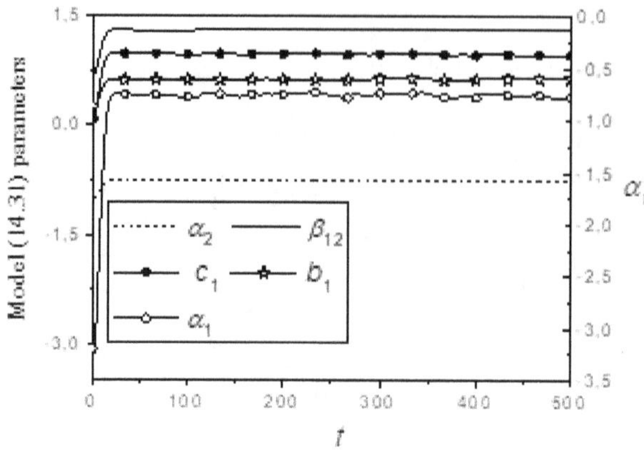

Show the adequacy estimation of the of models (14.31), (14.32) in Figure 3.

Figure 3. Adequacy of models (14.31), (14.32)

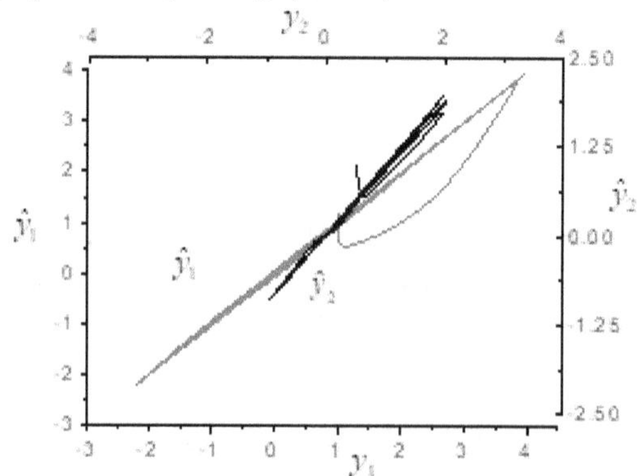

Figure 4. Tuning model (14.32) parameters

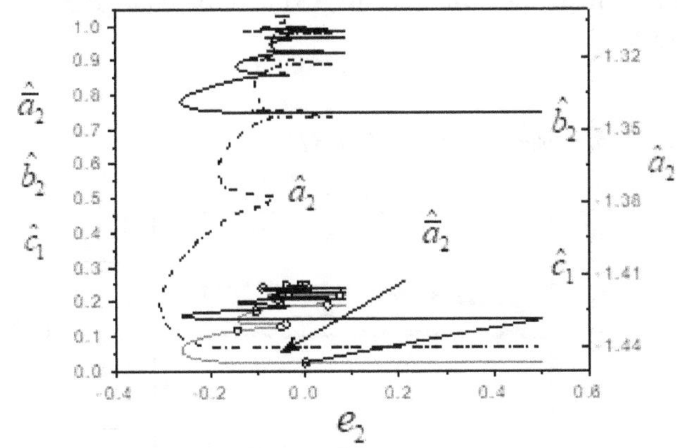

Figure 4 is presented the tuning process dynamics of the model (14.32) parameters depending on e_2. We see that the processes in the ASI for S_2 are nonlinear. The tuning process is more regular in the ASI for the S_1 subsystem.

Models (14.31) and (14.32) adaptation processes have different speeds (Figure 5).

Figure 5. ASI phase portrait in error space

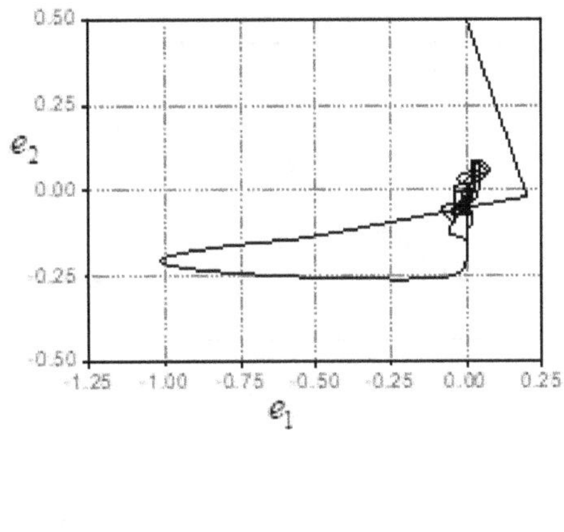

Consider the ASI with signal adaptation for S_2. Apply the model

$$\dot{\hat{y}}_2 = -k_2 e_2 + \hat{a}_2 y_2 + \hat{\bar{a}}_2 y_1 + \hat{b}_2 u_2 + u_{s,2} \tag{14.32a}$$

where $u_{s,2} = -d_2 e_2 x_2$.

We show results for the ASI in Figures 6-9. Show the adaptation of the model parameters (14.31) and the adequacy of the models in the output space in Figures 6 and 7. Figure 8 reflects the dynamics in ASI and the change in SA as e_2 function of the subsystem S_2.

Figure 6. Tuning parameters of model (14.31)

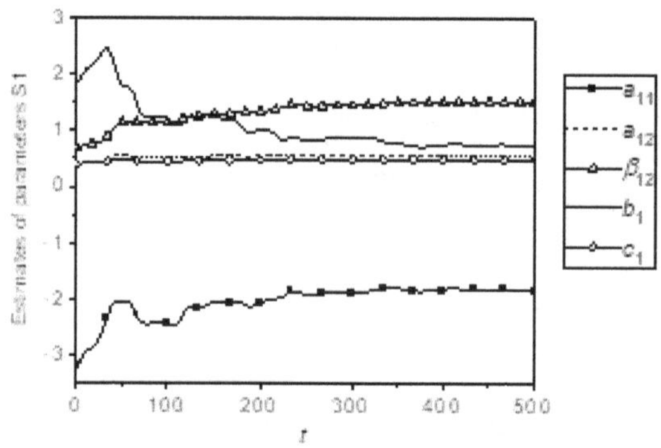

Figure 7. Adequacy of models (14.31) and (14.32a)

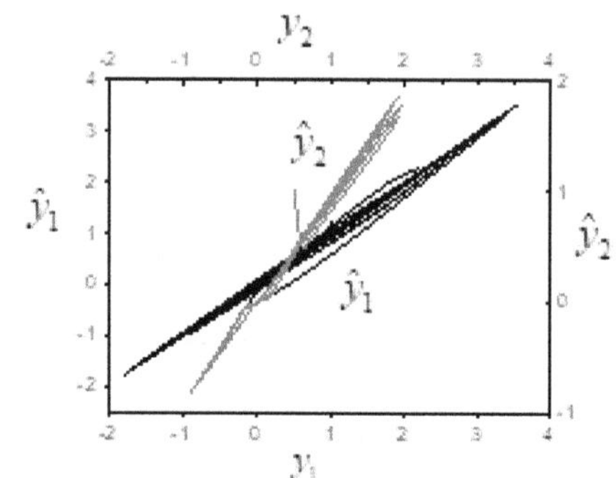

Figure 8. Phase portraits of the ASI and the output of the signal adaptation

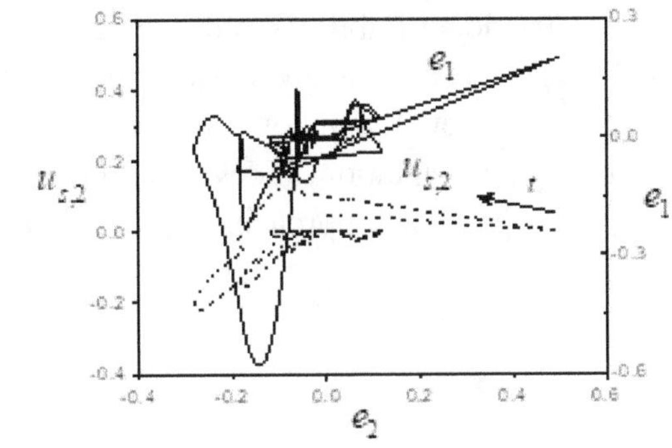

Figure 9. Tuning model (14.32a) parameters

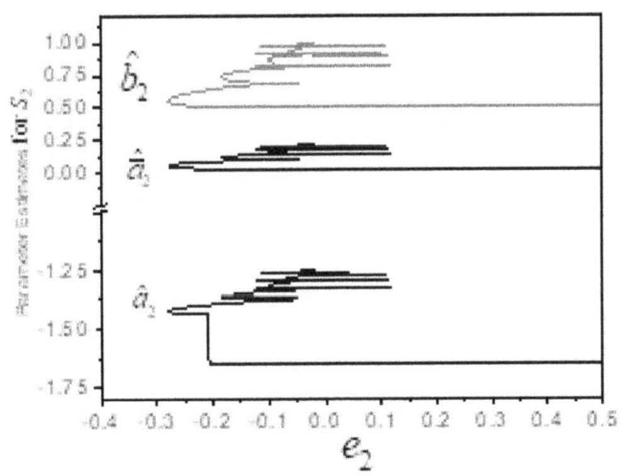

We see that the S_2-subsystem output with SA effects on the ASI adaptation of the S_1-subsystem. Therefore, ASI with SA should be applied considering the quality requirements of the identification process. Despite the compensating properties of CA, CA can lead to more complicated processes in ASI. The simulation of the algorithms from section "Adaptive algorithm as dynamic system" confirmed their work.

7. CONCLUSION

We consider nonlinear decentralized control systems in which the nonlinearity is under a quadratic condition. The problem of DS identifiability is investigated.

We analyze the influence of the constant excitation condition on solving of DS parametric identifiability problem. Quadratic estimates are got for the nonlinear part of the S_i-subsystem. Parametric identifiability conditions are offered. Algorithms of parametric and signal adaptation are designed. The exponential dissipativity of the AIS is proved. We propose a new Lyapunov function to obtain and verify the properties of a system with signal adaptation, and we present the simulation results confirming the efficiency of the proposed approach. Appendix contains proof results of the obtained results.

REFERENCES

Cho, S., Park, J.-W., & Sim, S.-H. (2015). Decentralized System identification using stochastic subspace identification for wireless sensor networks. *Sensors (Basel), 15*(4), 8131–8145. DOI: 10.3390/s150408131 PMID: 25856325

Gromyko, V. D., & Sankovsky, E. A. (1974). *Self-adjusting systems with a model.* Energy.

Gudi, R. D., Rawlings, J. B., Venkat, A., & Jabbar, N. (2004). Identification for decentralized MPC. *IFAC Proceedings, 37*(9), 299-304.

Hua, C., Guan, X., & Shi, P. (2004). Decentralized robust model reference adaptive control for interconnected time-delay systems. *Proceeding of the 2004 American Control Conference Boston,* Massachusetts June 30 - July 2, 2004, 4285-4289.

Karabutov, N. (2018). Structural-parametrical design method of adaptive observers for nonlinear systems. *International Journal of Intelligent Systems and Applications, 2*(2), 1–16. DOI: 10.5815/ijisa.2018.02.01

Karabutov, N. N. (2021). *Introduction to structural identifiability of nonlinear systems.* Lenand.

Karabutov, N. N. (2023). On adaptive identification of systems having multiple nonlinearities. *Russ. Technol. J., 11*(5), 94–105. DOI: 10.32362/2500-316X-2023-11-5-94-10

Lamara, A., Colin, G., Lanusse, P., Chamaillard, Y., & Charlet, A. (2012). Decentralized robust control-system for a non-square MIMO system, the air-path of a turbocharged Diesel engine. 2012 Workshop on Engine and Powertrain Control, Simulation and Modeling. The International Federation of Automatic Control Rueil-Malmaison, France, October 23-25, 130-137

Mao, X., & He, J. (2022). Decentralized System Identification Method for Large-Scale Networks. 2022 *American Control Conference* (ACC), Atlanta, GA, USA, 5173-5178.

Mei, H., Cai, W.-J., & Xiong, Q. (2005). Decentralized closed-loop parameter identification for multivariable processes from step responses. *Mathematics and Computers in Simulation, 68*(2), 171–192. DOI: 10.1016/j.matcom.2004.12.008

Podval'ny, S. L., Vasil'ev, E. M. (2022). Multilevel signal adaptation in non-stationary control systems. Bulletin of Voronezh state technical university, *18*(5), 38-45.

Wu, H. (2012). Decentralized adaptive robust control of uncertain large-scale non-linear dynamical systems with time-varying delays. *IET Control Theory & Applications*, *6*(5), 629–640. DOI: 10.1049/iet-cta.2011.0015

Yang, Q., Zhu, M., Jiang, T., He, J., Yuan, J., & Han, (2016). Decentralized Robust Adaptive output feedback stabilization for interconnected nonlinear systems with uncertainties. *Journal of control science and engineering*, article ID 3656578, 12.

APPENDIX A

Lemma 1 proof. As $\pi_1 X \leq F(X) \leq \pi_2 X$, then

$$\left(\pi_2 X - F(X)\right)^T \left(F(X) - \pi_1 X\right) \geq 0 \quad \text{(A1)}$$

where $\pi_1 > 0$, $\pi_2 > 0$.
After simple transformations, we get:

$$\begin{aligned}\chi &= \pi_2 X^T F(X) - \pi_1 \pi_2 X^T X - F^T(X)F(X) + \pi_1 X^T F(X) = \\ &\pi X^T F(X) - \pi_1 \pi_2 X^T X - F^T(X)F(X) \geq 0,\end{aligned} \quad \text{(A2)}$$

where $\pi = \pi_1 + \pi_2$. Let $X(t) \in EC_{\underline{\alpha}_X, \overline{\alpha}_X}$. Denote $X^T X = \|X\|^2$, $\underline{\alpha}_X \leq \|X\|^2 \leq \overline{\alpha}_X$, $\overline{\pi} = \pi_1 \pi_2$, where $\underline{\alpha}_X$, $\overline{\alpha}_X$ are positive numbers.
Transform χ to the form

$$\chi = \overline{\pi} X^T X - \left(\left[0.5 F^T(X)F(X) + \frac{(2\sqrt{0.5}\,\pi X^T F(X))}{2\sqrt{0.5}} \pm \frac{1}{4 \cdot 0.5}\pi^2 X^T X\right]\right) -$$

$$- 0.5 F^T(X)F(X) \geq 0$$

and get

$$\chi = \overline{\pi} X^T X - \left[\sqrt{0.5}\|F(X)\| + \frac{1}{2\sqrt{0.5}}\pi\|X\|\right]^2 - 0.5 F^T(X)F(X) + \frac{1}{2}\pi^2 X^T X \geq 0$$

or

$$F^T(X)F(X) \leq \underbrace{\left(2\overline{\pi} + \pi^2\right)}_{\eta}\|X\|^2 \leq \eta \overline{\alpha}_X.$$

APPENDIX B

Lemma 2 proof. As $\widehat{F} = F - \Delta F$, then we obtain from (14.7): $\left(\pi_2 X - \widehat{F}\right)^T \left(\widehat{F} - \pi_1 X\right) \geq 0$. Transform this inequality to the form:

$$(f_2 + \Delta F)^T (f_1 - \Delta F) \geq 0 \text{ (B1)}$$

where $f_1 = F - \pi_1 X, f_2 = \pi_2 X - F$. Then:

$$f_2^T f_1 - f_2^T \Delta F + \Delta F^T f_1 - \Delta F^T \Delta F,\ f_2^T f_1 - f_{12} \Delta F - \Delta F^T \Delta F \geq 0, \text{ (B2)}$$

where $f_{12} = f_2 - f_1$
Change (B2):

$$f_2^T f_1 - \left(\sqrt{0.5\Delta F^T \Delta F} + \sqrt{0.5 f_{12}^T f_{12}}\right)^2 - 0.5\Delta F^T \Delta F + 0.5 f_{12}^T f_{12} \geq 0,$$
$$0.5\Delta F^T \Delta F \leq f_2^T f_1 + 0.5 f_{12}^T f_{12}.$$

According to Lemma 1$f_2^T f_1 \leq \eta \overline{\alpha}_X$. Then:

$$\Delta F^T \Delta F \leq 2\eta \overline{\alpha}_X + f_{12}^T f_{12}$$

Obtain for $f_{12}^T f_{12}$

$$f_{12}^T f_{12} = (\pi X - 2F)^T (\pi X - 2F) = \pi^2 X^T X - 4\pi X^T F + 4F^T F,$$
$$f_{12}^T f_{12} = \|\pi X - 2F\|^2 \leq \delta_F,$$

where $\delta_F \geq 0$. So,
$$\Delta F^T \Delta F = \|\Delta F\|^2 \leq 2\eta \overline{\alpha}_X + \delta_F.$$

APPENDIX C

Theorem 1 proof. Derivative V_i

$$\dot{V}_i = -E_i^T Q_i E_i + E_i^T R_i \left(\Delta A_i X_i + \Delta B_i u_i + \sum_{j=1, j\neq i}^{m} \Delta \bar{A}_{ij} X_j + \Delta F_i(X_i) \right), \quad (C1)$$

where $K_i R_i + K_i^T R_i = -Q_i, Q_i = Q_i^T > 0$. Transform (C1):

$$\begin{aligned} \dot{V}_i - \underbrace{E_i^T Q_i E_i}_{-\lambda V_i} + \underbrace{|E_i^T R_i|}_{\sqrt{k_i V_i}} \left| \Delta A_i X_i + \Delta B_i u_i + \sum_{j=1, j\neq i}^{m} \Delta \bar{A}_{ij} X_j + \Delta F_i(X_i) \right| \le \\ \le -\lambda_i V_i + k_i V_i + 0.5 \left(\Delta A_i X_i + \Delta B_i u_i + \sum_{j=1, j\neq i}^{m} \Delta \bar{A}_{ij} X_j + \Delta F_i(X_i) \right)^2, \end{aligned} \quad (C2)$$

where $\lambda_i > 0$ is the minimum eigenvalue of the matrix Q_i.

We apply the Cauchy-Bunyakovsky-Schwarz inequality and Titus's lemma to the last term in (C2):

$$\begin{aligned} 0.5 \left(\Delta A_i X_i + \Delta B_i u_i + \sum_{j=1, j\neq i}^{m} \Delta \bar{A}_{ij} X_j + \Delta F_i(X_i) \right)^2 \le \\ \le 2 \left(\|\Delta A_i\|^2 \|X_i\|^2 + \|\Delta B_i\|^2 |u_i|^2 + \sum_{j=1, j\neq i}^{m} \|\Delta \bar{A}_{ij}\|^2 \|X_j\|^2 + \|\Delta F_i(X_i)\|^2 \right). \end{aligned} \quad (C3)$$

Since the conditions of Theorem 1 are fulfilled, then:

$$\dot{V}_i \le -\bar{\lambda}_i V_i + 2 \left(\bar{\alpha}_{X_i} \|\Delta A_i\|^2 + \bar{\alpha}_{u_i} \|\Delta B_i\|^2 + \sum_{j=1, j\neq i}^{m} \bar{\alpha}_{X_j} \|\Delta \bar{A}_{ij}\|^2 + \|\Delta F_i(X_i)\|^2 \right), \quad (C4)$$

where $\bar{\lambda}_i = \lambda_i - k_i$.

Apply lemmas 1, 2. Then:

$\Delta F_i^T \Delta F_i = \|\Delta F_i\|^2 \le 2\eta \bar{\alpha}_{X_i} + \delta_{F_i}$, (C5)

where $\eta = 2\bar{\pi} + \pi^2, \pi = \pi_1 + \pi_2, \bar{\pi} = \pi_1 \pi_2, \delta_i \ge 0$.

Get estimation for (C4):

$$\dot{V}_i \le -\bar{\lambda}_i V_i + 2 \left(\bar{\alpha}_{X_i} \|\Delta A_i\|^2 + \bar{\alpha}_{u_i} \|\Delta B_i\|^2 + \sum_{j=1, j\neq i}^{m} \bar{\alpha}_{X_j} \|\Delta \bar{A}_{ij}\|^2 + 2\eta \bar{\alpha}_{X_i} + \delta_i \right). \quad (C6)$$

As follows from (C6), if state variables have the property CE and

$2 \left(\bar{\alpha}_{X_i} \|\Delta A_i\|^2 + \bar{\alpha}_{u_i} \|\Delta B_i\|^2 + \sum_{j=1, j\neq i}^{m} \bar{\alpha}_{X_j} \|\Delta \bar{A}_{ij}\|^2 + 2\eta \bar{\alpha}_{X_i} + \delta_i \right) \le \bar{\lambda}_i V_i$,

then subsystem (14.1) is identifiable on the set or PI_{X_i}- identifiable.

APPENDIX D

Theorem 3 proof. Derivative $V_i = E_i^T P_i E_i$:

$$\dot{V}_i = -E_i^T Q_i E_i + E_i^T R\left(\Delta A_i X_i + \Delta B_i u_i + \sum_{j=1, j\neq i}^{M} \Delta \bar{A}_{ij} X_j + \Delta F_i(X_i)\right) \leq$$
$$-E_i^T Q_i E_i + |E_i^T R_i| \left|\Delta A_i X_i + \Delta B_i u_i + \sum_{j=1, j\neq i}^{M} \Delta \bar{A}_{ij} X_j + \Delta F_i(X_i)\right|. \quad \text{(D1)}$$

Apply the inequality

$$-az^2 + bz \leq -\frac{-az^2}{2} + \frac{b^2}{2a}, \quad a > 0, \ b \geq 0, \ z \geq 0. \ \text{(D2)}$$

As
$\Delta F_i = \tilde{F}_i^T \Delta N_{i,2} + \Delta \tilde{F}_i^T N_{i,2} = \tilde{F}_i^T \Delta N_{i,2} + D_{\Delta F_i}$,
then apply the conditions of Theorem 3 and get:

$$\dot{V}_i \leq -\mu_i V_i + \frac{2}{\mu_i}\left(\bar{\alpha}_{X_i} \|\Delta A_i\|^2 + \bar{\alpha}_{u_i} \|\Delta B_i\|^2 + \sum_{j=1, j\neq i}^{m} \bar{\alpha}_{X_j} \|\Delta \bar{A}_{ji}\|^2 + \right.$$
$$\left. + \left\|\tilde{F}_i^T(X_i, \hat{N}_{i,1}) \Delta N_{i,2} + D_{\Delta F_i}\right\|^2\right), \quad \text{(D3)}$$

where $E_i^T Q_i E_i = \mu_i E_i^T R_i E_i$, $\|\Delta A_i\|^2 = \mathrm{Sp}(\Delta A_i^T \Delta A_i)$. $\|D_{\Delta F_i}\|^2 \leq \bar{\upsilon}_{\Delta F_i}$ follows from construction of the system, $\bar{\upsilon}_{\Delta F_i} \geq 0$. Get the estimate for $\|\Delta A_i\|^2$:

$$\|\Delta A_i\|^2 = \mathrm{Sp}\left(\Delta A_i^T \Gamma_{A_i}^{-1} \Gamma_{A_i} \Delta A_i\right) \leq \bar{\lambda}_{\Gamma_{A_i}} \mathrm{Sp}\left(\Delta A_i^T \Gamma_{A_i}^{-1} \Delta A_i\right), \ \text{(D4)}$$

where $\bar{\lambda}_{\Gamma_{A_i}}$ is the maximum eigenvalue of the matrix Γ_{A_i}. Estimates for $\|\Delta B_i\|^2$, $\|\Delta \bar{A}_{ji}\|^2$ are obtained similarly. The estimate of the last term in (D3)

$$\|\tilde{F}_i^T \Delta N_{i,2}\|^2 \leq \|\tilde{F}_i\|^2 \|\Delta N_{i,2}\|^2 \leq \vartheta_{\tilde{F}_i} \Delta N_{i,2}^T \Gamma_{i,2}^{-1} \Gamma_{i,2} \Delta N_{i,2} \leq \vartheta_{\tilde{F}_i} \bar{\lambda}_{\Gamma_{i,2}} \Delta N_{i,2}^T \Gamma_{i,2}^{-1} \Delta N_{i,2},$$

where $\bar{\lambda}_{\Gamma_{i,2}}$ is the maximum eigenvalue of the matrix $\Gamma_{i,2}$, and $\tilde{F}$ is limited in construction, i.e. $\|\tilde{F}_i\|^2 \leq \vartheta_{\tilde{F}_i}$. Therefore, M⁺-form for $\dot{V}_i$

$$\dot{V}_i \leq -\mu_i V_i + \frac{2}{\mu_i} \kappa_i V_{\Delta,i} + \frac{2}{\mu_i} \bar{\upsilon}_{\tilde{F}_i}, \ \text{(D5)}$$

where $\kappa_i = \max\left(\bar{\alpha}_{X_i} \bar{\lambda}_{\Gamma_{A_i}}, \bar{\alpha}_{u_i} \bar{\lambda}_{\Gamma_{B_i}}, \bar{\alpha}_{X_j} \bar{\lambda}_{\Gamma_{\bar{A}_{ij}}}, \bar{\alpha}_{\tilde{F}_i} \bar{\lambda}_{\Gamma_{N_{i2}}}\right)$.
Then

$$\dot{V}_{\Delta,i} = -\mathrm{Sp}\left(\Delta A_i^T (\varphi_{A_i} \Delta A_i + R_i E_i X_i^T)\right) - \sum_{j=1}^{m} \mathrm{Sp}\left(\Delta \bar{A}_{ij}^T \left(\varphi_{\bar{A}_{ij}} \Delta \bar{A}_{ij} + R_i E_i X_j^T\right)\right) -$$
$$-\Delta B_i^T \left(\varphi_{B_{ij}} \Delta B_i + R_i E_i u_i\right) - \Delta N_{i,2}^T \left(\varphi_{F_i} \Delta N_{i,2} + \tilde{F}_i R_i E_i\right) = \omega_{1,i} + \omega_{2,i}. \quad \text{(D6)}$$

Consider the first component in (D6) $r_1 = \omega_{1,i(1)} + \omega_{2,i(1)}$:

$$r_1 = -\varphi_{A_i}\mathrm{Sp}(\Delta A_i^T \Delta A_i) - \mathrm{Sp}(\Delta A_i^T R_i E_i X_i^T) = -0.5\varphi_{A_i}\mathrm{Sp}(\Delta A_i^T \Delta A_i) -$$

$$-\left(0.5\varphi_{A_i}\mathrm{Sp}(\Delta A_i^T \Delta A_i) + 2\frac{\sqrt{0.5\varphi_{A_i}}}{2\sqrt{0.5\varphi_{A_i}}}\mathrm{Sp}(\Delta A_i^T R_i E_i X_i^T) \pm \frac{1}{4\cdot 0.5\varphi_{A_i}} X_i^T X_i E_i^T R_i R_i E_i\right) \le \tag{D7}$$

$$\le -0.5\varphi_{A_i}\mathrm{Sp}(\Delta A_i^T \Delta A_i) + \frac{1}{4\cdot 0.5\varphi_{A_i}} X_i^T X_i E_i^T R_i R_i E_i.$$

Using the transformations performed above for $\dot{V}_i$, we obtain:

$$r_1 \le -0.5\varphi_{A_i}\mathrm{Sp}(\Delta A_i^T \Delta A_i) + \frac{1}{4\cdot 0.5\varphi_{A_i}} X_i^T X_i E_i^T R_i R_i E_i \le$$
$$\le -0.5\underline{\lambda}_{A_i}\underline{\chi}_{A_i}\mathrm{Sp}\left(\Delta A_i^T \Gamma_{A_i}^{-1}\Delta A_i\right) + \underline{\chi}_{A_i}^{-1}\overline{\alpha}_{X_i}\rho_i V_i, \tag{D8}$$

where $\varphi_{A_i} \ge \underline{\chi}_{A_i}$, $\|R_i\| \le \rho_i$, $\varphi_{A_i}^{-1} \le \underline{\chi}_{A_i}^{-1}$.

Estimate $r_2 = \omega_{1,i(2)} + \omega_{2,i(2)}$, using the approach to get the estimate for r_1:

$$r_2 = -\sum_{j=1}^{m}\varphi_{\bar{A}_{ij}}\mathrm{Sp}\left(\Delta\bar{A}_{ij}^T\Delta\bar{A}_{ij}\right) - \sum_{j=1}^{m}\mathrm{Sp}\left(\Delta\bar{A}_{ij}^T R_i E_i X_j^T\right) = -0.5\sum_{j=1}^{m}\varphi_{\bar{A}_{ij}}\mathrm{Sp}\left(\Delta\bar{A}_{ij}^T\Delta\bar{A}_{ij}\right) \le$$

$$\le -0.5\sum_{j=1}^{m}\varphi_{\bar{A}_{ij}}\mathrm{Sp}\left(\Delta\bar{A}_{ij}^T\Delta\bar{A}_{ij}\right) + 0.5E_i^T R_i R_i E_i\sum_{j=1}^{m}\varphi_{\bar{A}_{ij}}^{-1}X_j^T X_j. \tag{D9}$$

Let $\underline{\chi}_{\bar{A}_{ij}} \le \varphi_{\bar{A}_{ij}}(t) \le \overline{\chi}_{\bar{A}_{ij}}$. Then

$$r_2 \le -0.5\sum_{j=1}^{m}\underline{\lambda}_{\bar{A}_{ij}}\underline{\chi}_{\bar{A}_{ij}}\mathrm{Sp}\left(\Delta\bar{A}_{ij}^T\Gamma_{\bar{A}_{ij}}^{-1}\Delta\bar{A}_{ij}\right) + \rho_i V_i\sum_{j=1}^{m}\underline{\chi}_{\bar{A}_{ij}}^{-1}\overline{\alpha}_{X_{ij}}. \tag{D10}$$

Estimate for $r_3 = -\varphi_{B_i}\Delta B_i^T\Delta B_i - \Delta B_i^T R_i E_i u_i$:

$$r_3 \le -0.5\underline{\lambda}_{B_i}\underline{\chi}_{B_i}\Delta B_i^T\Gamma_{B_i}^{-1}\Delta B_i + \underline{\chi}_{B_i}^{-1}\overline{\alpha}_{u_i}\rho_i V_i, \tag{D11}$$

where $\varphi_{B_i}(t) \ge \underline{\chi}_{B_i}$, $\varphi_{B_i}^{-1}(t) \le \underline{\chi}_{B_i}^{-1}$.

Consider $r_4 = -\varphi_{F_i}\Delta N_{i,2}^T\Delta N_{i,2} - \Delta N_{i,2}^T\tilde{F}_i R_i E_i - N_{i,2}^T\Delta\tilde{F}_i R_i E_i$:

$$r_4 \le -0.5\underline{\lambda}_{N_{i,2}}\underline{\chi}_{F_i}\Delta N_{i,2}^T\Gamma_{N_{i,2}}^{-1}\Delta N_{i,2} + \underline{\chi}_{F_i}^{-1}\overline{\alpha}_{\tilde{F}_i}\rho_i V_i + \left|N_{i,2}^T\Delta\tilde{F}_i R_i E_i\right| \le$$

$$\le -0.5\underline{\lambda}_{N_{i,2}}\underline{\chi}_{F_i}\Delta N_{i,2}^T\Gamma_{N_{i,2}}^{-1}\Delta N_{i,2} + \underline{\chi}_{F_i}^{-1}\overline{\alpha}_{\tilde{F}_i}\rho_i V_i + 0.5N_{i,2}^T\Delta\tilde{F}_i\Delta F_i^T N_{i,2} + 0.5E_i^T R_i R_i E_i,$$

where $\varphi_{F_i}(t) \ge \underline{\chi}_{F_i}$, $\|\tilde{F}_i\|^2 \le \overline{\alpha}_{\tilde{F}_i}$. Considering the designations introduced above and

$$0.5N_{i,2}^T\Delta\tilde{F}_i\Delta\tilde{F}_i^T N_{i,2} \le \delta_{N,i},\ \delta_{N,i} \ge 0,$$

we obtain

$$r_4 \le -0.5\underline{\lambda}_{N_{i,2}}\underline{\chi}_{F_i}\Delta N_{i,2}^T\Gamma_{N_{i,2}}^{-1}\Delta N_{i,2} + \left(\underline{\chi}_{F_i}^{-1}\overline{\alpha}_{\tilde{F}_i} + 1\right)\rho_i V_i + 0.5\delta_{N_i}. \tag{D12}$$

Let

$$\min_{\underline{\lambda}_{A_i}\underline{\chi}_{A_i}}\left(\underline{\lambda}_{A_i}\underline{\chi}_{A_i},\ \underline{\chi}_{\bar{A}_{ij}}\underline{\lambda}_{\bar{A}_{ij}},\underline{\lambda}_{B_i}\underline{\chi}_{B_i},\ \underline{\lambda}_{N_{i,2}}\underline{\chi}_{F_i}\right) = \beta_{\lambda\chi,i},$$

$$\max_{\chi_{A_i}^{-1}\overline{\alpha}_{X_i}}\left[\chi_{A_i}^{-1}\overline{\alpha}_{X_i},\ \sum_{j=1}^{m}\underline{\chi}_{\bar{A}_{ij}}^{-1}\overline{\alpha}_{X_{ij}},\underline{\chi}_{B_i}^{-1}\overline{\alpha}_{u_i},\left(\underline{\chi}_{F_i}^{-1}\overline{\alpha}_{\tilde{F}_i} + 1\right)\right] \le \vartheta_{\chi\alpha,i}.$$

APPENDIX E

Theorem 4 proof. Consider $\dot{V}_i$:

$$\dot{V}_i = -E_i^T Q_i E_i + E_i^T R_i \left(\Delta A_i X_i + \Delta B_i u_i + \sum_{j=1, j\neq i}^{m} \Delta \bar{A}_{ij} X_j + \Delta_{U_i} F_i \right). \text{ (E1)}$$

Apply the approach from Appendix D and get:

$$\dot{V}_i \leq -\mu_i V_i + \frac{2}{\mu_i} \left(\|\Delta A_i X_i\|^2 + \|\Delta B_i u_i\|^2 + \left\| \sum_{j=1, j\neq i}^{M} \Delta \bar{A}_{ij} X_j \right\|^2 + \|\Delta_{U_i} F_i\|^2 \right). \text{ (E2)}$$

where $\|\Delta A_i\|^2 = \mathrm{Sp}(\Delta A_i^T \Delta A_i)$, $\|\Delta A_{ij}\|^2 = \mathrm{Sp}(\Delta A_{ij}^T \Delta A_{ij})$.

According to section 3, U_i has the form (14.19), where is a diagonal matrix with positive elements. We select the matrix D_i from the condition $\|D_i\| \geq d_i \geq \eta \bar{\alpha}_{X_i}$, and $F_i(X_i) \in N_F(\pi_1, \pi_2)$ satisfies conditions Lemma 1, 2 conditions. According to the choice U_i, we get $\|\Delta_{U_i} F_i\|^2 \leq \sigma_{F_i}$, where $\sigma_{F_i} \geq 0$. Apply the mean theorem:

$$\|\Delta_{U_i} F_i\|^2 = \tau^{-1} \underbrace{\int_t^{t+\tau} \|\Delta_{U_i} F_i(\tau)\|^2 d\tau}_{I_{\Delta F_i}} = \tau^{-1} I_{\Delta F_i},$$

where $\tau > 0$, $I_{\Delta F_i} \leq \varphi$, $\varphi > 0$ Let $\tau\varphi \triangleq \phi$. Then (E2):

$$\dot{V}_i \leq -\mu_i V_i + \frac{2}{\mu_i} \left(\|\Delta A_i X_i\|^2 + \|\Delta B_i u_i\|^2 + \left\| \sum_{j=1, j\neq i}^{M} \Delta \bar{A}_{ij} X_j \right\|^2 + \tau^{-1} I_{\Delta F_i} \right). \text{ (E3)}$$

Using the proof scheme from Appendix D, we obtain:

$$\bar{\alpha}_{X_i} \|\Delta A_i\|^2 + \bar{\alpha}_{u_i} \|\Delta B_i\|^2 + \sum_{j=1, j\neq i}^{m} \bar{\alpha}_{X_j} \|\Delta \bar{A}_{ji}\|^2 + \tau^{-1} I_{\Delta F_i} \leq \kappa_{\Delta_s, i} V_{\Delta_s, i}, \text{ (E4)}$$

where $\kappa_{\Delta_s, i} > 0$, and

$\dot{V}_i \leq -\mu_i V_i + \frac{2}{\mu_i} \kappa_{\Delta_s, i} V_{\Delta_s, i}$. (E5)

Consider

$$\begin{aligned} \dot{V}_{\Delta_s, i} = & -\mathrm{Sp}\left(\Delta A_i^T (\varphi_{\bar{A}_{ij}} \Delta A_i + R_i E_i X_j^T)\right) - \sum_{j=1}^{m} \mathrm{Sp}\left(\Delta \bar{A}_{ij}^T (\varphi_{\bar{A}_{ij}} \Delta \bar{A}_{ij} + R_i E_i X_j^T)\right) - \\ & -\Delta B_i^T \left(\varphi_{B_{ij}} \Delta B_i + R_i E_i u_i\right) + \Delta_{U_i} F_i^T \Delta_{U_i} F_i. \end{aligned} \text{ (E6)}$$

Lemma 3. *Estimate*

$\Delta_{U_i} F_i^T \Delta_{U_i} F_i \leq -\bar{v} \|\Delta_{U_i} F_i\|^2 + 2\theta_i \rho_i d_i V_i + \sqrt{0.125}\, v \eta_i \bar{\alpha}_{X_i}$

is valid for $\Delta_{U_i} F_i^T \Delta_{U_i} F_i$.

Proof. $\Delta_{U_i} F_i^T \Delta_{U_i} F_i$ has the form

$\Delta_{U_i} F_i^T \Delta_{U_i} F_i = -E_i^T R_i D_i \Delta_{U_i} F_i - F_i^T \Delta_{U_i} F_i$.

Estimating $-E_i^T R_i D_i \Delta_{U_i} F_i$. Let in the domain $O_\nu(O)$, the equality holds:

$-E_i^T R_i D_i \Delta_{U_i} F_i = -\pi_i \left(\|E_i^T R_i D_i\|^2 + \|\Delta_{U_i} F_i\|^2 \right)$,

where $\pi_i > 0$, , is zero matrix, O_ν is some neighborhood of the point 0, .

Then

$$-\pi_i\left(\|E_i^T R_i D\|^2 + \|\Delta_{U_i} F_i\|^2\right) = -0.5\pi \|\Delta_{U_i} F_i\|^2 -$$
$$-\pi_i\left(0.5 \|\Delta_{U_i} F_i\|^2 + \|E_i^T R_i D_i\|^2 \pm \tfrac{2}{2}\sqrt{0.5}\|\Delta_{U_i} F_i\| \|E_i^T R_i D_i\|\right) \leq \text{(E7)}$$
$$\leq -0.5\pi_i \|\Delta_{U_i} F_i\|^2 + 0.5\sqrt{0.5}\,\pi_i \|\Delta_{U_i} F_i\| \|E_i^T R_i D_i\|.$$

Apply the inequality (D2) to (14.E7)

$$-E_i^T R_i D_i \Delta_{U_i} F_i \leq -\pi_i \|\Delta_{U_i} F_i\|^2 + \pi_i \|E_i^T R_i D_i\|^2.$$

As $\|D_i\| \leq d_i$, $\|R_i\| \leq \rho_i$, then

$$-E_i^T R_i D_i \Delta_{U_i} F_i \leq -\pi_i \|\Delta_{U_i} F_i\|^2 + 2\pi_i \rho_i d_i V_i. \text{ (E8)}$$

$\|\Delta_{U_i} F_i\|^2 = \tau^{-1} I_{\Delta F_i}$. Therefore, (E8)
$-E_i^T R_i D_i \Delta_{U_i} F_i \leq -\pi_i \tau^{-1} I_{\Delta F_i} + 2\pi_i \rho_i d_i V_i$. (E9)

Estimating $-F_i^T \Delta_{U_i} F$. Let $\nu_i > 0$ exist such that:
$-F_i^T \Delta_{U_i} F_i = -\nu_i \|F_i\|^2 - \nu_i \|\Delta_{U_i} F_i\|^2$
is fulfilled on $t >> t_0$.
Then

$$-F_i^T \Delta_{U_i} F_i = -\nu_i \|F_i\|^2 - \nu_i \|\Delta_{U_i} F_i\|^2 = -0.5\nu_i \|\Delta_{U_i} F_i\|^2 - 0.5\nu_i \|\Delta_{U_i} F_i\|^2 - \nu_i \|F_i\|^2 \leq$$
$$-0.5\nu_i \|\Delta_{U_i} F_i\|^2 + \sqrt{0.5}\,\nu_i F_i^T \Delta_{U_i} F_i \leq -0.25\nu_i \|\Delta_{U_i} F_i\|^2 + \sqrt{0.125}\,\nu_i \|F_i\|^2.$$

(E10)

Use (E9) and (E10) and get:
$\Delta_{U_i} F_i^T \Delta_{U_i} F_i \leq -(\pi_i + 0.25\nu_i)\|\Delta_{U_i} F_i\|^2 + 2\pi_i \rho_i d_i V_i + \sqrt{0.125}\,\nu_i \eta_i \overline{\alpha}_{X_i}$,
where $\|F_i\|^2 \leq \eta_i \overline{\alpha}_{X_i}$. ■
Use the estimates obtained in Appendix D. Introduce notations:
$\min\left(\lambda_{A_i}\chi_{A_i},\ \chi_{\overline{A}_{ij}}\lambda_{\overline{A}_{ij}}, \lambda_{B_i}\chi_{B_i},\ (\pi_i + 0.25\nu_i)\tau^{-1}\right) = \beta_{\Delta_s,i}$,
$\max_{\chi_{A_i}^{-1}\overline{\alpha}_{X_i}}\left[\chi_{A_i}^{-1}\overline{\alpha}_{X_i}, \sum_{j=1}^{m}\chi_{\overline{A}_{ij}}^{-1}\overline{\alpha}_{X_{ij}}, \chi_{B_i}^{-1}\overline{\alpha}_{u_i},\ \pi_i d_i\right] \leq \vartheta_{\Delta_s,i}$,
and apply Lemma 3. Then (E6)

$$\dot V_{\Delta_s,i} \leq -\beta_{\Delta_{s,i}} V_{\Delta_{s,i}} + \vartheta_{\Delta_{s,i}} \rho_i V_i + \sqrt{0.125}\,\nu_i \eta_i \overline{\alpha}_{X_i}, \text{ (E11)}$$

where $F_i^T \Delta_{U_i} F_i = -\nu_i \|F_i\|^2 - \nu_i \|\Delta_{U_i} F_i\|^2$, $\nu_i \geq 0$.
We obtain the system of inequalities for AS_{AS_i}

$$\underbrace{\begin{bmatrix} \dot V_i \\ \dot V_{\Delta_{s,i}} \end{bmatrix}}_{\dot W_{S_i}} \leq \underbrace{\begin{bmatrix} -\mu_i & \frac{2}{\mu_i}\kappa_{\Delta_{s,i}} \\ \vartheta_{\Delta_{s,i}}\rho_i & -\beta_{\Delta_{s,i}} \end{bmatrix}}_{A_{W_{S_i}}} \underbrace{\begin{bmatrix} V_i \\ V_{\Delta_{s,i}} \end{bmatrix}}_{W_{S_i}} + \underbrace{\begin{bmatrix} 0 \\ \sqrt{0.125}\,\nu_i \eta_i \overline{\alpha}_{X_i} \end{bmatrix}}_{L_{s,i}}. \text{ (E12)}$$

The exponential dissipation condition for $A\,S_{AS_i}$

$\mu_i^2 \beta_{\Delta_{S,i}} \geq 2\vartheta_{\Delta_{S,i}} \rho_i \kappa_{\Delta_{S,i}}$. (E13)

If we introduce the comparison system (see appendix D), then we get the estimate for $A\,S_{AS_i}$

$$W_{S_i}(t) \leq e^{A_{W_{S_i}}(t-t_0)} S_{W_{S_i}}(t_0) + \int_{t_0}^{t} e^{A_{W_{S_i}}(t-\sigma)} L_{S,i}\, d\sigma.$$

Chapter 15
Adaptive Observers for Nonlinear Systems

ABSTRACT

In this chapter, we consider the adaptive observer synthesis method for nonlinear dynamic systems (NDS) under uncertainty. The design introduces of two stages. The structural stage based on the application of geometric frameworks is allowed identifying of nonlinearity class and its structural parameters. We propose the method and criteria for the structural identifiability of NDS. It is believed that the excitation constancy condition is satisfied for system variables. We note that requirements to the input at stages of structural and parametric design of AO differ. The AO parametric design stage bases the results obtained at the first stage of the AO synthesis. We consider two cases of the structural information application. We focus on the incompleteness of structural information. Adaptive algorithms are proposed for tuning of AO parameters. The uncertainty estimation procedure is proposed. The adaptive system stability is proved.

1. INTRODUCTION

The design of adaptive observers (AO) is one of the most topical directions of control theory. The AO theory for the linear dynamic systems has been developed an end of the past century (Carrol and Lindorff, 1973; Carrol Monopoli, 1975; Kreisselmeier, 1986; Kudva and Narendra, 1973; Narendra and Kudva, 1974). Results (Carrol and Lindorff, 1973; Carrol Monopoli, 1975; Kreisselmeier, 1986; Kudva and Narendra, 1973; Narendra and Kudva, 1974) are the basis for the development and generalization of the theory to new system classes. Many works give to generalization of got results on nonlinear systems (see, for example, Baang et

DOI: 10.4018/979-8-3373-0822-7.ch015

al., (2007); Bastin, and Gevers (1988); Garimella and Yao (2003), Krstić M., and Kokotović (1996); Marino and Tomei (1992)).

So, AO are widely applied to the solution of unique problems. Often the nonlinearity is smooth and satisfies the Lipschitz condition. The structure of nonlinearity structure is set a priori. If the nonlinearity form is unknown, then it is approximated on the specified class of functions. But there is a class of systems for which a priori information is incomplete or unknown. Application of approaches considered above demands fulfilment of labor-consuming studies. The AO design is associated with the structural identifiability (SI) of nonlinear systems. This problem was set by Karabutov (2016) and was generalized by Karabutov (2011, 2018). SI provides a solution to the nonlinearity problem based on the information.

This chapter contains the solution of AO structural-parametric identification for nonlinear dynamic systems under uncertainty. AO has the canonical form (Bastin and Gevers (1988); Kudva and Narendra (1973).

2. PROBLEM STATEMENT

Consider system

$$\begin{aligned} \dot{\widetilde{X}} &= \widetilde{A}\widetilde{X} + \widetilde{A}_{\varphi}\varphi + \widetilde{B}_{r}r, \\ y &= C^{T}\widetilde{X}, \end{aligned} \tag{15.1}$$

where $\widetilde{X} \in \mathbb{R}^m$ is state vector; $r \in \mathbb{R}$ is input; $y \in \mathbb{R}$ is output; $\varphi(y) \in \mathbb{R}$ is nonlinear function;$\widetilde{A} \in \mathbb{R}^{m\times m}$is Frobenius matrix

$$\widetilde{A} = \begin{bmatrix} 0 & I_{m-1} \\ & \widetilde{A}_1 \end{bmatrix}$$

$\widetilde{A}_1 \in \mathbb{R}^m$ is the vector of unknown parameters belonging to a limited area $\mathbb{G}_A$; $\widetilde{A}_{\varphi} \in \mathbb{R}^m$, $\widetilde{A}_{\varphi} = \left[0,\ldots,0,\widetilde{a}_{\varphi}\right]^T$, $\widetilde{B}_r \in \mathbb{R}^m$, $\widetilde{B}_r = \left[0,\ldots,0,b_r\right]^T$ are unknown parameter vector; $I_{m-1} \in \mathbb{R}^{(m-1)\times(m-1)}$ is identity matrix; $C \in \mathbb{R}^m$, $C = [1\ 0 \ldots 0\ ]^T$. $\widetilde{A}$ is Hurwitz matrix.

Assumption 1. The input $r(t)$ is constantly excited and limited.

Assumption 2. Function $\varphi(y)$ is smooth, single-valued, and

$$\varphi \in F_{\varphi} = \left\{\kappa_1\xi^2 \leq \varphi(\xi)\xi \leq \kappa_2\xi^2,\ \xi \neq 0,\ \varphi(0)=0,\ \kappa_1 \geq 0,\ \kappa_2 < \infty\right\} \tag{15.2}$$

The data set

$$\mathbb{I}_o = \{u(t), y(t) t \in \mathbb{J} = [t_0, t_k]\}. \tag{15.3}$$

Apply to y, r, φ the auxiliary system

$$\dot{\tilde{P}}_v = \Lambda \tilde{P}_v + \tilde{H} v, v = y, r, \varphi \tag{15.4}$$

where $\Lambda \in \mathbb{R}^{(m-1)\times(m-1)}$ is a stable diagonal matrix; $\tilde{H} \in \mathbb{R}^{m-1}$, $\tilde{h}_i = 1$ ($i = \overline{1, m-1}$). Pair (Λ, H) is controllable.

Apply the mapping $\tilde{X} = TX$ to the system (15.1) and get:

$$\begin{aligned} \dot{X} &= AX + A_\varphi \varphi + B_r r, \\ y &= C^T X, \end{aligned} \tag{15.5}$$

where $T \in \mathbb{R}^{m\times m}$, $A_\varphi = T^{-1}\tilde{A}_\varphi$, $B_r = T^{-1}\tilde{B}_r$, $A = T^{-1}\tilde{A}T$,

$$A = \begin{bmatrix} A_y & \tilde{H}^T \\ & \Lambda \end{bmatrix}$$

Apply model for the identification system (15.5)

$$\begin{aligned} \dot{\hat{X}} &= A_M(\hat{X} - CC^T\hat{X}) + \hat{A}_y y + \hat{A}_\varphi \varphi + \hat{B}_r(t) r, \\ \hat{y} &= C^T \hat{X}, \end{aligned} \tag{15.6}$$

where $A_M \in \mathbb{R}^{m\times m}$ is the stable matrix of the form:

$$A_M = \begin{bmatrix} -k & \tilde{H}^T \\ 0 & \Lambda \end{bmatrix}$$

$k > 0$; $\hat{A}_y \in \mathbb{R}^m$, $\hat{A}_\varphi \in \mathbb{R}^m$, $\hat{B}_r \in \mathbb{R}^m$ are vectors of tuned parameters; $\hat{X} \in \mathbb{R}^m$ is state vector;$\hat{y} \in \mathbb{R}$ is model output.

Use the system (15.4) and transform equations (15.5) and (15.6) to the form

$$\dot{y} = A_y^T P_y + A_\varphi^T P_\varphi + B_r^T P_r \tag{15.7}$$

$$\dot{\hat{y}} = -k(\hat{y} - y) + \hat{A}_y^T P_y + \hat{A}_\varphi^T P_\varphi + \hat{B}_r^T P_r \tag{15.8}$$

where $P_y = \left[y\ \tilde{P}_y^T\right]^T, P_\varphi = \left[\varphi\ \tilde{P}_\varphi^T\right]^T, P_r = \left[r\ \tilde{P}_r^T\right]^T$.

Problem: for the system (15.7) satisfying the assumptions A1 and A2, define laws of tuning $\hat{A}_\varphi(t), \hat{A}_\varphi(t)$ and $\hat{B}_r(t)$ model (15.8) that:

$$\lim_{t\to\infty}\left|\hat{y}(t) - y(t)\right| \le \pi_y, \pi_y \ge 0$$

Stages of the AO design.

1. Structural stage (SS).
2. Parametric stage.

SS gives to nonlinearity localization $\varphi(y)$. The result localization is the class F_φ definition and structural parameter nonlinearity estimation. The parametric stage is based on SS and is given to tuning algorithms for AO.

3. SS

SS is based on structural identification $\varphi(y) \in F_\varphi$. We apply the approach and methods proposed by Karabutov (2016, 2017c). SI is the first step, ensures the construction of framework S_{ey}. The construction S_{ey} and estimate methods of nonlinearity structure for AO are described in the previous sections and is not considered here.

4. PARAMETRICAL STAGE

This stage is based on localization results for the nonlinearity $\varphi(y)$ obtained in section 3. Consider two cases of information on $\varphi(y)$.

Case 1. We have the complete information about the structure $\varphi(y)$. Subtract (15.7) of (15.8) and get the equation for the error $\varepsilon = \hat{y} - y$:

$$\dot{\varepsilon} = -k\varepsilon + \Delta A_y^T P_y + \Delta A_\varphi^T P_\varphi + \Delta B_r^T P_r \tag{15.9}$$

where $\Delta A_y = \hat{A}_y - A_y, \Delta A_\varphi = \hat{A}_\varphi - A_\varphi, \Delta A_r = \hat{A}_r - A_r$.

Apply Lyapunov function (LF) $V_\varepsilon(t) = 0.5\,\varepsilon^2(t)$. Determine $\dot{V}_\varepsilon$:

$$\dot{V}_\varepsilon = -k\varepsilon^2 + \varepsilon\left(\Delta A_y^T P_y + \Delta A_\varphi^T P_\varphi + \Delta B_r^T P_r\right)$$

and obtain parameter tuning algorithms of the model (15.8)

$$\Delta \dot{A}_y = \dot{\hat{A}}_y = -\Gamma_y \varepsilon P_y \tag{15.10}$$

$$\Delta \dot{A}_\varphi = \dot{\hat{A}}_\varphi = -\Gamma_\varphi \varepsilon P_\varphi \tag{15.11}$$

$$\Delta \dot{A}_r = \dot{\hat{A}}_r = -\Gamma_r \varepsilon P_r \tag{15.12}$$

from the condition $\dot{V}_\varepsilon \leq 0$, where $\Gamma_y \in \mathbb{R}^{m\times m}$, $\Gamma_\varphi \in \mathbb{R}^{m\times m}$, $\Gamma_r \in \mathbb{R}^{m\times m}$ are diagonal matrixes with positive diagonal elements.

Designate the obtained adaptive system (15.9) - (15.12) as AS_A.

Case 2. A posteriori information about $\varphi(y)$ has an approximate character.

Assumption 3. The structure $\varphi(y)$is specified as the set (Karabutov, 2018)

$$\varphi \in F_{ov} = \{W \in R^n : \varphi(y) = F^T(y, W_1)\, W_2, W$$
$$= [W_1^T, W_2^T]^T, W \in \mathbb{W}_a\}, \tag{15.13}$$

where $\mathbb{W}_a = \{W \in R^n : \underline{W} \leq W \leq \overline{W}\}$is the area formed posterior; $\underline{W}, \overline{W}$ are vector borders for W. Some elements $\underline{W}, \overline{W}$ can be unknown.

Then $\varphi(y)$ we describe as

$$\varphi(y) = F^T(y, W_1)\, W_2 \tag{15.14}$$

where $W \in \mathbb{W}_a \subset \mathbb{R}^n$ is the vector of parameters for $\varphi(y)$, $F(y, W_1) \in \mathbb{R}^n$.

We suppose that $\mathbb{W} = \mathbb{W}_1 \cup \mathbb{W}_2$. The set$\mathbb{W}_1 \subset \mathbb{R}^{n_1}$ $(W_1 \in \mathbb{W}_1)$ contains elements which are not tuned. Element $W_2 \in \mathbb{W}_2 \subset \mathbb{R}^{n_2}$ is estimated based on the tuning of parameters for AO. The vector $F(y, W_1) \in \mathbb{R}^{n_2}$ is formed at the structural synthesis stage of AO. Explain representation (15.14) as the proposed parametric concept for $\varphi(y)$.

Therefore, the major difference from the case 1 is need of the current estimations obtaining for the vector W_2 and for the nonlinearity $\varphi(y)$. The equation (15.8) is not applicable and is required for its modification.

Present model for the estimation $\varphi(y)$ as:

$$\widehat{\varphi}(y) = F^T(y, \widehat{W}_1)\widehat{W}_2 \tag{15.15}$$

where $\widehat{W}_2 \in \mathbb{R}^{n_2}$ is the tuned vector of parameters; $\widehat{W}_1 \in \mathbb{R}^{n_1}$ is the estimation for W_1 which is adjusted on iterative constraint. Next, we propose the constraint algorithm for $\widehat{W}_1$.

Apple (15.14), (15.15) and modify the adaptive model (15.8). Obtain the current estimation for the vector $\overline{P}_\varphi$:

$$\dot{\overline{P}}_\varphi = \Lambda \overline{P}_\varphi + \overline{H}\widehat{\varphi}(y) \tag{15.16}$$

and rewrite the model (15.8) as:

$$\dot{\hat{y}} = -k(\hat{y} - y) + \widehat{A}_y^T P_y + \widehat{A}_\varphi^T \widehat{P}_\varphi + \widehat{B}_r^T P_r$$

where $\widehat{P}_\varphi = \left[\widehat{\varphi}\ \overline{P}_\varphi^T\right]^T$. Transform this equation that to perform association with the equation (15.7). Obtain:

$$\dot{\hat{y}} = -k(\hat{y} - y) + \widehat{A}_y^T P_y + \widehat{A}_\varphi^T P_\varphi + \widehat{B}_r^T P_r + \omega, \tag{15.17}$$

where $\omega = \widehat{A}_\varphi^T\left(\widehat{P}_\varphi - P_\varphi\right) = \widehat{A}_\varphi^T \Delta P_\varphi$.

We see that the equation (15.17) contains uncertainty ω. Use the approach proposed by Karabutov (2017a) to the estimation ω. Apply numerical differentiation and get the estimation for $\dot{y}$. Designate the obtained variable as $\dot{y}_d(t)$. Apply model to the estimation $\dot{y}_d$

$$\dot{\hat{y}}_d = \widehat{A}_y^T P_y + \widehat{B}_r^T P_r \tag{15.18}$$

and determined by the misalignment $\sigma = \dot{\hat{y}}_d - \dot{y}_d$. Rewrite the equation (15.17) as:

$$\dot{\hat{y}} = -k(\hat{y} - y) + \widehat{A}_y^T P_y + \widehat{A}_\varphi^T P_\varphi + \widehat{B}_r^T P_r + \sigma \tag{15.19}$$

where σ is the current estimation ω. The equation for the error ε writes as:

$$\dot{\varepsilon} = -k\varepsilon + \Delta A_y^T P_y + \Delta A_\varphi^T P_\varphi + \Delta B_r^T P_r + \sigma$$

Design adaptive algorithms for the model (15.19).

Algorithms for $\widehat{A}_y$ and $\widehat{A}_r$ have the form (15.10), (15.12). Apply the LF V_ε and obtain the algorithm for $\widehat{A}_\varphi$ from the condition $\dot{V}_\varepsilon \leq 0$:

$$\Delta \dot{A}_\varphi = \dot{\widehat{A}}_\varphi = -\Gamma_\varphi \varepsilon \widehat{P}_\varphi \tag{15.20}$$

where $\Gamma_\varphi \in \mathbb{R}^{m\times m}$ is the diagonal matrix with positive diagonal elements.

Consider the Lyapunov function $V_\sigma(t) = 0.5\sigma^2(t)$. As $\sigma = \omega$, and

$$\Delta \dot{\bar{P}}_\varphi = \Lambda\Delta \bar{P}_\varphi + \bar{H} F^T(y, \widehat{W}_1)\Delta W_2 \tag{15.21}$$

that we obtain the algorithm for tuning $\widehat{W}_2$

$$\dot{\widehat{W}}_2 = -\Gamma_2 \sigma \widehat{A}_\varphi^T \bar{H} F \tag{15.22}$$

where $\Delta \bar{P}_\varphi = \bar{P}_\varphi - \widetilde{P}_\varphi, \Delta W_2 = \widehat{W}_2 - W_2, \Gamma_2 = \Gamma_2^T > 0$ is the matrix ensuring convergence of the algorithm (15.22).

Designate the adaptive system (15.10), (15.12), (15.15) - (15.22) as AS_φ.

Iterative constraint algorithm for the vector $\widehat{W}_1$.

1. $i = 0$.
2. Let $\widehat{W}_{1,i} = W_1^*$, where $W_1^* \in \mathbb{W}_a$.
3. Perform tuning of parameters for the system AS_{AP_φ}.
4. Check the condition $\lim_{t\to\infty} \left|\hat{y}(t) - y(t)\right| \leq \pi_y$. If it is satisfied, then finish the algorithm work, otherwise go to the step 5.
5. $i = i + 1$.
6. Suppose $\widehat{W}_{1,i} = \widehat{W}_{i-1} + \delta W_1$, where δW_1 chooses from the condition $\widehat{W}_{1,i} \in \mathbb{W}_a$.
7. Go to step 3.

5. STABILITY ANALYSIS

5.1. System AS_A

Let $\Delta K(t) \triangleq \left[\Delta A_y^T(t), \Delta A_\varphi^T(t), \Delta B_r^T(t)\right]^T$, $V_K(t) \triangleq 0,5\Delta K^T(t)\Gamma^{-1}\Delta K(t)$, $V(t) = V_\varepsilon(t) + V_K(t)$, where $\Gamma \triangleq \Gamma_y \dotplus \Gamma_\varphi \dotplus \Gamma_r$, $\dotplus$ is the direct sum of matrixes.

Theorem 1. *Let 1) functions $V_\varepsilon(t) = 0,5\,\varepsilon^2(t)$, $V_K(t) \triangleq 0,5\Delta K^T(t)\Gamma^{-1}\Delta K(t)$ are positive definite and satisfies conditions $\inf_{|\varepsilon|\to\infty} V_\varepsilon(\varepsilon) \to \infty$, $\inf_{\|\Delta K\|\to\infty} V_K(\Delta K) \to \infty$; 2)* *assumptions 1, 2 for the system (15.1) are satisfied. Then all trajectories of system* AS_A *are limited, lie in the area*

$$G_t = \{(\varepsilon, \Delta K) : V(t) \leq V(t_0)\}$$

and the estimation is fair

$$\int_{t_0}^{t} 2k V_\varepsilon(\tau) d\tau \leq V(t_0) - V(t)$$

The Theorem 1 proof is like the proof given in Chapter 12 Appendix A.

Theorem 1 shows that all trajectories of the adaptive system AS_A are limited. Ensuring the asymptotic stability demands the imposing of additional conditions on the system. Consider these conditions.

Let $P(t) \triangleq \left[P_y^T(t) \ \ P_\varphi^T(t) \ \ P_r^T(t)\right]^T, l = 3m$

Definition 1. The vector $P \in \mathbb{R}^{3m}$ is constantly excited (CE) with level α or has the property PE_α if

$$PE_\alpha : P(t) P^T(t) \geq \alpha I_l$$

is true for $\exists \alpha > 0$ and $\forall t \geq t_0$ on some interval $T > 0$.

If the vector $P(t)$ has property PE_α then we will write $P(t) \in PE_\alpha$.

The system (15.1) is stable, and the input $r(t)$ is restricted. Therefore, property PE_α is presented for the matrix $B_P(t) = P(t) P^T(t)$ as:

$$PE_{\alpha,\overline{\alpha}} : \ \alpha I_l \leq B_P(t) \leq \overline{\alpha} I_l$$

$$\forall t \geq t_0 \tag{15.23}$$

where $\overline{\alpha} > 0$ some number.

Let the estimation for $V_K(t)$ is fair

$$0.5 \ \beta_l^{-1}(\Gamma) \|\Delta K(t)\|^2 \ \leq \ V_K(t) \ \leq 0.5 \ \beta_1^{-1}(\Gamma) \|\Delta K(t)\|^2 \tag{15.24}$$

where $\beta_1(\Gamma)$, $\beta_l(\Gamma)$ are minimum and maximum eigenvalues of the matrix Γ.

Proof of the exponential stability is based on ensuring M^{+}-property for functions $\dot{V}_{\varepsilon}(t)$, $\dot{V}_{K}(t)$ (Karabutov, 2017b).

Definition 2. The non-positive quadratic form $W(Y,X)$ has M^{+}-property or $W(Y,X)\in M^{+}$, if it is representable, as:

$$W(Y,X) = -c_{y}\|Y\|^{2} + c_{x}\|X\|^{2}$$

for any $Y \in \mathbb{R}^{m}$, $X \in \mathbb{R}^{n}$ in the restricted area Ω_{D}, where $c_{y} \geq 0,\ c_{x} \geq 0$.

Lemma 1. *$\dot{V}_{\varepsilon}(t)$ have M^{+}-property*

$$\dot{V}_{\varepsilon} \leq -kV_{\varepsilon} + \frac{\overline{\alpha}\beta_{l}(\Gamma)}{k}V_{K} \tag{15.25}$$

The proof of Lemma 1 is given in appendix A.

Lemma 2. *$\dot{V}_{K}(t)$ have M^{+}-property*

$$\dot{V}_{K} \leq -\tfrac{3}{4}\vartheta\alpha\beta_{1}(\Gamma)V_{K} + \tfrac{8}{3}\vartheta V_{\varepsilon} \tag{15.26}$$

The proof of Lemma 2 is given in appendix B.

Application of Lemmas 1, 2 is the basis for exponential stability proof of the adaptive system AS_{A}.

Apply the method of Lyapunov vector functions (LVF). Consider LVF $V_{\varepsilon,K}(t) = \left[V_{\varepsilon}(t)\ \ V_{K}(t)\right]^{T}$. Let such functions $s_{\rho}(t) \geq 0$ exist that:

$$V_{\rho}(t) \leq s_{\rho}(t)$$

$$\forall\left(t \geq t_{0}\right) \,\&\, \left(V_{\rho}(t_{0}) \leq s_{\rho}(t_{0})\right)$$

, (15.27)

where $\rho = \varepsilon, K$.

The analysis of AS_{A} properties is based on the study of the inequality system

$$\begin{bmatrix}\dot{V}_{\varepsilon}\\ \dot{V}_{K}\end{bmatrix} \leq \underbrace{\begin{bmatrix}-k & \dfrac{\overline{\alpha}\ \beta_{l}(\Gamma)}{k}\\ \dfrac{8}{3}\vartheta & -\dfrac{3\alpha\vartheta\beta_{1}(\Gamma)}{4}\end{bmatrix}}_{A_{V}}\begin{bmatrix}V_{\varepsilon}\\ V_{K}\end{bmatrix} \tag{15.28}$$

Apply (15.27) and get comparisons vector system for (15.28):

$$\dot{S} = A_V S, \tag{15.29}$$

where $S = \left[s_\varepsilon \; s_K\right]^T, A_V \in \mathbb{R}^{2\times2}$ is the M-matrix (Gantmacher, 2000). The A_V matrix is an M-matrix if

$$k > 0,$$

$$k \geq \frac{4}{3}\sqrt{\frac{2\overline{\alpha}\beta_l(\Gamma)}{\alpha\beta_1(\Gamma)}}$$

Theorem 2 (Karabutov, 2018). *Let 1) positive definite of Lyapunov functions:*

$$V_\varepsilon(t) = 0.5\,\varepsilon^2(t), V_K(t) = 0.5\Delta\,K^T(t)\,\Gamma^{-1}\Delta K(t)$$

allow the infinitesimal highest limit at $\|\Delta K(t)\| \to 0, |\varepsilon(t)| \to 0$; 2) the vector $P(t)$ is piecewise continuous limited and $P(t) \in PE_{\alpha,\overline{\alpha}}$; 3) equality $e\Delta_N^T P_N = \vartheta\left(\Delta_N^T B_N \Delta^N + e^2\right)$ with $0 < \vartheta$ fairly in the area $O_\nu(O)$ where $\mathbb{O} = \left\{0, 0^{3m}\right\} \subset \mathbb{R} \times \mathbb{R}^{3m} \times \mathbb{J}_{0,\infty}$, O_ν is some neighborhood of the point O; 4) $\dot{V}_\varepsilon \in M^+, \dot{V}_K \in M^+$; 5) the estimation (15.24) is fair for the function $V_K(t)$; 6) the system of inequalities (15.28) is fair for $\dot{V}_\varepsilon$, $\dot{V}_K$; 7) the upper solution for $V_{\varepsilon,K}(t) = \left[V_\varepsilon(t) \;\; V_K(t)\right]^T$ satisfies to the equation (15.29) if inequality (15.27) for elements $V_{\varepsilon,K}$ is fair. Then the system AS_A is exponentially stable with the estimation:

$$V_{\varepsilon,K}(t) \leq e^{A_V(t-t_0)}\,S(t_0), \tag{15.30}$$

if

$$k > 0, k \geq \frac{4}{3}\sqrt{\frac{2\overline{\alpha}\beta_l(\Gamma)}{\alpha\beta_1(\Gamma)}} \tag{15.31}$$

The theorem 2 shows if the vector $P(t) \in PE$, then adaptive systems AS_A gives true parameter estimations of the system (15.7). System parameters must satisfy the condition (15.31).

5.2. System AS_{φ}

Theorem 3. Let 1) functions $V_{\varepsilon}(t)=0,5\,\varepsilon^2(t)$, $V_K(t)\triangleq 0,5\Delta\,K^T(t)\,\Gamma^{-1}\Delta K(t)$ are positive definite and satisfies conditions $\inf\limits_{|\varepsilon|\to\infty} V_{\varepsilon}(\varepsilon)\to\ \infty$, $\inf\limits_{\|\Delta K\|\to\infty} V_K(\Delta K)\to\ \infty$; 2) assumptions 1, 2 for the system (15.1) are performed. Then all trajectories of the system AS_{φ} are limited, belong the area $G_t\ =\ \{(\varepsilon,\Delta K,\Delta P)\ :V(t)\leq\ V(t_0)\}$, and the estimation:

$$\int_{t_0}^{t} 2k\,V_{\varepsilon}(\tau)\,d\tau\ \leq\ V(t_0)-V(t)$$

is fair.

Theorem 3 gives the stability estimate only for the trajectory part of the system AS_{φ}. It shows the limitation of processes at the top level of the adaptive system. At the bottom level, the tuning process of parameters is more difficult and depends on some indicators. Accounting of these indicators is impossible based on the applied criteria. The analysis of the relationships between these levels is acceptable if we consider restrictions on parameters. The AS_{φ} system is a hierarchical system, and we get constraints by applying LVF.

Obtain of exponential stability conditions for the system AS_{φ}. Consider the Lyapunov vector function:

$$V(t)=\left[V_{\varepsilon}(t)\ \ V_K(t)\ \ V_{P_{\varphi}}(t)\ \ V_{W_2}(t)\ \ V_{\sigma}(t)\right]^T \tag{15.32}$$

where " · " is Euclidean norm.

$$V_{P_{\varphi}}(t)=0.5\,\|\Delta\,P_{\varphi}(t)\|^2,\ V_K(t)=0.5\Delta\,K^T(t)\,\Gamma^{-1}\Delta\,K^T(t)$$

$$V_{W_2}(t)=\Delta\,W_2^T(t)\,\Gamma_2^{-1}\Delta\,W_2(t),V_{\varepsilon}(t)=0.5\,\varepsilon^2(t),V_{\sigma}(t)=0.5\,\sigma^2(t)$$

Let the following conditions be satisfied in the neighborhood $\mathbb{O}_{\nu}(O)$ of the equilibrium state O:

$$\begin{aligned}\varepsilon\Delta\,K^T P\ &=\ \vartheta_{K_1}(\varepsilon^2+\Delta\,K^T P\,P^T\Delta K),\\ \varepsilon\Delta\,K^T\Delta\,P_*\ &=\ \vartheta_{K_2}(\varepsilon^2+\Delta\,K^T\Delta\,P_*\Delta\,P_*^T\Delta K),\end{aligned} \tag{15.33}$$

$$\sigma\Delta W_2^T F A_\varphi^T \widetilde{H} = \vartheta_{g_1}\left(\Delta W_2^T F F^T \Delta W_2 + \sigma^2 A_\varphi^T \widetilde{H}\widetilde{H}^T A_\varphi\right),$$

$$\sigma\Delta W_2^T F \Delta A_\varphi^T \widetilde{H} = -\vartheta_{g_2}\left(\Delta W_2^T F F^T \Delta W_2 + \sigma^2 + \Delta A_\varphi^T \widetilde{H}\widetilde{H}^T \Delta A_\varphi\right), \quad (15.34)$$

$$-\sigma\varepsilon\left(P_\varphi + \Delta P_\varphi\right)^T \Gamma_\varphi \Delta P_\varphi = -\vartheta_\sigma\left\{\sigma^2 + \varepsilon^2 + (P_\varphi + \Delta P_\varphi)^T \Gamma_\varphi \Delta P_\varphi\right\},$$

where $\Delta P_* = \left[0^T \ \Delta P_\varphi^T \ 0^T\right]^T$; $\vartheta_{K_1}, \vartheta_{K_2}, \vartheta_{g_1}, \vartheta_{g_2} \ \vartheta_\sigma$ are positive numbers.

Theorem 4. *Let (i) components of the positive definite vector-function $V(t)$ (15.32) have the infinitesimal highest limit; (ii) the vector $P(t)$ is piecewise continuous limited and $P(t) \in PE_{\alpha,\bar{\alpha}}$; (iii) conditions (15.33), (15.34) are satisfied; (iv) elements of the vector $\dot{V}$ have M^+-property; (v) the following conditions are true for the system AS_φ:*

$$F(y, \widehat{W}_1)F^T(y, \widehat{W}_1) \geq \alpha_2 I_{n_2}, A_\varphi^T \widetilde{H}\widetilde{H}^T A_\varphi \geq \upsilon$$

$$\|\widetilde{H}\widetilde{H}^T\| \leq \eta, \|\Delta A_\varphi\|^2 \leq \eta_\varphi \|\Delta K\|^2$$

$$\sigma\left(\Delta A_\varphi + A_\varphi\right)^T\left(\Lambda\Delta P_\varphi + \widetilde{H} F^T \Delta W_2\right) \leq c\sigma^2$$

$$\left(P_\varphi + \Delta P_\varphi\right)^T \Gamma_\varphi \Delta P_\varphi \geq c_\varphi \|\Delta P_\varphi\|^2$$

where $\upsilon > 0$, $\eta > 0$, $\eta_\varphi > 0$, $c > 0$, $c_\varphi > 0$, $\alpha_2 \geq 0$; vi) $\dot{V}$ satisfies to the vector system of inequalities

$$\dot{V} \leq A_{V,\varphi} V \quad (15.35)$$

where:

$$A_{V,\varphi} = \begin{bmatrix} -k & \mu_{\varepsilon,K} & \mu_{\varepsilon,P} & 0 & 0 \\ \mu_{K,\varepsilon} & -\mu_K & \mu_{K,\varphi} & 0 & 0 \\ 0 & 0 & -\beta_{\min}(\Lambda) & \dfrac{\mu\beta_l(\Gamma_2)}{\beta_{\min}(\Lambda)} & 0 \\ 0 & 2\vartheta_{g_2} d\beta_l(\Gamma) & 0 & -\eta_{W_2}\alpha_2\beta_1(\Gamma_2) & \frac{2}{3}\vartheta_{g_1}\upsilon \\ 8\vartheta_\sigma & 0 & 8\vartheta_\sigma c_\varphi & 0 & -\frac{1}{2}c_\sigma \end{bmatrix}$$

(vii) the upper solution for $V(t)$ satisfies to comparison equation

$$\dot{S} = A_{V,\varphi} S, \quad (15.36)$$

if functions $s_\rho(t) \geq 0$ exist such that

$$V_\rho(t) \leq s_\rho(t) \forall (t \geq t_0) \;\&\; \left(V_\rho(t_0) \leq s_\rho(t_0)\right) \tag{15.37}$$

where $S = \left[s_\varepsilon \; s_K \; s_{P_\varphi} \; s_{W_2} \; s_\sigma \;\right]^T$, $\rho = \varepsilon, K, P_\varphi, W_2, \sigma$. *Then the system* AS_φ *is exponentially stable with the estimation* $V(t) \leq \exp\left(A_{V,\varphi}(t - t_0)\right)S(t_0)$, *if* $k > 0$, $k \mu_K > \mu_{K,\varepsilon}\mu_{\varepsilon,K}$, $\beta_{\min}(\Lambda) > 0$,

$$\eta_{W_2}\alpha_2\beta_1(\Gamma_2)(\beta_{\min}(\Lambda))^2\left(k\mu_K - \mu_{\varepsilon,K}\mu_{K,\varepsilon}\right) > 2\vartheta_{g_2} d\beta_l(\Gamma)\mu\beta_l(\Gamma_2)\left(k\mu_{K,\varphi} + \mu_{K,\varepsilon}\mu_{\varepsilon,P}\right)$$

$$c_\sigma\left[\eta_{W_2}\alpha_2\beta_1(\Gamma_2)(\beta_{\min}(\Lambda))^2\left(k\mu_K - \mu_{\varepsilon,K}\mu_{K,\varepsilon}\right) + 4\vartheta_{g_2} d\beta_l(\Gamma)\mu\beta_l(\Gamma_2)\left(k\mu_{K,\varphi} + \mu_{K,\varepsilon}\mu_{\varepsilon,P}\right)\right] > \frac{32}{3}\vartheta_\sigma\mu\beta_l(\Gamma_2)\vartheta_{g_1}\upsilon\left\{\left(\mu_{\varepsilon,K}\mu_{K,\varphi} + \mu_K\mu_{\varepsilon,P}\right) - c_\varphi\left(k\mu_K - \mu_{\varepsilon,K}\mu_{K,\varepsilon}\right)\right\},$$

where:

$$\mu_K = \left(\frac{3}{4}\vartheta_{K_1} + \frac{9}{32}\vartheta_{K_2}\vartheta_{K_3}\right)\beta_1(\Gamma), \eta_{W_2} = \left(\frac{3}{4}\vartheta_{g_1} + \vartheta_{g_2}\right)$$

$$\mu_{\varepsilon,K} = \frac{1}{k}(2\overline{\alpha} + 1)\beta_l(\Gamma), \|A_\varphi\|^2 \leq \chi_\varphi, \chi_\varphi > 0$$

$$\mu_{\varepsilon,P} = \frac{1}{k}\left(2\chi_\varphi + 1\right), \mu = \|\overline{H}\|^2 \max_y \|F(y)\|^2, \mu_{K,\varepsilon} = \frac{4}{3}\left(2\vartheta_{K_1} + \vartheta_{K_2}\right), \mu_{K,\varphi} = \frac{8}{3}\vartheta_{K_2}\vartheta_{K_3}$$

So, we showed that the AS_φ-system is exponentially stable. The difference from the AS_A-system is in the contour introduction for the vector P_φ estimation. Such modernization the AS_A-system gives to the complication of the adaptive system. The matrix $A_{V,\varphi}$ shows the mutual effect of processes in the AS_φ-system. The theorem 4 reflects interdependencies between parameters of the AS_φ-system.

Limitation of the variable σ in (15.19) follows from the theorem 4.

The proof of appendix C 4 is given in appendix C.

6. SIMULATION RESULTS

Consider the system (15.1) of the second order with parameters:

$$\widetilde{A}_{\varphi} = \widetilde{B}_{r} = [0\ 1]^{T}, \Lambda = \lambda = -2, y(0) = 2, \dot{y}(0) = 1$$

$$\widetilde{A} = \begin{bmatrix} 0 & 1 \\ -3 & -4 \end{bmatrix}, \varphi(y) = \begin{cases} w_{23} & if\ y > w_{11} \\ w_{22}y & if\ y \leq w_{11}, r(t) = 3\sin(0.1\pi t) \\ -w_{21} & if\ y \leq w_{12} \end{cases} \quad (15.38)$$

where $w_{11} = 1, w_{12} = -0.5, w_{21} = -1, w_{22} = w_{23} = 2$
Parameters of system (15.4):

$$\widetilde{P}_{v} \in R, \widetilde{P}_{v} = p_{v}, p_{y}(0) = p_{\varphi}(0) = p_{r}(0) = 0 \quad (15.39)$$

The parameter k in (15.8) is 1.5.

Solve the system (15.1) and obtain the set $\mathbb{I}_{o}$. The phase portrait of the system (15.1) is showed in Figure 1.

Figure 1. Phase portrait of system (15.1)

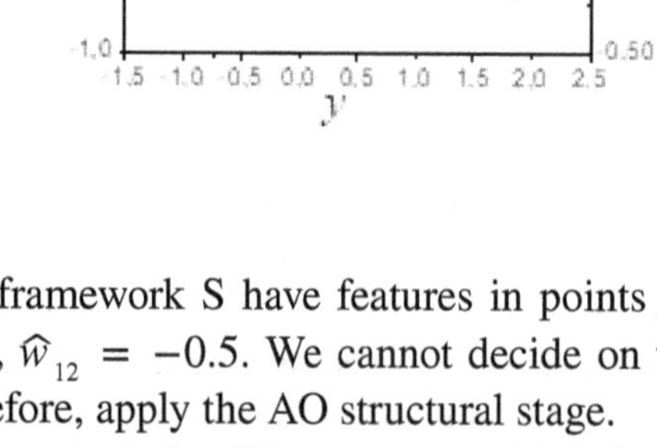

We see that the framework S have features in points $y_{1} = -0.5$ and $y_{2} = 1$. Therefore, $\widehat{w}_{11} = 1$, $\widehat{w}_{12} = -0.5$. We cannot decide on the form $\varphi(y)$ based on the analysis S. Therefore, apply the AO structural stage.

Construct the framework S_{ey}. We see what $\varphi(y)$ belongs to the class of non-linearities with saturation. The framework S analysis shows that the system is h_{δ_h}-identified (see Figure 2).

Figure 2. Framework S_{ey} with the selected fragments

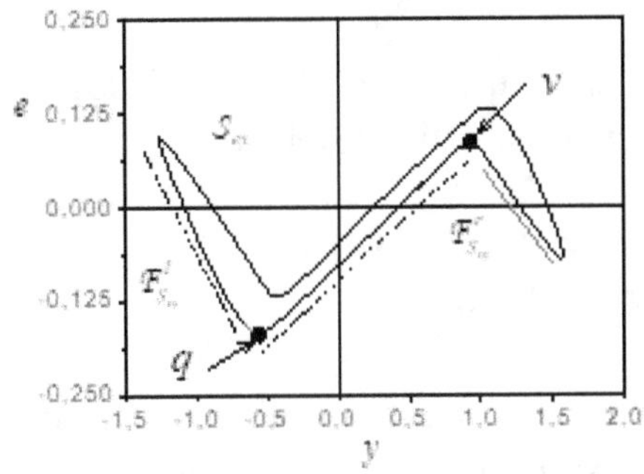

Algorithm for the estimation of saturation parameters on the line section S_{ey}.

1. Estimate h-identifiability of the system.
2. Construct the framework S_{ey}.
3. Estimate the SI of the system. If the system is h_{δ_h}-identifiable, then go to the step 4, otherwise, the end of the algorithm.
4. Go to S_{ey} and allocate the bottom-line section F_{gv}.
5. Calculate the coefficient of structural properties $k_s(t)$ on F_{gv}.
6. Consider the section $F^+_{gv} \subset F_{gv}$ on which $k_s(t)$ and $e(t)$ have identical signs.
7. Approximate the section $F^+_{gv} \subset S_{ey}$ the dependence

$$\hat{e}^+(t) = a_0^+ + a_1^+ y(t)$$

where a_0^+, a_1^+ are determined by the least squares method.

8. Calculate the average $\bar{e}$, $\bar{k}_s$ for $e(t)$, $k_s(t)$ on the time gap of definition F^+_{gv}.
9. Determine the estimation $\hat{a}_1^+$ of true parameter $\bar{a}_1^+$ as

$$\hat{a}_1^+ = \frac{\bar{e}\, a_1^+}{\bar{k}_s}$$

Apply this algorithm and calculate the angle of inclination estimation $\hat{b}_1^+$ of the upper linear part S_{ey} :

(i) lower line section: $\bar{a}_{+}^{+} = 0.18, \widehat{w}_2 = 1.9$;
(ii) upper line section: $\bar{\bar{b}}_1^{+} = 0.186,\ \widehat{w}_2 = 2.08$.

So, we have to parameter w_{22} the area $G_{22} = [1.9; 2.08]$. As transition points in the saturation area are known, saturation levels $\varphi(y)$:

1) if $y_1 = -0.5$ then $\varphi(y_1) \in [-1; -0.95]$;
2) if $y_1 = 1$ then $\varphi(y_2) \in [1.9; 2.08]$.

We average values in points y_1, y_2 and obtain estimations: $\widehat{w}_{21} = -0.975, \widehat{w}_{22} = 1.99$. The estimation for w_{22} is 1.99.

Parametrial Stage of Design AO

The case 1 considered in section 4 gives to model (15.8) and is well studied. Therefore, we will consider the case 2.

Let function $\varphi(y)$ have the form (15.14). Apply for identification $\varphi(y)$ the model:

$$\widehat{\varphi}(y) = \begin{cases} \widehat{w}_2 \widehat{w}_{11} & if\, y > \widehat{w}_{11} \\ \widehat{w}_2 y & if\, y \le \widehat{w}_{11} \\ -\widehat{w}_2 \widehat{w}_{12} & if\, y \le \widehat{w}_{12} \end{cases} \tag{15.40}$$

where $\widehat{w}_2 \triangleq \widehat{w}_{22}$. Estimations for $\widehat{w}_{11}, \widehat{w}_{12}$ are obtained at the stage of structural synthesis.

The adaptive algorithm for $\widehat{w}_2$

$$\dot{\widehat{w}}_2 = \begin{cases} -\gamma_2 \sigma y \hat{a}_{p_\varphi} & if (y \in (-0.5; 1])\ \&\ \left(\dot{\widehat{w}}_2 \notin G_{22}\right) \\ 0 & if (y \notin (-0.5; 1])\ \&\ \left(\dot{\widehat{w}}_2 \in G_{22}\right) \end{cases} \tag{15.41}$$

where $\gamma_2 > 0, \hat{a}_{p_\varphi} \in \mathbb{R}$ is the adjusted coefficient of the model (15.17).

Structure choice of the algorithm (15.41) is directed to compensation, not smoothness effect of the function $\varphi(y)$.

The input $r(t) \in PE_\alpha$ has the form

$$r(t) = 3\sin(0.1\pi t) + \sin(0.25\pi t) \tag{15.42}$$

Figure 3. Frameworks S, S_{ey} for system steady state with the input (15.41)

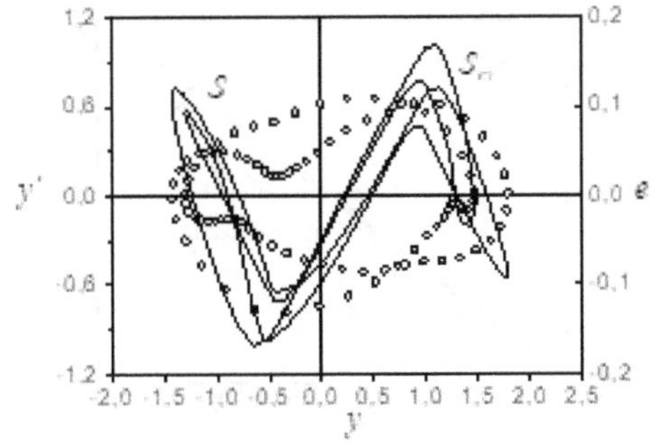

Remark 5. We applied different inputs at structural and parametric stages. Input (15.42) ensures adaptation process with given quality indicators. The input (15.38) is constantly excited and allows identifying the function $\varphi(y)$. Frameworks S, S_{ey} (Figure 3) reflect the effect of the input (15.42) and show the transformation of frameworks presented in Figure 1. SI process is complicated and demands preliminary processing of the set $\mathbb{I}_o$. The analysis S_{ey} does not allow obtaining acceptable estimations for W_1, W_2. These results explain the current status of the design problem AO for nonlinear systems, where the principle of linearization dominates.

Apply the method presented in section 4 to the uncertainty estimation ω in (15.17). The current estimation ω is determined as:

$$\sigma(t)=\begin{cases}\dot{\hat{y}}_d(t)-\dot{y}_d(t) & if \left|\frac{\varepsilon(t)}{y(t)}\right| > 0.01\\ 0 & if \left|\frac{\varepsilon(t)}{y(t)}\right| \leq 0.01\end{cases} \tag{15.43}$$

Structure choice of the algorithm for calculation $\sigma(t)$ considers remarks made for the adaptive algorithm (15.42).

Adaptive model (15.17) has the form

$$\dot{\hat{y}} = -k\varepsilon + \hat{a}_y y + \hat{a}_{p_y} p_y + \hat{a}_{p_\varphi} \hat{p}_\varphi + \hat{b}_{p_r} p_r + \sigma \tag{15.44}$$

Apply algorithms (15.10), (15.12), (15.5, (15.20), (15.22), (15.400, (15.41), (15.43). Matrixes Γ_i for adaptive algorithms

$$\Gamma_y = diag(0.015;0.0066), \Gamma_r = diag(0;0.05), \Gamma_\varphi = diag(0;0.00375), \Gamma_2 = 0.03$$

Figure 4. Model (15.15) adequacy estimation

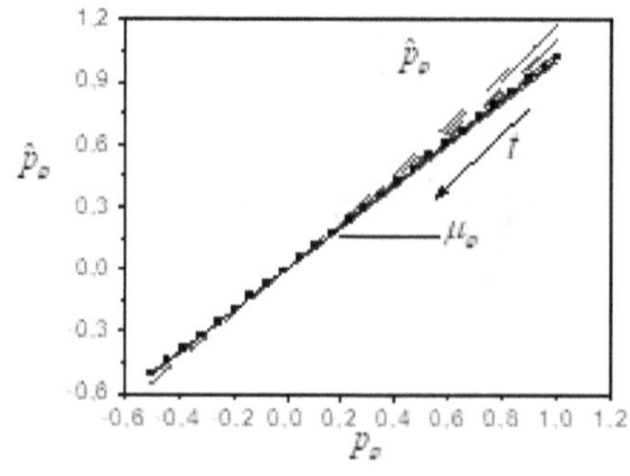

Adaption results of the AS_{φ}-system are shown in Figure 4– 12. Figures 4 and 5 represent the estimation process of the variable $\hat{p}_{\varphi}$ in two planes. Figure 4 shows the adequacy estimation of the model (15.16) in the space $(p_{\varphi}, \hat{p}_{\varphi})$. We represent the linear secant μ_{φ} for the dependence identification between $\hat{p}_{\varphi}$ and p_{φ}. The determination coefficient for μ_{φ} is 0.999. This inference is confirmed with changes $\hat{p}_{\varphi}$ and p_{φ} (Figure 5). We do not show the work results of the model (15.41) as they coincide with the presented charts.

Figure 5. Frameworks reflecting tuning process of the variable $\hat{p}_{\varphi}$

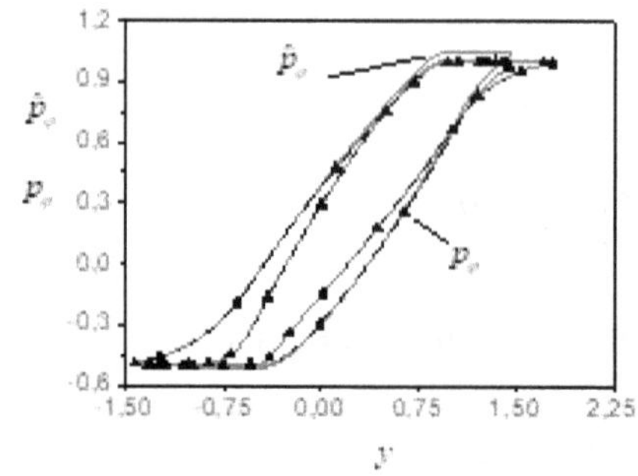

Frameworks showed in Figure 5 give the better understanding of obtaining estimation accuracy for the variable p_{φ}.

The change in the error ε and the uncertainty estimate σ are presented in Figures 6 and 7.

Figure 6. Error ε of AO work

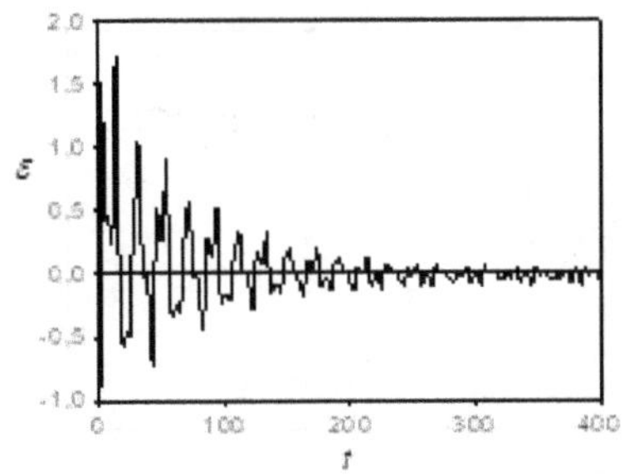

Figure 7. Estimation ω

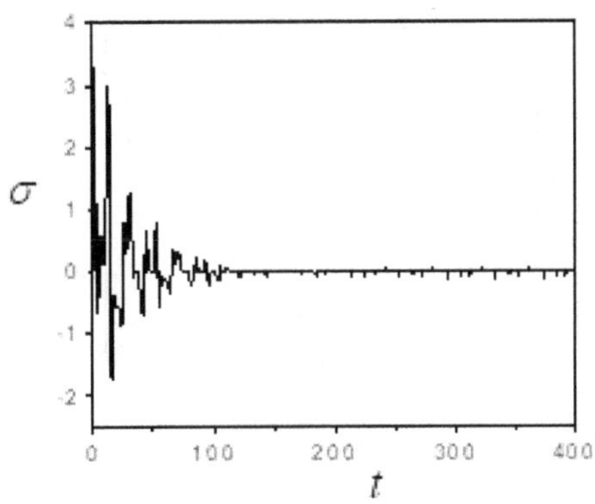

Tuning parameters is shown in Figures 8 - 14. Figure 8 presents to the tuning of parameters $\hat{a}_{p_y}$, $\hat{a}_{p_\varphi}$, $\hat{b}_{p_r}$, and Figure 9 shows tuning $\hat{a}_y$.

Figure 8. Tuning of parameters $\hat{a}_{n}$, $\hat{a}_{n}$, $\hat{b}_{n}$

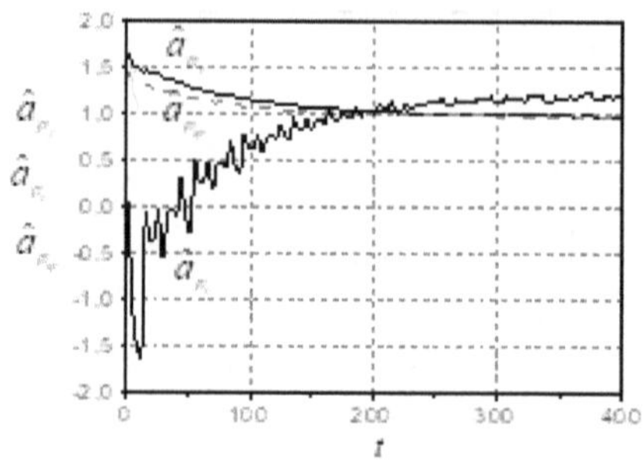

Figure 9. Tuning of parameter $\hat{a}_{n}$

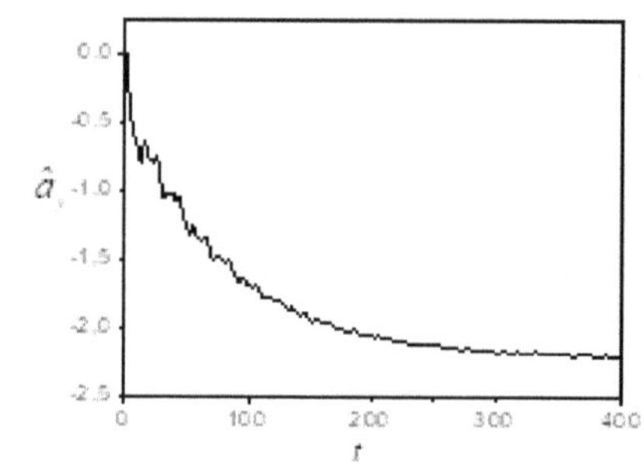

Tuning of the model (15.40) parameter $\hat{w}_2$ is based on the algorithm (15.41) and is shown in Figures 10 - 11. Figure 10 reflects of nonlinear process effect in the adaptive system on the tuning $\hat{w}_2$. We see that the tuning has the nonlinear character. This inference confirms Figure 11.

Figure 10. Effect of the variable y on tuning of the parameter $\widehat{w}_2$

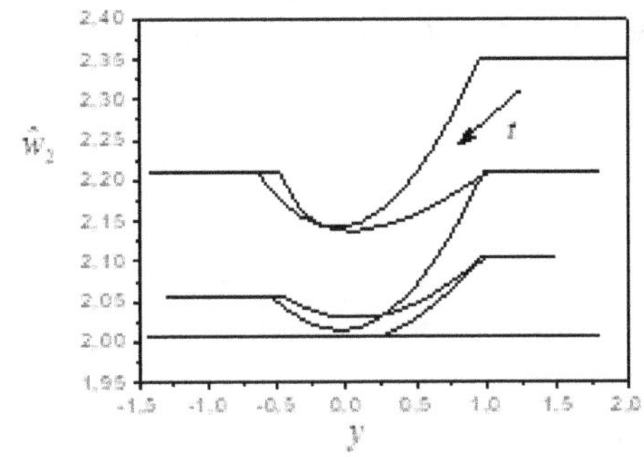

Figure 11. Tuning of parameter $\widehat{w}$,

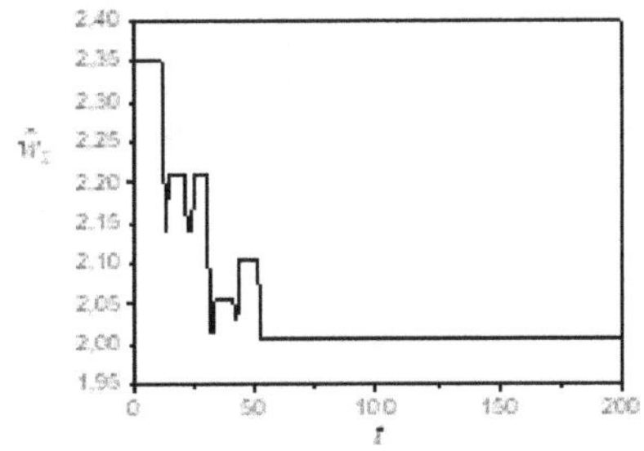

So, modelling results confirm the efficiency of the proposed approach to AO design. Uncertainty influences on properties of AO. The tuning process increases:

1) under the effect of local parameters when the basic parameters of AO are tunings;
2) at the estimation of the nonlinearity and an unknown auxiliary p_φ.

7. CONCLUSION

The adaptive observer design method for nonlinear dynamic systems under uncertainty is proposed. It is based on the application of structural and parametric approaches. The structural approach gives to identification problem solution for the nonlinear part of the system. It is based on the analysis of geometric frameworks S_{ey}, the analysis of which is described in previous parts. We decide on the identifiability of nonlinearity and estimate its parameters based on S_{ey}. The obtained structure and parameters of the nonlinearity are used at the parametric design stage of AO. The asymptotic stability of the adaptive system is proved.

REFERENCES

Baang, D., Stoev, J., & Choi, J. Y. (2007). Adaptive observer using auto-generating *B*-splines. *International Journal of Control, Automation, and Systems*, *5*(5), 479–491.

Barbashin, E. A. (1970). *Lyapunov function*. Nauka.

Bastin, G., & Gevers, R. (1988). Stable adaptive observers for nonlinear time-varying systems. *IEEE Transactions on Automatic Control*, *33*(7), 650–658.

Carrol, R. L., & Lindorff, D. P. (1973). An adaptive observer for single-input single-output linear systems. *IEEE Transactions on Automatic Control*, *18*(5), 428–435.

Carrol, R. L., & Monopoli, R. V. (1975). Model reference adaptive control estimation and identification using only and output signals. *Processings of IFAC 6th Word congress. Boston, Cembridge*, part 1, 58.3/1-58.3/10.

Gantmacher, F. R. (2000). *The Theory of Matrices*. AMS Chelsea Publishing: Reprinted by American Mathematical Society.

Garimella, P., & Yao, B. (2003). Nonlinear adaptive robust observer design for a class of nonlinear systems. *Proceedings of the American Control Conference Denver, Color ado* June 4-6, 2003, 4391-4396.

Karabutov, N. (2016). Structural identification of dynamic systems with hysteresis. *International Journal of Intelligent Systems and Applications*, *8*(7), 1–13.

Karabutov, N. (2017a). Adaptive observers for linear time-varying dynamic objects with uncertainty estimation. *International Journal of Intelligent Systems and Applications*, *9*(6), 1–14. DOI: 10.5815/ijisa.2017.06.01

Karabutov, N. (2017b). Adaptive observers with uncertainty in loop tuning for linear time-varying dynamical systems. *International Journal of Intelligent Systems and Applications*, *9*(4), 1–13.

Karabutov, N. (2017c). Frameworks in problems of structural identification systems. *International Journal of Intelligent Systems and Applications*, *9*(1), 1–19.

Karabutov, N. (2018). Structural-parametrical design method of adaptive observers for nonlinear systems. *International Journal of Intelligent Systems and Applications*, *10*(2), 1–13. DOI: 10.5815/ijisa.2018.02.01

Karabutov, N. N. (2011). *Structural identification of static objects: Fields, structures, methods*. Librokom.

Kreisselmeier, G. (1986). A robust indirect adaptive control approach. *International Journal of Control, 43*(1), 161–175.

Krstić, M., & Kokotović, P. V. (1996). Adaptive Nonlinear Output-Feedback Schemes with Marino-Tomei Controller. *IEEE Transactions on Automatic Control, 41*(2), 274–280.

Kudva, P., & Narendra, K. S. (1973). Synthesis of adaptive observer using Lyapunov direct method. *Jnt. J. Control, 18*(4), 1201–1216.

Marino, R., & Tomei, P. (1992). Global adaptive observers for nonlinear systems via filteres transformations. *IEEE Transactions on Automatic Control, 37*(8), 1239–1245.

Narendra, K. S., & Kudva, B. (1974). Stable adaptive schemes for system: Identification and control. *IEEE Transactions on Systems, Man, and Cybernetics, 4*(6), 542–560.

APPENDIX A

Lemma 1 proof. Write $\dot{V}_\varepsilon(t)$:

$$\dot{V}_\varepsilon = -k\varepsilon^2 + \varepsilon\Delta K^T P \leq -k\varepsilon^2 + |\varepsilon||\Delta K^T P| \quad (A1)$$

Apply the inequalities (15.24) and (Barbashin, 1970)

$$-az^2 + bz \leq -\frac{-az^2}{2} + \frac{b^2}{2a} \quad (A2)$$

where $a > 0,\ \ b \geq 0,\ z \geq 0$. Present (A1) as

$$\dot{V}_\varepsilon = -k\varepsilon^2 + \varepsilon\Delta K^T P \leq -k\varepsilon^2 + |\varepsilon||\Delta K^T P| \leq -kV_\varepsilon + \frac{\overline{\alpha}\beta_l(\Gamma)}{k}V_K \quad (A3)$$

So, we obtain M^+-property for $\dot{V}_\varepsilon$ from (A3).

APPENDIX B

Proof of Lemma 2. Write $\dot{V}_K(t)$

$$\dot{V}_K(t) = -\varepsilon \Delta K^T P \text{ (B1)}$$

Let $(\forall t \geq t^* > t_0)$ & $(\forall (e, \Delta K)) \in O_\nu(O))$ the condition be satisfied

$$\varepsilon \Delta K^T P = \vartheta(\Delta K^T B \Delta K + \varepsilon^2) \text{ (B2)}$$

where $\mathbb{O} = \{0, 0^{3m}\} \subset \mathbb{R} \times \mathbb{R}^{3m} \times \mathbb{J}_{0,\infty}$ is the system equilibrium position, $O_\nu(O)$ is some neighborhood of the point O, $0^{3m} \in \mathbb{R}^{3m}$ is the zero vector, $\vartheta > 0$ is some number, $t \in [0, \infty] = \mathbb{J}_{0,\infty}$.

Apply (B2) and present (B1) as:

$$\dot{V}_K(t) = -\vartheta(\Delta K^T B \Delta K + \varepsilon^2) \leq -\tfrac{3}{4}\vartheta \Delta K^T B \Delta K + \vartheta |\varepsilon| \sqrt{\Delta K^T B \Delta K} \text{ (B3)}$$

Apply inequalities (A2), (15.24) and obtain:

$$\dot{V}_K(t) \leq -\tfrac{3}{4}\vartheta \Delta K^T B \Delta K + \vartheta |\varepsilon| \sqrt{\Delta K^T B \Delta K} \leq -\tfrac{3}{4}\vartheta \alpha \beta_1(\Gamma) V_K + \tfrac{8}{3}\vartheta V_\varepsilon \text{ (B4)}$$

So, M^+-property for $\dot{V}_K$ has the form:

$$\dot{V}_K \leq -\tfrac{3}{4}\vartheta \alpha \beta_1(\Gamma) V_K + \tfrac{8}{3}\vartheta V_\varepsilon.$$

APPENDIX C

Proof of Theorem 4. The proof is based on ensuring M^+-property for elements $\dot{V}(t)$. Apply the approach described in Appendix A.

1. *The function* $V_{P_\varphi}(t) = 0.5\,\|\Delta P_\varphi(t)\|^2$. As:

$$\Delta \dot{P}_\varphi = \Lambda \Delta P_\varphi + \widetilde{H} F^T(y, \widehat{W}_1) \Delta W_2$$

that

$$\dot{V}_{P_\varphi} = \Delta P_\varphi^T \big(\Lambda \Delta P_\varphi + \widetilde{H} F^T \Delta W_2\big) \le -\beta_{\min}(\Lambda)\,\|\Delta P_\varphi(t)\|^2 + \|\Delta P\|\,\|\widetilde{H}\|\,\|F\|\,\|\Delta W_2\|, \quad \text{(C1)}$$

where $\beta_{\min}(\Lambda)$ is the minimum eigenvalue of the matrix Λ.

Apply the inequality (A2) and obtain:

$$\dot{V}_{P_\varphi} \le -\frac{\beta_{\min}(\Lambda)}{2}\|\Delta P_\varphi\|^2 + \frac{\mu}{2\beta_{\min}(\Lambda)}\|\Delta W_2\|^2 \quad \text{(C2)}$$

where $\mu = \|\widetilde{H}\|^2 \max_y \|F(y)\|^2$.

Let the estimation is fair for $V_{W_2}(t) = 0.5\Delta W_2^T(t)\,\Gamma_2^{-1} \Delta W_2(t)$:

$$0.5\beta_l^{-1}(\Gamma_2)\,\|\Delta W_2(t)\|^2 \le V_{W_2}(t) \le 0.5\beta_1^{-1}(\Gamma_2)\,\|\Delta W_2(t)\|^2 \quad \text{(C3)}$$

where $\beta_1(\Gamma_2), \beta_l(\Gamma_2)$ are minimum and maximum eigenvalues of the matrix Γ_2.

Apply (C3) and (C2) write as:

$$\dot{V}_{P_\varphi} \le -\beta_{\min}(\Lambda)\, V_{P_\varphi} + \frac{\mu \beta_l(\Gamma_2)}{\beta_{\min}(\Lambda)} V_2 \quad \text{(C4)}$$

Therefore, $\dot{V}_{P_\varphi} \in \mathrm{M}^+$.

2. *The function* $V_K(t) = 0.5\Delta K^T(t)\,\Gamma^{-1} \Delta K^T(t)$. $\dot{V}_K$ has the form:

$$\dot{V}_K = -\varepsilon \Delta K^T \breve{P}, \quad \text{(C5)}$$

where $\breve{P}(t) \triangleq \left[P_y^T(t)\ \ \widehat{P}_\varphi^T(t)\ \ P_r^T(t)\right]^T$.

The tuning algorithm of ΔA_φ has the form (15.20). Present it as:

$$\Delta \dot{A}_\varphi = \dot{\widehat{A}}_\varphi = -\Gamma_\varphi \varepsilon \widehat{P}_\varphi = -\Gamma_\varphi \varepsilon P_\varphi - \Gamma_\varphi \varepsilon \Delta P_\varphi \quad \text{(C6)}$$

where $\Delta P_\varphi = \widehat{P}_\varphi - P_\varphi$.

Consider (C6) and (C5) write as:

$$\dot{V}_K = -\varepsilon \Delta K^T P - \varepsilon \Delta K^T D_\varphi \Delta P_\varphi, \quad \text{(C7)}$$

where $D_\varphi \in \mathbb{R}^{3m\times 3m}, D_\varphi = 0_m + I_m + 0_m; 0_m \in \mathbb{R}^m, I_m \in \mathbb{R}^m$ are zero and unity matrixes.

Apply the approach described in Appendix B. Then present (C7) as:

$$\dot{V}_K = -\varepsilon \Delta K^T P - \varepsilon \Delta K^T D_\varphi \Delta P_* = \underbrace{-\vartheta_{K_1}(\varepsilon^2 + \Delta K^T P P^T \Delta K)}_{q_1} \underbrace{-\vartheta_{K_2}(\varepsilon^2 + \Delta K^T \Delta P_* \Delta P_*^T \Delta K)}_{q_2}, \quad \text{(C8)}$$

where $\Delta P_* = \left[0^T\ \Delta P_\varphi^T\ 0^T\right]^T$, $0 \in \mathbb{R}^m$ is zero vector, $\vartheta_{K_1} > 0$ and $\vartheta_{K_2} > 0$ are some numbers.

Transform q_1 to the form:

$$q_1 = -\vartheta_{K_1}(\varepsilon^2 + \Delta K^T P P^T \Delta K) \le -\frac{3}{4}\vartheta_{K_1} \alpha \beta_1(\Gamma) V_K + \frac{8}{3}\vartheta_{K_1} V_\varepsilon \quad \text{(C9)}$$

Present q_2 as:

$$q_2 = -\vartheta_{K_2}(\varepsilon^2 + \Delta K^T \Delta P_* \Delta P_*^T \Delta K) \leq -\frac{3}{4}\vartheta_{K_2} \Delta K^T \Delta P_* \Delta P_*^T \Delta K + \vartheta_{K_2}|\varepsilon|\sqrt{\Delta K^T \Delta P_* \Delta P_*^T \Delta K}. \quad \text{(C10)}$$

Apply the approach stated in appendix A and obtain

$$q_2 \leq -\frac{3}{4}\vartheta_{K_2} \Delta K^T \Delta P_* \Delta P_*^T \Delta K + \vartheta_{K_2}|\varepsilon|\sqrt{\Delta K^T \Delta P_* \Delta P_*^T \Delta K} \leq -\frac{3}{8}\vartheta_{K_2} \underbrace{\Delta K^T \Delta P_* \Delta P_*^T \Delta K}_{q_3} + \frac{4}{3}\vartheta_{K_2} V_\varepsilon. \quad \text{(C11)}$$

Transform q_3 to the form

$$q_3 = -\vartheta_{K_3}\left(\Delta K^T \Delta K + \Delta P_\varphi^T \Delta P_\varphi\right) \leq -\frac{3}{4}\vartheta_{K_3}\left\|\Delta K\right\|^2 - \vartheta_{K_3}\left(\frac{1}{2}\|\Delta K\| + \left\|\Delta P_\varphi\right\|\right)^2 + \vartheta_{K_3}\|\Delta K\|\left\|\Delta P_\varphi\right\|$$

where $\vartheta_{K_3} > 0$.

So, we obtain:

$$q_3 \leq -\frac{3}{4}\vartheta_{K_3}\|\Delta K\|^2 + \vartheta_{K_3}\|\Delta K\|\left\|\Delta P_\varphi\right\|$$

$$q_2 \leq -\frac{9}{32}\vartheta_{K_2}\vartheta_{K_3}\|\Delta K\|^2 + \frac{3}{8}\vartheta_{K_2}\vartheta_{K_3}\|\Delta K\|\left\|\Delta P_\varphi\right\| + \frac{4}{3}\vartheta_{K_2} V_\varepsilon.$$

The estimation for q_2 has the form after a simple transformation

$$q_2 \leq -\frac{9}{64}\vartheta_{K_2}\vartheta_{K_3}\|\Delta K\|^2 + \frac{4}{3}\vartheta_{K_2}\vartheta_{K_3}\left\|\Delta P_\varphi\right\|^2 + \frac{4}{3}\vartheta_{K_2} V_\varepsilon. \quad \text{(C12)}$$

Use (C9) and (C12) and obtain the estimation for $\dot{V}_K$:

$$\dot{V}_K \leq -\frac{3}{4}\vartheta_{K_1}\alpha\beta_1(\Gamma)V_K + \frac{8}{3}\vartheta_{K_1} V_\varepsilon - \frac{9}{32}\vartheta_{K_2}\vartheta_{K_3}\beta_1(\Gamma)V_K + \frac{4}{3}\vartheta_{K_2}\vartheta_{K_3}\left\|\Delta P_\varphi\right\|^2 + \frac{4}{3}\vartheta_{K_2} V_\varepsilon \leq -\left(\frac{3}{4}\vartheta_{K_1} + \frac{9}{32}\vartheta_{K_2}\vartheta_{K_3}\right)\beta_1(\Gamma)V_K + \frac{4}{3}\left(2\vartheta_{K_1} + \vartheta_{K_2}\right)V_\varepsilon + \frac{8}{3}\vartheta_{K_2}\vartheta_{K_3} V_{P_\varphi}.$$

So, M⁺-property for $\dot{V}_K$ has the form:

$\dot{V}_K \leq -\mu_K V_K + \mu_{K,\varepsilon} V_\varepsilon + \mu_{K,\varphi} V_{P_\varphi}$ (C13)

where $\mu_K = \left(\frac{3}{4}\vartheta_{K_1} + \frac{9}{32}\vartheta_{K_2}\vartheta_{K_3}\right)\beta_1(\Gamma)$, $\mu_{K,\varepsilon} = \frac{4}{3}\left(2\vartheta_{K_1} + \vartheta_{K_2}\right)$, $\mu_{K,\varphi} = \frac{8}{3}\vartheta_{K_2}\vartheta_{K_3}$.

3. *The function* $V_\varepsilon(t) = 0.5\,\varepsilon^2(t)$. The equation for ε write as:

$\dot{\varepsilon} = -k\varepsilon + \Delta K^T \breve{P} + A_\varphi^T \Delta P_\varphi$ (C14)

$\dot{V}_\varepsilon$ has the form:

$\dot{V}_\varepsilon = -k\varepsilon^2 + \varepsilon\left(\Delta K^T \breve{P} + A_\varphi^T \Delta P_\varphi\right)$ (C15)

As $\breve{P} = P + \Delta P_*$ present (C15) as:

$$\dot{V}_\varepsilon = -k\varepsilon^2 + \varepsilon\left(\Delta K^T P + \Delta K^T \Delta P_* + A_\varphi^T \Delta P_\varphi\right) \leq -kV_\varepsilon + \frac{\overline{\alpha}}{k}\left\|\Delta K\right\|^2 + \frac{\chi_\varphi}{k}\left\|\Delta P_\varphi\right\|^2 + \frac{1}{k}\left|\Delta K^T \Delta P_* \Delta P_*^T \Delta K\right|. \quad \text{(C16)}$$

where $\left\|A_\varphi\right\|^2 \leq \chi_\varphi, \chi_\varphi > 0$.

Apply inequality

$|\Delta K^T \Delta P_* \Delta P_*^T \Delta K| \leq 0.5\left(\|\Delta K\|^2 + \|\Delta P_\varphi\|^2\right)$

and rewrite (C16) as:

$\dot{V}_\varepsilon \leq -kV_\varepsilon + \frac{1}{k}(2\overline{\alpha} + 1)\beta_l(\Gamma)V_K + \frac{1}{k}\left(2\chi_\varphi + 1\right)V_{P_\varphi}.$

So, M$^+$-property for $\dot{V}_\varepsilon$ has the form:

$\dot{V}_\varepsilon \leq -kV_\varepsilon + \mu_{\varepsilon,K}V_K + \mu_{\varepsilon,P}V_{P_\varphi}$ (C17)

where $\mu_{\varepsilon,K} = \frac{1}{k}(2\overline{\alpha} + 1)\beta_l(\Gamma)$, $\mu_{\varepsilon,P} = \frac{1}{k}\left(2\chi_\varphi + 1\right)$.

4. *Lyapunov function* $V_{W_2}(t) = \Delta W_2^T(t)\Gamma_2^{-1}\Delta W_2(t)$. $\Delta W_2(t)$ is described by the equation $\Delta\dot{W}_2 = -\Gamma_2\sigma\widehat{A}_\varphi^T\overline{H}F$. Then:

$$\dot{V}_{W_2} = -\sigma\Delta W_2^T F\widehat{A}_\varphi^T\overline{H} = -\sigma\Delta W_2^T F\left(A_\varphi + \Delta A_\varphi\right)^T\overline{H} = \underbrace{-\sigma\Delta W_2^T F A_\varphi^T\overline{H}}_{g_1} - \underbrace{\sigma\Delta W_2^T F\Delta A_\varphi^T\overline{H}}_{g_2} \quad \text{(C18)}$$

Fulfil the transformation g_1 in (C18):

$g_1 = -\sigma\Delta W_2^T F A_\varphi^T\overline{H} = -\vartheta_{g_1}\left(\Delta W_2^T F F^T\Delta W_2 + \sigma^2 A_\varphi^T\overline{H}\,\overline{H}^T A_\varphi\right)$

where $\vartheta_{g_1} > 0$.

As the assumption 2 is fair and $y(t) \in PE_{\alpha'}$, the matrix $F\left(y, \widehat{W}_1\right)F^T\left(y, \widehat{W}_1\right)$ is constantly excited and

$F\left(y, \widehat{W}_1\right)F^T\left(y, \widehat{W}_1\right) \geq \alpha_2 I_{n_2}$

where $\alpha_2 > 0$ is some number.

Therefore,

$g_1 \leq -\vartheta_{g_1}\alpha_2\|\Delta W_2\|^2 - \vartheta_{g_1}\upsilon\sigma^2$

where $A_\varphi^T\overline{H}\,\overline{H}^T A_\varphi \geq \upsilon, \upsilon > 0$.

Obtain after simple transformations

$$g_1 \leq -\vartheta_{g_1}\alpha_2\left\|\Delta W_2\right\|^2 - \vartheta_{g_1}\upsilon\sigma^2 = -\tfrac{3}{4}\vartheta_{g_1}\alpha_2\left\|\Delta W_2\right\|^2 - \vartheta_{g_1}\left(\tfrac{1}{4}\alpha_2\|\Delta W_2\|^2 + \upsilon\sigma^2 \pm \sqrt{\alpha_2\upsilon}|\sigma|\|\Delta W_2\|\right) \leq$$

$$-\tfrac{3}{4}\vartheta_{g_1}\alpha_2\|\Delta W_2\|^2 + \vartheta_{g_1}\sqrt{\alpha_2\upsilon}|\sigma|\|\Delta W_2\| \leq -\tfrac{3}{4}\vartheta_{g_1}\alpha_2\beta_1(\Gamma_2)V_{W_2} + \tfrac{2}{3}\vartheta_{g_1}\upsilon V_\sigma.$$

So,

$g_1 \leq -\frac{3}{4}\vartheta_{g_1}\alpha_2\beta_1(\Gamma_2)V_{W_2} + \frac{2}{3}\vartheta_{g_1}\upsilon V_\sigma$ (C19)

Consider g_2 in (C18)

$$g_2 = -\sigma\Delta W_2^T F \Delta A_\varphi^T \widetilde{H} = -\vartheta_{g_2}\left(\Delta W_2^T F F^T \Delta W_2 + \sigma^2 + \Delta A_\varphi^T \widetilde{H}\widetilde{H}^T \Delta A_\varphi\right) \le$$

$$-\vartheta_{g_2}\tfrac{3}{4}\Delta W_2^T F F^T \Delta W_2 + \vartheta_{g_2}|\sigma|\sqrt{\Delta W_2^T F F^T \Delta W_2} - \vartheta_{g_2}\Delta A_\varphi^T \widetilde{H}\widetilde{H}^T \Delta A_\varphi \le$$

$$-\vartheta_{g_2}\tfrac{3}{4}\Delta W_2^T F F^T \Delta W_2 + \vartheta_{g_2}|\sigma|\sqrt{\Delta W_2^T F F^T \Delta W_2} - \vartheta_{g_2}\Delta A_\varphi^T \widetilde{H}\widetilde{H}^T \Delta A_\varphi \le$$

$$-\vartheta_{g_2}\tfrac{2}{4}\Delta W_2^T F F^T \Delta W_2 + \vartheta_{g_2}|\sigma|\sqrt{\Delta W_2^T F F^T \Delta W_2} +$$

$$\vartheta_{g_2}\sqrt{\Delta A_\varphi^T \widetilde{H}\widetilde{H}^T \Delta A_\varphi}\sqrt{\Delta W_2^T F F^T \Delta W_2} \le$$

$$\underbrace{\left\{-\vartheta_{g_2}\tfrac{1}{2}\Delta W_2^T F F^T \Delta W_2 + \vartheta_{g_2}|\sigma|\sqrt{\Delta W_2^T F F^T \Delta W_2}\right\}}_{g_{21}} +$$

$$\underbrace{\left\{-\vartheta_{g_2}\tfrac{1}{2}\Delta W_2^T F F^T \Delta W_2 + \vartheta_{g_2}\sqrt{\Delta A_\varphi^T \widetilde{H}\widetilde{H}^T \Delta A_\varphi}\sqrt{\Delta W_2^T F F^T \Delta W_2}\right\}}_{g_{22}},$$

where $\vartheta_{g_2} > 0$.

Transform g_{21}, g_{22} to the form

$$g_{21} = -\tfrac{1}{2}\vartheta_{g_2}\Delta W_2^T F F^T \Delta W_2 + \vartheta_{g_2}|\sigma|\sqrt{\Delta W_2^T F F^T \Delta W_2} \le$$

$$-\tfrac{1}{2}\alpha_2\vartheta_{g_2}\|\Delta W_2\|^2 + \vartheta_{g_2}|\sigma|\sqrt{\Delta W_2^T F F^T \Delta W_2} \le -\tfrac{1}{2}\alpha_2\vartheta_{g_2}\beta_1(\Gamma_2)V_{W_2} + \tfrac{1}{2}\vartheta_{g_2}V_\sigma,$$

$$g_{22} = -\tfrac{1}{2}\vartheta_{g_2}\Delta W_2^T F F^T \Delta W_2 + \vartheta_{g_2}\sqrt{\Delta A_\varphi^T \widetilde{H}\widetilde{H}^T \Delta A_\varphi}\sqrt{\Delta W_2^T F F^T \Delta W_2} \le$$

$$-\tfrac{1}{2}\alpha_2\vartheta_{g_2}\beta_1(\Gamma_2)V_{W_2} + \vartheta_{g_2}\Delta A_\varphi^T \widetilde{H}\widetilde{H}^T\|\Delta A_\varphi\|^2 \le$$

$$-\tfrac{1}{2}\alpha_2\vartheta_{g_2}\beta_1(\Gamma_2)V_{W_2} + \vartheta_{g_2}d\|\Delta K\|^2 \le -\tfrac{1}{2}\alpha_2\vartheta_{g_2}\beta_1(\Gamma_2)V_{W_2} + 2\vartheta_{g_2}d\beta_l(\Gamma)V_K,$$

where $\|\widetilde{H}\widetilde{H}^T\| \le \eta$, $\|\Delta A_\varphi\|^2 \le \eta_\varphi\|\Delta K\|^2$, $\eta > 0$, $\eta_\varphi > 0$.

So,

$$g_2 \le -\alpha_2\vartheta_{g_2}\beta_1(\Gamma_2)V_{W_2} + \tfrac{1}{2}\vartheta_{g_2}V_\sigma + 2\vartheta_{g_2}d\beta_l(\Gamma)V_K \quad \text{(C20)}$$

Apply (C19) and (C20), and obtain the estimation for $\dot V_{W_2}$

$$\dot V_{W_2} \le -\tfrac{3}{4}\vartheta_{g_1}\alpha_2\beta_1(\Gamma_2)V_{W_2} + \tfrac{2}{3}\vartheta_{g_1}\upsilon V_\sigma - \alpha_2\vartheta_{g_2}\beta_1(\Gamma_2)V_{W_2} + \tfrac{1}{2}\vartheta_{g_2}V_\sigma +$$

$$+2\vartheta_{g_2}d\beta_l(\Gamma)V_K \le -\left(\tfrac{3}{4}\vartheta_{g_1} + \vartheta_{g_2}\right)\alpha_2\beta_1(\Gamma_2)V_{W_2} + \tfrac{2}{3}\vartheta_{g_1}\upsilon V_\sigma + 2\vartheta_{g_2}d\beta_l(\Gamma)V_K.$$

So, M⁺-property for $\dot V_{W_2}$ has the form:

$$\dot V_{W_2} \le -\eta_{W_2}\alpha_2\beta_1(\Gamma_2)V_{W_2} + \tfrac{2}{3}\vartheta_{g_1}\upsilon V_\sigma + 2\vartheta_{g_2}d\beta_l(\Gamma)V_K \quad \text{(C21)}$$

where $\eta_{W_2} = \left(\tfrac{3}{4}\vartheta_{g_1} + \vartheta_{g_2}\right)$.

5. *The Lyapunov function* $V_\sigma(t) = 0.5\sigma^2(t)$. As σ is the estimation ω in (15.17) and $V_\sigma(t) = 0.5\omega^2(t)$, where:

$$\omega = \widehat{A}_\varphi^T\left(\widehat{P}_\varphi - P_\varphi\right) = \widehat{A}_\varphi^T\Delta P_\varphi = \left(A_\varphi + \Delta A_\varphi\right)^T\Delta P_\varphi$$

$\dot V_\sigma$ has the form:

$$\dot V_\sigma = \sigma\left(\widehat{A}_\varphi^T \Delta P_\varphi\right)' = \sigma\left(A_\varphi^T \Delta P_\varphi + \Delta A_\varphi^T \Delta P_\varphi\right)' =$$
$$\sigma \Delta \dot A_\varphi^T \Delta P_\varphi + \sigma\left(A_\varphi + \Delta A_\varphi\right)^T \Delta \dot P_\varphi =$$
$$-\sigma\varepsilon \widehat{P}_\varphi^T \Gamma_\varphi \Delta P_\varphi + \sigma\left(\Delta A_\varphi + A_\varphi\right)^T\left(\Lambda \Delta P_\varphi + \widetilde{H} F^T \Delta W_2\right) = \qquad (C22)$$
$$-\sigma\varepsilon \Delta P_\varphi^T \Gamma_\varphi \Delta P_\varphi - \sigma\varepsilon P_\varphi^T \Gamma_\varphi \Delta P_\varphi + \sigma\left(\Delta A_\varphi + A_\varphi\right)^T\left(\Lambda \Delta P_\varphi + \widetilde{H} F^T \Delta W_2\right).$$

Let positive numbers c, ϑ_σ and c_φ exist such that:

$$-\sigma\varepsilon\left(P_\varphi + \Delta P_\varphi\right)^T \Gamma_\varphi \Delta P_\varphi = -\vartheta_\sigma\left\{\sigma^2 + \varepsilon^2 + \left(P_\varphi + \Delta P_\varphi\right)^T \Gamma_\varphi \Delta P_\varphi\right\}$$

$$\sigma\left(\Delta A_\varphi + A_\varphi\right)^T\left(\Lambda \Delta P_\varphi + \widetilde{H} F^T \Delta W_2\right) \le c\sigma^2 \quad (C23)$$

$$\left(P_\varphi + \Delta P_\varphi\right)^T \Gamma_\varphi \Delta P_\varphi \ge c_\varphi \left\|\Delta P_\varphi\right\|^2$$

Apply (C23) to (C22), and obtain the inequality for $\dot V_\sigma$:

$$\dot V_\sigma \le -\vartheta_\sigma\left(\sigma^2 + \varepsilon^2 + c_\varphi \left\|\Delta P_\varphi\right\|^2\right) + c\sigma^2 = -c_\sigma \sigma^2 - \vartheta_\sigma \varepsilon^2 - \vartheta_\sigma c_\varphi \left\|\Delta P_\varphi\right\|,$$

where $c_\sigma = \left(\vartheta_\sigma - c\right) > 0$.

Apply the approach stated above and get M+-property for $\dot V_\sigma$:

$$\dot V_\sigma \le -\tfrac{1}{2} c_\sigma V_\sigma + 8\vartheta_\sigma V_\varepsilon + 8\vartheta_\sigma c_\varphi V_{P_\varphi} \quad (C24)$$

Use M+-properties for elements of the vector $\dot V$ and obtain

$$\begin{bmatrix} \dot V_\varepsilon(t) \\ \dot V_K(t) \\ \dot V_{P_\varphi}(t) \\ \dot V_{W_2}(t) \\ \dot V_\sigma(t) \end{bmatrix} \le \underbrace{\begin{bmatrix} -k & \mu_{\varepsilon,K} & \mu_{\varepsilon,P} & 0 & 0 \\ \mu_{K,\varepsilon} & -\mu_K & \mu_{K,\varphi} & 0 & 0 \\ 0 & 0 & -\beta_{\min}(\Lambda) & \dfrac{\mu\beta_l(\Gamma_2)}{\beta_{\min}(\Lambda)} & 0 \\ 0 & 2\vartheta_{g_2} d\beta_l(\Gamma) & 0 & -\eta_{W_2}\alpha_2\beta_1(\Gamma_2) & \tfrac{2}{3}\vartheta_{g_1} \upsilon \\ 8\vartheta_\sigma & 0 & 8\vartheta_\sigma c_\varphi & 0 & -\tfrac{1}{2}c_\sigma \end{bmatrix}}_{A_{V,\varphi}} \begin{bmatrix} V_\varepsilon(t) \\ V_K(t) \\ V_{P_\varphi}(t) \\ V_{W_2}(t) \\ V_\sigma(t) \end{bmatrix} \quad (C25)$$

The matrix $A_{V,\varphi}$ is M-matrix and stability conditions to have the form: $(-1)^i m_i\left(A_{V,\varphi}\right) > 0$, where $m_i(A_{V,\varphi})$ is ith the diagonal minor of the matrix $A_{V,\varphi}$, $i = \overline{1,5}$. Then:

$$k > 0, k\mu_K > \mu_{K,\varepsilon}\mu_{\varepsilon,K}, \beta_{\min}(\Lambda) > 0,$$

$$\eta_{W_2}\alpha_2\beta_1(\Gamma_2)\left(\beta_{\min}(\Lambda)\right)^2\left(k\mu_K - \mu_{\varepsilon,K}\mu_{K,\varepsilon}\right) > \qquad (C26)$$
$$2\vartheta_{g_2} d\beta_l(\Gamma)\mu\beta_l(\Gamma_2)\left(k\mu_{K,\varphi} + \mu_{K,\varepsilon}\mu_{\varepsilon,P}\right)$$

$$c_\sigma\left[\eta_{W_2}\alpha_2\beta_1(\Gamma_2)\left(\beta_{\min}(\Lambda)\right)^2\left(k\mu_K - \mu_{\varepsilon,K}\mu_{K,\varepsilon}\right) + 4\vartheta_{g_2} d\beta_l(\Gamma)\mu\beta_l(\Gamma_2)\left(k\mu_{K,\varphi} + \mu_{K,\varepsilon}\mu_{\varepsilon,P}\right)\right] >$$
$$\tfrac{32}{3}\vartheta_\sigma \mu\beta_l(\Gamma_2)\vartheta_{g_1} \upsilon\left\{\left(\mu_{\varepsilon,K}\mu_{K,\varphi} + \mu_K\mu_{\varepsilon,P}\right) - c_\varphi\left(k\mu_K - \mu_{\varepsilon,K}\mu_{K,\varepsilon}\right)\right\}.$$

Parameters choice AS_{φ}-system allows ensuring the fulfilment of conditions (C26). Construct the comparison system (SC) for (C25).

Let functions $s_{\rho}(t) \geq 0$ exist such that

$V_{\rho}(t) \leq s_{\rho}(t) \forall (t \geq t_0) \ \& \ (V_{\rho}(t_0) \leq s_{\rho}(t_0))$ (C27)

where $\rho = \varepsilon, K, P_{\varphi}, W_2, \sigma$.

SC has the form:

$\dot{S} = A_{V,\varphi} S$, (C28)

where $S = \left[s_{\varepsilon} \ s_K \ s_{P_{\varphi}} \ s_{W_2} \ s_{\sigma} \right]^T$.

The solution of system (C28)

$S(t) = \exp\left(A_{V,\varphi}(t - t_0)\right) S(t_0)$ (C29)

As (C27) and (C29) are fulfilled, the estimation is fair for $V(t)$

$V(t) \leq \exp\left(A_{V,\varphi}(t - t_0)\right) S(t_0)$

Chapter 16
AO for Linear Time-Varying Dynamic Systems

ABSTRACT

A method synthesis of adaptive observer (AO) is proposed for nonstationary linear dynamic systems (DS) when the constant excitation (CE) condition is not fulfilled. We consider AO synthesis as a solution to two problems. The solution to the first problem is to choose a constant matrix to reduce the CE impact. The procedures for obtaining this matrix are proposed. The matrix is considered constraints on AO parameters. The synthesis of adaptive multiplicative algorithms is the second problem solution. We propose procedures for estimating uncertainty in the system. Static model predicts uncertainty change. Uncertainty estimates are adjusted to minimize the error between the outputs of the system and the AO. We prove an exponential dissipative of the adaptive system.

1. INTRODUCTION

For a linear class of dynamical systems, the theory of adaptive observers (AO) (Carrol & Lindorff, 1973; Carrol & Monopoli 1975; Kreisselmeier, 1986; Kudva & Narendra, 1973; Narendra & Kudva1974; Nuyan & Carrol, 1979) is proposed at the end of the 20th century. The class of the adaptive systems having a special identification representation in space input-output was considered. Despite this, the research in this area continues. In particular, design methods of AO for time-varying systems (TVS) are proposed. Various approaches are offered to the AO construction for TVS (Bastin and Gevers, 1978; Feiler and Narendra2006; Karabutov, 2007; Jing et al., 2014; Zhang et al., 2001; Zhang et al., 2003).

The problem of TVS identification, as before, is actuated. The excitation constancy (EC) condition is a basis for the design of effective adaptive systems (AS). This condition is often not fulfilled. Estimates of the AO stability have not always

DOI: 10.4018/979-8-3373-0822-7.ch016

been studied under EC. The second important problem is the parametric disturbances estimation in AS. Parametric perturbations are assumed to be restricted. Methods for the parametric uncertainty estimation were not offered.

We propose the AO design method for a linear dynamic TVS, when EC condition non-fulfilment.

2. PROBLEM STATEMENT

Consider the system

$$x^{(m)} + \tilde{a}_m(t)x^{(m)} + \ldots + \tilde{a}_1(t)x = \tilde{b}(t)r, y = x \tag{16.1}$$

where $r \in \mathbb{R}$, $y \in \mathbb{R}$ are input and output, $\tilde{\Theta}(t) = [\tilde{a}_1(t)\ \tilde{a}_2(t)\ \ldots\ \tilde{a}_m(t)\ \tilde{b}(t)]^T$, $\tilde{\Theta} \in \mathbb{G}_\theta$, $\mathbb{G}_\theta \subseteq \mathbb{R}^{m+1}$ is restricted and a priori an unknown area.

Components of the vector $\tilde{\theta}(t)$ have the form

$$\tilde{\theta}_i(t) = \tilde{\theta}_i^0 + \Delta\tilde{\theta}_i(t), i = \overline{1, m+1}$$

where $|\Delta\tilde{\theta}_i(t)| \leq \bar{\theta}_i = \text{const} > 0$, $\tilde{\theta}_i^0 = \text{const}$.

The laws of change $\tilde{\theta}_i(t)$ are unknown.

Assumption 1. The $\Delta\tilde{\theta}_i(t)$ change velocity is restricted and unknown.

Assumption 2. The vector $\Delta\tilde{\Theta}(t) = \tilde{\Theta}(t) - \tilde{\Theta}^0$ belongs to set

$$\Delta\tilde{\Theta} \in \mathbb{G}_\theta \backslash \mathbb{G}_{\theta^0}$$

, (16.2)

where $\tilde{\Theta}^0 = [\tilde{a}_1^0\ \tilde{a}_2^0\ \ldots\ \tilde{a}_m^0\ \tilde{b}^0]^T$.

(16.2) is the condition of uncertainty. The experimental information:

$$\mathbb{I}_o = \{r(t), y(t), t \in \mathbb{J}\},$$

where $\mathbb{J} \subset \mathbb{R}$is the specified interval of the time.

The EC condition is not fulfilled for an input $r(t)$.

Problem: design adaptive observer for system (16.1) that the condition:

$$\lim_{t\to\infty} |\hat{y}(t) - y(t)| \leq \pi_y \tag{16.3}$$

is satisfied, where $\hat{y}(t)$is output AO, $\pi_y \geq 0$.

3. EQUATION FOR AO

Write the equation (16.1) in the form

$$\dot{y} = \Theta^T P + f(y, r, \Delta\widetilde{\Theta}) \tag{16.4}$$

where $P \in \mathbb{R}^{2m}$ is the generalized input representing an output of the auxiliary filter:

$$\dot{P}_1 = [\Lambda \dotplus \Lambda]\, P_1 + \left[I_1^T y \; I_1^T r\right]^T \tag{16.5}$$

$P = \left[y \; P_y^T \; r \; P_r^T\right]^T, P_1 = \left[P_y^T \; P_r^T\right]^T, P_1 \in \mathbb{R}^{2m-2}$ is the vector of auxiliary signals obtained on the basis of transformation $r(t)$ and $y(t)$; $\dotplus$ is the direct sum sign of matrixes; $\Lambda \in \mathbb{R}^{m-1}$ is the diagonal matrix with $\lambda_i < 0$ $(i = \overline{1, m-1})$; $I_1 \in \mathbb{R}^{m-1}$ is unit vector; $\Theta \in \mathbb{R}^{2m}$ is a vector corresponding $\widetilde{\Theta}^0$; $f(y, r, \Delta\widetilde{\Theta}) \in \mathbb{R}$ is a component depending on the parametric disturbance $\Delta\widetilde{\Theta}$ in (16.1).

Assumption 3. The input $r(t)$ does not satisfy the EC condition.

Assumption 4. $|f(y, r, \Delta\widetilde{\Theta})| < \infty$ at $|r(t)| < \infty$.

Apply to the system described by the equation (16.4) adaptive model (Karabutov, 2017):

$$\dot{\hat{y}} = -k(\hat{y} - y) + N^T H^T P + u \tag{16.6}$$

where $H \in \mathbb{R}^{2m \times l}$ is a matrix with constant parameters, $N \in \mathbb{R}^l$ is a vector of adjusted parameters, $u \in \mathbb{R}$ is a control, $l < 2m$.

Present $u(t)$ as

$u(t) = \overline{D}^T(t) P(t)$ (16.7)

where $\overline{D} \in \mathbb{R}^{2m}$ is some restricted vector.

Write the vector $\overline{D}$ as:

$$\overline{D}^T(t) = \mu(t) D \tag{16.8}$$

where $D \in \mathbb{R}^{2m}$, $\|\overline{D}(t)\| \leq d_0, d_0 > 0$ is some number, $\mu(t) \in \mathbb{R}$ is controlled variable.

We suppose that

$$\mu(t) \in \mathbb{G}_\mu = \{\mu \in \mathbb{R} : |\mu(t)| \leq 1 \forall t \geq t_0\} \tag{16.9}$$

The vector Dchoice ensures fulfilment of the condition (16.9).

4. DESIGN OF ADAPTIVE ALGORITHMS

Write the equation for an identification error $e(t)$ as:

$$\dot{e} = -ke + N^T H^T P - \Theta^T P - f(y,u,\Delta\Theta) + \mu z \quad (16.11)$$

where $z = D^T P$.

We see that the vector $HN(t)$ should approximate unknown parameters of the system (16.4). As parametric uncertainty exists, $e(t)$ cannot be reduced to zero. Apply control $u(t)$ to ensure fulfilment of the condition (16.3).

Linearize function $f(\cdot)$

$$f(y,u,\Delta\Theta,t) = \Delta\,\Theta^T(t)\,P(t)$$

where $\Delta\Theta(t)$ is some vector.

We assume that

$$|f(y,u,\Delta\Theta,t)| \leq \rho_f = \text{const} \geq 0 \quad (16.12)$$

Choose dimension l of the vector $N(t)$ based on frequency properties of the input $r(t)$. Designate an error of approximation $\Theta + \Delta\Theta(t)$ with the vector $HN(t)$ as $\delta_N(t)$:

$$\delta_N(t) = HN(t) - \Theta - \Delta\Theta(t)$$

$\delta_N(t)$ satisfies (16.2) and depends on the parametric uncertainty of the system (16.4). Uncertainty depends from $\delta_N(t)$ and satisfies the condition (16.12). Therefore, the condition for $\widehat{\Theta}(t) = \Theta + \Delta\Theta(t)$

$$\widehat{\Theta}(t) \in \mathbb{G}_A^k \Leftrightarrow \hat{\alpha}_\theta \leq \left\|\widehat{\Theta}(t)\right\| \leq \alpha_\theta.$$

is true.

Then

$$\|HN(t)\| \leq \alpha_N, \alpha_N < \infty$$

We obtain from $\delta_N(t)$ inequality

$$\nu = |\alpha_N - \alpha_\theta| \leq \|\delta_N(t)\| \leq |\alpha_N - \hat{\alpha}_\theta| = \nu_1 \tag{16.13}$$

Then the vector $N(t)$ belongs to set

$$N \in \mathbb{G}_N = \{N \in \mathbb{R}^l : |n_i(t)| \leq 1 i = \overline{1,l}\} \tag{16.14}$$

at the corresponding choice of matrix H.

Writhe (16.11) as

$$\dot{e} = -ke + \delta_N^T P + \mu z \tag{16.15}$$

Transform the equation (16.15) to the form:

$$\dot{e} = -ke + \sigma_N^T P - \sigma_\Delta^T P + \mu z$$

where $\sigma_N(t) = HN(t) - \Theta$, $\sigma_\Delta = \Delta\Theta(t)$. Let $h_\Delta = \sigma_\Delta^T P$.

Apply the method φ-algorithms (Karabutov, 2007) to the design of adaptation algorithms. Consider Lyapunov function (LF) $V_e(t) = 0.5 e^2(t)$. F-form is true for the derivative $\chi(t) = \dot{V}_e(t)$:

$$\chi = -ke^2 + \Psi_N^T(N)\Phi_N(e, N, H, P) + e\sigma_\Delta^T P + e\mu z \tag{16.16}$$

where $\Phi_N = eP^T(I^T D(|N|)I)^{-1}\sigma_N$, $\Psi_N(N) = I^T D(|N|)I$, $|N|$ is the absolute value from the vector $N(t)$, $D(|N|)$ is a diagonal matrix from the vector N, $I \in \mathbb{R}^l$ is the unit vector.

Obtain φ-algorithm of the vector $N(t)$ adaptation from D2 (16.16):

$$\dot{N} = -e\Gamma P(N) H^T P \tag{16.17}$$

where $\Gamma \in \mathbb{R}^{l \times l}$ is a symmetrical positive-definite matrix, $P(N)$ is a matrixed indicator belonging to $N \in \mathbb{G}_N$.

Various methods apply for the definition $P(N)$. If the indicator $P(N)$ is presented as:

$$P\left(N\right) = \begin{cases} I_l, N(t) \in \mathbb{G}_N \\ 0_l, N(t) \notin \mathbb{G}_N \end{cases}$$

that we have the multiplicative φ-algorithm for $N(t)$:

$$\dot{N} = \begin{cases} -e\Gamma H^T P, N(t) \in \mathbb{G}_N, \\ 0_l, N(t) \notin \mathbb{G}_N, \end{cases} \tag{16.18}$$

where I_l, 0_l unit and zero matrixes $l \times l$.

For accuracy raise, matrix H can be computed on each interval of a quasi-stationary system.

The control law is designed for the variable $\mu(t)$. Let conditions (16.8), (16.9) are satisfied. Determine the tuning law $\mu(t)$ from the condition:

$$\min_n \max_{\delta_N} \dot{V}_e(n, \delta_N, t) \tag{16.19}$$

$\max \dot{V}(t)$ on $\delta_N(t)$ is got on domain boundary (16.13), that is at such values $\delta_N(t)$, when $\|\delta_N(t)\| = \upsilon_1$. Therefore, transform $\chi(e, \mu, t)$to form

$$\chi \le -ke^2 + \Psi_\mu(e, \mu, z)\, \Phi_\mu(e, \mu, z) + |e|\, \|\delta_N\|\, \|P\| \tag{16.20}$$

where $\Psi_\mu = |e|\,|D^T P|$, $\Phi_\mu = \mu \operatorname{sgn}(e) \operatorname{sgn}(D^T P)$.

As$\mu \in \mathbb{G}_\mu$ the AO is robust to uncertainty $h_\Delta = \sigma_\Delta^T P$, if

$\|D\| > c(\upsilon_1)$ (16.21)

where $c(\upsilon_1) > 0$ is some number.

We obtain the following tuning φ-algorithms of the variable $\mu(t)$ based on (16.20) and (16.19):

(i) static $M_{s,s}$-algorithm

$$\mu(t) = -\gamma_s \operatorname{sgn}(e(t)) \operatorname{sgn}(D^T P(t)) \tag{16.22}$$

(ii) dynamic $M_{d,s}$-algorithm

$$\dot{\mu}(t) = -\gamma_{d,s} P(\mu) \operatorname{sgn}(e(t)) \operatorname{sgn}(D^T P(t)) \tag{16.23}$$

(iii) dynamic M_d- algorithm

$$\dot{\mu}(t) = -\gamma_d P(\mu)\, e(t)\, D^T P(t) \tag{16.24}$$

where $P(\mu)$ is the belonging $\mu(t) \in \mathbb{G}_\mu, \gamma_s > 0, \gamma_d > 0, \gamma_{d,s} > 0$.
Other algorithms for $\mu(t)$ can be obtained from the condition $\chi \leq 0$.

5. CHOICE OF MATRIX *H*

Consider the equation (16.4). Apply the differentiation operation and get $\dot{y}$. Designate the obtained variable as $\dot{y}_d(t)$. Choose time gap $\mathbb{J}_d \subset \mathbb{J}$ to the change y_d so that the system properties on it were homogeneous. It means that the trajectory $y_d(t)$, and hence, $y(t)$ should not have structural modifications.

Consider the equation

$$\dot{y}_d = \bar{H}^T P$$

on the interval $\mathbb{J}_d$, where $\bar{H} \in \mathbb{R}^{2m}, \bar{H} = HI, I \in \mathbb{R}^{2m}$.

Apply the least-squares method and obtain the vector $\bar{H}$ as:

$$\min_{\bar{H}} (\dot{y} - \dot{y}_d)^2 \rightarrow \bar{H}$$

Demands to the estimation $\bar{H}$.

H1. $\bar{H}$ ensures the following condition of dominance:

$\dot{y}(t) < \dot{y}_d(t)$ for almost $\forall t \in \mathbb{J}_* \subset \mathbb{J}$,

where $\mathbb{J}_*$ contains all t, since some $t \geq t_* > t_0$.

H2. Elements of vector $\bar{H}$ ensure the performance of the adaptive system.

Condition H2 is very important, as tuning of vector $N(t)$ in (16.6) depends on the elements of the vector $\bar{H}$. As $l \leq 2m$ we consider only such estimations $N(t)$, which ensure minimization of functional $Q(N) = (H(\bar{H})N(t) - \Theta)^2$.

Remark 1. The estimation $\bar{H}$ is obtained from a quasi-stationary condition of the system. This estimation has the approached character, sufficient for the implementation of the adaptive system.

Remark 2. If AS trajectories are not restricted on set $\bar{H}$, fulfil one of the following operations.

1. Obtain the estimation $\bar{H}$ for $t \in \mathbb{J}_{*,1}$, where $\mathbb{J}_{*,1} \neq \mathbb{J}_*$.
2. Change dimension l of the vector $N(t)$. It will lead to replacement of the vector $\bar{H}$ by the matrix H.

6. UNCERTAINTY ESTIMATION

Obtaining $\overline{D}$ is the complicated problem under uncertainty. Therefore, apply the following approach. It is based on obtaining of a variable for an estimation $h_\Delta(t)$.

Construct model

$$\hat{y}_d = \overline{H}_d^T P \tag{16.25}$$

for prediction of the change of the variable $\dot{y}_d(t)$ and compute values of the error $\varepsilon_d(t) = \hat{y}_d(t) - \dot{y}_d(t)$. $\varepsilon_d(t)$ is the uncertainty estimation in the system.

Remark 3. We have introduced the vector $\overline{H}_d$ to distinguish it from the vector $\overline{H}$ used for formation of the matrix H in model (16.6).

We present model (16.6) and the equation for the identification error (16.15):

$$\begin{aligned} SED{:}\dot{\hat{y}} &= -k(\hat{y} - y) + N^T H^T P + \mu_\varepsilon \varepsilon_d, \\ \dot{e} &= -ke + \sigma_N^T P + g_d, \\ \dot{\mu}_\varepsilon &= -\gamma_\varepsilon e \varepsilon_d, \end{aligned} \tag{16.26}$$

where $g_d = \mu_\varepsilon \varepsilon_d - \sigma_\Delta^T P$, μ_ε is the adjusted parameter.

μ_ε sets the level ε_d that disturbances to compensate in the adaptive system. Disturbances are the result of the application of the model $\hat{y}_d = \overline{H}^T P$ and action $h_\Delta(t)$.

Remark 4. Demands to the vector $\overline{H}$, considered in section 5, should be considered, when the variable $\varepsilon_d(t)$ is determined.

Another approach is based on obtaining of uncertainty estimations based on current adaptation results use. Apply the model:

$$\hat{y}_{d,N} = N^T H^T P \tag{16.27}$$

and determine the error $\varepsilon_N(t) = \hat{y}_{d,N}(t) - \dot{y}_d(t)$. ε_N is the current estimation of uncertainty $h_\Delta(t)$.

The model (16.27) use complicates the adaptation process. The AS operation quality can be improved under certain conditions if the model (16.27) is used. Here dimension of vector $N(t)$ has a significant effect. Therefore, the reasonable compromise should be observed between the vector $N(t)$ dimension and restrictions.

For the case (16.27), we pretend the model (16.6) and the equation for identification error (16.15) as:

$$\begin{aligned} SEN{:}\dot{\hat{y}} &= -k(\hat{y} - y) + N^T H^T P + \mu_N \varepsilon_N, \\ \dot{e} &= -ke + \sigma_N^T P + g_N, \\ \dot{\mu}_N &= -\gamma_N e \varepsilon_N, \end{aligned} \tag{16.28}$$

where $g_N = \mu_N \varepsilon_N - \sigma_\Delta^T P$, μ_N is the adjusted parameter.

7. PROPERTIES OF ADAPTIVE SYSTEMS

Consider properties of the adaptive system $AS_{N\mu}^{ss}$, which are described by equations (16.7), (16.8), (16.15), (16.17), (16.22). Boundedness of trajectories follows from the following statement.

Theorem 1. *Let: 1) function* $V_e(t) = 0,5\,e^2(t)$is positive definite and satisfies the condition $\inf\limits_{\|X\|\to\infty} V_e(e,t) \to \infty$; 2) assumptions 1, 2 are fulfilled for object (16.1); 3) parameters $N(t)$, $\mu(t)$of the model (16.6) belong to restricted areas $\mathbb{G}_N$, $\mathbb{G}_\mu$. Then all trajectories of system $AS_{N\mu}^{ss}$ are restricted and fair estimations:

$$\int_{t_0}^{t} \left(2V_e(\tau) - \gamma_s \varsigma \sqrt{2V_e(\tau)}\right) d\tau \le V(t_0) - V(t)$$

$$V_e(t) \le \frac{1}{k^2}\left(\upsilon_P^2 \max_t \|\delta_N(t)\|^2 + \gamma_s^2 \upsilon_z\right) \tag{16.29}$$

where $\varsigma = \min\limits_t |D^T P|$,

$\upsilon_P = \max\limits_t \|P(t)\|^2$

,

$\upsilon_z = \max\limits_t |z(t)|$

,

$$V(t) = V_e(t) + \int_{t_0}^{t} \delta_N^T(\tau)\, \Gamma^{-1} \Psi_N(N)\, \delta_N(\tau)\, d\tau \tag{16.30}$$

The proof of theorem 1 is given in Appendix A.

The theorem 1 shows that all system $AS_{N\mu}^{ss}$ trajectories are restricted. Limiting properties $AS_{N\mu}^{ss}$-system depends on properties of the adaptive algorithm (16.17) and the control law (16.22) for variable $\mu(t)$. $M_{s,s}$-algorithm ensures fulfilment of the condition (16.3). The error $e(t)$ decrease depends on the choice of the parameter γ_s of $M_{s,s}$-algorithm.

Consider $AS_{N\mu}^{ds}$-system described by equations (16.7), (16.8), (16.15), (16.17) and (16.23).

Theorem 2. *Let conditions of the theorem 1 are satisfied. Then all trajectories of the system* $AS^{ds}_{N\mu}$ are restricted and fair estimations (16.29) with $\gamma_s^2 = 1$, and:

$$\int_{t_0}^{t} 2V_e(\tau)d\tau \leq V(t_0) - V(t)$$

So $M_{s,s}$, $M_{d,s}$-algorithms ensure fulfilment of the target condition (16.3). Specific properties of algorithms do not allow to improve the AS work quality.

See the Theorem 2 proof in Appendix B.

Consider $AS^{d}_{N\mu}$-system described by the equations (16.7), (16.8), (16.15), (16.17) and (16.24).

Theorem 3. *Let conditions of the theorem 1 are satisfied. Then all trajectories of the system* $AS^{d}_{N\mu}$ are restricted and fair the estimation:

$$V_e(t) \leq \frac{\upsilon_P}{k^2} \max_t \|\delta_N(t)\|^2$$

The proof of theorem 3 is given in Appendix C.

In $AS^{d}_{N\mu}$-system, the condition (16.3) can satisfy with $\pi_y = 0$. It is possible if the vector $HP(t)$ is constantly excited.

Consider properties of the algorithm (16.18). The equation for the error:

$$\dot{e} = -ke + \sigma_N^T P - \sigma_\Delta^T P + \mu z \tag{16.31}$$

where $\sigma_N(t) = H(N(t) - N_\Theta) = H\Delta_N(t)$, $\Theta = HN_\Theta, \sigma_\Delta = \Delta\Theta(t)$.

We believe that LF has the M^+-property (see chapter 15).

Definition 1. The vector $P_H(t) = H^T P(t)$, $P_H \in R^l$ is constantly excited (CE) with level α or has property PE_α if:

$$PE_\alpha : P_H(t) P_H^T(t) \geq \alpha I_l$$

for some $\alpha > 0$ and $\forall t \geq t_0$ on some interval $T > 0$.

The system (16.1) is stable, and the input $r(t)$ is restricted. Therefore, property PE_α for the matrix $B_H(t) = P_H^T(t) P_H(t)$ present as:

$$PE_{\alpha,\bar{\alpha}} : \alpha I_l \leq B_H(t) \leq \bar{\alpha} I_l \forall t \geq t_0 \tag{16.32}$$

where $\bar{\alpha} > 0$ is some number.

Consider LF $V_N(t) = 0.5\Delta_N^T(t)\Gamma^{-1}\Psi_N(N)\Delta_N(t)$. Let

$$0.5\ \lambda_l^{-1}(\Gamma)\,\|\Delta_N(t)\|^2 \ \leq\ V_N(t)\ \leq 0.5\ \lambda_1^{-1}(\Gamma)\,\|\Delta_N(t)\|^2 \tag{16.33}$$

where $\lambda_1(\Gamma)$, $\lambda_l(\Gamma)$ are the minimum and maximum eigenvalues of the matrix Γ.

Lemma 1. Function $V_N(t)$ has M^+-property, if

$$\dot V_N \leq -\frac{3\alpha\vartheta\lambda_1(\Gamma)}{4} V_N + \tfrac{4}{3}\vartheta V_e$$

where $\vartheta > 0$ is some number.

The proof of lemma 1 is given in Appendix D.

Ensure the condition $\dot V_e \in M^+$.

Lemma 2. Function $V_e(t)$ has M^+-property if

$$\dot V_e \leq -kV_e + \frac{2\overline{\alpha}\ \lambda_l(\Gamma)}{k} V_N + \tfrac{1}{k}\left(\upsilon_P \max_t \|\sigma_\Delta\|^2 + \upsilon_z\right) \tag{16.34}$$

where $\|\sigma_N\|$ is the Euclidean norm of the vector σ_N.

The proof of lemma 2 is given in Appendix E.

Remark 5. M^+-property functions V_e and V_N are conditions of the exponential dissipativity (ED) for the AS. Passivity conditions of the adaptive system follow from these properties.

Show the ED of the AS (16.18), (16.31). Apply the method of Lyapunov vector functions (LVF).

Let functions $s_\rho(t) \geq 0$ exist such that

$$V_\rho(t) \leq s_\rho(t)\forall(t \geq t_0)\ \&\ \left(V_\rho(t_0) \leq s_\rho(t_0)\right), \rho = e, N \tag{16.35}$$

Apply lemmas 1 and 2 and obtain the system of inequalities for AS:

$$\begin{bmatrix}\dot V_e\\ \dot V_N\end{bmatrix} \leq \underbrace{\begin{bmatrix} -k_e & \dfrac{2\overline{\alpha}\ \lambda_l(\Gamma)}{k} \\ \tfrac{4}{3}\vartheta & -\dfrac{3\alpha\vartheta\lambda_1(\Gamma)}{4}\end{bmatrix}}_{A_V} \underbrace{\begin{bmatrix}V_e\\ V_N\end{bmatrix}}_{W_V} + \underbrace{\begin{bmatrix}\tfrac{1}{k}\left(\upsilon_P \max_t \|\sigma_\Delta(t)\|^2 + \upsilon_z\right)\\ 0\end{bmatrix}}_{F_V} \tag{16.36}$$

where $A_V \in \mathbb{R}^{2\times 2}$ is M-matrix.

Vector system of comparisons for (16.36)

$$\dot S = A_V S + F_V \tag{16.37}$$

Stability conditions of the matrix A_V (Gantmacher, 2000): $-m_1(A_V)> 0,\ m_2(A_V)> 0$, where m_1, m_2 are diagonal minors of the matrix A_V.

These conditions have the form:

$$k > 0, k \geq \frac{4}{3}\sqrt{\frac{2\bar{\alpha}\lambda_l(\Gamma)}{\alpha\lambda_1(\Gamma)}}$$

So, it's true the following statement.

Theorem 4. *Let 1) Lyapunov functions $V_N(t)= 0.5\Delta_N^T(t)\Gamma^{-1}\Psi_N(N)\Delta_N(t)\, V_e(t)= 0.5\, e^2(t)$ have the infinitely big limit at $|e(t)| \to \infty$, $\|\Delta_N(t)\| \to \infty$; 2) the vector $P_H(t)$ is piecewise continuous restricted and $P_H(t)\in PE_{\alpha,\bar{\alpha}}$; 3) equality $e\Delta_N^T P_N = \vartheta(\Delta_N^T B_N \Delta^N + e^2)$ with $0 < \vartheta$ fair in area $O_v(O)$; 4) $\dot{V}_e \in M^+$, $\dot{V}_N \in M^+$; 5) the estimation (16.33) is fair for the function $V_N(t)$; 6) the upper solution for Lyapunov vector function $V(t)= [\ V_e(t)\ \ V_N(t)\ ]^T$ satisfies the equation (16.37); 7) uncertainty $\sigma_\Delta(t)$ in (16.31) is restricted: $\|\sigma_\Delta(t)\| < \upsilon_\Delta\ \forall t \geq t_0$, where $\upsilon_\Delta \geq 0$ is some number. Then the system (16.18), (16.31) is exponentially dissipative and estimations are fair:*

$$V(t)\leq e^{A_V(t-t_0)} S(t_0)+\int_{t_0}^{\infty} e^{A_V(t-\tau)} F_V(\tau)\,d\tau$$

$$V(t)\leq \frac{1}{k}\left(\upsilon_P \max_t \|\sigma_\Delta(t)\|^2 + \upsilon_z\right)\Sigma_1(t)\big|_{t\to\infty}$$

where $\Sigma_1(t)$ is the first column of the matrix $\int_{t_0}^{t} e^{A_V(t-\tau)}\,d\tau$, if

$k > 0,$

$$k \geq \frac{4}{3}\sqrt{\frac{2\bar{\alpha}\lambda_l(\Gamma)}{\alpha\lambda_1(\Gamma)}}$$

The theorem 4 shows that the algorithm (16.18) is identifying if the vector $P_H(t)\in PE_{\alpha,\bar{\alpha}}$. System dissipative properties depend on a control algorithm for the variable $\mu(t)$. We consider control algorithm properties of the variable $\mu(t)$ above.

Consider AS properties with the uncertainty $h_\Delta(t)$ estimation. Designate system SED, (16.18) as $ASED_{N\mu_e}$. Boundedness of the variable $\varepsilon_d(t)$ follows from the method of its obtaining.

Theorem 5. *Let conditions 1), 2) theorem 1 are satisfied. Then all trajectories* $AS_{N\mu_e}$- system is restricted and the estimation fair:

$$V_e(t) \leq \frac{1}{k^2}\left(\overline{\alpha}\max_t \|\Delta_N(t)\|^2 + \max_t h_\Delta^2(t)\right)$$

The proof of theorem 5 is given in Appendix F.

Consider $ASEN_{N\mu_N}$-system: SEN and (16.18). Let number $\mu_{*,N}$ exists such that $h_\Delta(t) = \mu_{*,N}\varepsilon_N(t)$. Designate $\delta\mu_N = \mu_N - \mu_{*,N}$.

Theorem 6. *Let conditions 1), 2) of theorem 1 are satisfied and: (i)* $\underline{d}_N \leq \varepsilon_N^2 \leq d_N$, *where* $\underline{d}_N \geq 0, d_N > 0$*are some numbers; (ii) positive definite function* $V_{\mu_N}(t) = 0.5\gamma_N^{-1}(\delta\mu_N(t))^2$ *satisfies the condition* $\inf_{\|X\|\to\infty} V_e(e,t) \to \infty$. *Then all trajectories* $ASEN_{N\mu_N}$*-system are restricted with the estimation:*

$$V(t) \leq V(t_0) - 2k\int_{t_0}^{t} V_e(\tau)\,d\tau$$

where:

$$V(t) = V_e(t) + V_{\mu_N}(t) + V_{\sigma_N}(t) = 0.5e^2(t) + 0.5\gamma_N^{-1}(\delta\mu_N(t))^2 + 0.5\sigma_N^2(t)\Gamma^{-1}\sigma_N(t)$$

Let further conditions are satisfied: 3) M^+- properties for $\dot{V}_e$, $\dot{V}_N$ have the form:

$$\begin{bmatrix}\dot{V}_e\\ \dot{V}_{\mu_N}\end{bmatrix} \leq \underbrace{\begin{bmatrix}-k & \frac{2d_N}{k}\\ \upsilon & -\frac{3}{4}\upsilon\underline{d}_N\gamma_N^2\end{bmatrix}}_{A_{\mu_N}}\begin{bmatrix}V_e\\ V_{\mu_N}\end{bmatrix} + \underbrace{\begin{bmatrix}\frac{\overline{\alpha}}{k}\max_t\|\Delta_N\|^2\\ 0\end{bmatrix}}_{F_{\mu_*}} \tag{16.39}$$

4) functions $s_\rho(t) \geq 0$ exist such that

$$V_\rho(t) \leq s_\rho(t)\forall(t \geq t_0)\ \&\ \left(V_\rho(t_0) \leq s_\rho(t_0)\right), \rho = e,\ \mu_N$$

5) the upper solution of the system (16.39) satisfies to the vector system of comparison

$$\dot{S}_{e,\mu_N}(t) = A_{\mu_N} S_{e,\mu_N}(t) + F_{\mu_N}$$

where $S_{e,\mu_N} = \begin{bmatrix} s_e & s_{\mu_N} \end{bmatrix}^T$; 6) uncertainty $h_\Delta(t) = \sigma_\Delta^T(t) P(t)$in (16.31) is restricted. Then $ASEN_{N\mu_N}$-system is exponentially dissipative and estimations are fair:

$$V(t) \leq e^{A_{\mu_N}(t-t_0)} S_{e,\mu_N}(t_0) + \int_{t_0}^{t} e^{A_{\mu_N}(t-\tau)} F_{\mu_N}(\tau)\, d\tau$$

$$V(t) \leq \frac{\overline{\alpha}}{k} \max_t \|\Delta_N\|^2 \Sigma_{1,\mu}(t)\big|_{t\to\infty}$$

if

$$k > 0, k^2 \underline{d}_N \gamma_N^2 \geq \frac{8}{3} d_N$$

where $\Sigma_1(t)$ *is the first column of the matrix* $\int_{t_0}^{t} e^{A_\gamma(t-\tau)} d\tau$.

See proof of theorem 6 in Appendix H.

Theorem 6 shows that properties $ASEN_{N\mu_N}$-system depends on the adaptive algorithm (16.18) work quality. Therefore, we choose the vector $N(t)$ maximum dimension of the *SEN*-system to the misalignment $\Delta_N(t)$ decrease. Explain it to that quality of the tuning $N(t)$ effects on the magnitude of the error $e(t)$ and ε_N.

8. RESULTS OF MODELING

Consider the system (16.1) second order with parameters

$\tilde{a}_1(t) = -2 + 0.3\sin(0.04\pi t), \tilde{a}_2(t) = -3 + 0.5\sin(0.09\pi t). \tilde{b} = 1.5, \tilde{a}_2$ $(t) = -3 + 0.5\sin(0.09\pi t).\ \tilde{b} = 1.5, r(t)$

$$= 2 + \sin(0.3\pi t).$$

The equation (16.1) was integrated with the step 0.2 s. $x(0) = 2$, $\dot{x}(0) = 1$. Obtain the informational set $\mathbb{I}_o$. Parameter $\Lambda$is (16.5) is -1.5, $P_y(0) = 0$, $P_r(0) = 0$, $k = 1$.

Consider *SED*-system. Apply the following approach to the get of the variable ε_d. Fulfil segmentation of variable $y(t)$ values for $t \in \mathbb{J}$, based on structural homogeneity observance for modification $y(t)$. Apply the method described in section 4, and find the vector $\overline{H}_d$ on each subinterval $\mathbb{J}_i \subset \mathbb{J}$. Choose the value $\overline{H}_d$, ensuring the maximum value of determination coefficient for the model $\hat{y}_d = \overline{H}_d^T P$. So, $\overline{H}_d = [0.28;\ -1.19;\ 0.59]^T$ is on the interval [39.2; 79.2]s. Fulfil the prediction $\dot{y}(t)$, used by the model $\hat{\dot{y}}_d = \overline{H}_d^T P$, and determine ε_d. Next, apply the variable ε_d for implementation *SED*-system. Show change ε_d in Figure 1.

Figure 1. Uncertainty estimation based on application SED-system

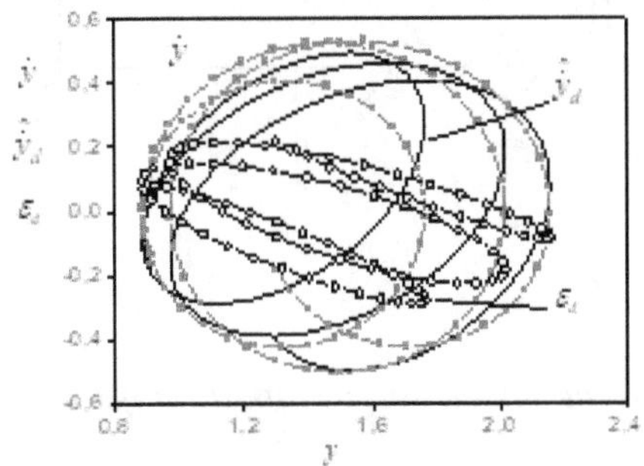

Determine the matrix H for implementation *SED*-systems. Apply the approach described in section 5. The vector $\overline{H} = [-1.75; 0.65; 1.8]^T$for (16.6) is obtained on the time gap [12.2; 19.2] s. The matrix H:

$$H = diag(\overline{H}) \tag{16.40}$$

The vector $N(t) \in R^2$ is adjusted by algorithm (16.18). Matrix Γ in (16.18) has the form $\Gamma = diag(0.005; 0.05)$. Factor γ_ε in (16.26) $\gamma_\varepsilon = 5$. Work results of *SED*-system are shown in Figures 2 - 6.

Figure 2. Vector N(t) tuning

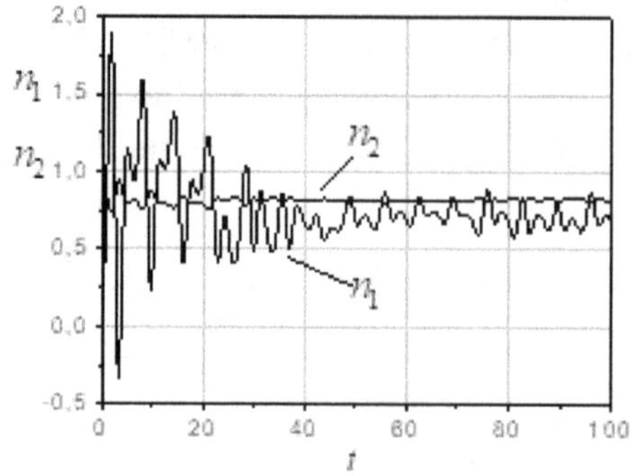

Figure 3. Errors e, ε_d and the uncertainty estimation η_d

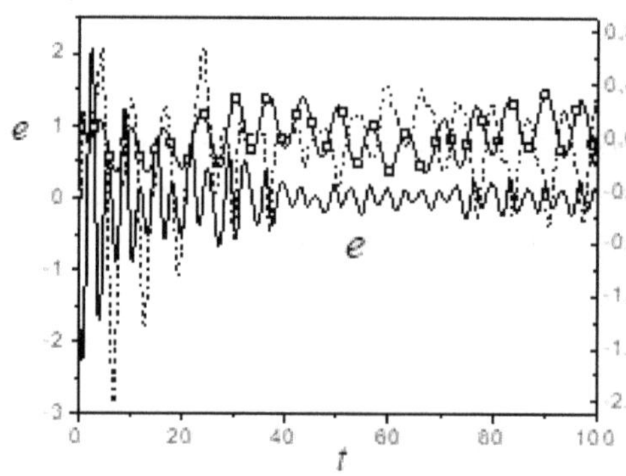

Figure 2 represents the vector $N(t)$ tuning. Errors $e(t)$, ε_d and the current value of uncertainty $\eta_d = \mu_\varepsilon \varepsilon_d$ are presented in the Figure 3.

Show change of parameters $\left[\hat{\theta}_1(t) \quad \hat{\theta}_2(t) \quad \hat{\theta}_3(t)\right]^T = HN(t)$ in Figure 4. Tuning of the variable μ_ε for system (16.26) is shown in Figure 5. System (16.4) parameters are varied in ranges:

$\theta_1(t) = [-1.95; -1.25]$, $\theta_2(t) = [0.4; 0.95]$,

$\theta_3 = 1.5$

.

Figure 4. Tuning of parameters for systems (16.4)

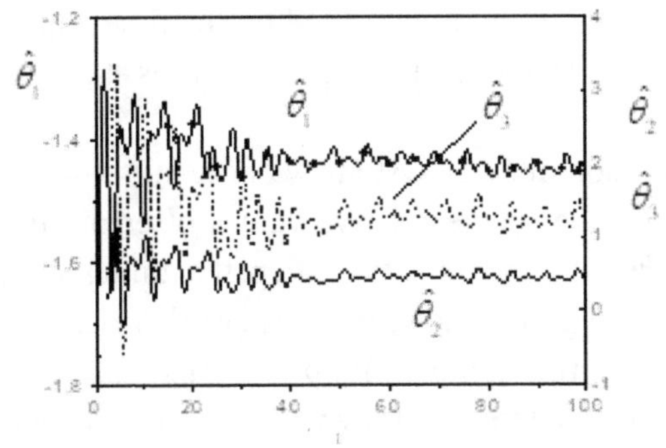

Figure 5. Change of parameter μ_c

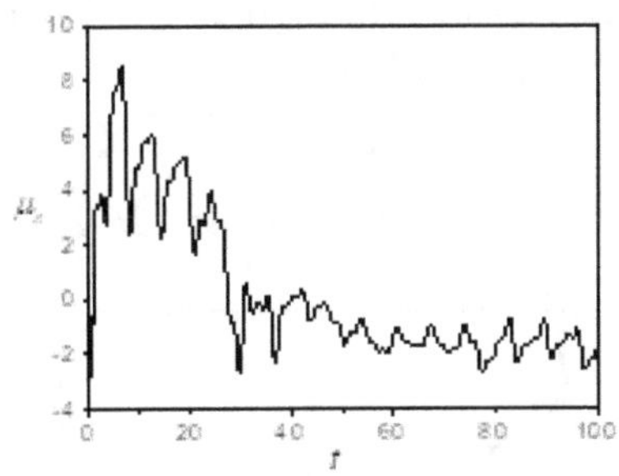

Figure 6. Work quality comparison of the system (16.26) with one and two adjusted parameters

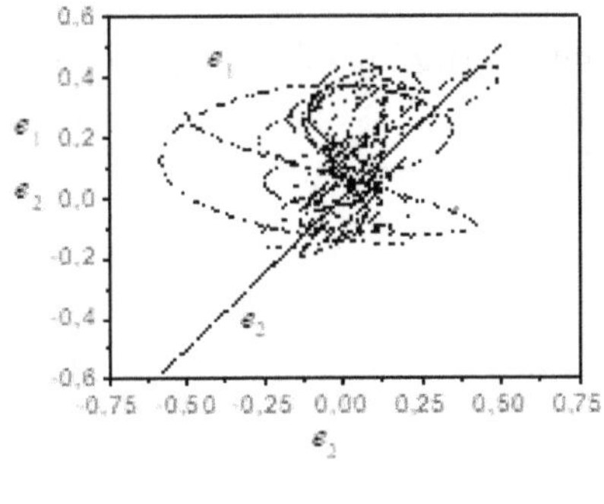

The analysis of obtained results confirms the workability of the adaptive *SED*-systems. Explain such change: (i) application of a static model (SM) for uncertainty identification; (ii) getting of parameters SM on the time gap. This note is genuine for the tuning of vector $N(t)$ elements. The AO work is stabilized at the tuning time increase of parameters.

We give a comparison of *SED*-systems work results with one and two adjusted parameters (Figure 6). Designate relative values of errors for *SED*-systems with $N \in \mathbb{R}$ and $N \in \mathbb{R}^2$ as $e_1(t)$ and $e_2(t)$. We except the tuning initial stage of the model (16.26) vector $N(t)$The Figure 6 shows the framework S_e described by the function $f(e_2): e_2(t) \rightarrow e_1(t)$. S_e given the comparative position $e_1(t)$ and $e_2(t)$, starting $t \geq 19.2$s. We see that the adaptive system accuracy with $N \in \mathbb{R}$ is lower than with $N \in \mathbb{R}^2$. The framework S_e is more informative than change $e_i(t)$ on the plane (t, e_i). We have made this conclusion based on the theorem 6 for the $ASEN_{N\mu_N}$-system. It is also true and for $ASED_{N\mu_e}$-system.

Consider the $ASEN_{N\mu_N}$-system. The matrix H has the form (16.40), $N(t) \in \mathbb{R}^2$. Matrix Γ in (16.18): $\Gamma = diag(0.004; 0.0007)$. The factor γ_N of *SEN*-system is 0.0006. Show results for the $ASED_{N\mu_N}$-system in Figures 7 - 10. Figure 7 represents tuning of the vector $N(t)$ for the system (16.28).

Figure 7. Tuning of model parameters for system (16.28)

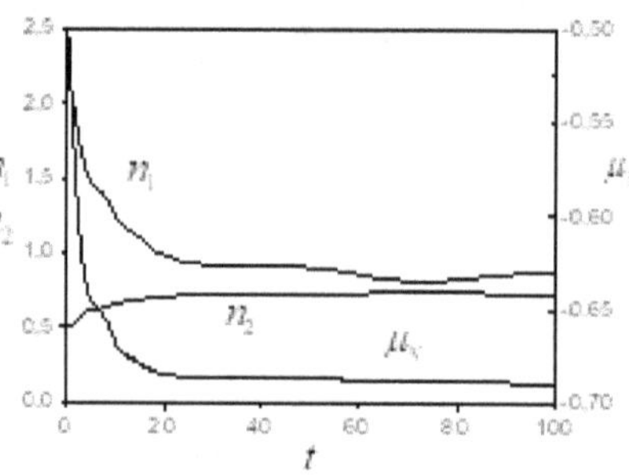

Compare these results with the tuning of system (16.26) parameters. We see that the adaptation process in the $ASEN_{N\mu_N}$-system has a monotonic character. Explain this the adequate estimation of the uncertainty $h_\Delta(t)$.

Figure 8. Estimations for system (16.4) parameters got by system (16.28)

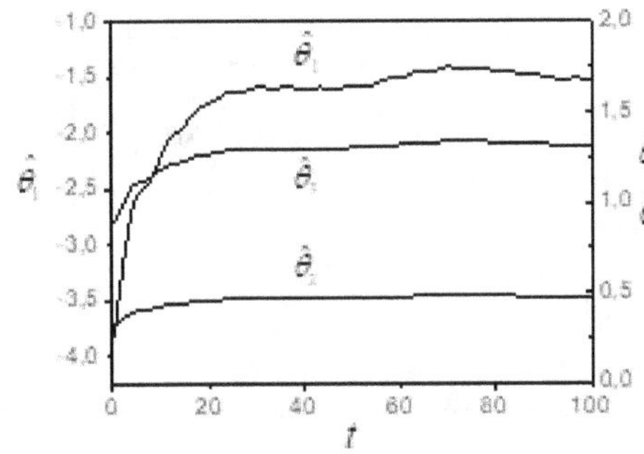

Figure 9 contains results for the work $ASED_{N\mu_N}$-system. We see that the AS is exponentially dissipative.

Figure 9. Adequacy estimation of work system (16.28)

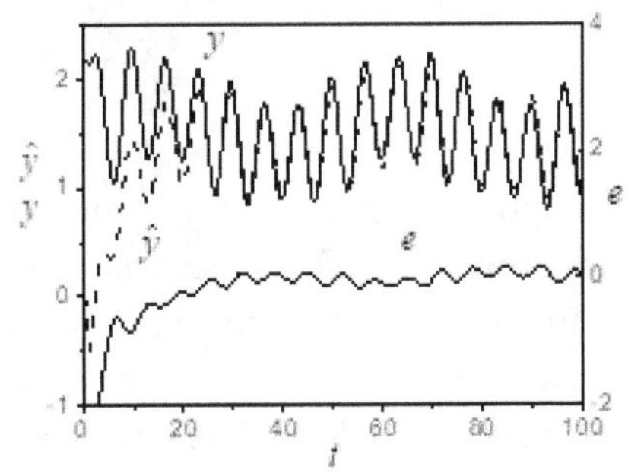

In Figure 10, we present uncertainty estimation results at different parameters of systems (16.26), (16.28). Designations in Figure 10: 1 are systems (16.26) with $N(t) \in \mathbb{R}$ and the static law $\eta_d = 0.9\varepsilon_d$ for the estimation $h_\Delta(t)$; 2 is systems (16.28) with $N(t) \in \mathbb{R}^2$ and the static law $\eta_N = \varepsilon_N$; 3 is systems (16.28) with $N(t) \in \mathbb{R}^2$ and the algorithm $\dot{\mu}_N = -\gamma_N e \varepsilon_N$; 4 is systems (16.26) with $N(t) \in \mathbb{R}$ and the algorithm $\dot{\mu}_\varepsilon = -\gamma_\varepsilon e \varepsilon_d$. Results are presented as structures described by functions $f_\eta : e(t) \rightarrow \eta_i(t),\ i = d, N$.

Figure 10. Uncertainty identification results

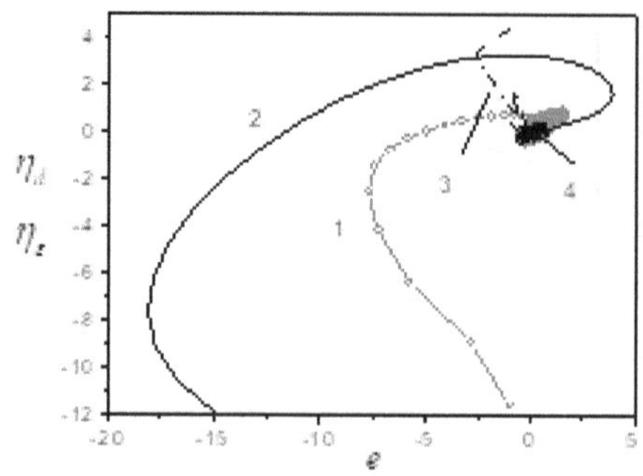

The *SEN*-system shows the best results (see Figure 10). Remaining variants of the considered systems have the tuning large time and a low accuracy of the uncertainty estimation.

10. CONCLUSION

The design method of adaptive observers (AO) for time-varying linear dynamic systems at non-fulfilment is proposed. AO synthesis is presented as the solution of two tasks. The solution of first a problem is the constant matrix choice decreasing the effect of excitation constancy (EC). Procedures are shown for obtaining of this matrix. The matrix sets restrictions for AO parameters. The design method of adaptive multiplicative algorithms results from the second problem solution under got restrictions. Procedures for the uncertainty estimation are proposed. They are based on obtaining of static models giving the uncertainty change forecast. Optimum estimations for the uncertainty are got, which minimizes the error between outputs of the system and AO. The adaptive system exponential dissipativity is proved. Modelling results are confirmed the effectiveness of designed methods and procedures.

REFERENCES

Carrol, R. L., & Lindorff, D. P. (1973). An adaptive observer for single-input single-output linear systems. *IEEE Transactions on Automatic Control, 18*(5), 428–435. DOI: 10.1109/TAC.1973.1100367

Carrol, R. L., & Monopoli, R. V. (1975). Model reference adaptive control estimation and identification using only and output signals. *Processings of IFAC 6th Word congress. Boston, Cembridge*, part 1, 58.3/1-58.3/10.

Feiler, M. J., & Narendra, K. S. (2006). Simultaneous identification and control of time–varying systems. In *Proceedings of the 45th IEEE Conference on Decision and Control,* San Diego CA, U.S.A. DOI: 10.1109/CDC.2006.377529

Gantmacher, F. R. (2000). *The Theory of Matrices.* AMS Chelsea Publishing: Reprinted by American Mathematical Society.

Jing, N., Juan, Y., Xuemei, R., & Gu, Y. (2014). Robust adaptive estimation of nonlinear system with time-varying parameters. *International Journal of Adaptive Control and Signal Processing, 29*(8), 1055–1072. DOI: 10.1002/acs.2524

Karabutov, N. (2017). Adaptive observers for linear time-varying dynamic objects with uncertainty estimation. *International Journal of Intelligent Systems and Applications, 9*(6), 1–14. DOI: 10.5815/ijisa.2017.06.01

Karabutov, N. N. (2007). *Adaptive identification of systems: information synthesis.* URSS.

Kreisselmeier, G. (1986). A robust indirect adaptive control approach. *International Journal of Control, 43*(1), 161–175. DOI: 10.1080/00207178608933456

Kudva, P., & Narendra, K. S. (1973). Synthesis of an adaptive observer using Lyapunov direct method. *International Journal of Control, 18*(4), 1201–1216. DOI: 10.1080/00207177308932593

Narendra, K. S., & Kudva, B. (1974). Stable adaptive schemes for system: Identification and control. *IEEE Transactions on Systems, Man, and Cybernetics, 4*(6), 542–560. DOI: 10.1109/TSMC.1974.4309363

Nuyan, S., & Carrol, R. L. (1979). Minimal order arbitrarily fast adaptive observer and identifies. *IEEE Transactions on Automatic Control, 24*(2), 496–499. DOI: 10.1109/TAC.1979.1101989

Zhang, Q., & Clavel, A. (2001). Adaptive observer with exponential forgetting factor for linear time varying systems. *Proceedings of the 40th IEEE Conference on decision and control (16.CDC '01)*, 4, 3886–3891. DOI: 10.1109/.2001.980478

Zhang, Y., Fidan, B., & Ioannou, P. A. (2003). Backstepping control of linear time-varying systems with known and unknown parameters. *IEEE Transactions on Automatic Control, 38*(11), 1908–1925. DOI: 10.1109/TAC.2003.819074

APPENDIX A

Proof of Theorem 1. Consider Lyapunov function $V(t)$ (16.30). The choice of the second component in the right part (16.30) is substantiated by Karabutov (2007). Find the derivative $V(t)$ on motions F2-system. After simple transformations, we obtain the inequality for $\dot{V}(t)$:

$$\dot{V} \leq -2V_e - \gamma_s \varsigma \sqrt{2V_e}, \text{ (A1)}$$

where $\varsigma = \min_t |D^T P|$.

Apply the condition 1) of theorem 1. As follows from (A1), the system $AS_{N\mu}^{ss}$ is stable. Integrate (A1) on the time and obtain:

$$V(t_0) - \int_{t_0}^{t} \left(2V_e(\tau) + \gamma_s \varsigma \sqrt{2V_e(\tau)}\right) d\tau \geq V(t).$$

As $V(e, \delta_N, \mu)$ satisfies the condition 1) theorem 1, all trajectories of the system $AS_{N\mu}^{ss}$ belong the area $\mathbb{G}_t = \{(e, N, \mu) : V(t) \leq V(t_0)\}$. Then:

$$\int_{t_0}^{t} \left(2V_e(\tau) + \gamma_s \varsigma \sqrt{2V_e(\tau)}\right) d\tau \leq V(t_0) - V(t).$$

Obtain the inequality (16.29). Consider $\dot{V}_e(t)$:

$$\dot{V}_e = -ke^2 + e\delta_N^T P - \gamma_s |e||z| \leq -ke^2 + |e|\left(\|\delta_N\| \|P\| + \gamma_s |e||z|\right). \text{ (A2)}$$

Use the inequality

$$-az^2 + bz \leq -\frac{-az^2}{2} + \frac{b^2}{2a}, \quad a > 0, \ b \geq 0, \ z \geq 0$$

and transform (A2) into the form:

$$\dot{V}_e \leq -\frac{k}{2}e^2 + \frac{1}{2k}\left(\|\delta_N\| \|P\| + \gamma_s |z|\right)^2. \text{ (A3)}$$

Write (A3) after simple transformations as:

$$\dot{V}_e \leq -kV_e + \frac{1}{k}\left(\upsilon_P \max_t \|\delta_N(t)\|^2 + \gamma_s^2 \upsilon_z^2\right), \text{ (A4)}$$

where $\upsilon_P = \max_t \|P(t)\|^2$, $\upsilon_z = \max_t |z(t)|$.

From (A4), we obtain the estimation (16.29) for $V_e(t)$. ■

APPENDIX B

Proof of Theorem 2. Consider LF

$$V(t) = V_e(t) + \int_{t_0}^{t} \delta_N^T(\tau)\Gamma^{-1}\Psi_N(N)\delta_N(\tau)d\tau + \int_{t_0}^{t} \dot{\mu}(\tau)\gamma_{d,s}^{-1}\Psi_\mu(\mu)\mu(\tau)d\tau.$$

Find the derivative $V(t)$ and transform $\dot{V}(t)$ on motions of the system $A\,S_{N\mu}^{ds}$ to the form

$\dot{V} \leq -2V_e$. (B1)

V_e satisfies to the condition 1) theorem 1. We see $\dot{V}(t) < 0$. $A\,S_{N\mu}^{ds}$-system is stable. Integrate (B1):

$\int^{t} 2V_e(\tau)d\tau \leq V(t_0) - V(t)$. (B2)

Obtain the boundedness of trajectories for $A\,S_{N\mu}^{ds}$-system. The estimation for $\dot{V}_e$ has the form:

$\dot{V}_e \leq -\frac{k}{2}e^2 + \frac{1}{2k}\left(\|\delta_N\|\,\|P\| + |z|\right)^2.$

Apply (A2) and obtain the estimation (16.29) with $\gamma_s^2 = 1$. ■

APPENDIX C

Proof of Theorem 3. Consider Lyapunov function

$$V_{e\mu}(t) = V_e(t) + \gamma_d^{-1} \int_{t_0}^{t} \dot{\mu}(\tau)\Psi_\mu(\mu)\mu(\tau)\,d\tau = V_e(t) + V_\mu(t).$$

Present $\dot{V}_{e\mu}$ as:

$$\dot{V}_e + \dot{V}_\mu = -ke^2 + e\delta_N^T P \le -\frac{k}{2}e^2 + \frac{\upsilon_P}{2k}\max_t \|\delta_N(t)\|^2.$$

Then

$$V_e(t) \le \frac{\upsilon_P}{k^2}\max_t \|\delta_N(t)\|^2 - \int_{t_0}^{t} \dot{V}_\mu(\tau)\,d\tau.$$

Let $\Psi_\mu(\mu) = 1$. Then:

$$\int_{t_0}^{t} \dot{V}_\mu(\tau)\,d\tau = \frac{1}{2\gamma_d}\left(\mu^2(t) - \mu^2(t_0)\right) > 0.$$

So

$$V_e(t) \le \frac{\upsilon_P}{k^2}\max_t \|\delta_N(t)\|^2. \blacksquare$$

APPENDIX D

Proof of Lemma 1. Consider $\dot{V}_N(t)$:

$$\dot{V}_N(t) = -e\Delta_N^T P_H. \text{ (D1)}$$

We get property $\dot{V}_N(t) \in M$. Let the condition:

$$e\Delta_N^T P_H = \vartheta\left(\Delta_N^T B_H \Delta_N + e^2\right)$$

is satisfied for $(\forall t \geq t^* > t_0)$ & $\left(\forall (e, \Delta_N)\right) \in O_\nu(O))$, where $\mathbb{O} = \{0, 0_m\} \subset \mathbb{R} \times \mathbb{R}^m \times \mathbb{J}_{0,\infty}$ is the equilibrium position of system, $O_\nu(O)$ is some neighborhood of a point $O, 0 \in R^l$ is the zero vector, $t \in [0, \infty] = \mathbb{J}_{0,\infty}$, $\vartheta > 0$ is some number.

Transform (D1) to the form (Karabutov, 2007):

$$\dot{V}_N(t) = -\vartheta\left(\Delta_N^T B_H \Delta_N + e^2\right).$$

Next, we use (16.32). Let $\Psi_N(N) = P(N) = I_l$. Then:

$$\dot{V}_N \leq -\tfrac{3}{8}a\vartheta\Delta_N^T\Delta_N + \tfrac{2}{3}\vartheta e^2 \leq -\frac{3a\vartheta\lambda_1(\Gamma)}{4} V_N + \tfrac{4}{3}\vartheta V_e \Leftrightarrow \dot{V}_\delta \in M^+.$$

APPENDIX E

Proof of Lemma 2. Apply (16.31) and present $\dot{V}_e$ as:

$$\dot{V}_e \leq -ke^2 + |e| \left|\delta_N^T P_H + \sigma_\Delta^T P + \mu z\right| \;\Rightarrow\; \dot{V}_e \in M. \text{ (E1)}$$

Ensure the condition $\dot{V}_e \in M^+$. Apply the proof of the theorem 1 and obtain:

$$\dot{V}_e \leq -\frac{k}{2}e^2 + \frac{1}{2k}\left(\Delta_N^T P_H + \sigma_\Delta^T P + \mu z\right)^2 \leq -kV_e + \frac{1}{k}\left(2\overline{\alpha}\ \lambda_l(\Gamma)\, V_N + \upsilon_P \max_t \|\sigma_\Delta(t)\|^2 + \upsilon_z\right)$$

.

So, property $\dot{V}_e \in M^+$ for $\dot{V}_e$ has the form:

$$\dot{V}_e \leq -kV_e + \frac{2\overline{\alpha}\ \lambda_l(\Gamma)}{k}V_N + \frac{1}{k}\left(\upsilon_P \max_t \|\sigma_\Delta(t)\|^2 + \upsilon_z\right). \blacksquare$$

APPENDIX F

Proof of Theorem 5. Consider LF:

$$V(t) = V_e(t) + 0.5\gamma_\varepsilon^{-1}\mu^2(t) = V_e(t) + V_\mu(t)$$

Write $\dot{V}$ as:

$$\dot{V} = -ke^2 + e\Delta_N^T P_H + e(\mu_e \varepsilon_d - h_\Delta) - e\mu_e \varepsilon_d$$

Apply the proof scheme of theorems 1, 2 and get

$$\dot{V} \leq -kV_e + \frac{1}{k}\left[\bar{\alpha}\max_t \|\Delta_N(t)\|^2 + \omega\right] \text{ (F1)}$$

where $\max_t h_\Delta^2(t) \leq \omega, \omega \geq 0$ is some number. The inequality (16.38) follows from (F1). ■

APPENDIX H

Proof of Theorem 6. Show boundedness of system $ASED_{N\mu_N}$ trajectories. Let number $\mu_{*,N}$ exists such that the equality $h_\Delta(t) = \mu_{*,N}\varepsilon_N(t)$ is true.

Consider LF:

$$V(t) = V_e(t) + V_{\mu_N}(t) + V_{\sigma_N}(t) = 0.5e^2(t) + 0.5\gamma_N^{-1}\left(\delta\mu_N(t)\right)^2 + 0.5\sigma_N^2(t)\Gamma^{-1}\sigma_N(t)$$

where $\delta\mu_N = \mu_N - \mu_{*,N}$.

Write $\dot{V}$ as:

$$\dot{V} = -ke^2 + e\Delta_N^T P_H + e\delta\mu_N\varepsilon_N - e\delta\mu_N\varepsilon_N - e\Delta_N^T P_H \text{ (H1)}$$

We obtain from (H1) that $\dot{V} \le -2kV_e$. Integrate this inequality and obtain:

$$V(t) \le V(t_0) - 2k\int_{t_0}^{t} V_e(\tau)\,d\tau \text{ (H2)}$$

Hence, all trajectories of the system $ASED_{N\mu_N}$ belong area:

$$\Omega = \left\{(e, \Delta_N, \delta\mu_N) : V(t) \le V(t_0)\right\}$$

Consider LF $V_e(t)$ and $V_{\mu_N}(t) = 0.5\left(\delta\mu_N(t)\right)^2$. Transform $\dot{V}_e$ to the form:

$$\dot{V}_e = -ke^2 + e\Delta_N^T P_H + e\delta\mu_N\varepsilon_N \text{ (H3)}$$

The vector P_H has property PE_α. Apply the proof scheme of the lemma 1 and ensure for $\dot{V}_e$ property $\dot{V}_e \in M^+$ in the form:

$$\begin{aligned}\dot{V}_e &= -ke^2 + e\Delta_N^T P_H + e\delta\mu_N\varepsilon_N \le -ke^2 + |e|\left|\Delta_N^T P_H + \delta\mu_N\varepsilon_N\right| \le \\ &\quad -ke^2 + \tfrac{1}{k}\left[\|\Delta_N\|^2\|P_H\|^2 + (\delta\mu_N)^2\varepsilon_N^2\right] \le \\ &\quad -kV_e + \tfrac{\overline{\alpha}}{k}\max_t\|\Delta_N\|^2 + \frac{2d_N}{k}V_{\mu_N},\end{aligned} \qquad \text{(H4)}$$

where $\varepsilon_N^2 \le d_N$, $d_N \ge 0$ is some number.

Ensure the condition $\dot{V}_{\mu_N} \in M^+$. By analogy with (H4) obtain:

$$\begin{aligned}\dot{V}_{\mu_N} &= -\gamma_N e\delta\mu_N\varepsilon_N = -\upsilon\left(\gamma_N^2\delta\mu_N^2\varepsilon_N^2 + e^2\right) \\ &\le -\tfrac{3}{4}\upsilon\gamma_N^2\delta\mu_N^2\varepsilon_N^2 + \upsilon\gamma_N\left|e\delta\mu_N\varepsilon_N\right| \le -\tfrac{3}{4}\upsilon\underline{d}_N\gamma_N^2 V_{\mu_N} + \upsilon V_e.\end{aligned} \qquad \text{(H5)}$$

As υ does not influence on properties of the system assume $\upsilon = 1$. So, we have for $\dot{V}_e$ and $\dot{V}_{\mu_N}$ following inequalities:

$$\begin{bmatrix}\dot{V}_e \\ \dot{V}_{\mu_N}\end{bmatrix} \le \begin{bmatrix}-k & \frac{2d_N}{k} \\ \upsilon & -\frac{3}{4}\upsilon\underline{d}_N\gamma_N^2\end{bmatrix}\begin{bmatrix}V_e \\ V_{\mu_N}\end{bmatrix} + \begin{bmatrix}\frac{\overline{\alpha}}{k}\max_t\|\Delta_N\|^2 \\ 0\end{bmatrix} \text{ (H6)}$$

Let $S_{e,\mu_N} = \begin{bmatrix} s_e & s_{\mu_N} \end{bmatrix}^T$,

$$A_{\mu_N} = \begin{bmatrix} -k & \frac{2d_N}{k} \\ \upsilon & -\frac{3}{4}\upsilon \underline{d}_N \gamma_N^2 \end{bmatrix}, F_{\mu_N} = \begin{bmatrix} \frac{\overline{\alpha}}{k} \max_t \|\Delta_N\|^2 \\ 0 \end{bmatrix}$$

We assume that functions $s_\rho(t) \geq 0$ exist such that

$$V_\rho(t) \leq s_\rho(t)\ \forall\left(t \geq t_0\right) \ \&\ \left(V_\rho(t_0) \leq s_\rho(t_0)\right), \rho = e,\ \mu_N$$

We obtain the vector system of comparison for (H6):

$$\dot{S}_{e,\mu_N}(t) = A_{\mu_N} S_{e,\mu_N}(t) + F_{\mu_N} \text{ (H7)}$$

Stability conditions of the system (H7) have the form:

$$k > 0, k^2 \underline{d}_N \gamma_N^2 \geq \frac{8}{3} d_N$$

Hence

$$V(t) \leq e^{A_{\mu_N}(t-t_0)} S_{e,\mu_N}(t_0) + \int_{t_0}^{t} e^{A_{\mu_N}(t-\tau)} F_{\mu_N}(\tau)\, d\tau$$

Limiting properties the $ASED_{N\mu_N}$-system estimate as

$$V(t) \leq \frac{\overline{\alpha}}{k} \max_t \|\Delta_N\|^2 \Sigma_{1,\mu}(t)\Big|_{t\to\infty}$$

where $\Sigma_1(t)$ is the first column of the matrix $\int_{t_0}^{t} e^{A_\nu(t-\tau)} d\tau$. ■

Conclusion

Two aspects of the identification theory are discussed: structural identification (SI) and adaptive parameter estimation. These problems we consider under uncertainty. A new approach to SI based on the application of geometric frameworks (GF) is proposed. GF reflects the system state in the structural space, allowing us to formalize SI problems. Until now, SI has been an intuitive field of knowledge, using indirect methods to evaluate the structural parameters of the system. The analysis and application of GF provides solutions to various structural identification problems. In particular, we propose the formalized approach to the structural identifiability (SID) of nonlinear dynamical systems and show that structural identifiability follows from SI.

We note the use of GF to evaluate such a structural parameter as the Lyapunov exponent (LE). This approach does not require the formation of time series, but uses available measurement information. We prove the identifiability and observability of LE and propose methods for their evaluation.

The information set (IS) characteristics of the system play a significant role in solving SI and SID problems. This is the constant excitation (CE) property. The verification of the CE condition is based on checking of the S-synchronizability for nonlinear dynamical systems.

Using the GF method provided the problem solution for estimating parametric constraints under uncertainty. We propose procedures and algorithms for estimating the domain of parametric constraints, which are based on the principle of preferences and adaptive algorithms.

The second direction is related to the design of adaptive identification systems for complex dynamic systems (DS) under uncertainty. Consideration of the operating conditions is presented as restrictions on the system. We use the algorithm method to get of adaptive algorithms. We obtain classes of both static and adaptive control and identification algorithms. The GF analysis provides an assessment of the system features that we use in the design of adaptive systems.

Compilation of References

Achho, L. (2013). Hysteresis Modeling and Synchronization of a Class of RC-OTA Hysteretic-Jounce-Chaotic Oscillators. *Universal journal of applied mathematics, 1*(2), 82-85. DOI: 10.13189/ujam.2013.010207

Acho, L. (2013). Hysteresis modelling and synchronization of a class of RC-OTA hysteretic-jounce-chaotic oscillators. *Universal Journal of Applied Mathematics, 1*(2), 82-85 14 41.

Aghakouchak, A. A., Ghafouripou, A., & Kiamehr, H. (2000). An overview of system identification methods and applications. Part I: Methods of system identification and dynamic tests. *Conference: 4th international conference on coasts, ports & marine structure at: Bandar Abbas, Iran, 1*, 1-31.

Aguirregabiria, V., & Mira, P. (2007). Dynamic Discrete Choice Structural Models: A Survey. *Working Paper 297*. University of Toronto.

Aguirre, L. A., & Letellier, C. (2009). Modeling nonlinear dynamics and chaos: A Review. *Mathematical Problems in Engineering, 2009*(1), 1024–1235. DOI: 10.1155/2009/238960

Aivazyan, S. A., Yenyukov, I. S., & Meshalkin, L. D. (1985). *Applied statistics. Study of relationships*. Moscow: Finansy i statistika.

Alessi, L., Barigozzi, M., & Capasso, M. (2008). A robust criterion for determining the number of static factors in approximate factor models. *Working paper series*, 903.

Ali, M., Chughtai, S. S., & Werner, H. (2010). Identification of spatially interconnected systems. *Proceedings of the 48h IEEE Conference on Decision and Control (CDC) held jointly with 2009 28th Chinese Control Conference*. DOI: 10.1109/CDC.2009.5399748

Ali, M., Abbas, H., Chughtai, S. S., & Werner, H. (2011). Identification of spatially interconnected systems using neural network. *49th IEEE Conference on decision and control (CDC)*. DOI: 10.1109/CDC.2010.5717080

Aliyeva, A. S. (2019). Identification of multiconnected dynamic objects with uncertainty based on neural technology and reference converters. *Informatics and control problems, 39*(2), 93-102.

Aljamaan, I. (2016). *Nonlinear closed-loop system identification in the presence of non-stationary noise source (Unpublished doctoral thesis)*. University of Calgary., http://hdl.handle.net/11023/3070

Aljanaideh, O., Rakheja, S., & Chun-Yi, S. (2014). Experimental characterization and modeling of rate-dependent asymmetric hysteresis of magnetostrictive actuators. *Smart Materials and Structures, 23*(3), 035002. DOI: 10.1088/0964-1726/23/3/035002

Almon, S. (1965). The distributed lag between capital appropriations and expenditures. *Econometrica, 33*(1), 178–196. DOI: 10.2307/1911894

Andersson, C., Ribeiro, A. H., Tiels, K., Wahlström, N., & Schön, T. B. (2019). Deep Convolutional Networks in System Identification. *2019 IEEE 58th Conference on Decision and Control (CDC), Nice, France*, 3670-3676.

Anishchenko, V. S., Astakhov, V., Neiman, A., Vadivasova, T., & Schimansky-Geier, L. (2007). *Nonlinear dynamics of chaotic and stochastic systems. nonlinear dynamics of chaotic and stochastic systems. Tutorial and modern developments* (2nd ed.). Springer., DOI: 10.1007/978-3-540-38168-6

Armstrong, B. (2006). Models for the relationship between ambient temperature and daily mortality. *Epidemiology (Cambridge, Mass.), 17*(6), 624–631. DOI: 10.1097/01.ede.0000239732.50999.8f PMID: 17028505

Audoly, S., D'Angio, L., Saccoman, M. P., & Cobelli, C. (1998). Global identifiability of linear compartmental models - a computer algebra algorithm. *IEEE Transactions on Automatic Control, 45*, 36–47. PMID: 9444838

Audoly, S., D'Angio, L., Saccomany, M. P., & Cobelli, C. (1998). Global identifiability of linear compartmental models – a computer algebra algorithm. *IEEE Transactions on Automatic Control, 45*, 36–47.

Avdeenko, T. V. (2013). Identification of linear dynamic systems with use parametrical space separators. *Automatics and program engineering*, 1(3), 16-23,

Avdeenko, T. V. (2013). Identification of linear dynamical systems using concept of parametric space separators. *Automated Software Engineering, 1*(3), 16–23.

Aykan, M., & Özgüven, H. N. (2012). Parametric Identification of Nonlinearity from Incomplete FRF Data Using Describing Function Inversion. In D. Adams, G. Kerschen, A. Carrella (Eds). *Topics in Nonlinear Dynamics,* Volume 3. *Conference Proceedings of the Society for Experimental Mechanics Series.*, New York: Springer. DOI: 10.1007/978-1-4614-2416-1_25

Ayvazyan, S. A., Enyukov, I. S., & Meshalkin, L. D. (1985). *Applied Statistics: Dependency Research. Reference edition.* Finansy i Statistika.

Azeez, M. A. F., & Vakakis, A. F. (1999). Numerical and experimental analysis of a continuously overhung rotor undergoing vibro-impacts. *International Journal of Non-linear Mechanics, 34*(3), 415–435. DOI: 10.1016/S0020-7462(98)00022-5

Baang, D., Stoev, J., & Choi, J. Y. (2007). Adaptive observer using auto-generating *B*-splines. *International Journal of Control, Automation, and Systems, 5*(5), 479–491.

Balonin, N. A. (2010). *Identifiability theorems.* St. Petersburg: Publishing house "Polytechnic".

Balonin, N. A. (2010). *Theorems of identifiability.* Politekhnika publishing house.

Barbashin, E. A. (1970). *Lyapunov function.* Nauka.

Bardsley, J. M. (2004). A bound-constrained Levenberg-Marquardt algorithm for a parameter identification problem in electromagnetics. https://www.researchgate.net/publication/255577468 (2005).

Bardsley, J. M. (2004). A bound-constrained Levenburg-Marquardt algorithm for a parameter identification problem in electromagnetics. http://www.math.umt.edu/bardsley/papers/EMopt04.pdf

Bastin, G., & Gevers, R. (1988). Stable adaptive observers for nonlinear time-varying systems. *IEEE Transactions on Automatic Control, 33*(7), 650–658.

Bellman, R. E. (2003). *Dynamic programming.* Dover Publications.

Benettin, G., Galgani, L., Giorgilli, A., & Strelcyn, J.-M. (1980). Lyapunov characteristic exponents for smooth dynamical systems and for Hamiltonian systems; a method for computing all of them. Part. I. *Theory. Pt. II: Numerical applications. Meccanica, 15*, 9–30. DOI: 10.1007/BF02128236

Berryman, J. G. (1978). Choice of operator length for maximum entropy spectral analysis. *Geophysics, 43*(7), 1384–1391. DOI: 10.1190/1.1440902

Besekersky, V. A., & Popov, E. P. (1975). *Theory of automatic control systems* (Third edition, revised). Nauka.

Bespalov, A. V., & Polyakhov, N. D. (2016). Comparative analysis of methods for estimating first Lyapunov exponent. *Modern problems of science and education, 6*, 7.

Billings, S. A. (1982). Structure Detection and Model Validity Tests in the Identification of Nonlinear Systems. Research Report. *ACSE Report 196*, Department of Control Engineering, University of Sheffield.

Binczak, S., Busvelle, E., Gauthier, J.-P., & Jacquir, S. (2000). Identification of Unknown Functions in Dynamic Systems. https://pageperso.lis-lab.fr/eric.busvelle/papers/bioid.pdf

Bodunov, N. A. (2012). Introduction to the theory of local parametrical identifiability. *Differential equations and management processes, 2*.

Bodunov, N. A. (2012). Introduction to theory of local parametric identifiability. *Differential equations and control processes*, 1-137.

Bodunov, N. A. (2012). Introduction to the theory of local parametrical identifiability. *Differential equations and management processes*, 2.

Boguslavsky, I. A. (2006). *Polynomial approximation for nonlinear estimation and control problems*. Fizmatlit.

Bohr, H. (2018). *Almost Periodic Functions*. Dover Publications, Incorporated.

Box, G. P., & Jenkins, G. M. (1976). *Time series analysis: Forecasting and control*. Holden-Day.

Bozhokin, S. V., & Parshin, D. A. (2001). *Fractals and Multifractals*. Moscow-Izhevsk: Scientific Publishing Centre \. *Regular and Chaotic Dynamics*.

Bozhokin, S. V., & Parshin, D. A. (2001). *Fractals and Multifractals*. Scientific Publishing Centre Regular and Chaotic Dynamics.

Bradley, E., & Kantz, H. (2015). Nonlinear time-series analysis revisited. *arXiv:1503.07493v1*. doi: DOI: 10.1063/1.4917289

Brandhofer, S., Myers, C. R., Devitt, S. J., & Polian, I. (2023). Multiplexed pseudo-deterministic photon source with asymmetric switching elements. arXiv:2301.07258.

Brian, D. O., & Anderson, R. R., Bitmead, C., Johnson R. Jr., Kokotovic, P. V. et al. (1986). *Stability of adaptive system: passivity and averaging analysis*. Cambridge: MIT Press.

Britenkov, A. K. & Dedus, F. F. (2012). Prediction of time sequences using a generalized spectral-analytical method. Mathematical modelling. Optimal control. *Bulletin of the Nizhny Novgorod University named after N.I. Lobachevsky*, 28-32.

Bukov, V. N., Maksimenko, I. M., & Ryabchenko, V. N. (1998). Control of multivariable systems. *Automation and Remote Control, 6*, 97–110.

Buzug, T., & Pfister, G. (1992). Optimal delay time and embedding dimension for delay-time coordinates by analysis of the global static and local dynamical behaviour of strange attractors. *Physical Review A, 45*(10), 7073–7084. DOI: 10.1103/PhysRevA.45.7073 PMID: 9906779

Bylov, B. F., Vinograd, R. E., Grobman, D. M., & Nemytsky, V. V. (1966). *Theory of Lyapunov indexes and its application to stability problems*. Nauka.

Bylov, F., Vinograd, R. E., Grobman, D. M., & Nemytskii, V. V. (1966). *Theory of Lyapunov exponents and its application to problems of stability*. Nauka.

Campos, J., Ericsson, N. R., Hendry D. F. (2005). General-to-specific modeling: An overview and selected bibliography. *International finance discussion papers, 2005*(838). DOI: 10.17016/ifdp.2005.838

Carino, J. A., Delgado-Prieto, M., Iglesias, J. A., Sanchis, A., Zurita, D., Millan, M., Ortega Redondo, J. A., & Romero-Troncoso, R. (2018). Fault Detection and Identification Methodology Under an Incremental Learning Framework Applied to Industrial Machinery. *IEEE Access : Practical Innovations, Open Solutions, 99*, 1–9. DOI: 10.1109/ACCESS.2018.2868430

Carrol, R. L., & Monopoli, R. V. (1975). Model reference adaptive control estimation and identification using only and output signals. *Processings of IFAC 6th Word congress. Boston, Cembridge*, part 1, 58.3/1-58.3/10.

Carrol, R. L., & Lindorff, D. P. (1973). An adaptive observer for single-input single-output linear systems. *IEEE Transactions on Automatic Control, 18*(5), 428–435.

Carter, R. A. L., & Zellner, A. (2004). The arar error model for univariate time series and distributed lag models. *Studies in Nonlinear Dynamics and Econometrics, 8*(1), 2–2. DOI: 10.2202/1558-3708.1132

Castle, J. L., Doornik, J. A., & Hendry, D. F. (2010). *Evaluating automatic model selection. (Discussion paper series, 474) (working Paper)*. Economists Online.

Chadeev, V. M., & Gusev, S. S. (2012). Identification with restrictions. determining a static plant parameter estimate. Proceedings of the VII International Conference System Identification and Control Problems" SICPRO 'OS Moscow January 28-31, 2008. V.A. Trapeznikov Institute of Control Sciences. Moscow: V.A. Trapeznikov Institute of Control Sciences, 261-269.

Chan-Zheng, C., Borja, P., & Scherpen, J. (2021). Passivity-based control of mechanical systems with linear damping identification. arXiv: 2105.04324v2.

Chen, L., & Narendra, K. S. (2004). Identification and Control of a Nonlinear Discrete-Time System Based on its Linearization: A Unified Framework. *IEEE Transactions on Neural Networks*, *15*(3), 663–673. DOI: 10.1109/TNN.2004.826206 PMID: 15384554

Chia, T. L., Chow, P. C., & Chizeck, H. J. (1991). Recursive parameter identification of constrained systems: An application to electrically stimulated muscle. *IEEE Transactions on Biomedical Engineering*, *38*(5), 429–442. DOI: 10.1109/10.81562 PMID: 1874525

Chis, O.-T., Banga, J. R., & Balsa-Canto, E. (2011). Structural identifiability of systems biology models: A critical comparison of methods. *PLoS One*, *6*(4), 1–16. DOI: 10.1371/journal.pone.0027755 PMID: 22132135

Choquet, G. (1964). *L' enseignement de la geométrie*. Hermann.

Cho, S., Park, J.-W., & Sim, S.-H. (2015). Decentralized System identification using stochastic subspace identification for wireless sensor networks. *Sensors (Basel)*, *15*(4), 8131–8145. DOI: 10.3390/s150408131 PMID: 25856325

Chow, G. C. (2011). Usefulness of adaptive and rational expectations in economics, *CEPS Working Paper No. 221*.

Corrêa, M. V., Aguirre, L. A., & Saldanha, R. R. (2002). Using steady-state prior knowledge to constrain parameter estimates in nonlinear system identification. *IEEE Transactions on Circuits and Systems. I, Fundamental Theory and Applications*, *49*(9), 1370–1381. DOI: 10.1109/TCSI.2002.802345

Demetriou, I. C., & Vassiliou, E. E. (2008). A distributed lag estimator with piecewise monotonic coefficients. *Proceedings of the World Congress on Engineering. 2008.* V. 2. *WCE 2008, July 2 - 4, 2008, London, U.K.*

Demetriou, I. C., & Vassiliou, E. E. (2009). An algorithm for distributed lag estimation subject to piecewise monotonic coefficients. *International Journal of Applied Mathematics*, *39*(1), 1–10.

Demetriou, I. C., & Vassiliou, E. E. (2014). On Distributed-Lag Modeling algorithms by r-convexity and piecewise monotonicity. In Rassias, T., Floudas, C., & Butenko, S. (Eds.), *Optimization in Science and Engineering. Springerб 115-140.* DOI: 10.1007/978-1-4939-0808-0_6

Deshpande, A., Chen, Q., Wang, Y., Lai, Y.-C. G., & Do, Y. (2010). Effect of smoothing on robust chaos. *Physical Review. E, 82*(2), 026209. DOI: 10.1103/PhysRevE.82.026209 PMID: 20866895

Dhrymes, P. J. (1971). *Distributed Lags: Problems of estimation and formulation.* Holden-Day.

Dhrymes, P. J. (1971). *Distributed Lags: Problems of Estimation and Formulation.* Holden-Day.

Dieci, L., Russell, R. D., & Van Vleck, E. S. (1997). On the computation of Lyapunov exponents for continuous dynamical systems. *SIAM Journal on Numerical Analysis, 34*(1), 402–423. DOI: 10.1137/S0036142993247311

Draper, N. R., & Smith, H. (2014). *Applied Regression Analysis.* John Wiley & Sons.

Dubrovsky, A. M., Mkhitaryan, V. S., & Troshin, L. I. (2003). *Multidimensional statistical methods.* Finance and Statistics.

Du, C., Yang, C., Li, F., & Gui, W. (2019). A novel asynchronouss control for artificial delayed Markovian jump systems via output feedback sliding mode approach. *IEEE Transactions on Systems, Man, and Cybernetics. Systems, 49*(2), 364–374. DOI: 10.1109/TSMC.2018.2815032

Duffee, G. (2006). R. (2005). Term structure estimation without using latent factors. *Journal of Financial Economics, 79*(3), 507–536. DOI: 10.1016/j.jfineco.2005.03.007

Egorov, I. N., & Umnov, V. P. (2022). *Control systems for electric drives of technological robots and manipulators: textbook.* Vladimir: Publishing house of Vladimir State University.

Egorov, I. N. (2010). *Positional force control of robotic and mechatronic devices.* Publishing House of Vladimir State University.

Eikhoff, P. (1974). *System identification: parameter and state estimation.* Wiley-Interscience.

Elgerd, O. I. (1967). *Control Systems Theory.* Mc Graw-Hill.

Evtushenko, Ju. G. (1982). Methods of solution of extremum problems and their application in optimisation problems. Moscow: Nauka.

Eykhoff, P. (Ed.). (1974). *Modern Identification Methods*. John Wiley and Sons.

Feder, J. (1988). *Fractals*. Plenum Press.

Feiler, M. J., & Narendra, K. S. (2006). Simultaneous identification and control of time–varying systems. In *Proceedings of the 45th IEEE Conference on Decision and Control,* San Diego CA, U.S.A. DOI: 10.1109/CDC.2006.377529

Filatov, V. V. (2013). Structural characteristics of anomalies of geophysical fields and their use in forecasting. *Geophysics, geophysical instrumentation*, 4(16), 34-41.

Fisher, I. (1937). Note on a Short-cut Method for Calculating Distributed Lags. *Bulletin de l'Institut International de Statistique*, *29*, 323–327.

Fomin, V. N., Fradkov, I. L., & Yakubovich, V. A. (1981). *Adaptive control in dynamic plants*. Nauka, Ru.

Forssell, U., & Ljung, L. (1999). Closed-loop identification revisited. *Automatica, 35*(7), 1215–1241. DOI: 10.1016/S0005-1098(99)00022-9

Fradkov, A. L. (1979). Shema of a high-speed gradient and its application in adaptive control problems. *Automatics and telemechanics, 9*, 90-102.

Furasov, V. D. (2005). *Tasks of the guaranteed identification. Digital systems.* Moscow: BINOMIAL. Laboratory of knowledge.

Gabor, D., Wiby, W. P. L., & Woodcock, R. A. (1961). A universal nonlinear filter, predictor and simulator which optimizes itself by learning processes. *Proceedings of the IEEE - Part B: Electronic and Communication Engineering*, *108*, 422–438.

Gagliardini, P., Gouriéroux, C., & Renault, E. (2010). Efficient Derivative Pricing by the Extended Method of Moments. *Swiss Finance Institute Research,* 10-07.

Gagliardini, P., Gourieroux, C., & Renault, E. (2011). Efficient derivative pricing by the extended method of moments. *Econometrica*. *Econometrica*, *79*(4), 1181–1232. DOI: 10.3982/ECTA7192

Gantmacher, F. R. (2000). *The theory of matrices*. AMS Chelsea Publishing.

Gantmakher, F. R. (2000). *The Theory of Matrices*. Chelsea Pub. Co.

Gao, H., Jézéquel, L., Cabrol, E., & Vitry, B. (2019). Characterization of a Bouc-Wen model-based damper model for automobile comfort simulation. *HAL* Id: hal-02188563.

Garimella, P., & Yao, B. (2003). Nonlinear adaptive robust observer design for a class of nonlinear systems. *Proceedings of the American Control Conference Denver, Color ado* June 4-6, 2003, 4391-4396.

Gasparrini, A., Armstrong, B. M., & Kenwardb, G. (2010). Distributed lag non-linear models. *Statistics in Medicine, 29*(21), 2224–2234. DOI: 10.1002/sim.3940 PMID: 20812303

Gedon, D., Wahlström, N., Schön, T. B., & Ljung, L. (2021). Deep State Space Models for Nonlinear System Identification, *arXiv:2003.14162v3* [eess.SY].

Gencay, R., & Dechert, W. D. (1992). An algorithm for the n Lyapunov exponents of an n-dimensional unknown dynamical system. *Physica D. Nonlinear Phenomena, 59*(1-3), 142–157. DOI: 10.1016/0167-2789(92)90210-E

Gershenfeld, N. (1999). *The Nature of Mathematical Modelling*. Cambridge University Press.

Ghafooripour, A., & Kiamehr, H. (2000). An overview of system identification methods and applications part ii: theory, type of tested structures, history and prospective of system identification. *4th International conference on coasts, ports & marine structure, Nov 2000 Shahid Rajaee Port Complex, Bandar Abbass*, 509-540.

Giacomini, R., & Rossi, B. (2009). Model comparisons in unstable environments, *ERID Working Paper 30*, Duke.

Giacomini, R., & Rossi, B. (2009). Model comparisons in unstable environments. *ERID Working Paper 30*, Duke.

Giannakis, G. B., & Serpedin, E. A. (2001). A bibliography on nonlinear system identification. *Signal Processing, 81*(3), 533–580. DOI: 10.1016/S0165-1684(00)00231-0

Glentis, G.-O., & Slump, C. H. (1995). A highly modular normalized adaptive lattice algorithm for multichannel least squares filtering. *1995 International Conference on Acoustics, Speech, and Signal Processing*. DOI: 10.1109/ICASSP.1995.480508

Goldenberg, L., Matiushkin, B. D., & Poliak, M. (1990). *Digital signal processing: Handbook*. Radio and Communications.

Golovko, V. A. (2005). Neural network methods of processing chaotic processes. Scientific session of MEPhI-2005. VII All-Russian Scientific and Technical Conference Neuro-Informatics-2005: Lectures on neuroinformatics. Moscow: MIPhI, 43-88.

Graupe, D. (1975). *Identification of systems*. Litton Educational Co.

Graupe, D., & Cline, W. K. (1975). Derivation of ARMA parameters and orders from pure AR models. *Int. J. Syst. ScL, 10*(2), 101–106.

Greene, B. R. (2004). *The fabric of the cosmos: space, time and the texture of reality*. Random House, Inc.

Gromiko, D., & Sankovsky, I. (1974). *Self-customized systems with model*. Energy.

Gromyko, V. D., & Sankovsky, E. A. (1974). *Self-adjusting systems with a model*. Energy.

Gudi, R. D., Rawlings, J. B., Venkat, A., & Jabbar, N. (2004). Identification for decentralized MPC. *IFAC Proceedings, 37*(9), 299-304.

Guillaume, J. H. A., Jakeman, J. D., Marsili-Libelli, S., Asher, M., Brunner, P., Croke, B., Hill, M. C., Jakeman, A. J., Keesman, K. J., Razavi, S., & Stigter, J. D. (2019). Introductory overview of identifiability analysis: A guide to evaluating whether you have the right type of data for your modeling purpose. *Environmental Modelling & Software, 119*, 418–432. DOI: 10.1016/j.envsoft.2019.07.007

Gupta, N., & Chopra, N. (2016). Stability analysis of a two-channel feedback networked control system. *Conference: 2016 Indian Control Conference (ICC)*. DOI: 10.1109/INDIANCC.2016.7441129

Haber, R. (1985). Nonlinearity tests for dynamic processes. *IFAC proceedings volumes, 18*(5), 409-414.

Hadamard, J. (1948). *Lecons de geometrie elementaire. T. 2: Geometrie dans l'espace*. Armand Colin & Cie.

Hametner, C., & Jakubek, S. (2011). Nonlinear identification with local model networks using GTLS techniques and equality constraints. *IEEE Transactions on Neural Networks, 22*(9), 1406–1418. DOI: 10.1109/TNN.2011.2159309 PMID: 21788188

Hamilton, J. D. (1994). *Time Series Analysis*. Princeton University Press. DOI: 10.1515/9780691218632

Hannan, E. J., & Quinn, B. G. (1979). The Determination of the order an autoregression. *Journal of the Royal Statistical Society. Series B, Statistical Methodology, 41*(2), 190–195. DOI: 10.1111/j.2517-6161.1979.tb01072.x

Hansen, B. E. (2022). *Econometric*. Princeton University Press.

Hao, L., Yang, H., Sun, Z., Xiang, C., & Xue, B. (2021). Modeling and compensation control of asymmetric hysteresis in a pneumatic artificial muscle. *Journal of Intelligent Material Systems and Structures*, *28*(19), 2769–2780. DOI: 10.1177/1045389X17698588

Hastie, T., Tibshirani, R., & Friedman, J. (2001). *The elements of statistical learning*. Springer. DOI: 10.1007/978-0-387-21606-5

Hayashi, C. (1964). *Nonlinear oscillations in physical systems*. McGraw-hill Book Company.

Holmes, P. (1982). The dynamics of repeated impacts with a sinusoidally vibrating table. *Journal of Sound and Vibration*, *84*(2), 173–189. DOI: 10.1016/S0022-460X(82)80002-3

Hołyst, J. A., & Urbanowicz, K. (2000). Chaos control in economic model by time-delayed feedback method. *Physica A*, *287*(3–4), 587–598. DOI: 10.1016/S0378-4371(00)00395-2

Hosseini, S. M., Johansen, T. A., & Fatehi, A. (2011). Comparison of nonlinearity measures based on time series analysis for nonlinearity detection. *Modeling, Identification and Control*, *32*(4), 123–140. DOI: 10.4173/mic.2011.4.1

Hsiao, C. (2003). *Analysis of Panel Data*. Cambridge University Press. DOI: 10.1017/CBO9780511754203

Hua, C., Guan, X., & Shi, P. (2004). Decentralized robust model reference adaptive control for interconnected time-delay systems. *Proceeding of the 2004 American Control Conference Boston,* Massachusetts June 30 - July 2, 2004, 4285-4289.

Iqbal, A., Aftab, S., Ullah, I., Saeed, M. A., & Husen, A. (2019). (). A classification framework to detect DoS attacks. *International Journal of Computer Network and Information Security*, *11*(9), 40–47. DOI: 10.5815/ijcnis.2019.09.05

Isermann, R., Baur, V., Bamberger, W., Kneppo, P., & Siebert, H. (1974). Comparison of six on-line identification and parameter estimation methods. *Automatica*, *10*(1), 81–103. DOI: 10.1016/0005-1098(74)90012-0

Isermann, R., & Münchhof, M. (2011). *Identification of dynamic systems*. Springer-Verlag Berlin Heidelberg. DOI: 10.1007/978-3-540-78879-9

Ivakhnenko, A. G. (1975). *Long-term forecasting and management of complex systems*. Technika.

Izobov, N. A. (2006). *Introduction in theory of Lyapunov indexes*. BGU.

Jenkins, G. M., & Watts, D. G. (1969). *Spectral analysis and its applications*. Holden-Day.

Jing, N., Juan, Y., Xuemei, R., & Gu, Y. (2014). Robust adaptive estimation of nonlinear system with time-varying parameters. *International Journal of Adaptive Control and Signal Processing*, *29*(8), 1055–1072. DOI: 10.1002/acs.2524

Johnston, J. (1972). *Econometric methods* (2nd ed.). McGraw-Hill Book Company.

Johnston, J. (1972). *Econometric Methods*. McGraw-Hill.

Jones, R. H. (1976). Autoregression order selection. *Geophysics*, *41*(4), 771–773. DOI: 10.1190/1.1440649

Jorgenson, D. W. (1964). Minimum variance, linear, unbiased seasonal adjustment of economic time series. *Journal of the American Statistical Association*, *59*(307), 681–724. DOI: 10.1080/01621459.1964.10480722

Kailath, T. (Ed.). (1977). *Linear Least-Squares Estimation,* Stroudsburg / Pennsylvania: Dowden, Hutchinson and Ross, Inc.

Kalman, R. E. (1961). On the general theory of control systems. *Proceeding first IFAC Congress on Automatic Control,* Moscow, 1960; Butterworths, London, 1, 481-492.

Kalman, R., Falb, P. L., & Arbib, M. A. (1969). *Topics in mathematical system theory*. McGraw-Hill.

Karabutov, N. (2012a). About structures of state systems identification of static object with hysteresis. *International journal of sensing, computing & control, 2*(2), 59-69.

Karabutov, N. (2012b). Structural identification of static processes with hysteresis nonlinearities in civil engineering. *Journal of civil engineering and science, 1*(4), 22-29.

Karabutov, N. (2023). On adaptive identification of systems having multiple nonlinearities. *Russian technological journal, 11*(5), 94–105.

Karabutov, N. N. (1993). Integral estimations of quality of transients in adaptive systems in the presence of restrictions and perturbations. *Reports of Academy of sciences, 331*(2), 158-160.

Karabutov, N. N. (2009a). Structural identification of systems: analysis of informational structures. Moscow: URSS: Book house Librokom.

Karabutov, N. N. (2010). Structures, fields and methods of identification of nonlinear static systems in the conditions of uncertainty," *Intelligent control and automation, 1*, 59-67.

Karabutov, N. N. (2010). Structures, fields and methods of identification of nonlinear static systems in the conditions of uncertainty. *Intelligent control and automation, 1*, 59-67. DOI: 10.4236/ica.2010.12007

Karabutov, N. N. (2011). Structural identification of static objects: Fields, structures, methods. Moscow: URSS: Book house Librokom.

Karabutov, N. N. (2012a). About structures of state systems identification of static object with hysteresis. International *journal sensing, computing and control, 2*(2), 59-69.

Karabutov, N. N. (2012b). Structural identification of static systems with distributed lags. *International journal of control science and engineering, 2*(6), 136-142.

Karabutov, N. N. (2020). S-synchronization structural identifiability and identification of nonlinear dynamic systems. *Mekhatronika, avtomatizatsiya, upravlenie, 21*(6), 323-336.

Karabutov, N. N. (2024). On identification of interconnected systems. *Russian technological journal, 12*(5), 63-76.

Karabutov, N. (2012). About structures of state systems identification of static object with hysteresis. *Int. J. Sensing. Computing and Control*, 2(2), 59–69.

Karabutov, N. (2013). Identification of parametrical restrictions in static systems in conditions of uncertainty. *International Journal of Intelligent Systems and Applications*, *5*(4), 43–54. DOI: 10.5815/ijisa.2013.04.04

Karabutov, N. (2013). Structural identification of systems with distributed lag. *International Journal of Intelligent Systems and Applications*, *5*(11), 1–10. DOI: 10.5815/ijisa.2013.11.01

Karabutov, N. (2015a). Structural identification of nonlinear dynamic systems. *International Journal of Intelligent Systems and Applications*, *7*(9), 1–11. DOI: 10.5815/ijisa.2015.09.01

Karabutov, N. (2015b). Structural methods of estimation Lyapunov exponents linear dynamic system. *International Journal of Intelligent Systems and Applications*, *7*(10), 1–11. DOI: 10.5815/ijisa.2015.10.01

Karabutov, N. (2016). Structural Identification of dynamic systems with hysteresis. *International Journal of Intelligent Systems and Applications*, *8*(7), 1–13.

Karabutov, N. (2016a). Structural identification of dynamic systems with hysteresis. *International Journal of Intelligent Systems and Applications*, *8*(7), 1–13. DOI: 10.5815/ijisa.2016.07.01

Karabutov, N. (2016b). System with distributed lag: Adaptive identification and prediction. *International Journal of Intelligent Systems and Applications*, *8*(6), 1–13. DOI: 10.5815/ijisa.2017.06.01

Karabutov, N. (2017). Adaptive observers with uncertainty in loop tuning for linear time-varying dynamical systems. *International Journal of Intelligent Systems and Applications*, *9*(4), 1–13. DOI: 10.5815/ijisa.2017.04.01

Karabutov, N. (2017). Frameworks in Problems of Structural Identification Systems. *International Journal of Intelligent Systems and Applications*, *9*(1), 1–19.

Karabutov, N. (2017). Structural methods of design identification systems. In Uvarova, L. A., Nadykto, A. B., & Latyshev, A. V. (Eds.), *In Nonlinearity problems, solutions and applications* (Vol. 1, pp. 233–274). Nova Science Publishers, Inc.

Karabutov, N. (2017). Structural Methods of Design Identification Systems. In Uvarova, L., Nadykto, A. B., & Latyshev, A. V. (Eds.), *Nonlinearity: Problems, Solutions and Applications* (Vol. 1, pp. 233–274). Nova Science Publishers Inc.

Karabutov, N. (2017c). Frameworks in problems of structural identification systems. *International Journal of Intelligent Systems and Applications*, *9*(1), 1–19.

Karabutov, N. (2018). *Frameworks in identification problems: Construction and analysis*. Lenand.

Karabutov, N. (2018). Structural-parametrical design method of adaptive observers for nonlinear systems. *International Journal of Intelligent Systems and Applications*, *10*(2), 1–16. DOI: 10.5815/ijisa.2018.02.01

Karabutov, N. (2018a). About Lyapunov exponents identification for systems with periodic coefficients. *International Journal of Intelligent Systems and Applications*, *10*(11), 1–10. DOI: 10.5815/ijisa.2018.11.01

Karabutov, N. (2018b). About structural identifiability of nonlinear dynamic systems under uncertainty. *Global Journal of Science Frontier Research: (A). Physics and Space Science*, *18*(11), 51–61.

Karabutov, N. (2018c). *Frameworks in identification problems: Design and analysis*. URSS/Lenand.

Karabutov, N. (2021). Geometrical Framework Application Directions in Identification Systems. [Review]. *International Journal of Intelligent Systems and Applications, 13*(2), 1–20. DOI: 10.5815/ijisa.2021.02.01

Karabutov, N. (2021). S-synchronization and excitation constancy in structural identifiability problem of nonlinear systems. *EPJ Web of Conferences, 248*, 01004.

Karabutov, N. (2021). Structural Identifiability of Systems with Multiple Nonlinearities. *Contemporary Mathematics, 2*(2), 103–172. DOI: 10.37256/cm.222021763

Karabutov, N. (2021a). Geometrical frameworks in identification problem. *Intelligent Control and Automation, 12*(2), 17–43. DOI: 10.4236/ica.2021.122002

Karabutov, N. (2022). Structural Identifiability of Feedback Systems with Nonlinear Adulterating. *Contemporary Mathematics, 3*(2), 257–269. DOI: 10.37256/cm.3220221258

Karabutov, N. (2023). Identifiability and detectability of Lyapunov exponents in robotics. In Mellal, M. A. (Ed.), *Design and Control Advances in Robotics* (pp. 152–174). IGI Globalss Publisher of Timely Knowledge.

Karabutov, N. (2024). Bouc-Wen hysteresis identification: structural-parametric approach. In *Advances in Mathematics Research* (Vol. 36, pp. 99–144). Nova Science Publishers, Inc.

Karabutov, N. N. (2006). *Adaptive identification of systems. Information synthesis.* URSS, Librokom.

Karabutov, N. N. (2006). *Adaptive identification of systems: informational synthesis.* URSS.

Karabutov, N. N. (2006). *Adaptive identification of systems: Informational synthesis.* URSS.

Karabutov, N. N. (2006). *Adaptive system identification: information synthesis.* URSS.

Karabutov, N. N. (2007). *Adaptive identification of systems: information synthesis.* URSS.

Karabutov, N. N. (2008). *Adaptive identification of systems: Information synthesis.* URSS.

Karabutov, N. N. (2009). *Structural identification of systems. analysis of information structures* [Russian language]. URSS/ Librokom.

Karabutov, N. N. (2009). *Structural Identification of Systems. The Analysis of Information Structures.* URSS/ Librokom.

Karabutov, N. N. (2009). Structural identification of systems: analysis of information portraits. URSS. DOI: 10.13140/2.1.1843.2324

Karabutov, N. N. (2009). *Structural identification of systems: the analysis of informational structures*. URSS, Librokom.

Karabutov, N. N. (2009b). Structural identification of a static object by processing measurement data. *Measurement Techniques*, *52*(1), 7–15. DOI: 10.1007/s11018-009-9219-1

Karabutov, N. N. (2011). *Structural identification of static plants: Fields, structures, methods*. URSS, Librokom.

Karabutov, N. N. (2012). Adaptive algorithms for structural parameters identification of single-valued nonlinearities of static systems with a vectorial input. *Int. J. Sensing. Computing and Control*, *2*(1), 1–11.

Karabutov, N. N. (2012). Structural identification of static systems with distributed lags. *International Journal of Control Science and Engineering*, *2*(6), 136–142. DOI: 10.5923/j.control.20120206.01

Karabutov, N. N. (2013). Method of φ-algorithms of control. In Nadykto, A. B., Uvarova, L., & Latyshev, A. V. (Eds.), *Mathematical models of non-linear phenomena, processes, and systems: from molecular scale to planetary atmosphere* (pp. 259–396). Published by Nova Science Publishers.

Karabutov, N. N. (2013). Structural identification of nonlinear static system on basis of analysis sector sets. *International Journal of Intelligent Systems and Applications*, *6*(1), 1–10.

Karabutov, N. N. (2016). *Adaptive Identification of Systems: Information Synthesis*. URSS.

Karabutov, N. N. (2016a). *Structural identification of static objects: Fields, structures, methods*. URSS.

Karabutov, N. N. (2016b). *Structural identification of systems: Analysis of information structures*. URSS.

Karabutov, N. N. (2017). Structural methods of design identification systems. In Uvarova, L. A., Nadykto, A. B., & Latyshev, A. V. (Eds.), *Nonlinearity problems, solutions and applications* (Vol. 1, pp. 233–274). Nova Science Publishers Inc.

Karabutov, N. N. (2018). *Structures in identification problems: Construction and analysis*. URSS.

Karabutov, N. N. (2020). Frameworks application for estimation of Lyapunov exponents for systems with periodic coefficients. *Mekhatronika, Avtomatizatsiya. Upravlenie*, *21*(1), 3–13.

Karabutov, N. N. (2020). S-synchronization, structural identifiability and identification of nonlinear dynamic systems. *Mekhatronika, Avtomatizatsiya. Upravlenie*, *21*(6), 323–336.

Karabutov, N. N. (2021). *Introduction to structural identifiability of nonlinear systems*. URSS/Lenand.

Karabutov, N. N. (2021b). *Introduction to the structural identifiability of nonlinear systems*. URSS/LENAND.

Karabutov, N. N. (2022). Identifiability and detectability of Lyapunov exponents for linear dynamical systems. *Mekhatronika, Avtomatizatsiya. Upravlenie*, *23*(7), 339–350.

Karabutov, N. N. (2023). On adaptive identification of systems having multiple nonlinearities. *Russ. Technol. J.*, *11*(5), 94–105. DOI: 10.32362/2500-316X-2023-11-5-94-10

Karabutov, N. N. (2024). Structural Identifiability Evaluation of System with Non-symmetric Nonlinearities. *Mekhatronika, Avtomatizatsiya. Upravlenie*, *25*(2), 55–64.

Karabutov, N., & Feklin, V. (2016). Adaptive identification of systems with distributed lags. *Journal of Mathematical Sciences*, *216*(5), 649–666. DOI: 10.1007/s10958-016-2925-8

Kashyap, R. L. (1980). Inconsistency of the aic rule for estimation the order of autoregressive models. *IEEE Transactions on Automatic Control*, *25*(5), 996–998. DOI: 10.1109/TAC.1980.1102471

Kashyap, R., & Rao, A. (1976). *Dynamic stochastic models from empirical data*. Academic Press.

Katkovnik, V. J. (1976). *Linear estimations and stochastic problems of optimization*. Nauka, Ru.

Kay, S. M. (1999). *Modern spectral estimation: Theory and application*. Prentice-Hall, Inc., Englewood Cliffs.

Kerschen, G., Worden, K., Vakakis, A., & Golinval, J.-C. (2006). Past, present and future of nonlinear system identification in structural dynamics. *Mechanical systems and signal processing*, Elsevier, *20*(3), 505-592.

Kerschen, G., Worden, K., Vakakis, A., & Golinval, J. (2006). Past, present and future of nonlinear system identification in structural dynamics. *Mechanical Systems and Signal Processing*, *20*(3), 505–592. DOI: 10.1016/j.ymssp.2005.04.008

Kholmatov, U. (2023). The possibility of applying the theory of adaptive identification to automate multi-connected objects. *The American journal of engineering and technology*, 31-38.

Kogan, M. M., & Nejmark, J. I. (1987). Adaptive locally-optimum control. *Automatics and telemechanics, 8*, 126-136.

Koh, C. G., & Perry, M. J. (2010). *Structural identification and damage detection using genetic algorithms*. CRC Press.

Ko, S., & Bitmead, R. R. (2007). Identifiability analysis for nonlinear feedback limit-cycle systems using harmonic balance methods. *Proceedings of the IEEE Conference on Decision and Control, 39*(1), 2506 – 2511. DOI: 10.1109/CDC.2006.377326

Koyck, L. M. (1954). *Distributed Lags and Investment Analysis*. North-Holland Publishing Company.

Krasovsky, A. A. (1957). On two-channel automatic control systems with antisymmetric connections. *Automation and telemechanics, 18*(2), 126–136.

Krasovsky, A.A. (1975). Optimal algorithms in problems of identification with adaptive model. *Automatics and telemechanics, 8*, 75-82.

Krasovsky, A. A. (Ed.). (1987). *Handbook of the theory of automatic control*. Nauka.

Krasovsky, A. A. (Ed.). (1987). *Reference book on automatic control theory*. Nauka.

Krasovsky, A. A., Bukov, V. N., & Shendrik, V. S. (1977). *Universal algorithms of optimum control of continuous processes*. Nauka.

Kreisselmeier, G. (1986). A robust indirect adaptive control approach. *International Journal of Control, 43*(1), 161–175.

Krstić, M., & Kokotović, P. V. (1996). Adaptive Nonlinear Output-Feedback Schemes with Marino-Tomei Controller. *IEEE Transactions on Automatic Control, 41*(2), 274–280.

Kudva, P., & Narendra, K. S. (1973). Synthesis of adaptive observer using Lyapunov direct method. *Jnt. J. Control, 18*(4), 1201–1216.

Kudva, P., & Narendra, K. S. (1973). Synthesis of an adaptive observer using Lyapunov direct method. *International Journal of Control, 18*(4), 1201–1216. DOI: 10.1080/00207177308932593

Kulchitsky, O. J. (1983). Algorithms of type of stochastic approximation in a contour of adaptive discrete stochastic linear dynamic system. I. *Automatics and telemechanics, 9*, 102-119.

Kuntsevich, V. M., & Lichac, M. M. (1985). *Synthesis of optimum and adaptive control systems*. Naukova dumka, UA.

Kuntsevich, V. M., & Lychak, M. M. (1985). *Synthesis of optimal and adaptive control systems: A game approach*. Naukova dumka.

Kurtoglu, T., & Tumer, I. Y. (2008). A Graph-Based Fault Identification and Propagation Framework for Functional Design of Complex Systems. *Journal of Mechanical Design, 130*(5), 051401–051408. DOI: 10.1115/1.2885181

Kushner, H. J., & Clark, D. S. (1978). *Stochastic approximation methods for constrained and unconstrained systems*. Springer. DOI: 10.1007/978-1-4684-9352-8

Kwakernaak, H., & Sivan, R. (1972). *Linear optimal control systems*. Wiley Interscience.

Lamara, A., Colin, G., Lanusse, P., Chamaillard, Y., & Charlet, A. (2012). Decentralized robust control-system for a non-square MIMO system, the air-path of a turbocharged Diesel engine. 2012 Workshop on Engine and Powertrain Control, Simulation and Modeling. The International Federation of Automatic Control Rueil-Malmaison, France, October 23-25, 130-137

Lancaster, P. (1969). *Theory of Matrices*. Academic Press.

Lansing, K. J. (2002). Real-time estimation of trend output and the illusion of interest rate smoothing. *FRBSF economic. RE:view*, •••, 18–34.

Lauer, F., Bloch, G., & Vidal, R. (2011). A continuous optimization framework for hybrid system identification. *Automatica, 47*(3), 608–613. DOI: 10.1016/j.automatica.2011.01.020

Leboutet, Q., Roux, J., Janot, A., Guadarrama-Olvera, J. R., & Cheng, G. (2021). Inertial parameter identification in robotics: A Survey. *Applied Sciences (Basel, Switzerland), 11*(9), 4303. DOI: 10.3390/app11094303

Lee, R. (1964). *Optimal estimation, identification, and control*. MIT Press.

Li, C. (2021). Closed-loop identification for a class of nonlinearly parameterized discrete-time systems. *Automatica, 131*, 109747. DOI: 10.1016/j.automatica.2021.109747

Li, C., Min, F., & Li, C. (2018). Multiple coexisting attractors of the serial–parallel memristor-based chaotic system and its adaptive generalized synchronization. *Nonlinear Dynamics*, *94*(7191), 1–22. DOI: 10.1007/s11071-018-4524-3

Liebert, W., Pawelzik, K., & Schuster, H. G. (1991). Optimal embedding of chaotic attractors from topological considerations. *Europhysics Letters*, *14*(6), 521–526. DOI: 10.1209/0295-5075/14/6/004

Ljung, L. (1999). *System identification: Theory for the User*. Prentice Hall PTR.

Ljung, L. (2000). *System Identification Toolbox User's Guide. Computation. Visualization. Programming. Version 5*. The MathWorks Inc.

Lü, J., & Zhang, S. (2007). Controlling Chen's chaotic attractor using backstepping design based on parameters identification. *Physics Letters. [Part A]*, *32*(4), 148–156. DOI: 10.1016/S0375-9601(01)00383-8

Lyapunov, A. M. (1992). *General problem about a movement stability*. CRC Press.

Lyapunov, A. M. (1992). *General problem of motion stability*. CRC Press. DOI: 10.1080/00207179208934253

Macek, W. M., & Redaelli, S. (2000). Estimation of the entropy of the solar wind flow. *Physical Review. E*, *62*(5), 6496–6504. DOI: 10.1103/PhysRevE.62.6496 PMID: 11101986

Magnusson, L., & Mavroeidis, S. (2012). Identification using stability restrictions. https://econ.sciences-po.fr/sites/default/files/SCident32s.pdf

Malinetskiy, G. G., & Potapov, A. B. (2000). *Modern problems of nonlinear dynamics*. Editorial URSS.

Malinvaud, E. (1969). Méthodes statistiques de l'économétrie. Deuxième édition. London-Paris.

Malinvaud, E. (1980). *Statistical methods in econometrics* (3rd ed.). North-Holland Publishing Co.

Malinvaud, E. (1980). *Statistical Methods in Econometrics*. North-Holland., DOI: 10.1604/9780444854735

Mao, X., & He, J. (2022). Decentralized System Identification Method for Large-Scale Networks. 2022 *American Control Conference* (ACC), Atlanta, GA, USA, 5173-5178.

Marino, R., & Tomei, P. (1992). Global adaptive observers for nonlinear systems via filteres transformations. *IEEE Transactions on Automatic Control*, *37*(8), 1239–1245.

Marple, S. L.Jr. (1987). *Digital spectral analysis: With applications*. Prentice-Hall.

Martinyuk, A. A. (1980). Method of integral inequalities in the theory of stability of motion. *Soviet Applied Mechanics*, *16*(4), 267–281. DOI: 10.1007/BF00884242

Martinyuk, A. A., Lanshmikantam, V., & Lila, S. (1989). *Stability of movement: a method of integral inequalities*. Naukova dumka, UA.

Mavroeidis, S., & Magnusson, L. (2014). Identification using stability restrictions. *Econometrica*, *82*(5), 1799–1851. DOI: 10.3982/ECTA9612

Mazunin, V. P., & Dvoinikov, D. A. (2010). Parametric constraints in nonlinear control systems of mechanisms with elasticity. *Electrical Engineering*, *5*, 9–13.

Mazunin, V. P., & Dvojnikov, D. A. (2010). Parametrical restrictions in nonlinear control systems of mechanisms with elasticity. *Russian Electrical Engineering*, *81*(5), 227–231. DOI: 10.3103/S1068371210050020

Mead, J. L., & Renaut, R. A. (2010). Least squares problems with inequality constraints as quadratic constraints. *Linear Algebra and Its Applications*, *432*(8), 1936–1949. DOI: 10.1016/j.laa.2009.04.017

Meerov, M. V., & Litvak, B. L. (1972). *Optimization of multicommunicated control systems*. Nauka.

Mehra, R. K. (1974). Optimal input signals for parameter estimation in dynamic systems. A survey and new results. *IEEE Transactions on Automatic Control*, *19*(6), 753–768. DOI: 10.1109/TAC.1974.1100701

Mei, H., Cai, W.-J., & Xiong, Q. (2005). Decentralized closed-loop parameter identification for multivariable processes from step responses. *Mathematics and Computers in Simulation*, *68*(2), 171–192. DOI: 10.1016/j.matcom.2004.12.008

Mejari, M., Piga, D., & Bemporad, A. (2018). A bias-correction method for closed-loop identification of linear parameter-varying systems. *Automatica*, *87*, 128–141. DOI: 10.1016/j.automatica.2017.09.014

Michel, A., Wang, K., & Hu, B. (2001). *Qualitative theory of dynamical systems*. CRC Press., DOI: 10.1201/9780203908297

Morozovsky, V. E. (1970). *Multi-connected automatic control systems*. Energiya.

Mosteller, F., & Tukey, J. (2019). *Data analysis and regression: A second course in statistics* (1st ed.). Pearson.

Mosteller, F., & Tukey, J. W. (1977). *Data Analysis and Regression: A Second Course in Statistics*. Addison-Wesley.

Nagarajaiah, S. (2017). Sparse and low-rank methods in structural system identification and monitoring. *X International Conference on Structural Dynamics (EURODYN 2017). Procedia Engineering*, *199*, 62–69. DOI: 10.1016/j.proeng.2017.09.153

Narendra, K. S., & Kudva, B. (1974). Stable adaptive schemes for system: Identification and control. *IEEE Transactions on Systems, Man, and Cybernetics*, *4*(6), 542–560.

Nelson, C. R., & Schwert, G. W. (1974). Estimating the parameters of a distributed lag model from cross-section data: The Case of hospital admissions and discharges. *Journal of the American Statistical Association*, *69*(347), 627–633. DOI: 10.1080/01621459.1974.10480179

Noël, J., & Schoukens, J. (2018). Grey-box state-space identification of nonlinear mechanical vibrations. *International Journal of Control*, *91*(5), 1118–1139. DOI: 10.1080/00207179.2017.1308557

Nuyan, S., & Carrol, R. L. (1979). Minimal order arbitrarily fast adaptive observer and identifies. *IEEE Transactions on Automatic Control*, *24*(2), 496–499. DOI: 10.1109/TAC.1979.1101989

Ortega, J., & Rheinboldt, W. (1970). *Iterative solution of nonlinear equations in several variables*. Academic Press.

Pai, P. F., & Palazotto, A. N. (2008). Detection and identification of nonlinearities by amplitude and frequency modulation analysis. *mechanical systems and signal processing*, *22*(5), 1107-1132.

Palanthandalam-Madapusia, H. J., Van Peltb, T. H., & Bernstein, D. S. (2010). Parameter consistency and quadratically constrained errors-in-variables least-squares identification. *International Journal of Control*, *83*(4), 862–877. DOI: 10.1080/00207170903470666

Papadopoulos, P. N., Guo, T., & Milanović, J. V. (2017). Probabilistic Framework for Online Identification of Dynamic Behavior of Power Systems with Renewable Generation. *IEEE Transactions on Power Systems*, *33*(1), 45–54. DOI: 10.1109/TPWRS.2017.2688446

Parsen, E. (1974). Some recent advances in time series modelling. *IEEE Transactions on Automatic Control*, *19*(6), 723–730. DOI: 10.1109/TAC.1974.1100733

Pavlov, Y. N., Nedashkovskii1, V. M., & Tihomirova, E. A. (2015). Identification of nonlinear dynamic systems possessing some nonlinearities. *Science and education of the Bauman MSTU*, *07*, 217–234.

Pecora, L. M., Moniz, L., Nichols, J., & Carroll, T. L. (2007). A Unified approach to attractor reconstruction. *Chaos (Woodbury, N.Y.)*, *17*(1), 013110. DOI: 10.1063/1.2430294 PMID: 17411246

Perederiy, Y. A. (2012). Method for calculation of Lyapunov exponents spectrum from data series. *Izvestiya VUZ. Izvestiya Vyssih Ucebnyh Zavedenij. Prikladnaya Nelinejnaya Dinamika*, *20*(1), 99–104.

Perelman, I. I. (1983). Model structure selection Methodology in control objects identification. *Automation and telemechanics, 11*, 11 5–29.

Pershin, Y. V., & Slipko, V. A. (2018). Dynamical attractors of memristors and their networks. *arXiv: 1808.07947.*

Piga, D., & Tyth, R. (2014). A bias-corrected estimator for nonlinear systems with output-error type model structures. *Automatica*, *50*(9), 2373–2380. DOI: 10.1016/j.automatica.2014.07.021

Pillonetto, G., Dinuzzo, F., Chen, T., De Nicolao G., & Ljung, L. (2014). Kernel methods in system identification, machine learning and function estimation: A survey. *Automatica, 50*(3), 657-682. doi.: DOI: 10.1016/j.automatica.2014.01.001

Pintelon, R. (2002). Frequency-domain subspace system identification using non-parametric noise models. *Automatica*, *38*(8), 1295–1311. DOI: 10.1016/S0005-1098(02)00036-5

Pintelon, R., & Schoukens, J. (2012). *System Identification. A Frequency Domain Approach* (2nd ed.). Published by John Wiley & Sons, Inc., DOI: 10.1002/9781118287422

Podval'ny, S. L., Vasil'ev, E. M. (2022). Multilevel signal adaptation in non-stationary control systems. Bulletin of Voronezh state technical university, *18*(5), 38-45.

Poljak, B. T. (1976). Convergence and a velocity of convergence iterative rocedures I. *Automatics and telemechanics, 4*, 101-107.

Poljak, B. T. (1983). *Introduction in optimization*. Nauka, Ru.

Polyak, B. T., & Tsypkin, J. Z. (1980). Optimum pseudogradients algorithms of adaptation. *Automatics and telemechanics, 8*, 74-84.

Polyak, B. T., & Tsypkin, J. Z. (1983). Optimum algorithms of criteria optimisation in the conditions of uncertainty. *Reports AS USSR*, *273*(2), 315–318.

Popov, E. P. (1988). *Theory of nonlinear automatic control and control systems.* Nauka.

Porcher, R., & Thomas, G. (2001). Estimating Lyapunov exponents in biomedical time series. [R]. *Physical Review. E, 64*(1), 010902. DOI: 10.1103/PhysRevE.64.010902 PMID: 11461218

Prawin, J., Rao, A. R. M., & Sethi, A. (2020). Parameter identification of systems with multiple disproportional local nonlinearities. *Nonlinear Dynamics, 100*(1), 289–314. DOI: 10.1007/s11071-020-05538-1

Prokhorov, S. A., & Grafkin, V. V. (2010). *Structural and spectral analysis of random processes*. SNC RAS.

Raibman, N. S. (1978). Identification of control objects (Review). *Automation and telemechanics, 6*, 80–93.

Raibman, N. S., & Terekhin, A. T. (1965). Dispersion methods of random functions and their application for the study of nonlinear control objects. *Automation and telemechanics, 26*, 500–509.

Raibman, N. S., & Chadeev, V. M. (1975). *Construction of models of production processes*. Energiya.

Roettgen, D., Allen, M. S., Kammer, D., & Mayes, R. L. (2017). Substructuring of a Nonlinear Beam Using a Modal Iwan Framework, Part I: Nonlinear Modal Model Identification. *Dynamics of Coupled Structures, 4*, 165–178. DOI: 10.1007/978-3-319-54930-9_15

Rosenstein, M. T., Collins, J. J., & De Luca, C. J. (1993). A practical method for calculating largest Lyapunov exponents from small data sets Source. *Physica D. Nonlinear Phenomena, 65*(1-2), 117–134. DOI: 10.1016/0167-2789(93)90009-P

Rossi, B. (2005). Optimal tests for nested model selection with underlying parameter instability. *Econometric Theory, 21*(5), 962–990. DOI: 10.1017/S0266466605050486

Saccomani, M. P. (2013). Structural vs Practical identiability in system biology. International Work-Conference on Bio informatics and Biomedical Engineering. IWBBIO 2013. *Proceedings. Granada, 18-20*(March), 305–313.

Saccomani, M. P., & Thomaseth, K. (2010). Structural vs practical identifiability of nonlinear differential equation models in systems biology. Bringing mathematics to life. In Rogato, A., Zazzu, V., & Guarracino, M. (Eds.), *Dynamics of mathematical models in biology* (pp. 31–42). Springer.

Saccomani, M. P., & Thomaseth, K. (2010). Structural vs practical identifiability of nonlinear differential equation models in systems biology. In Rogato, A., Zazzu, V., & Guarracino, M. (Eds.), *Dynamics of mathematical models in biology: Bringing mathematics to life* (pp. 31–42). Springer.

Safari, S., & Monsalve, J. M. L. (2021). Direct optimization-based model selection and parameter estimation using time-domain data for identifying localised nonlinearities. *Journal of Sound and Vibration, 501*, 116056. DOI: 10.1016/j.jsv.2021.116056

Saha, B., Goebel, K., Poll, S., & Christopherson, J. (2009). Prognostics methods for battery health monitoring using a Bayesian framework. *IEEE Transactions on Instrumentation and Measurement, 58*(2), 291–296. DOI: 10.1109/TIM.2008.2005965

Salyga, V. I., Karabutov, N. N., & Kusmina, O. I. (1988). Identification of nonlinear plants by the method of φ-algorithms. *Identification and system parameter estimation. 8th IFAC/IFOS symposium,* Beijing, China, *2*. 661–665.

Shakiba, S., Ourak, M., Poorten, E. V., Ayati, M., & Yousefi-Koma, A. (2021). Modeling and compensation of asymmetric rate-dependent hysteresis of a miniature pneumatic artificial muscle-based catheter. *Mechanical Systems and Signal Processing, 154*(107532), 1–23. DOI: 10.1016/j.ymssp.2020.107532

Shilnikov, L. P., Shilnikov, A. L., Turaev, D. V., & Chua, L. O. (2001). *Methods of qualitative theory in nonlinear dynamics. Part II.* World Scientific., DOI: 10.1142/4221

Shilov, G. (1973). *Mathematical analysis*. MIT Press.

Shin, M.-J., Guillaume, J. H., Croke, A. B. F. W., & Jakeman, A. J. (2015). A review of foundational methods for checking the structural identifiability of models: Results for rainfall-runoff. *Journal of Hydrology (Amsterdam), 520*, 1–16. DOI: 10.1016/j.jhydrol.2014.11.040

Singh, A., & Moore, K. J. (2021). Identification of multiple local nonlinear attachments using a single measurement case. *Journal of Sound and Vibration, 513*, 116410. DOI: 10.1016/j.jsv.2021.116410

Sirca, G. F.Jr, & Adeli, H. (2012). System identification in structural engineering. *Scientia Iranica, 19*(6), 1355–1364. DOI: 10.1016/j.scient.2012.09.002

Skokos, Ch. (2010). *The Lyapunov Characteristic Exponents and Their Computation* (Vol. 790). Lect. Notes Phys. DOI: 10.1007/978-3-642-04458-8_2

Skorospeshkin, M. V. (2008). Adaptive two-channel correction device for automatic control systems. *Bulletin of the Tomsk Polytechnic University, 312*(5), 52–57.

Smirnov, V. I. (1968). A course of higher mathematics.: Vol. 5. *Integration and functional analysis* [Russian language]. Nauka.

Söderstrom, T. (1975). Comments on order assumption and singularity of information matrix for pulse transfer function models. *IEEE Transactions on Automatic Control, 20*(3), 445–447. DOI: 10.1109/TAC.1975.1100949

Solow, R. (1960). On a family of lag distributions. *Econometrica, 28*(2), 393–406. DOI: 10.2307/1907729

Song, G., Xu, L., & Ding, F. (2019). Recursive least squares algorithm and stochastic gradient algorithm for feedback nonlinear equation-error systems. international journal of modelling, identification and control, *32*(3-4), 251-257.

Song, J., & Der Kiureghian, A. (2006). Generalized Bouc–Wen model for highly asymmetric hysteresis. *Journal of Engineering Mechanics, 6*(6), 610–618. DOI: 10.1061/(ASCE)0733-9399(2006)132:6(610)

Stepin, V. S. (Ed.). (2001). *New Philosophical Encyclopedia: In 4 volumes* / Edited by. Moscow: Mysl.

Stigter, J. D., & Peeters, R. L. M. (2007). On a geometric approach to the structural identifiability problem and its application in a water quality case study. *Proceedings of the European Control Conference 2007 Kos,* Greece, July 2-5, 2007, 3450-3456. DOI: 10.23919/ECC.2007.7068560

Stoica, P., & Söderstrom, T. (1982). On non-singular information matrices and local identifiability. *International Journal of Control, 35*(2), 323–329. DOI: 10.1080/00207178208932896

Takens, F. (1980). Detecting strange attractors in turbulence," *Dynamical systems and turbulence. Lecture notes in mathematics* / Eds. D.A. Rand, L.-S. Young. Berlin: Springer-Verlag, *898*, 366-381.

Takens, F. (1980). Detecting strange attractors in turbulence. In D. A. Rand, L.-S. Young (Eds.), *Dynamical Systems and Turbulence. Lecture Notes in Mathematics* (366–381). Berlin: Springer-Verlag, 898.

Teh, C. Y., Kerk, Y. W., Tay, K. M., & Lim, C. P. (2018). On modelling of data-driven monotone zero-order TSK fuzzy inference systems using a system identification framework. *IEEE Transactions on Fuzzy Systems, 26*(6), 3860–3874. DOI: 10.1109/TFUZZ.2018.2851258

Thamilmaran, K., Senthilkumar, D. V., Venkatesan, A., & Lakshmanan, M. (2006). Experimental realization of strange nonchaotic attractors in a quasiperiodically forced electronic circuit. *Physical Review. E, 74*(9), 036205. DOI: 10.1103/PhysRevE.74.036205 PMID: 17025727

Theil, H., & Stern, R. M. (1960). A simple unimodal lag distribution. *Metroeconomica, 12*(2-3), 111–119. DOI: 10.1111/j.1467-999X.1960.tb00277.x

Toth, R., Sanandaij, B. M., Poolla, K., & Vincent, T. L. (2012). Compressive system identification in the linear time-invariant framework," *50th IEEE Conference on Decision and Control and European Control Conference (CDC-ECC) Orlando, FL, USA, December 12-15, 2011*, 783-790.

Tsay, W.-J. (2005). The long memory autoregressive distributed lag model and its application on congressional approval. *Institute of Economics, Academia Sinica.* DOI: 10.1016/j.electstud.2009.06.004

Tsypkin, J. Z. (1995). *Bases of identification informational theory*. Nauka.

Van den Hof, P. (1997). Closed-loop issues in system identification. In Y. Sawaragi & S. Sagara (Eds.), System Identification (SYSID'97) (1547-1560*) Proc. 11th IFAC Symp. System Identification, Kitakyushu, Japan, 8-11 July 1997. IFAC Proceedings Series, 3*, Van den Hof, P. (1998). Closed-loop issues in system identification. *Annual reviews in control, 22*(0), 173 – 186.

Van Gig, J. (1978). *Applied general systems theory*. Harper & Row.

Van Peltb, H. D., & Bernstein, S. (2001). Quadratically constrained least squares identification. *Proceedings of the American Control Conference,* Arlington, VA June 25-27, 2001, 3684-3689. DOI: 10.1109/ACC.2001.946207

Vanli, O. A., & Del Castillo, E. (2007). Closed-Loop System Identification for Small Samples with Constraints. *Technometrics, 49*(4), 382–394. DOI: 10.1198/004017007000000146

Varna, A. L., Swaminathan, A., & Wu, M. (2008). A Decision Theoretic Framework for Analyzing Binary Hash-based Content Identification," *DRM '08 Proceedings of the 8th ACM workshop on Digital rights management*, 67-76. DOI: 10.1145/1456520.1456532

Veluvolu, K., & Soh, Y. (2011). Fault reconstruction and state estimation with sliding mode observers for Lipschitz non-linear systems. *IET Control Theory & Applications, 5*(11), 1255–1263. DOI: 10.1049/iet-cta.2010.0171

Vinogradov, I. M. (Ed.). (1979). *Mathematical Encyclopedia* (Vol. 2). Soviet Encyclopedia.

Voronov, A. A. (1985). *Introduction in dynamics of complex controlled systems.* Nauka.

Wagenmakers, E.-J. (2007). A practical solution to the pervasive problems of p values. *Psychonomic Bulletin & Review, 14*(5), 779–804. DOI: 10.3758/BF03194105 PMID: 18087943

Walter, E. (1982). *Identifiability of state space models.* Springer-Verlag. DOI: 10.1007/978-3-642-61823-9

Wei, Z., & Wang, D. (2012). Non-symmetrical Bouc–Wen model for piezoelectric ceramic actuators. *Sensors and Actuators. A, Physical, 181*, 51–60. DOI: 10.1016/j.sna.2012.03.048

Wei, Z., Xiang, B. L., & Ting, R. X. (2014). Online parameter identification of the asymmetrical Bouc-Wen model for piezoelectric actuators. *Precision Engineering, 38*(4), 921–927. DOI: 10.1016/j.precisioneng.2014.06.002

Wen-Ming, S. (2010). Parameter Estimation with constraints based on variational method. *Journal Marine. Sci. Appl, 9*, 105–108.

Wen-Ming, S. (2010). Parameter Estimation with Constraints Based on Variational Method. *Journal Marine. Sci., 9*, 105–108.

Wiggins, S. (2003). Introduction to applied nonlinear dynamical systems and chaos. Springer Science & Business Media, 4(5). .DOI: 10.1007/b97481

Wolf, A., Swift, J. B., Swinney, H. L., & Vastano, J. A. (1985). Determining Lyapunov exponents from a time series. *Physica, 16D*(16), 285–301.

Woodside, C. M. (1971). Estimation of the order of linear systems. *Automatica, 7*(6), 727–733. DOI: 10.1016/0005-1098(71)90010-0

Wu, B., Han, X., & Hui, N. (2021). System identification and controller design of a novel autonomous underwater vehicle. *Machines (Basel), 9*(109), 1–15. DOI: 10.3390/machines9060109

Wu, H. (2012). Decentralized adaptive robust control of uncertain large-scale non-linear dynamical systems with time-varying delays. *IET Control Theory & Applications, 6*(5), 629–640. DOI: 10.1049/iet-cta.2011.0015

Xie, S.-L., Liu, H.-T., Mei, J.-P., & Guo-Ying, G. (2018). Modeling and compensation of asymmetric hysteresis for pneumatic artificial muscles with a modified generalized Prandtl–Ishlinskii model. *Mechatronics, 52*, 49–57. DOI: 10.1016/j.mechatronics.2018.04.001

Yang, Q., Zhu, M., Jiang, T., He, J., Yuan, J., & Han, (2016). Decentralized Robust Adaptive output feedback stabilization for interconnected nonlinear systems with uncertainties. *Journal of control science and engineering*, article ID 3656578, 12.

Yang, H., Chen, Y., Sun, Y., & Hao, L. (2018). A novel kriging-median inverse compensator for modeling and compensating asymmetric hysteresis of pneumatic artificial muscle. *Smart Materials and Structures, 27*(11), 115019. DOI: 10.1088/1361-665X/aad758

Yang, H., Zhang, Y., Chen, Y., Sun, Y., & Hao, L. (2020). A novel learning adaptive hysteresis inverse compensator for pneumatic artificial muscles. *Smart Materials and Structures, 29*(1), 015035. DOI: 10.1088/1361-665X/ab4b84

Yang, Q., Zhu, M., Jiang, T., He, J., Yuan, J., & Han, J. (2016). Decentralized Robust Adaptive output feedback stabilization for interconnected nonlinear systems with uncertainties. *Journal of Control Science and Engineering, 2016*, 1–12. DOI: 10.1155/2016/3656578

Yoder, J. (2007). Autoregressive distributed lag models. *WSU Econometrics, II*, 91–115.

Young, P. C., Jakeman, A. J., & McMurtrie, R. (1980). An instrumental variable method for model order identification. *Automatica, 16*(3), 281–294. DOI: 10.1016/0005-1098(80)90037-0

Zhang, Q., & Clavel, A. (2001). Adaptive observer with exponential forgetting factor for linear time varying systems. *Proceedings of the 40th IEEE Conference on decision and control (16.CDC '01)*, 4, 3886–3891. DOI: 10.1109/.2001.980478

Zhang, W., Zhang, Z., & Wang, H. (2019). Adaptive state feedback control f or nonlinear systems with multiple interval time-varying delays and non-symmetric dead-zone input: A Lyapunov-Krasovskii approach. *Transactions of the Institute of Measurement and Control, 41*(14), 1–12. DOI: 10.1177/0142331219842284

Zhang, X., Chen, X., Zhu, G., & Su, C.-Y. (2020). Output feedback adaptive motion control and its experimental verification for time-delay nonlinear systems with asymmetric hysteresis. *IEEE Transactions on Industrial Electronics, 67*(8), 6824–6834. DOI: 10.1109/TIE.2019.2938460

Zhang, Y., Fidan, B., & Ioannou, P. A. (2003). Backstepping control of linear time-varying systems with known and unknown parameters. *IEEE Transactions on Automatic Control*, *38*(11), 1908–1925. DOI: 10.1109/TAC.2003.819074

Zhao, Z., Li, C., Ahlin, K., & Du, H. (2016). Onlinear system identification with the use of describing functions – a case study. *VibroenginIeering PROCEDIA*, *8*, 33–38. DOI: 10.1016/j.proeng.2016.01.076

Zyryanov, G. V. (2010). *Control systems for multi-connected objects: a textbook.* SUSU Publishing Center.

About The Author

Nikolay Karabutov graduated from Kharkiv Institute of Radio Electronics (Ukraine) in 1972. He got his PhD in 1982. He got his Doctor of Engineering in 1993. He obtained an academic title of the Professor in 1998. Head of the Department of Informatics and Computer Technologies of the Moscow State Academy of the Water Transport (1995-2002). He is Professor of Problem Control Department at MIREA - Russian Technological University since 2002. Professor Karabutov got the State Award in Science and Technology of the Russian Federation in 1995 for researches on adaptive systems. He was the Professor of Department of Applied Informatics under the Moscow State Industrial University (2002-20014), professor Mathematics department of the Moscow Financial University (2011-2014), professor Department of Applied Informatics of Moscow Polytechnic University.

Index

A

D

E

H

I

L

M

N

P

Q

S

T

U

www.ingramcontent.com/pod-product-compliance
Lightning Source LLC
LaVergne TN
LVHW080609300926
849162LV00005B/1093